1 MONTH OF
FREE
READING

at

www.ForgottenBooks.com

By purchasing this book you are eligible for one month membership to ForgottenBooks.com, giving you unlimited access to our entire collection of over 1,000,000 titles via our web site and mobile apps.

To claim your free month visit:

www.forgottenbooks.com/free95929

ISBN 978-1-5283-7300-5
PIBN 10095929

TRANSACTIONS

OF THE

AMERICAN INSTITUTE OF MINING AND METALLURGICAL ENGINEERS

(INCORPORATED)

WITH WHICH IS CONSOLIDATED THE

AMERICAN INSTITUTE OF METALS

VOL. LXV

CONTAINING PAPERS AND DISCUSSIONS ON PETROLEUM AND GAS

NEW YORK, N. Y.
PUBLISHED BY THE INSTITUTE
AT THE OFFICE OF THE SECRETARY
29 WEST 39TH STREET
1921

PREFACE

In this volume are the papers and discussions on Petroleum and Gas that were presented at the Chicago meeting, September, 1919, the Lake Superior and St. Louis meetings, August and September, 1920, the New York meetings of 1920 and 1921, and the Wilkes-Barre meeting, September, 1921; also proceedings of the St. Louis meeting.

Petroleum and Gas Meeting at St. Louis

A SPECIAL meeting arranged by the Petroleum and Gas Committee of the American Institute of Mining and Metallurgical Engineers was held on Tuesday and Wednesday, Sept. 21 and 22, 1920, in the Assembly Room of the American Annex Hotel, St. Louis, Mo.   Those in attendance were guests of the St. Louis Local Section.

Preceding the first session on Tuesday morning, the members and guests were registered and presented with the usual Institute badges. The morning session was opened at 10:30 by Ralph Arnold of Los Angeles, Calif., chairman of the Petroleum and Gas Committee.   In his opening remarks, he suggested that the petroleum section specialize more and more on the technical problems of the oil industry and that an effort be made to enlarge the membership of the Institute among the technical men of the industry.   The following program was presented:

Oil Fields of Russia, by A. Beeby Thompson and.T. G. Madgwick, of London, England.   Presented by H. A. Wheeler; discussed by Arthur Knapp and R. Van A. Mills.

This most comprehensive paper was one of what is hoped to be a series to be presented by some of our foreign members.   It is by far the best description in English of the world-famous Baku and other fields of Russia.

Extended Life of Wells Due to Rise in Price of Oil, by Willard W. Cutler, Jr., of Chevy Chase, Md.   Presented by the author; discussed by J. L. Darnell.

This paper brought graphically before the audience the fact that the economic life of a well lengthens as the price of oil goes up.

Urgency for Deeper Drilling on the Gulf Coast, by A. F. Lucas, of Washington, D. C.   Presented by Mowry Bates; discussed by David White, W. E. Pratt, Mowry Bates, R. Van A. Mills, J. L. Henning, Arthur Knapp and E. DeGolyer.

. This paper opened up the always interesting subject of salt domes on the Gulf Coast and the possibility of the occurrence of oil at great depth in these structures.

Petroleum Industry of Trinidad, by George A. Macready, of Los Angeles, Calif· Presented by R. A. Conkling; discussed by Arthur Knapp, R. A. Conkling, E. De-Golyer, Ralph Arnold and R. Van A. Mills.

Oil Shales and Petroleum Prospects in Brazil, by H. E. Williams, of Rio de Janeiro, Brazil.   Presented by J. Elmer Thomas; discussed by David White, Mowry Bates, B. O. Mahaffy, J. Elmer Thomas and Ralph Arnold.

The latter paper brought out the point that there are possibilities of developing oil from oil shales, and in addition that there are certain localities along the eastern flanks of the Andes in Brazil that may eventually yield commercial quantities of oil.

TUESDAY AFTERNOON SESSION

The afternoon session was opened at 2.30 and was presided over by Vice-chairman E. DeGolyer. The following papers were given:

Determination of Pore Space in Oil and Gas Sands, by A. F. Melcher of Washington, D. C. Presented by W. E. Pratt; discussed by R. Van A. Mills, Walter M. Small, W. W. Cutler, Jr. and David White.

This paper brought out the point that there are other determining factors affecting the oil saturation of rocks than the size and shape of the grains.

Application of Taxation Regulations to Oil and Gas Properties, by Thomas Cox, of Oakland, Calif. Presented by E. B. Hopkins; discussed by Ralph Arnold, J. L. Henning and W. E. Pratt.

Oil Possibilities of Northern Alabama, by D. M. Semmes, of University, Ala. Presented by Walter M. Small; discussed by David White and Mowry Bates.

Efficiency in Use of Oil and Gas as Fuel, by W. N. Best, of New York. Presented by James H. Hance; discussed by S. O. Andros, H. P. Mueller, J. L. Henning, I. N. Knapp, and C. H. Matthews.

Industrial Representation in the Standard Oil Co. of N. J., by C. J. Hicks, New York. Presented by John L. Henning; discussed by Ralph Arnold, W. E. Pratt, and Mr. Trowbridge.

Valuation Factors in Casinghead Gas Industry, by O. U. Bradley, Muskogee, Okla. Presented by W. B. Wilson; discussed by W. E. Pratt, E. DeGolyer, J. L. Henning, W. M. Small and Mr. Reeves.

Nature of Coal, by J. E. Hackford of London, England. Presented by David White; discussed by Ralph Arnold, David White, W. E. Pratt and E. DeGolyer.

In the evening an informal smoker was given to the visiting members and guests at the American Annex Hotel. Moving pictures were shown; some short speeches were given and suitable refreshments provided. The evening was greatly enjoyed by those present.

WEDNESDAY SESSIONS

The Wednesday morning session was called to order at 10 o'clock by Ralph Arnold, who presided. The following program was presented:

Analysis of Oil-field Water Problems, by A. W. Ambrose, Bartlesville, Okla. Presented by C. E. Beecher; discussed by R. A. Conkling, R. Van A. Mills, E. De-Golyer, Mr. Reilly and Mr. Compton.

Contribution of Oil Geology to Success in Drilling, by F. G. Clapp, of New York. Presented by W. E. Wrather; discussed by E. DeGolyer.

Ultimate Source of Kentucky Crudes, by W. R. Jillson, of Frankfort, Ky. Presented by title, as manuscript was not received in time for preparing abstract.

Oil-field Brines, by C. W. Washburne of New York. Presented by Walter M. Small; discussed by R. Van A. Mills, E. DeGolyer, W. M. Small, R. A. Conkling, Mr. Reilly, W. E. Pratt and W. E. Wrather.

The last paper brought out further discussion regarding the theories of origin of salt domes.

The above list completed the formal papers. Following the formal meeting the ensuing papers were presented without discussion:

Gulf Cretaceous Oil Fields, by Julius Fohs. Presented by the author.
Oil Resources of Illinois, by Mr. Mylius, of Urbana, Ill. Presented by the author.
Influence of Faults in the Illinois Fields, by H. A. Wheeler, of St. Louis, Mo. Presented by the author.

Prior to the adjournment of the meeting a resolution was passed extending the thanks of those present to the St. Louis Local Section for its hospitality and for the courtesies extended during the meeting, with a special vote of thanks to Dr. H. A. Wheeler for his untiring efforts in making the meeting a success.

The afternoon of the twenty-second was spent in a trip to interesting points about St. Louis, in automobiles provided by the St. Louis Section.

PAPERS

they could not remain in the government if such legislation was passed as a government measure. The compromise reached was that a bill should be passed declaring that no one could sink a test well for oil or gas in Great Britain without a license from the government, and the question of royalty and ownership would be dealt with after the war. The government gave an undertaking to Parliament that it would not recognize the payment of royalties on oil until Parliament had acted. This legislation was passed in October, 1918. The government then took the land necessary for nine well sites (seven in Derbyshire and two in Staffordshire) under the powers given it by the Defence of the Realm Acts. This gave the right of occupancy, but not of ownership. Later, two additional sites were taken in Scotland; as one of these was taken after the signing of the armistice the validity of the action is now the subject of a lawsuit. The present condition is, therefore, that while the government may still legally, for the time being, have the power to take sites under the Defence of the Realm Acts, it cannot justify the expediency of so doing; it cannot acquire such sites by agreement, because this would involve the payment of a royalty to the landlord, or the recognition of his ownership of the oil, and it cannot grant a license to anyone else because this also would involve the same recognition indirectly.

The first well sunk by the government found commercial oil, and while it would have been relatively easy to pass legislation giving the ownership of the oil to the government when the majority of the landlords had no belief in its existence, the laborites and extreme radicals have now been furnished with the politically effective argument that the oil was found with government money. Even the utilization of the oil found in the test wells, which will be limited to the ones already started, is subject to the serious handicap that whenever the government starts to remove the oil from the tankage at the well site the landlord will immediately start injunction proceedings.

FUTURE COMMERCIAL PROSPECTS

In the center of England the Mountain limestone (Mississippian) is exposed along the axis of the Pennine fold. Like the similar carboniferous limestones in Kentucky and Missouri, it is cut by spar and lead veins, but unlike these, it contains numerous important seepages of petroleum. The upper 100 to 150 ft. (30 to 45 m.) of this limestone is dolomitic. Overlying the Mountain limestone are the Yoredale shales and sandstones, which in the important area to the east have a thickness of from 400 to 700 ft. (121 to 213 m.) and in the area to the west, 2000 to 2500 ft. The Yoredale shales are followed by the Millstone grits series of shales and important porous sandstones with a total thickness on the east of 700 to 900 ft., and on the west of about 300 ft.; these, in turn, are succeeded by the productive coal measures. On each side of the main Pennine fold, subsidiary folds produce a whole series of local domes, anticlines, and terraces in the regions where the limestone is overlaid by the Yoredale and succeeding rocks. There is considerable faulting, but the character of the oil produced in the limestone is such that, while it is of a paraffin base, it oxidizes even more rapidly than an asphaltic oil. There are no surface exudations of oil of importance on either side of the main limestone mass, but for the last century the coal mines on either side have encountered important flows of oil on fault planes.

The discovery well is located on a faulted dome at Hardstoft, Derbyshire, where none of the coal mines had found oil in the fault planes. It started in the coal measures, found wax in drilling through a fault, a commercial supply of gas in the Millstone grits, which was muddied off, and oil in the top of the limestone at a depth of 3078 ft. (938 m.). This well has been flowing at the rate of 12 bbl. per day since June of this year, and is estimated to have a pumping capacity in excess of 50 bbl. The well has not been "shot;" first, because the transportation of nitroglycerine on the roads of England is not permitted, and, second, because the war emergency being over, the question of the ownership of the oil has become acute, and when the present tankage is filled the removal of the oil will undoubtedly involve a legal fight.

Two wells, located on domes south of Hardstoft, both started in the coal measures, penetrated the Millstone grits without finding gas in any considerable quantities, showed a little oil in the top of the limestone, and are now drilling in the limestone, where they have encountered a little gas. It is planned to "shoot" these wells whenever conditions permit. Three wells on different structures to the north of Hardstoft have encountered commercial gas in the Millstone grits, but have not yet reached the limestone. The two wells which have been started on the west side of the Pennine axis in Staffordshire have not yet reached a sufficient depth to be interesting. The area in the center of England that has important petroleum possibilities is between 20,000 and 30,000 square miles.

The two wells that are being drilled in Scotland are in an entirely different category. They are merely "wildcat" wells, with a moderate chance of being successful. One is located at West Calder, on a dome in the oil-shale fields, 16 mi. southwest of Edinburgh, and the other on a dome at Darcy, 10 mi. southeast of Edinburgh—both in Edinburghshire. They both start in what is considered the northern equivalent of the lower part of the Mountain limestone, which is here for the most part the oil-shale series. They will both penetrate between 2000 and 2500 ft. (609 and 761 m.) important untested sandstones underlying the oil shales, and are expected to reach the old red sandstone (Devonian) at from 3300 to 4000 ft. A certain amount of free oil and wax has been found in connection with the shale mining—sometimes in the associated sandstones; sometimes on the faces of the igneous sills. This free oil has always been considered as due to destructive distillation of the shale by heat from the igneous rocks, but Mr. J. E. Hackford finds that it has many things which distinguish it from an oil that could be produced by the destructive distillation of the shales, and reaches the conclusion that it has come from below after the igneous rocks had cooled. This, taken in connection with the fact that the Devonian sandstones show some oil in the north of Scotland and in the Orkneys, has led to the location of the two test wells in Scotland.

The present work in Great Britain had its inception in 1914, when the outbreak of the war enabled the writer and his associates to carry out a long deferred desire to see just what the numerous indications of petroleum in Great Britain really meant. Thanks to the great mass of fundamental geological information which the Geological Survey of Great Britain had collected and published, and particularly to the detail work carried out in certain coal fields, it was possible in·a short time to present to Lord Cowdray the conclusion that the petroleum possibilities of the Midlands of England were of a most amazing and striking character. Lord Cowdray, after a momentary hesitation, shared our enthusiasm. With the increase of the submarine menace, he offered to place the services of his firm and his petroleum staff at the disposal of the nation, free of cost, for carrying this work forward as a war measure. This was a gift made to the nation without any commitment of any kind on the part of the British Government to Lord Cowdray.

Special mention should be made of the work of Mr. Eugene L. Ickes, a graduate of the University of California and an American geologist of marked ability, Mr. Roderic Crandall, of Stanford University, who was in charge of the technical administration of the work, and Mr. Victor L. Conaghan, drilling superintendent, who was very kindly supplied as a war measure by the United States Bureau of Mines.

The oil from the Hardstoft well has the following characteristics: Specific gravity, 0.823; sulfur, 0.26 per cent.; gasoline, 7.5 per cent.; kerosene, 39.0 per cent.; wax, 6.0 per cent.; gas oil, 20.0 per cent.; lubricating oil, 30.0 per cent. The oil is particularly rich in very high-grade lubricants.

DISCUSSION

CHESTER W. WASHBURNE, New York, N. Y. (written discussion). The work of Mr. Veatch and his associates in directing the work that promises to add England to the list of oil-producing countries indicates the value of science, as well as their ability to apply it. Englishmen long have been searching the corners of the earth for oil, without recognizing the possibilities at home. I would like to ask Mr. Veatch whether any chemists have ascertained what constituents in the oil are responsible for its susceptibility to oxidation. He says that, "while it is of paraffin base, it oxidizes even more rapidly than an asphaltic oil." Can he give us the percentage of unsaturated hydrocarbons, or any similar information concerning the chemical nature of the oil? This experience in England indicates the possibility of oil in other parts of the world that have been neglected in explorations.

Oil Fields of Persia

By Campbell M. Hunter, London, Eng.

(New York Meeting, February, 1920)

Petroleum is found in almost every province in Persia. On the northern frontier, along the southern shore of the Caspian Sea, it is found near Enzelli and Shakhtesar and gas at Khoremabad. Oil is also found at Gumish Tepe, northwest of Astrabad, on the southeastern shore of the Caspian Sea. Further inland, to the south of Astrabad, oil is found at Dchahkuh-i-balae, also on the margin of the Khorasan desert at Semnan, 115 mi. (35 m.) east of Teheran.

Along the western frontier, from northwest to southeast, oil is encountered at Ouschachi, north of Lake Urumieh; in the province of Azerbaijan, at Zohab, Khanikin, and other places in the district of Kormanshah in the province of Ardelan. Further south, in the province of Luristan, oil is found east of Mendeli and in the Pusht-i-kuh districts. Considerable quantities of oil are also obtained from Schuster, Maidan-i-Naphtun, Ram-Hormuz, Beheban, which are almost on a straight line, running northwest and southeast along the foothills of the Bakhtiari Mountains. At Ahwaz, in the province of Arabistan, oil has been found along another range of hills whose axis also lies in a northwest-southeast direction.

In the Fars province, boring for oil has taken place at Daliki, and indications of oil are found at Kheri, Fasa, Darab, and other places. A gas show is also recorded at Kuhi-Sung-Atush in this province, 30 mi. (48 km.) east of Darab. In the south of Persia, oil is encountered on Qishm Island, also at Ahmedi and other places north of Bunder Abbas. On the southeastern frontier, oil is found on the Sarhad range of hills. Thus, oil indications have been noted over a distance of approximately 1100 mi. along the western and 700 mi. along the northern frontiers of Persia.

History

The first working of oil in Persia of which there is any record took place at Kir-ab-us, Susiana, now known as Kirab, about 57 mi. northwest of Schuster. Herodotus (about B. C. 450) reported a well near Ardericca that produced three different substances; namely, asphalt, salt, and oil. The oil, which was black and had a strong smell, was called Rhadamance by the Persians.

At Daliki, many years ago, a well sunk to a depth of 124 ft. pierced hard sandstone and blue clay and encountered semi-solid bitumen and liquid petroleum in small quantities. The same company, the Persian Bank Mining Rights Corpn., also drilled on the island of Qishm, though unsuccessfully. Later, surveys at Zohab and near Schuster indicated more favorable conditions. In the former district, oil has been exploited for centuries from primitive, shallow, hand-dug wells, some being reported to have yielded oil in undiminished quantities for upwards of 50 years.

In 1903, W. K. D'Arcy, prompted by rumors of oil in Persia, started a systematic investigation of the country, and in 1903–4 drilled two wells at Kasr-i-Shirin, one to a depth of 800 ft. (243 m.) and the other to 2100 ft. (640 m.). Drilling was conducted in various districts, but without any great success. After about £200,000 ($1,000,000) had been spent in this way and there were serious thoughts of abandoning the whole project, D'Arcy heard of oil seepages and springs in the neighborhood of Schuster and had these examined. After overcoming much opposition from the natives, a concession was secured and drilling begun. The first bore hole, at a depth of 1100 ft., pierced the oil sands and the oil spouted to a height of 70 ft., carrying away the derrick.

In 1909, the Anglo-Persian Oil Co. was formed with the object of working the concession obtained by Mr. D'Arcy from the Persian Government in 1901. This concession runs for 60 years, from May, 1901, and gives the exclusive right to drill for, produce, buy, and carry away oil and petroleum products throughout the Persian Empire, except in the provinces of Azer, Badjan, Gilan, Mazanderan, Astarbad and Khorasan.

Before the formation of this company, preliminary examinations and tests had been carried out in compliance with the terms of the concession, by the First Exploitation Co. The concession to the company provided for the allotment to the Persian Government of £20,000 fully paid shares, as well as a cash payment of £20,000, and a royalty of 16 per cent. of the net yearly profits.

On the inception of the Anglo-Persian Oil Co., the actual holding of the First Exploitation Co. was limited to 1 sq. mi. in the Maidan-i-Naphtun field, which is situated in a territory belonging to the Bakhtiari Khans. The agreement with the latter tribes provides that they shall receive 3 per cent. of the shares in any company formed to work oil in their country; and to facilitate the working of the agreement, it was decided to form a second subsidiary company, known as the Bakhtiari Oil Co., Ltd., to work the remainder of the oil-bearing lands in the Bakhtiari country. All the shares of these two companies not held in Persia are the property of the Anglo-Persian Company.

The concession taken over by the Anglo-Persian Oil Co. covers an are

of some 500,000 sq. mi., only a small part of which has been examined.
In 1914, the British Government decided to take an interest in the development of the Persian oil fields, and to this end, entered into an agreement
with the Anglo-Persian Oil Co. under which they took up 1000 £1 preferred and 2,000,000 £1 ordinary shares.

—PERSIA—
REFERENCE
■ Oil Wells Proved Areas
● Reported Oil Shows
////// Anglo Persian Oil Company's Concession
++++ Russo Persian Petroleum Concession
SCALE:
Miles 20 10 0   100   200 Miles

In 1917, the Russo-Persian Petroleum Co. obtained from the Persian
Government an exclusive concession for prospecting for oil in the district
of Ardebil, and in the provinces of Gilan, Mazanderan, and Astrabad.
In the same year, the same company purchased a number of oil-carrying
steamers and sent out a party of geologists under Prince Ameradzhebe.

GEOLOGY

For the purpose of this paper, Persia may be divided into three areas: Northern Persia, embracing the provinces of Azerbaijan, Gilan and Mazanderan; western Persia, in which lie the provinces of Ardelan, Luristan, Bakhtjari, Arabistan, Fars, Laristan and the Island of Qishm; southeastern Persia, comprising the district of Mekran.

The oil-bearing region in northern Persia lies between Lake Urumieh and the Caspian Sea, a distance of about 200 mi. in breadth, and belonging chiefly to the Tertiary period. In the north of this region at Ahar, natural shows of petroleum are seen in a stratum of apparently foraminiferous sandstone, which gives off petroleum emanations a few feet below the surface. There are also several mud volcanoes in this district. Between Ahmenabad and Ahar, the region is terraced in asymmetrical folds, the principal axis of folding lying roughly due east and west with synclines about 2 mi. apart, the dip on the one side being between 65° and 75° and on the other between 12° and 15°.

To the south of Ahar, the greater part of the formations belong to the upper Carboniferous period; and in the Savalian Kuh Mountains in the southeast, rock salt and gypsum are found in large quantities. Faulting is very prevalent in this range, associated with numerous highly petroliferous mud volcanoes.

East of Ahar, at Ardebil on the Mugan steppe, extensive shell beds resting on rocks of Pliocene age similar to those found at Baku exist; it is thought that the oil fields of northern Persia are a continuation of those of Baku. Similar shell beds exist near Marand almost due west of Ahar, and near Sofian and Tabris, which is built on alluvial beds of Miocene age. The country east of Tabris belongs to the Mesozoic period and contains very considerable deposits of rock salt and gypsum. In a depression close to Sirab, traces of oölite are found; and north and south of this site Carboniferous shales are met with.

The regional tectonics of the Belfathemar divide, which lies to the south of Sirab, consist of a lengthy anticlinal fold along which, at several places, oil and gas escape; in warm weather fumes of sulfuretted hydrogen and sulfur dioxide are found in the gulleys. It is the belief of Charles Bouvard, Sir Boverton Redwood, and many others that the petroleum of northern Persia is of organic origin. Toward the end of 1917, a geological survey of Gilan and Mazanderan was in contemplation on behalf of the Russo-Petroleum Co., which has acquired concessions in these provinces.

The geology of southern and western Persia, especially to the north of the Persian Gulf, has been investigated on a comprehensive scale by Doctor Pilgrim who gives the following geological formations in descending order:

Recent and sub-recent, .	Shelly conglomerates and dead coral reef of littoral; red sand hills of coast of Trucial Oman, alluvium of Mesopotamia, etc.
Pleistocene.............	Foraminiferal, oolite.
Pliocene................	Bakhtiari series grits and conglomerates.
Miocene................	Fars series marls, clays, and sandstones with limestones and interbedded strata of rock gypsum.
Miocene................	Urumieh series limestones.
Oligocene..............	Nummulitic limestone of Khamir.
Eocene	Nummulitic limestone of Persia, Muscat, and Bahrain series.
Upper Cretaceous......	Hormuz series, lavas and tuffs, etc.

In connection with the oil geology of western and southern Persia the most important of these series are the Fars, Bakhtiari, and Hormuz. The Fars series, which is by far the most important and widespread of these three, is subdivided into three divisions, viz.: Basal or gypsum beds, plateau beds, and coastal beds. According to Doctor Pilgrim, the basal beds of the Fars series consist of layers of rock about 10 ft. thick, containing a large amount of gypsum interbedded with impure limestone, clays, and shales. The thickness of these basal beds is stated to be 1000 ft. and they appear to be represented from Lake Urumieh, in the north, to Bunder Abbas, in the south. They are of reddish color, due to iron oxide. Conformably overlying the basal beds, but with no definite line of demarcation, are the plateau beds, which extend from Kaniarij plain to Bunder Abbas. These, Doctor Pilgrim considers to be the thickest of the Fars series with a thickness of from 14,000 to 15,000 ft. (4270 to 4575 m,) at Kotal Malo, but rapidly thinning to 8000 ft. between Konartakha plain and the Kotal Kamarij. They consist of blue and red clays, or marls, alternating with sandstones and thinly bedded fossiliferous limestones. These limestones are most frequently found in the lower beds and the sandstones predominate in the upper.

The plateau beds seem to merge into the coastal beds, which vary from 300 to 1000 ft. in thickness and are composed of pale gray clays, or marls, passing sometimes into soft argillaceous limestone. Interbedded with these are thin calcareous bands crowded with shells and grit. Resting with great unconformity on the Fars series is the Bakhtiari series, which consists of detrital deposits of every degree of coarseness, is about 1000 ft. thick, and has as its most characteristic rock a conglomerate of red and green chert pebbles. This series is practically unfossiliferous and is not more recent than the Pliocene age.

In a paper read before the Institute of Petroleum Technologists in London (in 1918), Messrs. Busk and Mayo, describing the Bakhtiari country and dealing with these series, divide the Fars series into a lower gypsiferous group, varying in thickness from 1000 to 3500 ft. (304 to 762 m.); a middle, or passage, group, similar to Doctor Pilgrim's plateau group, but with a thickness rarely exceeding 1000 ft.; and an upper, or argillaceous, group consisting of purple and red shales and clays with intercalated massive sandstones, consisting of chert grains cemented by a calcareous matrix.   Due to the presence of plant remains and the absence of marine fossils, they considered this upper group to be of lacustrine origin and about 2700 ft. thick.

The Bakhtiari series, which they consider a sediment of non-marine origin, being deposited during a period of earth movement producing uplift and depression along well-defined lines, attains its greatest thickness of about 15,000 ft. (4575 m.) in the synclines thus formed.   The lower group of this series, about 12,000 ft. thick, consists of clays, shales, and intercalated sandstones and conglomerates at odd places, particularly toward the top of the group, which are often 1500 ft. thick and are of deltaic form.   Although oil and oily residues are found in parts of this group, these are thought to be simply due to redeposition from the adjacent Fars series.   The upper group of this series rests unconformably upon all rocks below, except in the synclinal troughs, and consists of conglomerate of well-rounded limestone and chert pebbles well cemented together and about 2000 ft. thick.   The typical syncline in the Bakhtiari district measures about 7 mi. (11 km.) across and in the trough has 15,000 ft. of the Bakhtiari series overlying 5000 ft. of Fars series.   They consider that further earth movements, continuing to the present, have produced a very complicated series of fanlike structures with thrust faults coming up to the surface.

Oil has been found in this district to be contained in the detrital limestones forming the base of the Fars series and has been flowing at Maidan-i-Naphtun, under strong pressure, for the last 10 years.   The accompanying table shows the difference between the thickness of the various formations, as calculated by Doctor Pilgrim and Messrs. Busk and Mayo:

	Dr. Pilgrim, Feet	Thickness According to Messrs. Busk and Mayo, Feet
Pliocene = Bakhtiari series	1,000	15,000
Miocene = Fars series	Basal, 1,000	Lower or gypsiferous group, 3,500
	Plateau beds, 15,000	Middle or passage group, 1,000
	Coastal beds, 1,000	Upper or argillaceous group 2,700

In describing the Ahwaz Pusht-i-kuh country, Messrs. Busk and Mayo state that there is an anticlinal structure running for 100 mi. (160 km.) in a west-northwest and east-southeast direction through Ahwaz.   This structure forms the furthest outlying fold of the Iranian Mountain chain and is asymmetrical, having a steep vertical, or inverted, dip on the southwestern face and a gentle slope on the northeastern limb.   In the neighborhood of Ahwaz, the crest of the anticline is in the form of elongated domes, and denudation has shown that the lowest 200 to 300 ft. (60 to 91 m.) of exposed beds belong to the middle group of the Fars series.

There is one main oil horizon in the central field, which has been proved at depths varying from 1200 to 1300 ft. (365 to 396 m). This horizon is responsible for the greater part of the production of this field; and as the oil is found in a hard porous limestone, a steady production is obtained with little necessity for cleaning out the wells.

At the White Oil Springs, two seepages occur on the crest of the fold, from which a colorless oil resembling kerosene is obtained. The production amounts to about 20 gal. (75 l.) per day, and is used by the natives for domestic purposes.

The Ahwaz anticline is about 36 mi. (58 km.) to the southwest of the Maidan-i-Naphtun fold; and although no evidence of petroleum appears at the surface, the White Oil Springs horizon is expected to exist at no great depth.

At Qishm, oil issues from the lowest exposed beds at two places, about ½ mi. apart. The seepages are not considerable.

TECHNOLOGY

There is little published information relating to drilling methods on the Persian oil fields. D'Arcy, in his early exploration work, employed Canadian drillers and Canadian drilling rigs. This system, but with the wire rope taking the place of the old-fashioned drilling rods, is now extensively used on the field, though rotary drilling has been tried. Inserted joint casing is generally used, as formations encountered give little trouble through caving. Considerable gas is yielded by the wells and is used under the boilers.

TABLE 1

Location	Specific Gravity	Flash Point of Crude, Degrees F.	Benzine, Per Cent.	Kerosene, Per Cent.	Lubricating Oils, Per Cent.	Sulfur	Odor	Color
				Fractions				
Schuster District.	0.927			27.0	45			Dark green
White Oil Springs (Ahwaz)........	0.773					Present	Inoffensive	Light straw
Tchiah Sourlch (Near Kasr-i-Shirin).	0.815	Low	9.4	57.6		0.4 per cent. present	Inoffensive	Brown, strongly fluorescent
Daliki..........	1.016	170				Present	Strongly of sulfuretted hydrogen	Dark brown
Qishm..........	0.837	190					Pleasant	Brownish red

No recent production figures have been published, but it is understood that the wells come in as gushers and continue to flow for a considerable time.   One well, at least, is reported to have yielded over 100,000 tons of oil by flowing.   Early in 1919, it was stated that the wells already drilled were estimated to be capable of producing 5,000,000 tons per annum.   Table 1 gives brief particulars of some of the oils.

From Maidan-i-Naphtun, which is situated about 800 ft. above sea level, the oil is conveyed to the refinery at Abadan through two pipe lines of 6-in. and 10-in. (15 and 25 cm.) diameter, the distance being about 145 mi. (233 km.).   The diameter of the former pipe line is increased to 8 in. about 53 mi. from the field to enable the production from White Oil Springs and Ahwaz to be pumped to the refinery.   Upon the completion of certain additional pumping stations, the joint pipe lines will have a total carrying capacity of about 3,000,000 tons—say 22,000,000 bbl. The refinery at Abadan, which is an island at the head of the Persian Gulf, was completed in 1913, with an estimated annual throughput capacity of about 240,000 tons (1,750,000 bbl.).   Since then the refinery has been considerably extended, and is now capable of treating the bulk of the company's production.

Initially, considerable difficulty was experienced in eliminating the sulfur present in the Maidan-i-Naphtun oil.   Various processes were tried, and it is only within the last year or two that a satisfactory treatment has been evolved.

Production Statistics and Future Possibilities

Up to 1916, about thirty wells had been drilled at Maidan-i-Naphtun, all of which were gushers; no wells had at that time been put to pump. The production for the year ending March, 1912, was about 600,000 bbl.; during the ensuing six months the yield had been increased to 1,000,000 bbl.   Since then a considerable number of additional wells have been brought in while a still larger number have been drilled to the oil sand, but not completed pending the development of increased marketing facilities.

While little information has been published by the Anglo-Persian Oil Co. on the development of its concession, there can be no question that Persia is destined to furnish enormous quantities of oil, and to take a leading position among the world's great oil-producing countries. From the outset, the company's production has been greatly in excess of its transporting, treating, and marketing facilities.

DISCUSSION

The Chairman (E. DeGolyer, New York, N. Y.).—The Persian fields are being operated by the Anglo-Persian Oil Co., of which the British Government has control, under what amounts to a monopoly.   According to the company reports, some of the wells discovered are among the largest in the world. The country is becoming extremely important in the production of petroleum at the present time. According to the estimate of the United States Geological Survey in 1918, Persia produced 7,200,000 bbl. of oil, and was fifth in importance among the producing nations.

The Chairman of the Board of the Anglo-Persian Oil Co., in reporting to the stockholders in 1918, mentioned a well that had produced 1,500,000 tons with no apparent diminution in pressure and no apparent diminution in productive capacity. I think that American petroleum geologists and technologists have been overlooking the importance of the Persian fields as a source of supply.

DAVID WHITE,* Washington, D. C.—In connection with the description of the oil indications of Persia and Mesopotamia, mention should be made of the recent publication by the Hamburg Colonization Institute of a rather extensive memoir by Walter Schweer. This report, which was evidently compiled for German consumption when the war should be over, contains many details concerning the distribution and character of the oil indications, with something of the geology and the concessions held by various countries in Turkey, Palestine, Arabia, Syria, Persia, and Armenia. This report will be found very valuable and helpful by Americans interested in the great potential oil fields of the near East.

* Chief Geologist, U. S. Geol. Survey.

Oil Fields of Russia

By A. Beeby Thompson and T. G. Madgwick, London, Eng.

(St. Louis Meeting, September, 1920)

For more than 2500 years, natural gas issues in the Surakhany district of the Apsheron peninsula were the object of pilgrimages by fire worshippers and Hindoos from Burma and India. Even as late as 1890, Hindoo priests conducted ceremonies in a temple at Surakhany, which probably replaced a more ancient one; but later, the visits of the pilgrims were prohibited in order to check the spread of Asiatic diseases in that region.

For centuries, limited supplies of oil have been abstracted from shallow excavations in the Caspian oil belt and dispatched into the interior of Asia and elsewhere for medicinal and industrial purposes. Statistics show a yield of 37,400 bbl. in 1863, but only since 1869 has there been serious development; in that year the yield was 203,000 bbl. At that time, hand digging was supplanted by drilling, and the enormous wells that resulted from tapping sources hitherto beyond the reach of operators completely demoralized the industry for a time, owing to inadequate outlets for the products.

The early activities in this area were greatly hindered by annoying taxation, monopolies, imperial land grants, etc., but when these were revoked or adjusted, in 1877, the industry sprang into prominence and, between 1898 and 1901, the Baku fields produced practically one-half of the world's supply of oil.

Within a few miles of Baku lie the two richest oil fields in the world; viz., the Balakhany-Saboontchy-Romany and the Bibi-Eibat, the latter constituting almost a suburb of the city. For many years the gasoline obtained in the refineries of the Baku area was burned in pits, being considered an undesirable product, and until 1870 the residue also was destroyed, its value as a fuel not being recognized. Kerosene was the main product sought by the refiners. It was shipped across the Caspian Sea and up the Volga to the industrial centers of Russia. Only on the completion of the Baku-Batoum railway did the Baku oil fields secure important commercial communication with the outside world through the medium of the Black Sea. The first tank steamer was successfully launched on the Caspian in 1879, by Messrs. Nobels, for transporting oil in bulk instead of in barrels. In 1905, an 8-in. pipe line to Batoum was completed; this was capable of transporting to seaboard 8,000,000 bbl. of kerosene per annum.

In 1903, the important Grozny oil field was proved by a great flowing well sunk by an enterprising Englishman, who, however, was ruined by the claims for compensation made by peasants whose habitations and lands were destroyed by the deluge of oil, which could not be controlled for years.   The property on which the well was drilled has since given over 300,000 bbl. of oil per acre.

In 1901, general interest was directed to the Binagadi oil field by the bringing in of a 10,000-bbl. well.   The field lies close to Baladjari railway station and only a few miles from Baku and the refineries, to which a pipe line was subsequently laid.   In this year, also, an important oil field was located at Berekei; but after a few years' work hot sulfurous waters flooded the oil sands and, as no suitable means for its exclusion were devised, the field was practically abandoned, although some wells continued to yield for years.   Berekei lies on the Caucasian railway near the port of Derbent.   The oil from the field was piped to the railway and taken in tank wagons to its destination.

Another interesting field is Holy Island, off the north coast of the Apsheron peninsula, where 400-bbl. wells have been struck and a considerable area has been proved to be oil-bearing.   Oil is shipped direct to the Volga by tankers proceeding from Baku.   In 1908, the Surakhany district a few miles southeast of the main Saboontchy oil field was developed by deep wells, and large gushers of the typical Baku oil resulted from drilling beneath the upper light oil and gas-yielding beds that until then had been exclusively worked.

For many years the island of Cheleken, off the Asiatic coast of the Caspian Sea, near Krasnovodsk, had been the scene of some moderate operations; but from 1911 onwards large yields were obtained from wells sunk in the Ali Tepe district in the southwestern part.   These wax-containing oils were generally shipped to Baku for treatment at Black-town, the refinery suburb of Baku.

In 1909, the Maikop oil field attracted considerable attention as the result of a large gusher of light oil being struck by almost the first trial well in the Shirvansky district.   Since that time, a fair production has been obtained although the very prolific area was proved to be small. This field lies on the northern flanks of the western, and sinking, end of the Caucasian range, over which a pipe line was laid to the port of Touapse.   Pipe lines were also laid to Ekaterinodar, where a refinery was erected, as well as Shirvansky.

A promising oil field was developed, about 1910, in the Emba district north of the Caspian sea and inland from the port of Gurieff.   Around Dossor, large flowing wells were struck and, prior to the war, extensive arrangements were being made to dispose of the product.   Pipe lines were laid to Bolshaya Rakashka, where refinery operations were conducted, and submarine pipe lines were carried through the shallow-water belt to facilitate shipment of the products up the Volga.

A single Turkestan field, in Fergana on the Trans-Caspian Railway at Chimion, has yielded substantial supplies of oil that finds a ready local market. It is said that at Maili Sai good productions resulted from trial wells sunk by the government.

LEASING OF OIL LANDS

Many original grants of oil lands were gifts bestowed on court favorites, but when some system was introduced terms were based solely on the unique Baku conditions, and prospecting licenses of about 100 acres were granted from which 27 acres could be selected when oil had been found. The original annual rentals of $2 per acre for the first ten years, increasing ten times each ten years, were soon superseded by percentage royalties, which varied from 25 per cent. upwards, with minimum annual payments. At one time tenders on a royalty basis were publicly solicited, but speculation led to such absurd offers that the government abandoned the practice. For instance, at times, operators tendered royalties of 75 per cent. with large minimum payments merely to protect their boundaries from aggressive competitors.

The Cossack lands of the Terek-Kuban provinces were subject to a rental of $5 per acre per annum and about 4 c. per bbl. royalty for the first 120,000 bbl. and 2 c. per bbl. afterwards; but rights were reserved to revise the royalties after 12 years. Insistence in perpetuating the old leasing laws based on the unique oil fields of Baku has been a great hindrance to prospecting in Russia, and it is to be hoped that some more rational policy will be introduced in the near future.

DISPOSAL OF RUSSIAN OILS

The products of the Baku oil fields go largely to supply the internal demands of Russia through the medium of the Caspian Sea and the river Volga, although the freezing of the northern Caspian and the Volga in the winter months restricts movements of oil to about 8 months in the year. A pipe line and railway to Batoum are available for the conveyance of lamp oils and other oil products to the Black Sea, where ocean vessels can approach via the Dardanelles. The large refineries are situated at Blacktown, a suburb of Baku.

Oil from the Grozny field and refineries is either piped to the port of Petrovsk on the Caspian Sea or sent by rail to Novorossisk on the Black Sea. From Holy Island and Cheleken, oils are mainly sent to

Baku for treatment; while the North-Caspian (Emba) oils are shipped at Bolshaya Rakashka for transmission up the Volga. Central Asian oils find a ready Asiatic market and are useful for the Trans-Caspian railway service. Maikop oils can either be pumped to Touapse on the Black Sea or to Ekaterinodar, a large town that feeds a wide, fertile, agricultural region. Extensive tank farms are situated at Baku, Grozny, Batoum, and Novorossisk; also up the Volga, where the winter supplies are accumulated during the summer months.

OIL MANIFESTATIONS

Probably no country in the world exhibits a greater display of oil-field surface phenomena than Russia. There are thousands of square miles flanking the Caucasus Mountains and encircling the Caspian Sea that justify an investigation. Difficulties of language, inaccessibility, danger to life, indifference of the authorities to their mineral resources, and irritating restrictions have contrived to suppress any initiative that existed. For miles around the Baku oil fields, the oil series lie spread out like the leaves of a book under the nearly desert-like surroundings of that devastated region. Mud volcanoes on a gigantic scale in every stage of activity may be witnessed, as well as perpetual fires fed by incessant issues of natural gas. Acres of asphaltic residues and streams of viscous oils oozing from immense thicknesses of oil-soaked sands are common. These phenomena, mingled with sulfurous waters, present problems for study that are nowhere else reproduced on such a vast scale Over extensive areas, shallow hand-dug wells sunk into the outcropping inclined or vertical strata yield appreciable, and often considerable, quantities of oil.

Many equally imposing exhibits of oil phenomena may be seen on Holy Island and Cheleken, where for miles numerous oil residues, gas exudations, and sulfurous- and salt-water issues may be examined along the outcropping beds. Fierce outbursts of oil and gas occasionally startle the inhabitants, cause damage to property, and loss of life. Twice within the writer's knowledge, such outbursts in the Yasmal Valley have caused conflagrations that illuminated the sky for miles around each night. Big outflows of oil have also been recorded; and during earthquakes, considerable alarm has been occasioned by the ignition of gas that issued from cracks in the earth. Little less interesting are the great mud volcanoes of the Taman peninsula, which area has not received the attention its manifestations merit.

An interesting phenomenon is the submarine gas issue. Prior to the development of the Baku oil field, several places in the Caspian Sea were known where the ebullition caused by escaping gas was sufficient to capsize boats.

GEOLOGY

As is the case in many other oil fields, structure is the dominating feature of the chief oil fields of Russia. Comparatively simple partial domes characterize the two great fields of Baku, but both have flanks on one side where the oil-bearing series outcrop and display those surface phenomena usually associated with oil. The whole series of Tertiary strata in which the oil is secreted consists of unconsolidated clays, sandy clays, and sands of all grades of fineness that readily break down and crumble when pierced by the drill. Their fragile nature is the cause of unusual difficulties in drilling, as throughout a thickness of over 3000 ft. there are constant irregular and ill-defined alternations of sands and clays that merge into one another in a way that makes a log very unreliable when prepared from collected samples. Some sands are charged with oil, some with gas alone, and others with oil and gas. Many of the water-bearing quicksands run freely on penetration and fill the hole to a depth of hundreds of feet. At times, too, oil-saturated clays continue to ooze into the well, rendering progress very difficult.

Certain sandy horizons can occasionally be traced for some distance and definite water and oil horizons have been located within restricted areas. Generally, however, the pliable beds have been so contorted and crushed that no single bed can be recognized for any considerable distance. Geological study was always made more difficult by the further disruption the beds sustained when rich oil sands were struck. Masses of surrounding strata were expelled on penetrating a rich oil sand; in addition, thousands of tons of sand was either ejected with oil during flows or removed with the oil during its abstraction. As much as 50 per cent. sand (by weight) has been suspended in the oil for a time, and often 10,000 tons of sand have been ejected daily, for some weeks, from a well piercing a virgin and prolific sand body.

Stratigraphy

The oil-bearing rocks of the Russian oil fields are of Miocene and, to a less extent, of Oligocene age; that is to say, occur in the deposits of the old Caspian-Mediterranean sea that surrounded the Caucasus and generally lie unconformably upon the more highly disturbed Cretaceous beds. The large area covered by these deposits, particularly along the foot of the northeastern slopes of the mountains, has enabled very many occurrences of oil to be noticed. It is these Tertiary rocks alone that present any interest, although considerable quantities of gas have occurred in the Cretaceous, a noteworthy instance being during the construction of the tunnel on the Novorossisk line. Oil seepages are likewise known in the Cretaceous rocks.

Detailed geological study of this great area has not been attempted, in fact, published maps have been confined to the districts in which development has taken place. Tabular columns are appended of the Neftianaia-Shirvanka area in the Kuban, or northwest area; the Grozny, or northeast field; and the Apsheron Peninsula, the most eastern portion and seat of by far the most important development, the Baku fields. These tables show that the Miocene rocks present similar characteristics throughout the northern flanks of the mountains but that at Baku there is a distinct facies, which is largely concealed by the characteristic Pliocene and Post-Pliocene formations of the Caspian Sea. Nevertheless, the Apsheron Peninsula as a whole presents a very complete exposure of the succession, notably in its northwest corner and in the Yasmal valley and adjacent hills farther south, while the Pliocene and younger rocks are seen around the fields and farther

TABLE 1.—*Section of Apsheron Peninsula (after Golubiatnikov).*

STAGE		FORMATION	THICKNESS, METERS	LITHOLOGICAL CHARACTER
Post Pliocene	Upper	Coastal deposits of present Caspian extending to 10 m. above present sea-level.	5-10	Sands, clay, and shell fragments.
		Older Caspian deposits forming conglomerates at a height of 12 and 26 m. and reaching to a height of 34 m. above present sea-level.	14	Sands, clays, boulders and shell conglomerates.
		Aralo-Caspian Terraces at a height of 96 and 186.5 m. (beds not disturbed.)	3-6	Limestones, sands with boulders and conglomerates.
	Middle	Bakunian (disturbed).	46	Limestones, sandstones, sands, clays and conglomerates.
Pliocene		Apsheronian.	453	Limestones, boulder limestones, oolites, shell beds, sandy limestones, calcareous sandstones, sands, marls, sandy clays, and clays; limestones predominate in the upper beds, sands in the middle, and clays in the lower; the thick clay series (110 m.) contain layers of tufaceous sands at base.
	Lower	Pontian (?)	76	Dark colored clays interbedded with sand and marl; contain gas at Bibi-Eibat.
		Transition beds.	11.3	Dark clays with interbedded gas sands; sands gas bearing at Bibi-Eibat.

TABLE 1.—*Section of Apsheron Peninsula (after Golubiatnikov).* (*Continued*).

Stage		Formation	Thickness, Meters	Lithological Character
Miocene	Upper	Akchagylian.	49.4	Dark colored clay shales and shaley clays interbedded with limestones and white tufaceous sands; the sands are gas and oil containing in Bibi-Eibat.
		Fresh water formation.	490	Clays, sandy clays, sands with clay and sand; clays predominate.
	Middle?	Unfossiliferous series:		
		First series, sand oil bearing at Bibi-Eibat.	434	Sandy clay series; sands predominate.
		Second series, sand oil bearing in Yasmal Valley.	185	Sands and sandstone with interbedded clays.
		Break in the series until the Spirialis beds		
	Lower	Spirialis beds.	98	Siliceous, calcareous and sandy clay rocks with interbedded ferruginous sandstones; in places oil bearing.
		Cedroxylon beds.		Dark colored, laminated shales with concretions of siliceous sandy rocks.
Oligocene?		Amphisyle beds.		Shales, dark, and chocolate colored, weathering yellow.
		Lamna beds.		Green, sandy clay shales with interbedded siliceous sandy rocks and white marls; oil bearing in places.

TABLE 2.—*Section of Grozny Field (after Charnotsky)*

Stage		Formation	Thickness, Meters	Lithological Character
Miocene	Meotic.	Akchagylian	Up to 425	Limestones, conglomerate of limestone pebbles, calcareous sandstone, and clayey sands, calcareous clays.
		Unconformity.		
	Middle Sarmatian formation.	Beds with fish and remains of Cetacea.	43	Calcareous (and shaley) clays with numerous limestone beds.
	Transition from Sarmatian to Mediterranean.	Spaniodontella beds.	50	Shales, sandy clays, clay sandstones, calcareous clays and sandstones, pure sandstones, water sands.
		Chokrakian.	370	Shales, sandy clays, clay sandstones, pure sandstones, calcareous sandstones, limestones, (often nodular), dolomite; sandstones oil bearing.
	Mediterranean	Spirialis beds.	?	Black shales, limestones, black nodular limestones, dolomite.

TABLE 3.—*Section of Neftianaia-Shirvanka*

	STAGE	LOCAL HORIZON	THICKNESS, METERS	LITHOLOGICAL CHARACTER
Miocene	Meotic.	Congeria Panticapea beds.		Dolomitic limestone, at base dark marls.
			Unconformity.	
	Upper Sarmatian.	Mactra Caspia beds.	25–30 +	Thin clay beds, a few thin partings of ferruginous sandstone, shell beds, gypsum.
	Middle Sarmatian.	Beds with typical Middle Sarmatian fauna.	thin	Dark gray clays, at top beds of shells.
		Cryptomactra beds.		Dark gray marls, at base thinly laminated gray marls.
	Lower Sarmatian.	Beds with Lower Sarmatian fauna.	400–500	Shell limestones.
		Beds with fish and plant remains.	in E	Dark gray marls with beds of gray thinly laminated marls.
	Middle Miocene	Spaniodon beds.	10–15	Compact marly limestones with porous partings.
	Mediterranean	Spirialis beds.	200–400	Dark marls and yellow gray marls.
		Chokrakian.	20–25	Sands and limestone, shell beds, dark marls.
	Lower Miocene.	Oil formation.	225 (Neftianaia)	Dark shaley muds not effervescing in HCl; beds of coarse sands and sandstones becoming gravels in places or conglomerates; most beds contain oil.
Oligocene	Upper		480 (r. P. shekh)	
	Middle	Foraminifera beds.	800	White clays and marls, beds of shaly bituminous marls.
	Lower	Beds with Pecten Bronni.		Green-gray marls.
			Unconformity.	
Cretaceous	Senonian.			White chalk marl with few beds of dark marls and coarse sandstones.
			Unconformity.	
	Aptian	Beds with Ammonites.		Dark sandy clays with beds of coarse sandstone.

east.   The Grozny field, forming as it does a very complete instance of a subsidiary fold, may well be referred to as a classic example of the asymmetrical anticline; it yields no exposures of the older rocks.   Still farther westward, in the Kuban area, the development has been along the margin of the Tertiaries where they are creeping around the final Cretaceous anticlinorium of the Main Range before it disappears beneath nearly level younger formation in the Taman Peninsula to reappear under simi_ lar conditions in the Crimea.

The same formations occur south and southwest of Baku and are recognizable across the Caspian. It is impossible to draw any geo_ logical limit to the oil province of Southern Russia, it must be studied in the future as part of the Eurasian Fields.

Structure

Regionally, the structure is that of folding parallel to the main ridge of the Caucasus with development of asymmetric folds along the north. east side, more disturbed conditions to the south, and gentle plunging of the Tertiaries and Pleistocene at the ends, with quite considerable local folding in the thick series of plastic rocks composing the earlier Tertiaries of these districts. Two well marked directions of folding northeast and northwest, of which the former is the older, are recognizable. The intersection of these two lines at the Taman and Apsheron Peninsulas leads to local development of great pressure, which, acting on the plastic material, gives rise to the phenomenon of "salses," or "mud volcanoes" as they are often called, in which the softer underlying strata are squeezed out in the form of mud, associated with much salt water and gas, the latter being composed of hydrocarbons and at times emitted on a grandi- ose scale. The exuded material forms considerable hills, at the top of which a crater shows activity, even in times of relative quiescence. The rather complex structures resulting from the intersection of these folds are often hidden beneath the Pleistocene rocks, which partake of no folding, and the Pliocene which have suffered slight deformation. This is especially the case in the Taman peninsula. It is also the case, to a certain extent, in the Apsheron peninsula though good exposures exist.

The Apsheron peninsula is built up of Pleistocene and Tertiary rocks and both the Pliocene and the Miocene rocks are well developed. It is the Upper Miocene that carries the pay so far developed. The shell limestones of the Pliocene, which form the Baku building stone, are less easily denuded than the Miocene rocks and so form escarpments around the oil fields, usually bordering plateaus, while the Miocene rocks, where exposed, form gently undulating country.

The Akchagylian (Upper Miocene) contain abundant fish remains and thin beds of volcanic ash. The only known occurrence of pay in them is at Surakhany; it is of the filtered type and, therefore, secondary. The fresh-water formation contains shell remains and algæ. It covers large areas in the older fields but is concealed at Surakhany. The underlying unfossiliferous series forms with it the source of the bulk of the oil hitherto won in Russia.

The unconformable Lower Miocene, capped by the hard siliceous limestones with very characteristic casts of Spirialis, which form a useful mapping horizon, is petroliferous; especially noticeable are the Oligocene fish shales, but these strata have yielded no pay.

Four directions of folding occur, northwest, meridional, latitudinal, and north-northeast. The first, that of the main Caucasian ridge, prevails at Bibi-Eibat, Holy Island, and over much of the northeast part of the peninsula; the second at Surakhany. Many folds are subject to change, and complex structures ensue, in the formation of which faulting stands in close relation. Faults are usually of small throw and coincide with the axial crests, but three important lines must be noted as they dominate the Peninsular structure. The first is the circular uplift following the ridges Kabiriadig-Puta, Atashka-Shaban-Dagh, and Kobi-Bos-Dagh; this lies west of any important developed area. The second runs along Atashka-Shaban-Dagh, Gyokmabj-Khurdalan, and Binagady; and the third through Fatmagi-Dygia, Kir-Maku, Balakhany-Saboontchy-Romany, and Surakhany-Zykh. These two lines form, with the Bibi-Eibat fold, a horseshoe line open to the sea; it is along this line that the big production has been obtained. The last important tectonic movements took place at the close of the Upper Pliocene period, the next before was at the commencement of the Upper Pliocene, while between the Middle and the Lower Oligocene much greater dislocation took place.

Oil Occurrence

The following horizons are known to carry oil:

1. Middle Pliocene: lower Apsheronian at Surakhany and Romany.
2. Lower Pliocene: Pontian at Bibi-Eibat.
3. Akchagylian: at Bibi-Eibat, Surakhany, Romany.
4. Fresh-water beds: at Bibi-Eibat, Surakhany, Romany, Saboontchy.
5. Unfossiliferous beds: Puta, Atashka, Khurdalan, Binagady, Kir-Maku, Balakhany, Holy Island.
6. Lower Miocene: (Spirialis) Puta, Atashka, Kobi, Gyokmabj, Khurdalan, Binagady, Holy Island.
7. Amphisyle beds: Sumgait.
8. Lamna beds: Western hills from Kobi mud volcano onwards.

The impregnation is sporadic, thus the phenomenally rich fresh-water beds beneath Bibi-Eibat and Surakhany show no signs of oil on their outcrops in the adjacent Yasmal valley and Zykh, respectively. These facts lend weight to the hypothesis that their oil is in secondary accumulation, which would have been facilitated by the downward increase of permeable strata and the big unconformity above the Lower Miocene. The temperatúres of the oil at Bibi-Eibat have been acquired in accordance with the geothermals, hence the invasion must be mainly a long accomplished fact.

Of the eight horizons, the third has importance in Surakhany for gas and "white oil;" the fourth is the most prolific and is of value in Surakhany, Saboontchy, Romany, and Bibi-Eibat; the fifth in Balakhany, Binagady, Holy Island, Khurdalan, on Atashka, and near Puta Station.

Surakhany.—The Pliocene and Miocene rocks are folded into a broad, flat anticline striking north and south. Eastern dips are 10° to 20°; the western 4° to 10°. The visible fold continues to the north end of Surakhany Lake and southward toward the faulted area at Zykh. The crest is much faulted and in the salt lake are numerous fissures filled with inspissated oil, while over a wide area fissures emit gas. The oil zone having now been proved, by drilling, to continue through to Romany, it is probable that the two areas are on one curving fold.

Gas springs occur over the whole central part of the district, in the Lake, on the hill Atashka ("eternal Fires"), at the Temple, and in the depression of Karatchkhur. To sink 20 to 30 ft. is enough to strike gas. The upper gas beds cover an area of 4800 m. by 1500 m. (1780 acres) and represent the apex of the fold. Outside this area, the same beds contain no gas. Drilling has shown that all porous beds in the upper strata contain gas; the lower, gas and oil. From 36 m. (118 ft.) to 480 m. (1575 ft.), twenty-three gas sands were struck, the pressure at times reading 30 atmospheres.

White oil begins at 200 m. (656 ft.), at the top of the Akchagylian, and increases downwards. Only the central region is oil bearing. From 200 to 335 m. (656 to 1099 ft.) five white oil sands were struck. The specific gravity is 0.785.

Black oil was first struck at 480 m. (1575 ft.) in the fresh-water formation, as previously determined by Golubiatnikov. During recent years, great development has taken place below this. The first horizons yielded oil having a specific gravity of 0.820.

Balakhany, Saboontchy, Romany.—A wide, faulted anticline represents the continuation, southeast of the mud volcano Bog-Boga, of two folds, that of Kir-Maku to the northwest and Binagady more westerly. The fold pitches to the southeast, disappears beneath the Pliocene beyond Romany, but is probably continuous with Surakhany. The fresh-water formation covers most of the field, the unfossiliferous series appearing in the salt marsh of Kir-Maku. The oil in the upper beds at Romany is lighter, in the lower beds at Balakhany heavier.

Bibi-Eibat.—This is a dome plunging north-northwest and possibly beneath the sea. At its crest are two subsidiary parallel domes, one with its apex on Group XIX, the other roughly on Group XX. Between them, the shallow minor syncline forms the low hill running out to sea at Cape Naftalan. Much minor faulting, usually parallel to the major axis, occurs. The fresh-water beds are exposed around Naftalan where the best production has been obtained. The beds are very uniform lithologically, the sands being fine to medium grained, the former like dust and known as "gas sands" to the driller, while alternating clay and fine sand is termed "gas clay." "Water sands" are cemented with lime, "oil sands" are loose and when saturated with gas and oil are thought to resemble caviar. The occurrence of hard concretions of sandstone probably accounts for the removed cementing material.

Ignoring the beds of inspissated oil and gas and oil shows in upper beds, the first important pay was struck at 280 ft. Other oil sands were struck at intervals but were exhausted down to 700 ft. before any study of the field was made. Between here and 640 m. (2100 ft.) were twenty workable sands with a total thickness of 120 m. (390 ft.). Water sands were few and it was only below 2600 ft. that the predominance of sands made water-shut-off of such importance. Water occurs in all sands and made the marginal plots unprofitable.

Binagady, Khurdalan, and Puta.—These areas are geologically much alike. Oil is obtained from the unfossiliferous series, which are folded into much faulted anticlines. The oil is heavy. Binagady became prominent as a producer during the war; the other fields are classed in the "hand dug" production.

Holy Island.—The Pleistocene beds here rest directly on the unfossiliferous beds, very much disturbed, with the Spirialis horizon just showing to indicate the succession. The fold is an elongated dome with the major axis northwest; it is asymmetrical with overfolded side in places. It is much faulted and there is evidence of other domes outside the developed area (in the northern part of the Island).

Seepages occur at the southern end of the anticline and in the central salt marsh and mud volcanoes, etc. along the crest. The oil has infiltrated into the Pleistocene and has formed Kir deposits in the sandstones of the northeast part of the fold. This part has been developed with the drill, the wells giving 800 poods daily from 1300 ft. The specific gravity is 0.944.

Grozny.—Outside the Apsheron Peninsula, the most important field is Grozny. Here are two folds. The old field is an elongated asymmetrical dome, slightly bulging outwards on its steep side, accompanied with dip faulting, which marks out distinct provinces as regards water and richness of pay. Dips on the north vary from 40° to 90°; on the south, from 20° to 30° and flatten out at the ends to 6° to 15°. The length is 9 mi. west northwest.

The new Bellik field is a nearly symmetrical fold and lies to the east of the old field.

As shown by the columnar section, Table 2, the beds are of Miocene age, the oil occurring in the Chokrakian (transition Mediterranean-Sarmatian), and they do not outcrop. Just at the apex of the old field, the overlying Spaniodontella beds are exposed. The oil occurs in sandstones associated with shales, sandy clays, limestone, and dolomites; whether *in situ* appears doubtful.

Maikop.—Here oil occurs in beds of Upper Oligocene age, in a succession of shales, marls, and sandy beds. Beneath is a thick mass of foraminiferal marls, above are the Mediterranean-Sarmatian beds of the Chokrakian limestone. The Tertiaries lie unconformably upon denuded Cretaceous rocks in the oil-field region of Shirvansky, no Eocene beds being interposed, but farther west these latter appear. The oil sand is a narrow strip down the dip and probably represents a former river bed. It is sealed, by overlapping against some of the Cretaceous islands penetrating the foraminiferal marls; the dip of the Tertiaries is 10°. The Pioneer well was able to show a yield of 375,000 bbl. from the shallow depth of 281 feet.

About 140 km. (87 mi.) westward lies Kudako where the first Russian gusher was struck, in February, 1866, at a depth of 70 ft. with a reported yield of 1,000,000 poods (120,500 bbl.). Subsequent wells reached 700 to 1050 ft. under conditions similar to those at Maikop. The specific gravity of the oil was 0.840 to 0.865. Another small pool was opened by Tweddle in the early eighties, by the river Il, with production from both the oil sand and the overlying Chokrakian, the latter being a heavy oil.

Berekei.—Here a much faulted anticline occurs involving much the same horizons as at Maikop, but oil occurs in many horizons and, being associated with hot water, may come from some depth. Its specific gravity is 0.868.

Cheleken.—The actual productive area is in the southwest corner of the island, where there is a dome with its major axis northwest, but with dips southwest of 15° to 50° and northeast of 18° to 20°. It is much faulted parallel to the axis and the steep side is involved in a trough fault, whence the best production of late years has been obtained. The dome itself has produced for many years. Toward the center of the island, the larger dome of Chokrak has many oil indications but the productive Pliocene beds have been denuded and the underlying continental formation of unknown age is exposed.

Emba.—Here the surface is entirely covered by Pleistocene beds and the subsoil structure can only be explored by the drill. Salt masses occur and the detailed geology has not yet been worked up.

Ferghana.—The Syr-Darya (Jaxartes) valley is a Cretaceous-Tertiary basin lying between the western continuation of the Tienshan mountains and the Zarafshan-Chain. Oil indications occur in the margins of the Cretaceous rocks and are associated with rather complex secondary anticlinal structure often partly concealed by Pleistocene rocks. The worked fields of Chimion are on the southern margin.

DRILLING OPERATIONS

Owing to the highly disturbed and unconsolidated sediments in the Baku oil fields it has been found impossible to adopt the standard American cable system or even the rotary. The need for wells having exceptionally large initial and completed diameters is due to the necessity of excluding waters and penetrating swelling and caving ground during progress, as well as to permit of the extraction of oil by bailers; consequently, the "stove-pipe" system is mostly employed. Initial diameters of 36 to 40 in. (91 to 101 cm.) are usual when ultimate diameters of 12 to 14 in. are desired at a depth of 2000 ft. Massive surface gear is necessary to manipulate columns of such size and tools of such weight. Engines or motors of 50 to 60 hp. are usually employed to drive the rigs.

Because of the enormous volume of sand expelled or raised with the oil, the drilling difficulties of the Baku oil fields rather increased with the development of the field, thereby neutralizing the favorable influence of natúral improvements that were gradually introduced. Usually from 1 to 3 years were occupied in drilling the deeper wells, and their cost, in pre-war days, was not less than $25 a foot; nearly 50 per. cent. of this sum was for the casing alone.

Amid such disturbed and loose sediments, no water-flush system was permissible as the water freely entered partly exhausted sands and found access to all surrounding wells, from which it was bailed. Away from the old fields, as at Surakhany, where a considerable thickness of more consolidated, non-petroliferous (or slightly so) beds have to be penetrated before the normal loose, oil-yielding facies is reached, rotary drilling has proved successful and greatly accelerated progress has been made.

The system in vogue is the free-fall which, being operated by rods from the walking beam, transmits a positive action to the drill, enabling tools to be rotated against a resistance and the motion of underreamers to be positive. Wire-rope cable drilling has been successfully performed in the Baku oil field under skilled direction, but the risks are great and the ultimate speeds never exceeded those of the free-fall drilling system. Rarely could more than a few feet be left unlined without danger, and often 70 per cent. of the time was occupied in the maintenance. of the freedom of the column of casing to insure its descent of only a few hundred feet.

Unlike Baku, the Grozny strata are much more compact, and although many of the productive beds are unconsolidated sands which are freely expelled with the oil, the intervening beds hold up sufficiently to permit the employment of standard wire-line cable drilling, consequently quicker work results. The pre-war cost of wells 3000 ft. deep did not exceed $12 per ft., of which about 50 per cent. was for the casing. In the Ural-Emba oil field, where great thicknesses of gypsum and salt have been pierced and the ordinary sediments are fairly consolidated, the rotary drill has proved highly successful.

Many of the companies operating in the Maikop oil field used the old free-fall system, others used Galician rigs, but in the shallow field portable Star rigs were found to accomplish just as fast work as the others, while making unnecessary the use of expensive derricks, and lengthy dismantling and re-erection of plant on the completion of each well.

Cheleken conditions resemble those of Baku and wells entail long and costly work to complete.

The main feature distinguishing Russian from American practice is the design and employment of positive tools, owing to the great difficulties of dealing with loose sediments and the enormous financial losses sustained by the abandonment of a well that has required several years to make and on which perhaps $50,000 or more has been expended. Of exquisite design and workmanship, many of the fishing tools cost thousands of dollars, and they were invariably used on solid or hollow fishing rods, which permitted the most delicate handling and certain release if they failed in their object. All fishing rods had a loose collar joint and feather so that they could be rotated right- or left-handed at will; reliance was never placed on a trip movement, as is the case with many American fishing tools. Owing to the heaving nature of the beds, it was often essential to employ powerful water flushes to free the material around a lost bit; for this purpose 3-in. pipes were customary.

The enforced use of riveted stove-pipe casing rendered cementations for water exclusion lengthy and delicate operations, as the failure to hold water or resist pressure without leakage prevented the simpler American circulation systems from being adopted. Anything beyond a shoe cementation made it imperative to fill the casing with earthy matter and its subsequent extraction was often as difficult as drilling a new well in most countries.

Production Methods

During the early history of the Baku oil field, practically all wells penetrating a virgin sand body of any importance flowed so violently that their effective control was practically impossible. Enormous masses of sand mixed with boulders and pieces of rock were often ejected for days and weeks, rendering approach to the well dangerous. Single wells have given for weeks as much as 10,000 tons daily of sand mixed with an equal weight of oil, and all objects placed to obstruct or deflect discharge were destroyed or perforated in a few hours. Usually hardwood blocks or chilled, cast-iron plates, 12-in. thick, were pushed over the mouth of the well some distance above the ground, and the vertical jet was thereby deflected horizontally. These "fountain shields," as they were termed, were replaced as they became destroyed.

The wells themselves did not escape damage as the casing was often torn to shreds, each soft rivet causing the initiation of a vertical rifling that extended upwards as the sand-blast action continued. In certain regions, when excessive gas pressures were encountered, well after well was sunk and destroyed after a few days, eruption before the pressure was relieved sufficiently to permit normal development. Occasionally, sand only or oil-soaked clay would be expelled for days, or even weeks, before oil entered or deepening could be resumed. On the cessation of flowing, the wells were often in a delicate condition and remunerative yields, free from water that entered the damaged casing, could only be secured by the maintenance of a high head of oil that, usually, exceeded the static head of upper water sources. Such a condition could only be effected by keeping the well clear of sand at the bottom and so facilitating the entry of oil. A little water that practically always gained admission nd collected near the base of the well not only served to compact the sand, but greatly impeded the entrance of oil; consequently, the water had to be abstracted at regular intervals. The only method of handling such a condition was by bailing, and the scientific application of this principle reached a high degree of efficiency in the Russian oil fields.

Bailing drums 16 ft. in circumference were driven by engines developing up to 150 hp. each, and velocities of 1500 ft. per min. were common. Single bailers up to 7.5 bbl. capacity were used in large-diameter wells of great yields and productions up to 2000 bbl. daily were raised by this method. The mean cost of bailing Baku wells in pre-war days, averaging 120 bbl. a day each, was under 20 c. per bbl. but the large yielding wells individually would cost but a fraction of this to bail.

The only other process for raising oil that met with success under the early Baku conditions was the air-lift. In wells of small diameter with a high level of liquid where bottom bailing had to be frequent to remove water and sand accumulations, the air-lift proved very successful. Although the cost of operating the air-lift greatly exceeded that of bailing, the excess costs were often repaid many times by the augmented yield. Emulsions in some cases gave trouble, but usually sandy water and oil were alternately discharged at more or less regular intervals during continuous or intermittent working as the case might be. In low-level wells all discharges were intermittent in operation; in high-level wells, the discharge was continuous.

With the gradual reduction of gas pressure on the Baku oil fields, opportunities arose for the use of pumps in the less sandy wells, but the constant need for renewals of cup leathers, barrels, or plungers has caused the system to be unpopular. The oil is never quite free from sand and, as the density of the oil is light and sand quickly sinks in the column, there is a tendency for the plunger and valve to become choked up if left idle for a few minutes.

There are fewer objections to the use of pumps in most of the other oil fields of Russia but while the period of intermittent flowing continues bailing is preferred in order to keep the well clear of sand. Deep well pumps are used in the Grozny and other fields when the wells have settled down.

OIL-FIELD YIELDS

The Baku oil fields are by far the richest yet discovered. Not merely are the loose uncemented sands capable of high absorption, but they are plentifully distributed throughout a depth of several thousand feet of sediments and often reach a considerable thickness. Thus, within the confines of a single plot, a dozen highly productive sands may be struck aggregating several hundred feet in thickness. A selected plot at Bibi-Eibat has yielded nearly 2,500,000 bbl. per acre, and the whole operated area of 250 acres in that field has produced over 1,500,000 bbl. per acre. Even the greater Balakhany-Saboontchy field of Baku, aggregating about 2600 acres, has yielded fully 500,000 bbl. per acre and is still capable of enormous collective production, though the individual output of wells is now small. Enough oil has been abstracted from this field to cover the whole area to a depth of 63 ft. neglecting entirely the many millions of cubic feet of gas with its contained gasoline that has been lost.

The influence of interference and the process of exhaustion is, perhaps, best illustrated by the steady decline of initial productions of new wells. Between 1892 and 1896, the first half yearly output of new wells was around 108,000 bbl. (600 bbl. per day). In 1912, this had fallen to 15,000 bbl. (80 bbl. per day) and during the same interval the ultimate yield had sunk from 675,000 bbl. per well to about 225,000 bbl. In 1895, the average annual production of wells at Baku was 75,000 bbl.; in 1909, the average had been depressed to 30,000 bbl. although in the same period wells were on an average 60 per cent. deeper. In the Bibi-Eibat field, footage ceased to increase or even sustain production after 1904 when the zenith of production was attained in that region.

Civil disturbances for some years prior to the war and general disorganization since make any estimates and predictions of little value. The fields are still capable of giving enormous quantities of oil, and their present potential capacity is probably between 25,000,000 and 20,000,000 bbl. a year. Much local speculation is aroused as to the results that will attend. the drilling of reclaimed plots in Bibi-Eibat bay, the Great Lake of Romany, and large reserved areas surrounded by old producing plots.

An unusual amount of scientific interest surrounds the obsequies of these famous Baku fields, and it would be a world's loss if trustworthy data were not kept for the benefit of our successors. The final phase is apparently in sight, as what appears to be basal water has been penetrated beneath the great oil-bearing series. Considerable thicknesses of water-bearing sands exist, but whether these are underlain by other oil-bearing sands it is difficult to say. One would surmise not, and there is just the possibility that the upper riches may be partly due to the expulsion of the former oil contents of these beds by water.

At present, the dregs of these vast oil fields are being mainly secured through the medium of water which has percolated into the disrupted and badly disturbed beds from which sufficient solid matter alone has been flung by thousands of wells to raise the oil-field surface many feet. From the thousands of wells a mixture of oil and water is constantly being raised, but both the oil and the water contents are being reduced. Some areas no longer yield water at all where formerly expensive measures had to be undertaken for its exclusion; here the least oil is now obtained as natural filtration without the aid of gas or water is insignificant. At other points the static head of the liquid is gradually falling, and unless the lower water is admitted the whole field may be eventually dried up. No synclinal or edge water that cannot be overcome by bailing encroaches on the exhausted upper oil strata, so that the chief migration of oil may have been vertical rather than lateral.

The Surakhany field to the east of the Balakhany-Saboontchy area has now become the most interesting in the Baku zone. Years ago, the deep development of the oil series was predicted by geologists and their predictions have been verified. Enormous volumes of natural gas were obtained from shallow fine sands in the area and the product was piped to the oil fields for fuel. At increased depths, gushers of white oil, specific gravity 0.785, were struck in similar sands, and there is now little doubt that they represented a filtration product of migration from the underlying normal series.

Large gushers of the typical Baku grade oil have been struck in the Surakhany area where an active development was in progress until the war.

In the Grozny oil field, wells have given very substantial productions along a belt of many miles. No area has excelled in productivity the original plot on which the first well was drilled, about 1897. This point corresponded with the maximum elevation on the pitching anticline and attracted attention by its surface manifestations. This plot has yielded over 320,000 bbl. of oil per acre. In 1914, Grozny yielded 10,500,000 bbl. from about 8000 acres. The field has given about 150,000 bbl. per acre and is still far from exhausted. After an initial flow, wells continue to yield normally. A typical well yielding an ultimate production of 80,000 bbl. gives about 50 per cent. of its total production in the first year, 22 per cent. in the second, 15 per cent. in the third, and 7½ per cent. in the fourth.

Exceptionally good results were obtained on the Bellik oil field, discovered in 1912. It is really an extension of the old Grozny oil field or a parallel fold. Pioneer wells gave large and sustained flows, and an important field is likely to result, possessing the Grozny characteristics.

No detailed statistics are available concerning the important Emba oil field of the Uralsk. Large flowing wells were struck in some number near Dossor and considerable shipments of oil were made to the Volga. There is every indication of a large and useful field being opened up. The Island of Cheleken, off Krasnovodsk, gave its maximum output in 1912, when 1,500,000 bbl. were reported. In 1913, the production was under 1,000,000 bbl. and the fall continues. Activity was confined to the Ali Tepe sector, which area seems to have passed its best days. Holy Island has attracted sporadic attention and appears to justify more. One company operates and produces about 800,000 bbl. a year when conditions are normal.

The small field near Shirvansky, known as the Maikop, has yielded over 4,000,000 bbl. of oil; a production of about 250,000 bbl. a year is still maintained despite the fact that the area has not been greatly extended.

Little information is forthcoming about Ferghana oil field of Turkestan. At one time the production reached 450,000 bbl. a year and the oil found a ready local sale in that part of the world. At Maili Sai, no further drilling has been undertaken.

Innumerable abortive or uncertain tests have been made in the Caucasian oil belt. Some were undertaken at a time when nothing short of 500-bbl. wells attracted any interest at all. At many spots productions have been obtained that would pay well in any other part of the world. All along the Caspian Sea littoral, from Baku to Derbent, there are frequent and encouraging indications of oil. At Berekei, a little northwest of Derbent, about 500,000 bbl. of oil were taken from a field in which 2000- to 5000-bbl. wells were struck; but water troubles eventually drove away nearly all operators. At Kaikent, a large gusher was struck in an initial effort, although all subsequent wells failed. Around Baku at Binagadi, a much despised region left for years to peasants to develop by primitive methods gave, in 1913, 1,750,000 bbl. of oil and has since exceeded 4,000,000 bbl. a year. At Puta and Khardalan, a thriving industry developed as a result of hand-dugs sunk into the outcropping oil sands and an output of 1,700,000 bbl. resulted in 1914. Drilling would yield very different results.

Miles of territory flanking the Caucasus foot-hills that fringe the valley of the Kura are capable of remunerative development. In some places small wells, still flowing, testify to past efforts, and at many places the peasants satisfy all local wants by sinking shafts into outcropping oil sands.

At Ildohani in the Tionct valley, near Tiflis, productive wells were sunk that flowed oil of light density in which crystals of wax separated; at other places in the same district fruitless experiments were made with antiquated plants where modern methods might have succeeded. At Chatma, near the River Jora, there are numerous indications of oil and interesting structures.

Miles of the Black Sea littoral and Taman peninsula are potential oil fields; indeed, many dozens of productive wells have been sunk in that region where such stupendous mud volcanoes are in evidence. Russian geologists have estimated that there are fully 30,000 sq. mi. of interesting undeveloped oil land in Russia and this is probably no exaggeration.

Chief Russian Oil Fields

The main oil fields of Russia, in the order of their relative importance, are as follows:

				Approximate Production, in Barrels	Approximate Proved Area, Acres
North Caucasian	Baku	Balakhany-Saboontchy field, Romany, Bibi-Eibat field,	to 1918	1,597,690,000	2,600
				....	1,000
		Surakhany field, to 1918		54,920,000	
	Grozny	Old Field, Bellik field,	to 1917	139,858,000	8,000
	Binagadi, to 1918			22,620,000	
	Khurdalan Puta Berekei	to 1917		9,082,000	
	Holy Island, to 1918			5,562,000	
Emba Ural North Caspian	Dossor, etc., to 1917			8,575,000	
Tcheleken East Caspian	Cheleken Island, to 1917			7,317,000	
	Ferghana field, to 1917			3,620,000	
West Caucasian	Maikop, to 1917			4,750,000	

TABLE 4.—*Approximate Production, in Thousands of Barrels (8.3 Poods to a Barrel)*

	Previous to 1914	1914	1915	1916	1917	1918
Baku Oil Fields						
Balakhany-Saboontchy	⎫ 1,427,950	32,000	31,800	29,000	24,600	12,500
Romany..............	⎭		.			
Bibi-Eibat...........		8,700	9,550	10,800	7,350	3,440
Surakhany...........	12,400	6,200	7,270	11,600	11,150	6,300
Binagadi.............	7,380	2,650	3,930	4,160	3,560	940
Holy Island..........	1,980	712	844	820	844	362
Khurdalan, Puta, etc...	5,300	840	930	1,180	832	
Terek						
Grozny field..........	⎫ 93,800 ⎧	10,600	9,260	8,280	6,620	5 gushers
Bellik field..........	⎭	1,028	1,350	4,100	4,820	burning 9 months
North Caspian						
Emba oil field........	845	2,000	1,990	1,870	1,870	
Kuban						
Maikop..............	2,870	482	915	241	242	
Asiatic Russia						
Cheleken............	5,670	602	482	362	201	
Ferghana............	2,680	217	241	241	241	

DISCUSSION

ARTHUR KNAPP, Shreveport, La.—From this paper one would be led to believe that the American system of drilling was not a success in Russia; I spent two years in the Baku field and know that this is not true. The first rotary rig that I know of was sent over there in 1913. The Russian engineers were so opposed to the use of the rotary that it was not until 1914 that a well was drilled using American methods throughout.

The first rotary hole was drilled 1800 ft. (548 m.) in about three months and offset a well that took 2½ years to drill by the Russian method. The Russian method used about 120 tons of casing while the American method used only two strings, 10 and 6 in., with a saving of from $75,000 to $80,000. During the next two years, between eighteen and twenty rotary wells were finished in the field. In every case there was a saving of from 20 to 60 per cent. on casing and from 30 to 50 per cent. on time and labor.

The fundamental difference between the Russian and American systems of producing oil is that in America we try to keep the oil formation from moving, by the use of screens, where necessary, while the Russian engineer does everything he can to produce a large quantity of sand. The drilling with the rotary is about the same as the drilling in the Midway field in California.

The Russian engineers have opposed the use of deep well pumps as a means of producing oil. I installed an ordinary 2-in. pump in a well that produced about 12 bbl. of oil and increased the production to 24 bbl. on the beam. Our statistics showed a saving of 50 per cent. over bailing. Another company under English control put about twelve wells to pumping on jacks. They were run for some time and the cost compared with the same kind and number of wells being produced by bailing. The pumping wells used about 25 per cent. of the steam that was required by the bailing wells. The labor costs were 12 per cent. and the repairs and upkeep 5 per cent. of the bailing costs.

The deepest wells in the Baku fields at the time that I was there were about 3000 ft. deep. They took at least three years to drill by the Russian system and cost about $125,000. Only one out of three of the Russian wells at this depth was a successful producer. The uniform success of the American system was very much in its favor. Out of the fifteen or twenty wells drilled during 1914–15, only two of the American wells were lost.

The American rotary system is being very rapidly adopted. When the war stopped imports the Russians tried with good success to make rotaries of their own. The rotary may never entirely supplant the Russian rig but it has been a great success and has come to stay in Russia.

Mr. Thompson's paper is the only one that I know of that brings our knowledge of the Russian oil fields up to date. It is a valuable addition to our literature.

A. BEEBY THOMPSON (author's reply to discussion).—Mr. Arthur Knapp's remarks, without some qualification, are apt to be misleading. In the past, the leading oil companies of Baku have spent large sums in experimenting with American plant and have offered high rewards to any successful operator who could increase speeds and reduce costs but a small percentage. The best operators were sought and every facility granted them but until just before the war no improved results had been achieved; indeed, until the modern heavy type rotary was introduced the problem appeared hopeless. Any driller who could have saved but 20 per cent. in time or costs of drilling wells in the rich Baku oil field could have made contracts that would have yielded him a fortune in a few years, as the time factor in such congested areas meant so much to operators. Where wells are drilled within 100 ft. or less of each other and a few square miles are perforated by thousands of wells, it is almost inconceivable that rotary flush drilling could be uniformly successful, for the mud enters the loose, partly exhausted sands and follows channels of flow to neighboring producing wells. Attempts in Bibi-Eibat many years ago caused many wells to turn to mud and water when the rotary penetrated one of the main sands from which the wells were drawing oil.

Rotary drilling has proved successful only in areas outside the congested fields where great thicknesses of uninteresting beds have to be pierced before the productive oil series is reached or where oil occurs in more compacted strata. Any operator able to save $75,000 worth of casing and nine-tenths of the time of drilling in proved areas could within a few years be a wealthy man.

The Russian engineers have not opposed pumping on principle, as in Grozny and Maikop pumping has been conducted for years. As far back as 1900, the writer made persistent efforts to use pumps in the Baku oil fields but the large quantities of sand accompanying the oil made their use impossible. In no case could a highly productive well be pumped for more than a few hours without the pump being choked by sand and the cups or plungers being cut to pieces. Induced flows through the pump sometimes brought in sufficient quantities of sand to fill hundreds of feet of the tubing. At that time no 12 or 24 bbl. well was accepted and probably 100 bbl. was the minimum payable yield. Conditions are quite different today and there are many Baku properties where the slowly infiltrating oil is sufficiently free from sand to enable grouped pumping to be conducted. In the past, the output of a well fell off to an unpayable yield unless the sand that entered the well was constantly being removed by bottom bailing.

Petroleum in the Argentine Republic

BY STANLEY C. HEROLD,* TULSA, OKLA.

(New York Meeting, February, 1920)

AT THE present time five localities in the Argentine Republic are known to bear direct evidences of the presence of petroleum. The segregation of these localities is more or less arbitrary inasmuch as minor indications may be found to extend from one locality to the other at no regular distance apart, especially in the northern and western part of the republic. These localities are listed as follows: North Argentine-Bolivian region, Salta-Jujuy district, provinces of Mendoza and Neuquen, Comodoro Rivadavia, and the Gallegos-Punta Arenas region.

Economic conditions attract us to the possibilities of developing these and other regions of countries in the southern hemisphere. Development work will, naturally, be undertaken first in such localities as present direct manifestations of the presence of petroleum; "hidden fields" may exist, but, unless discovered by accident, their development will be left to the last.

The problems to be solved in the development of the petroleum resources of the Argentine republic are mainly of stratigraphy, structure, and transportation. We are not here concerned with the unfavorable climate of the countries to the north in the tropics where, for us of the "far north," life hangs by a thread ready to be severed by a mosquito, gnat, or tropical germ.

NORTH ARGENTINE-BOLIVIAN REGION

The North Argentine-Bolivian region has already been described by the author.[1] Geographically and geologically this is admittedly one field extending from Argentine into Bolivia. It is not necessary to repeat here the various conditions pertaining to this field, though the summary may be quoted as follows:

Extending from northern Argentine northward into central Bolivia is a belt of petroleum seepages. On account of the remoteness of the district it has, heretofore, been little considered by oil operators. The regional geology is comparatively well understood but the local features have not been carefully detailed.

Development work in the past has been done on an unscientific basis and has led to failures. At the present time, access to the region is somewhat difficult but no

* Chief Geologist, Tulsa District, Sinclair Oil and Gas Co.
[1] *Trans.* (1919) **61**, 544.

serious problem would be encountered in improving the conditions. The nearest railroad terminal is at Embarcación, 114 mi. (183 km.) south of the Bolivian border, or 72 mi. (116 km.) from the nearest manifestation of petroleum in natural springs.

The oil is of high quality and the seepages occur in creek beds along the Sierra de Aguaragüe fault, and at other isolated places.

Native labor is good and government policies are sympathetic toward foreign exploitation.

Though the structural features of the region, as a unit, have been worked out by reconnaissance surveys, there still remain many local sections upon which no detailed study has been made.

Several small areas have been proved unfavorable for production, though the region as a whole cannot be condemned on this account.

Since writing the above there has been no further development in this district, to the author's knowledge, though individuals have had their geologists there.

THE SALTA-JUJUY DISTRICT

The Salta-Jujuy district lies to the west of the foregoing area, northwest and north of the town of Salta, extending into Bolivia. There may be no logical reason for separating these fields except that the latter lies in the mountainous and somewhat inaccessible part of the country. The stratigraphy of one is closely related to that of the other. Seepages are small and widely scattered, of high quality oil, and not of continuous flow, for heavy rains may either temporarily efface or bring to light slight showings of petroleum, depending on local conditions.

The structure of the district is rather complex due to the folding and faulting of the strata lying on the side of igneous formation protruded in the Andes uplift. As the surface is made up of steep mountains and narrow gorges largely, there is small probability of extensive development. The seepages occur along the exposures of beds dipping at high angle and along faults.

PROVINCES OF MENDOZA AND NEUQUEN

These two provinces lie on the eastern flank of the Andes Mountains due west from Buenos Aires and adjoining the Republic of Chile. The province of Mendoza is traversed by the trans-Andean railway which extends from coast to coast. Seepages, generally of tar or asphalt and heavy oil, extend in a north-and-south line along the frontal ranges, paralleling the main trend of the range. The author made but a casual observation in this district and is therefore not competent to enter a detailed discussion of stratigraphic conditions. The main features are beds of steep dip and numerous faults. Some development has been undertaken in the past but up to autumn of the year 1917 no success had been met. The area has its possibilities.

COMODORO RIVADAVIA

At Comodoro Rivadavia is situated the only successfully developed oil field in the Argentine Republic. It is located in the southeast corner of the territory of Chubút along the Atlantic seacoast, immediately north of the town of Comodoro Rivadavia, on the Gulf of St. George (San Jorge) at approximately latitude south 45° 45′ and longitude west 67° 20′ from Greenwich. From Buenos Aires, the field lies in a direction of south 30° west, 1164 mi. (1875 km.) distant, as the ships sail.

The area includes the reserved land of the Argentine Government of 5000 ha. (12,050 acres), 5000 by 10,000 m. along the coast covering the town of Comodoro Rivadavia itself. Furthermore, it includes various areas adjoining this reservation to the north, west, and south. Three properties were producing petroleum in the latter part of the year 1917: namely, the government reservation, the Compañia Argentina de Comodoro Rivadavia, and the Astro property, the latter situated about 20 km. north. Many claims or concessions have been taken up by local parties, some of which appear to be favorable for production.

Previous to the accidental discovery of oil by drilling for water there were said to be absolutely no signs of the existence of oil or gas under the surface in this region. The domestic water supply of the town of Comodoro Rivadavia was very poor and in such condition as to render the district unhealthy. In the year 1908, the Argentine Government sent a drilling outfit there to prospect for water, a site having been chosen opposite the bank building in town. No water was encountered so the drill was removed to a place 3½ km. north. Drilling proceeded until a strong flow of gas was encountered and later a gusher of oil at 1770 ft. (540 m.) below sea level. The well was probably not over 70 ft. above the sea.

Since discovery, drilling has been continued with more or less regularity until, in the latter part of 1917, about sixty wells had been put down on the reservation, a fair proportion of which proved successful. Water is now brought from the hills to the west through pipe and supplies all requirements.

Within the government area, at least three sedimentary series exposed at the surface have been differentiated. The lowest stratum exposed is that of a white, soft, tufaceous formation lying at the base of the hills immediately to the north of the oil field. Traces of carboniferous matter have been found in this formation but no fossils capable of recognition were on record at the time of the author's visit. Its age was considered to be Lower Eocene or possibly Upper Cretaceous. About 50 ft. of the series is exposed.

The next younger formation is a series of sandstones and shales. The sandstone is light brown in color, soft, and easily eroded. Sand grains are of medium size. Beds vary in thickness from 10 to 50 ft., bedding planes well defined. The shale is also light brown where exposed and very soft. The entire thickness of the series is approximately 200 ft. Fossils of this formation were considered to be of Eocene Age.

The third and youngest stratified series is the so-called "Patagonia" series, a formation composed largely of soft, light brown, thinly bedded sandstone. Some shale occurs. It is this formation that stands in high cliffs to the west of the field. At least 300 ft. of the series is exposed in the immediate vicinity. Fossils are very abundant. They are probably of Oligocene age.

In addition to these stratified deposits there is a great amount of chert, water-worn pebbles lying loosely on the ground above the Patagonia series and particularly along the beach. These pebbles are predominantly of yellow, red, green, and black colors and undoubtedly were transported from a distance.

The formations below the tuff penetrated by the drill are mostly gray shales and sandstones, the hardness of which varies somewhat in the different strata. They may be Lower Eocene or Upper Cretaceous.

The beds at the surface lie at a very low angle, somewhat similar to conditions in the Mid-Continent fields. The normal dip is toward the southeast with sufficient undulation to produce flat dome structures with closures in contours on the northwest. At the close of Patagonia time, a gentle but extensive uplifting took place throughout the entire region, leaving the strata almost horizontal, producing the great "pampas," or high plains, to the west toward the Andes. Evidently little, if any, lateral pressure was exerted upon the strata, for they appear little disturbed except for their elevation. The sea and rains have ravaged the coast line, leaving a shelf with low relief along the coast; it is on this shelf that we find the development of the field.

The oil is found on the above-mentioned domes in a sand that lies conformably to the series at the surface at a depth close to 530 m. (1740 ft.) below sea level. The texture of the sand seems to vary considerably, producing non-porous parts sufficiently tight to exclude the oil. For this reason oil is not always encountered as soon as the oil formation is struck. In some wells it was reported that oil was not encountered until a depth of 580 m. (1900 ft.) had been reached. Overlying the oil series is a soft bluish-gray shale. The age of the series was thought to be Upper Cretaceous, though this was not certain. Water has been encountered but no difficulty is experienced in shutting it off.

The wells of this field have been drilled by a combination of the rotary drill to a depth of 462 m. (1515 ft.) and the Fauck system with rods to the oil strata. The rigs are of the closed-in type, covered with sheet iron on four sides to the top. They must be heavily guyed to prevent damage from strong winds prevailing during certain seasons. At the time of the writer's visit, two American rigs had recently been built to use cable tools in combination with the rotary outfit. As far as they had been used, they were making an admirable record compared with the rods. It seems quite necessary to use the rotary for the upper part of the hole, as the walls are subject to caving. Strong flows of briny water are encountered at 350 m. (1150 ft.) and at 435 m. (1428 ft.). Gas is often struck at 150 m. (492 ft). and at 400 m. (1312 ft.).

In August, 1917, twenty-five wells were producing 4000 bbl. of oil per day and an average of 60 m. (195 ft.) of new hole per day was being drilled. The oil is heavy, about 18° Bé. on an average, black in color, with a low content of gasoline. There is a small refinery on the ground for extracting the gasoline. The refuse is returned to the storage tanks for shipping to Buenos Aires, where it is used by industrial plants as fuel. The government maintained a fleet of four tankers at that time. Loading was often done with difficulty on account of lack of harbor facilities, since the Gulf of St. George is quite open to the Atlantic. Although but twenty-five wells were producing, about thirty-five others had been drilled and had either been dry, lost holes, or abandoned as no longer profitable to operate. It is understood that this record has been considerably improved since that time.

A fair proportion of the territory can be surveyed in detail, as is being done in the Mid-Continent field. The most favorable localities may therefore be selected for the drill with a minimum of failures. Stratigraphic conditions are favorable for a considerable extension of present known producing area. Transportation presents little difficulty, since the field adjoins the sea and market conditions are capable of great expansion; all oil not needed by the industries in Buenos Aires may be devoted to use in oil-burning vessels, which would call if fuel were available.

GALLEGOS-PUNTA ARENAS REGION

At the southernmost section of the continent is located the Gallegos-Punta Arenas region. In addition to the well-known gas springs near the town of Punta Arenas, the manifestations extend northward to a district due west of the port of Gallegos. In the latter locality, the streams are reported by competent observers to be carrying small quantities of crude oil toward Lake Blanca. The author has not studied this region, so is unable to give information regarding conditions and possibilities of development.

DISCUSSION

THE CHAIRMAN (E. DeGOLYER, New York, N. Y.).—The production of petroleum in the Argentine is entirely in the hands of the government. Some years ago, in attempting to develop a water supply for the Patagonian region, the department in charge of drilling brought in a gushing well. The president immediately withdrew a considerable reserve around this well; subsequently, the government reduced the size of the reserve to 5000 hectares and proceeded to develop the property. I have studied several of the reports of the Commission that has the matter in charge, and am not able to determine whether or not it is a profitable venture for the government, but it seems to have developed a distinct policy of exclusion. The mining laws, like the mining laws in most Latin-American countries, practically made no provision for petroleum. They were a development of the old Spanish mining codes when petroleum was not recognized as a mineral. In most of these Latin-American countries, some sort of special legislation has been required to make it possible for one to go in to develop the petroleum resources. As the tendency in the Argentine seems to be to keep the thing in the hands of the government, there is the peculiar condition that a nation that has no coal fuel, and where fuel is the utmost importance, seems to be determined to have no oil development either.

S. C. HEROLD (author's reply to discussion).—At the present time, development work is carried on in various parts of the Argentine Republic by several distinct companies largely financed by foreign capital, though one or two have included a fair proportion of local capital. While some of these operating companies have not brought in producing wells, the possibilities are particularly good in some instances.

The Argentine Government now has two reserved zones; namely, that referred to by Mr. DeGolyer at Comodoro Rivadavia at the southeast corner of the Territory of Chubut adjacent to the coast line along the Gulf of St. George, and that at Station Plaza Huincul in the Territory of Neuquen, both of 5000 hectares area. It is understood that a third zone may be set aside near the Cerro Negro region, Neuquen Territory. The favored location for work by the private companies has been, so far, near the zones reserved by the Government. These zones, with their adjoining lands, are within "Territorial" jurisdiction; as the territories are controlled from the national seat of government in Buenos Aires, the national laws providing for petroleum development are the only laws prevailing. The provinces may have their own departments of mining, etc., and may detail their laws respecting oil claims so long as such details do not conflict with the spirit of the national laws; within the territories no such detail can be worked out.

It cannot be properly stated that there is any intentional exclusion policy on the part of the country. Any exclusion prevailing is due rather to that which the Government has failed to do than to what it has done. We can hardly expect a country that has only reached the stage of development in petroleum which the Argentine has experienced to date to have all matters worked out in detail. Capital invested in that country will, therefore, be exposed to a risk when the day for the interpretation of the present laws is at hand. Assurances that are now generously given will avail little in the interpretation of the law when there are invested possibly several millions and the coveted fluid is gushing from the wells.

The Argentine Government's venture at Comodoro Rivadavia would appear to be profitable. The price of Comodoro crude at Buenos Aires last November was 100 pesos per metric ton, approximately $6.50 per bbl. It is improbable that the rate has changed, for the Government has practically a monopolistic control over the production to date. As the production of the field is about 5000 bbl. per day, with the actual delivery of that amount to the market, it is quite obvious that there would be "something wrong" if there is no profit. As a matter of fact, the books do show a handsome gain over expenditures. The property is not handled in a most efficient manner, nor do the men in charge think it is, for they admit the difficulties of Government control.

Petroleum in the Philippines

BY WARREN DUPRÉ SMITH,* PH. D., EUGENE, ORE.

(New York Meeting. February, 1920)

IT HAS been 5 years since the writer left the Philippine Islands and while in that country his chief work did not lie in this field, though he has visited all but one of the localities mentioned in this article. The principal field studies relating to oil were made by his colleague, Mr. Wallace E. Pratt. The writer's investigations dealt with the general stratigraphy and paleontology of the Philippines. With the exception of the investigations made by Mr. Pratt and the writer very little information on this subject is available.

A number of geologists in the employ of large oil companies have visited the Islands from time to time but, following the general rule, the public has scarcely ever been permitted to learn anything of the results. The writer is indebted to the Bureau of Insular Affairs, Washington, for late information regarding recent legislation in the Islands relating to petroleum.

All the known oil seeps, petroleum residues, such as ozocerite, and natural-gas emanations are associated with Tertiary sediments. The chief seeps and most promising prospects are located as follows:

1. Bondoc Peninsula (lower end), Tayabas Province, Southeast Luzon.
2. The west coast of the island of Cebú from Alegria north to, and perhaps beyond, Toledo.
3. Central Mindanao not far from Lake Lanao.
4. The ozocerite veins on the Island of Leyte in the northwestern part in the vicinity of the town of Villaba.
5. Natural gas from some deep wells in Tertiary shale formations on the eastern flank of the Cordillera and extending out under the plain on the island of Panay.

The first mention in geological literature, to the writer's knowledge, of gas or petroleum in the Philippines is found in the description of the Island of Panay by the Spanish geologist, Abella, in the year 1890. In 1898, an oil well was being dug on the estate of Smith Bell & Co., an English concern, near the town of Toledo on the west coast of Cebu. In

---

* Formerly Chief, Division of Mines, Bureau of Science, Manila.

TABLE 1.—*Provisional Table of Philippine Stratigraphy*

Period	Formation	Name	Type Locality	Distribution	Economic Deposits	Characteristic Fossil
Recent	Piedmont deposits	......	Pangasinan.	General.	Gravel	
	Fluviatile deposits High level	......	......	General.	Bench placers, gold	
	Lower level	......	Cagayan, Mindanao. Paracale, Luzon.	General.	Deep placers, gold and platinum	
	Spring deposits	......	Mountain Province.	Central Luzon.	Salt and silica	
	Talus deposits	......	Mountain Province.	General.	Building sand	
	Littoral deposits	......	Sangley Point.	General.	Building stone and	
	Coral reefs	......	......	General.	Building stone....... lime.	Leaves of Euphorbiaceæ (?) deer's and sharks' teeth.
	Pyroclastic deposits	"Guadalupe stone"	Pasig River.	Rizal, Laguna, and Batangas Provinces.		
	Raised coral reefs	......	......	Many parts of west coast.	Lime.	
Unconformity (not marked)	Mesa conglomerates and	......	NE of Ilocos Norte	Manila Bay.		
	Basalt and andesite flows and ejecta.	......	El Fraile Island. Mt Arayat. Mnt Apo.	Southern Luzon and Mindanao.	Road metal.	
Pleistocene and Pliocene	Raised coral reefs	......	Cu, out east.	Many parts of archipelago.	......	*Hinsia diyki* Mart.
	Marls	Laoag.	Ilocos Norte.	Many parts of Islands. Well developed. Agusan Valley, Mindanao.	......	
	Piles	......	Mnt Mariveles.	Luzon, Panay, Mindanao, etc.	Construction material. Road metal.	
Unconformity	Limestone, upper.	Licos.	Cbu.	Cebu, Northern Luzon, Panay.	Burned for lime........	Small lepidocyclines. Shells very similar to recent forms; corals abundant. *Orbitolites martini.*
Miocene	Marl	Alpaco.	Cebu.	Cebu and Luzon.	Iron.	
	Iron formation?	Hison.	Bulacan, Luzon.	Eastern Luzon.	Gold, silver, manganese, lead, copper, road metal.	
	Intrusive andesite and diorite, quartz diorites dacites.		Cebu.	Cebu, Luzon, Masbate.	Copper.	
	Quartz porphyry	Mancayan.	Lepanto.	Luzon.	Gold.	
Unconformity	Granite and granite gneiss	......	Camarines.	Luzon.	Gold.	
	Crystalline schists	Zamboanga.	Mindanao.	Mindanao and Luzon.	Gold.	

					Large lepidocyclines. Lepidocyclina richthofeni. Lithothamnium, etc.
Middle limestone......	Binangonan......	Luzon........	Luzon, Cebu, etc.....	For construction material.	
Sandstone with conglomerate phases. Shale... (Two facies probably contemporaneous.)	Suague......	Panay........	Throughout the Archipelago.	Abrasives.	Vicarya ikia and Callianassa dijki.
	(1) Batan......	Batan Island...	Throughout the Visayan Islands, Luzon and Mindanao. ..., Gu, Leyte..... Batan, Gu, etc.....	Coal..... Light oil.	ikes niasi.
Lower limestone......	(2) Vigo...... Camansi...	Tayabas Peninsula.. Cebu........		Coal.	
Limestone (covered and underlain by shale.)			Luzon, Palawan, etc....		Radiolaria and sponge spicules. Radiolaria: Radiolaria: Cenosphaera and Dictyomitra.
Cherts........	Baruyen......	Ilocos Norte....	Luzon, Panay, Palawan.		
"Slates"........	Ulion........	Panay.			

Lower Miocene or Oligocene. — Oligocene.... Unconformity........ Eocene........ — Pre-Tertiary, Jurassic?....

* Or late Pleistocene.
† Doubt as to true position.

this year, an insurrection broke out against Spanish rule and the drillers were driven from the well, which was abandoned. It is now practically as the insurrectos left it; that is, choked with rubbish which would have to be cleared before operations could be renewed.

About 1910, a number of Americans became interested in the petroleum deposits known to occur on the peninsula of Tayabas in the southeastern part of Luzon, and which were apparently unknown to the Spaniards. Two wells 117 and 300 ft. (35 and 91 m.) respectively, were dug from which a few gallons of oil were pumped, but nothing has been done since then, as far as we know. At the end of 1917, there was considerable excitement in the Islands over the alleged discoveries of oil in the Lanao region of the Island of Mindanao. The existence of petroleum on that island was known to the writer as early as 1908, but he could not visit the localities owing to the hostility of the natives.

In 1919, an Act was passed by both branches of the Philippine Legislature, and approved by the Governor-General on Mar. 4, providing for the creation of the National Petroleum Co. This Act is as follows:

Section 1.—A company is hereby organized, which shall be known as the National Petroleum Company, the principal office of which shall be in the city of Manila, and which shall exist for a period of fifty years, from and after the date of the approval of this Act.

Section 2.—The said corporation shall be subject to the provisions of the Corporation Law in so far as they are not inconsistent with the provisions of this Act, and shall have the general powers mentioned in said Law and such other powers as may be necessary to enable it to drill wells for the development of petroleum deposits, and to work said deposits and sell the output thereof.

Section 3.—The capital of said corporation shall be five hundred thousand pesos, divided into five thousand shares of stock having a par value of one hundred pesos each, and no stock shall be issued at less than par nor except for cash.

Section 4.—The Governor-General, on· behalf of the Government of the Philippine Islands, shall subscribe for not less than fifty-one per cent. of said capital stock, and the remainder may be offered to the provincial and municipal governments or to the public at a price not below par which the board of directors shall from time to time determine. Ten per centum of the value of all stock subscribed shall be paid at the time of the subscription, and the balance thereof shall be paid at such time as shall be prescribed by the board of directors. The voting power of all such stock owned by the Government of the Philippine Islands shall be vested exclusively in a committee consisting of the Governor-General and the presiding officers of both Houses of the Legislature.

Table 1 gives the provisional stratigraphic column of the Philippines. Table 2 furnishes a tentative correlation of the Far Eastern Tertiary stratigraphy including that of the Philippines. Some of the shales referred here to the Miocene may belong to the Oligocene. The oil horizon (there may be more than one) is probably in the Oligocene or the lower Miocene shales.

Attention is called to an error in the stratigraphic column given by the writer in one table in an earlier paper and incorporated by Pratt in one of his.[1] In that table, the Oligocene is given as resting directly upon a pre-Tertiary complex. The Eocene is well developed in the Philippine coal fields and is doubtless to be found in the oil fields as well as in the coal fields, though perhaps not so extensively.

In view of a prevailing opinion that these islands are largely volcanic, it should be pointed out that there are large areas where the only surface rocks are sediments and other areas where volcanic rocks form a veneer over the underlying Tertiary sandstones, shales, and limestones. Intrusive diorite is found near the center of many of the large land masses and more rarely intrusive granite, andesitic and diabasic intrusive are also found in many places near the borders of the masses.

The section made by the streams flowing eastward from the cordillera of the island of Panay affords, perhaps, as clear a view of the sequence of strata comprising a part of the Tertiary as can be obtained anywhere in the Philippines. The dominant formation is shale with thin-bedded, intercalated sandstones of which there are some 15,000 ft. (4572 m.) along the Tigum river. These shales belong to the same horizon as those in the Bondoc Peninsula, known as the Vigo series, and are Lower Miocene or Oligocene. This shale yields small amounts of natural gas, which may or may not have any relation to small coal seams.

Apparently there are three principal shale horizons; lowermost, the Eocene, which is associated with sandstones and coal seams; next, the Oligocene or Lower Miocene, in which the oil seeps are found; and uppermost, the Miocene with more coal seams.

---

[1] Occurrence of Petroleum in the Philippines. *Econ. Geol.* (1918) **11**, 247.

TABLE 2.—*Tentative Correlation of the Far Eastern Tertiary*

Series	Philippines	Borneo	Java	Formosa	Japan
Pliocene	Raised coral reef limestone.	....	Limestone.	Raised coral reef limestone.	Marine this with Mytilus, Conchocele.
Pliocene					
Pre-Pliocene	Limestone and marl beds; fossil elephants' teeth reported from two localities in Luzon and one in Mindanao; foraminiferal marls of Karrer probably belong here also.	....	Etage calcareux.	....	Beds containing fossil plants and those containing numerous specimens of Carcharodon megalodon.
Miocene Upper	{Upper limestone, with small Lepidocyclines. Limestones with Cycloclypeus.	Group H with small Lepidocyclines and Lithothamnium.	Limestones with small Lepidocyclines. Volcanic tuff and sands, with Cycloclypeus; limestones with Lepidocyclines and Cycloclypeus.	Limestones with Lepidocyclines and Lithothamnium.	Limestones with Lepidocyclines, Lithothamnium.
Middle	Sandstones and shales with Vicarya callosa coal seams.	....	Sandstones.	Sandstones, shale, and coal seams.	Sandstone; shales; coal seams with Vicarya callosa in the overlying beds.
Lower	Middle limestone, with large Lepidocyclines, with Nummulites; coal seams.	Group G. Groups D and C with Nummulites.	Etage brecheux. Limestone with Nummulites.	....	
Oligocene	Lower limestone with Nummulites.				
Eocene	Some coal seams; exact position doubtful.	Group B with Operculina. Group A with Orthophragmina omphalus.	Quartzose argillites with coal seams.	....	Tuffs with Nummulites (Bonin series).
Paleocene	Uncertain.	....	....	....	Sandstone, shales, coal seams.

The typical oil (Vigo) shale on Bondoc Peninsula may be described as (quoting from report of Pratt and Smith, p. 331) "consisting of fine-grained shale and sandy shale interstratified in thin regular beds from 5 to 10 cm. in thickness. Occasional beds of sandstone occur varying from 10 cm. to 1 m. in thickness. The fine-grained shale is gray, blue, or black and is made up almost entirely of clay. The blue or black fine-grained shale in the Vigo formation usually emits a slight odor of light oils upon fresh fracture and in some outcrops is highly petroliferous. The material loses this odor and assumes a light-gray color after it has been exposed to the air and becomes thoroughly dry." These shales contain numerous foraminifera of the genus *Globigerina*, which may be the source of the oil. Although present numerously, these organisms did not appear to comprise any large percentage of the volume of these shales.

On the island of Cebu there is a similar shale series, dark blue in color and fine-grained, in which the oil seeps are found.

Another important feature of the Philippine Tertiary is the presence of several limestone horizons in striking contrast to the American west coast Tertiary. These limestones, in places, attain thicknesses of several hundred feet.

In general, we may say that the Philippine Islands consist of a series of anticlinal regions, which are marked by the island masses, and synclines, which are occupied by the narrow straits between the islands. The principal folding has been east and west, with minor flexures north and south. The anticlines are generally sharp, as is the case in Sumatra, Java, Burma, and other parts of the Far East. In South Sumatra, Tobler has shown that the anticlines are only 2000 ft. across the crest and that wells must be located on the crest in every case. In Tayabas, the structure can easily be understood by a study of the map accompanying the paper by Pratt and the writer. This shows that the anticlines are generally sharp and the dips are quite steep. The Maglihi anticline on Bondoc Peninsula is a typical example and less than $\frac{1}{2}$ mi. wide. The axis of the principal structures coincides with the general direction of the principal tectonic lines of the archipelago; *i.e.*, north and south with minor departures from this. Accompanying this folding there has been more or less faulting. Just how great the throws are, not enough detailed work has been done to determine. There is enough evidence to indicate a considerable amount of faulting throughout the archipelago.

On Cebu, Pratt has shown that the seeps near Alegria are located at the crest of a very sharp anticline. The well near Toledo, which shows some oil, is apparently on a monocline in which the beds dip 50° to 60° to the northwest.

In Leyte, the petroleum seeps are along the outcrops of steeply dipping strata (Vigo shale series).

On Panay, the shale beds yielding natural gas are generally monoclinal, but there is one well-defined anticline, known as the Maasin anticline, which might be a favorable location for a test well.

In Tayabas and Cebu, there is a sandstone member, to which the local name "Canguinsa sandstone" has been given, which lies unconformably upon and overlapping the great shale series. In some cases, as in Leyte, residual bitumens are found in this formation.

Very probably, transportation will depend on inter-island boats as most of the 750 mi. of railroad in the archipelago does not tap the petroleum localities. In the transportation of machinery and the location of docks, great care will have to be exercised because of typhoons.

Labor is plentiful but unskilled. However, the Filipinos show a great aptitude along mechanical lines and, under competent white foremen, make very excellent workmen. The prices for labor vary with the different tribes and localities. Present prices are commensurate with those in other parts of the world. The Filipino's fondness for holidays ne-

TABLE 3.—*Physical and Chemical Properties of Philippine Petroleum* *

Description of Petroleum	Crude Oil		Distillation Products						Remarks
	Color by Transmitted Light	Specific Gravity	Gasoline, to 150° C.		Kerosene, 150° to 300° C.		Heavy Oils, 300° to 400° C.	Residue Above 400° C.	
			Volume	Specific Gravity	Volume	Specific Gravity	Volume	Specific Gravity	
Tayabas, Bahay well I., depth 40 m. Sampled by division of mines 24 hr. after well had been drained.	Brown to wine red.	0.8325	39.0	0.770	44.5	0.850	16.5		Flash point 0° C. (32° F.). sulfur absent; initial boiling point, 91° C., paraffin, 8.1 per cent; specific gravity at 15° C.
Cebu, Philippines: Well at Toledo, Cebu...	Dark brown.	0.885	6.2	0.762	42.32	0.832	38.3	0.901 13.17	Residue above 375° C.; specific gravity at 15°C.
Oil seep at Alegria, Cebu	Dark brown.		17.5		30.5		35.0		Residue contained foreign sediment.
Leyte, Philippines........		0.926							Flash point 74° C. (166° F.).

*Analyses by Richmond, and other chemist of Philippine Bureau of Science.

cessitates a larger pay-roll than is usually warranted by the size of the job. The manager will have to take this into account. ·

Philippine petroleum has a paraffine base and is usually reddish to violet in color. It is quite clear, and closely resembles oil from Burma and Sumatra. Table 3 gives a fairly complete analysis made by Richmond, former chemist of the Bureau of Science, and others. Table 4 contains an analysis of the petroleum residue from the Island of Leyte. The paraffine content of the Philippine petroleum is very high; a beer bottle full of oil from the Toledo well in 1908, which the writer collected and put imperfectly sealed into his saddle bags, on unpacking three days later, contained no oil but was half full of solid paraffine.

The residual bitumens from Villaba, Leyte, are found in lenses and pockets in the Canquinsa sandstone. These have been fully investigated by Pratt. Physically they more nearly correspond to ozocerite than to any other of the natural bitumens.

TABLE 4.—*Physical Properties of Natural Bitumens from Villaba, Leyte*

PROPERTY	OUTCROP A AND B	PROPERTY	OUTCROP A AND B
Specific gravity...	1.05	Luster............	Brilliant
Hardness........	2.00	Structure.........	Columnar
Color............	Jet black	Fracture..........	Conchoidal
Streak...........	Black	Flows............	Intumesces, softens, and flows imperfectly at 150° C.

The writer agrees with Pratt in the belief that there is a small commercial supply of oil in the Philippines, very much as in Formosa. He seriously doubts, however, that petroleum exists in large enough quantities to attract large capital from America, considering the distance and the many unfavorable conditions to be encountered. The size, steepness of dip, and broken nature of many of the structures are not favorable to large production.

All the oil which the writer has seen in the Islands is in the shallow wells mentioned and in seeps in shales, and these seeps have been small. He has seen no oil in beds either below or above these shales. In the petroliferous shales are a number of forminifera with Globigerina predominating. It may be that all the oil has come from the decomposition of these organisms.

DISCUSSION

WALLACE E. PRATT,[*] Houston, Tex. (written discussion).—Doctor Smith's correction of my statement in *Economic Geology* that the Philippine Oligocene rests directly upon a pre-Tertiary basement of crystalline

---

[*] Chief Geologist, Humble Oil & Refining Co.

rocks is well taken. I concede that some of the older indurated shales and limestones may properly be classed as Eocene. My statement should have been limited to the petroleum-bearing areas that I described and in which it applies by reason of the absence of the older rocks.

In this connection, I may record my impression that the typical Philippine section is not as thick as the 15,000-ft. (4572 m.) section exposed on the eastern flank of the cordillera of the Island of Panay and mentioned by Doctor Smith. I think the average thickness for the Philippine sedimentary column would be about one-third of the figure quoted. The Panay section is unusual in another respect; at its base there are thousands of feet of unfossiliferous conglomerates and coarse sands which appear to be of extreme shallow-water origin, whereas in the typical column this basal member is not more than 200 ft. (61 m.) thick.

A recent press dispatch states that Cebu, the second largest port in the Philippine Islands, and a city of about 100,000 population, is now paving its streets with rock asphalt secured from a quarry in the vicinity of the town of Villaba, on the neighboring island of Leyte. I believe this work marks the first commercial use of the petroliferous deposits of the Philippines. The circumstance is interesting not only as a first step in making this natural resource serve industry, but as an evidence of the extent of the residual deposits that constitute the surface indications of petroleum in one of the possible oil fields in the Philippines.

As a matter of fact, not only is the stratigraphic column in the Philippines—dominantly shale with interbedded sandstone as described by Doctor Smith—of suitable character and adequate thickness to yield commercial petroleum, but the surface evidences of the existence of petroleum on some of the islands are quite remarkable. These various seepages and asphalt deposits are described briefly in my paper in *Economic Geology* to which Doctor Smith refers.[2]

The situation in the Philippines, in so far as the geologic conditions are concerned, is certainly one that would lead many of the geologists engaged in the present fervid search for new petroleum fields to recommend drilling exploration on some of the islands, provided their clients commanded adequate capital. The possibilities in the Philippines are the more impressive when one reflects that in Borneo, Sumatra, and Java, as well as in Formosa and Japan, commercial production of a petroleum similar in character to that which comes to the surface at places in the Philippines is obtained from beds of the same geologic age and composition as those in the Philippines.

---

[2] More detailed descriptions with maps and geologic sections will be found in the following references: Wallace E. Pratt and Warren D. Smith: Geology and Petroleum Resources of the Southern Part of Bondoc Peninsula, Tayabas Province, Philippines. *Phil. Jnl. Sci.*, Bur. Sci., Manila (1913) Sec. A, 5, 301–376; Wallace E. Pratt; Occurrence of Petroleum in Cebu, *Idem.* (1915) Sec. A, 4; Wallace E. Pratt: Petroleum and Residual Hydrocarbons in Leyte, *Idem.* (1915) Sec. A, 4.

Under different conditions, a prospect like that in the Philippines would evoke an active drilling campaign. I have had opportunity to make direct comparison in the field between conditions in parts of Central America, for instance, and in the Philippines and I can conceive no possible contention but that the geologic conditions in the Philippines are decidedly the more promising. Yet these same regions in Central America have interested dozens of large petroleum corporations; concessions there have been sought eagerly for years and are at present, indeed, being exploited.

I am convinced that adequate exploration of the petroleum deposits in the Philippines has been prevented, not by unfavorable geologic conditions but by prohibitive regulations of the local mining laws. Practically all the possibly petroleum-bearing territory in the Philippines is government land. It can be acquired only under laws similar to the mining laws of the United States. Petroleum lands are subject to "location" as placer-mining claims. An individual may obtain a single claim of 8 hectares (20 acres) in any one field while a corporation comprising eight individuals can secure only one claim of 64 hectares (160 acres) in any one field. Except by a direct evasion of the law, therefore, it is impossible to control the acreage requisite for large operation, such as must be contemplated by any enterprise that looks as far afield as the Philippine Islands.

It is unlikely that successful exploration will result from the efforts of the Government owned corporation, the formation of which is recorded in the legislation quoted by Doctor Smith. This corporation, like any other, is subject to the laws that prevent the acquisition of suitably large holdings and its capitalization of 500,000 pesos ($250,000) is not adequate. If the Filipinos were to grant concessions of hundreds of thousands of acres, as some of the Central and South American republics have done, I believe their possible petroleum resources would be promptly and thoroughly tested by the drill.

DAVID WHITE,* Washington, D. C.—This paper is very timely since the Philippine Islands are presumably open to the enterprise of the American driller, whereas much of the territory in that part of the world is closed to us.

The United States has ambitious plans for the operation of a great merchant marine, which is to be oil burning in the main, and it takes but a glance at the world map to see the strategic advantage of oil supplies in the Philippines for such marine operations. It is a little difficult to understand why more attention has not been given to the Philippines, in spite of the difficulties attending development in these islands.

---

* Chief Geologist, U. S. Geol. Survey.

A number of American oil companies are, I believe, at the present moment taking an interest in the possibilities of the Philippines. Important factors in the formulation of opinion regarding the importance of the oil deposits of the Philippines, as brought out by Mr. Pratt, are the point of view and the breadth of experience of the examining geologist. The average oil geologist, whose experience has been mainly in the Appalachian region, the Mid-Continent, or Louisiana, or possibly even in California, on seeing the narrowness of the basins, the closeness of the folding, and the presence of igneous rocks here and there, would be likely to draw unfavorable conclusions as to the possibilities of the Philippine Islands. Geologists visiting that region should be familiar with the geological conditions of oil occurrence in Japan, Formosa, the East Indies, or in the Baku district, and their conclusions should be formulated with the knowledge and understanding of the occurrence of oil in those regions rather than in the United States.

Petroleum Industry of Trinidad

By George A. Macready, Los Angeles, Calif.

(St. Louis Meeting, September, 1920)

Trinidad, British West Indies, is an island near the north coast of South America, situated between latitudes 10° and 11° N., and opposite the numerous outlets of the Orinoco River Delta. It is separated from Venezuela by the Gulf of Paria (salt water) and straits over 5 mi. (8 km.) wide. The area of the island is approximately 1750 sq. mi. (453,-250 hectares) and the population is approximately 400,000. The climate is tropical with an annual rainfall of from 45 to 60 in. (114 to 152 cm.).

The oil fields consist of several units, or fields, located in the southern half of the island. Approximately 90 per cent. of the total production has been yielded by fields situated within 7 mi. (11.3 km.) of the famous asphalt lake and on the southwest peninsula.

The most important producing fields, or units, are the following, which are shown on the accompanying map:

Brighton, or Pitch Lake Field, operated by the Trinidad Lake Petroleum Co., Ltd., is situated beside the famous Pitch Lake; it even encroaches on the lake.

Vessigny Field, operated by the Trinidad Lake Petroleum Co., Ltd., is situated 2 mi. (3.2 km.) south of Pitch Lake.

Lot One Field, operated by the Petroleum Development Co., Ltd., the United British Oilfields of Trinidad, Ltd., and Stollmeyer, Ltd., is situated 3 mi. south of Pitch Lake upon Lot One of Morne l'Enfer Forest Reserve and adjoining properties.

Parry Lands Field, operated by the United British Oilfields of Trinidad, Ltd., and the Petroleum Development Co., Ltd., is situated 3½ mi. south of Pitch Lake on Lot Three of Morne l'Enfer Forest Reserve and adjoining properties.

Point Fortin Field, operated by the United British Oilfields of Trinidad, Ltd., is situated at Point Fortin, 6 mi. southwest of Pitch Lake.

Fyzabad Field, operated by Trinidad Leaseholds, Ltd., is situated several miles southwest of Fyzabad Village and 6 mi. south-southeast of Pitch Lake.

Barracpore Field, operated by Trinidad Leaseholds, Ltd., is situated several miles south of San Fernando and 15 mi. (24.14 km.) east of Pitch Lake.

Tabaquite Field, operated by Trinidad Central Oilfields, Ltd., is situated 4 mi. southeast of Tabaquite Railroad Station, and 30 mi. (48.28 km.) northeast of Pitch Lake.

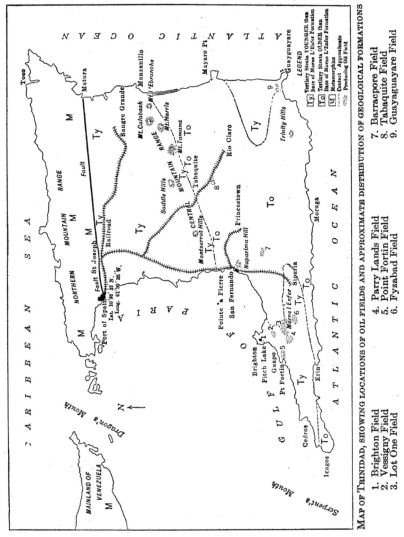

MAP OF TRINIDAD, SHOWING LOCATIONS OF OIL FIELDS AND APPROXIMATE DISTRIBUTION OF GEOLOGICAL FORMATIONS

LEGEND

1. Brighton Field
2. Vessigny Field
3. Lot One Field
4. Parry Lands Field
5. Point Fortin Field
6. Fyzabad Field
7. Barracpore Field
8. Tabaquite Field
9. Guayaguayare Field

Guayaguayare Field, operated by Trinidad Leaseholds, Ltd., is situated in the extreme southeast corner of the island 45 mi. (72.42 km.) from Pitch Lake.

From 1870 to 1900, several attempts were made to obtain oil on Trinidad but although small quantities of oil were encountered, no commercial production resulted, and most of the wells were abandoned. An attempt was also made to obtain oil from the crude lake asphalt, probably by a cracking process, but without commercial success.

The present industry can be said to commence with wells drilled since 1900 near Guayaguayare, in the extreme southeast corner of the Island. Several years later wells were drilled near Point Fortin, southwest of Pitch Lake, which yielded commercial quantities of oil but not sufficient for export.

In 1908, the New Trinidad Lake Asphalt Co., Ltd., commenced drilling at the Pitch Lake and encountered an excellent flow of oil in its second well. Other wells were drilled and, in 1910, this company exported the first steamship cargo of oil from Trinidad. Since then, the quantity of oil produced and the number of companies exporting has increased. The production in 1908 was 169 bbl., in 1912, 436,805 bbl.; and in 1917, 1,599,455 bbl.

Geology

Stratigraphy

All petroleum produced by Trinidad has been yielded by strata of Tertiary age. In general, the Tertiary strata consist of clays, shales, marls, and sandstones; conglomerate is extremely rare and limestone is uncommon. The sandstone is usually composed of small quartz grains uniformly sorted. Cretaceous and metamorphic rocks underlie the Tertiary. The most important portion of the Tertiary strata consists of sandstone and shale, which grades upward into marl and shale containing marine organic material and evidences of petroleum. The organic material in this shale is probably the primary, or "mother," source from which Trinidad petroleum is derived. The upper portion of the shale contains sandy strata into which petroleum has migrated and accumulated in quantities sufficient for commercial exploitation. Eocene fossils occur in the lower part, but the upper part may extend into the Oligocene. This includes the Naparima clay, Cruse oil zone, and Stollmeyer oil zone.

The Morne l'Enfer formation unconformably overlies the above-mentioned, and consists of sandstone and clay shale in approximately equal proportions. The lower sandstones are often heavily impregnated with asphalt and often outcrop as "pitch sand" cliffs. The author believes that this asphalt has migrated from the underlying shales and marls. Near Pitch Lake, some oil may be produced from this formation. Strata younger than the Morne l'Enfer have not yielded commercial quantities of oil and are unimportant.

The accompanying tabulation describes the geological formations of

Trinidad in more detail. The areal distribution of the formations is shown approximately on the map.

Structure

The areal geology of the island is separated into two parts by the great east-west fault passing near Port of Spain and Matura, and extending from the Atlantic Ocean into Venezuela. North of the fault is the area of Metamorphics, forming the Northern Mountain Range. South of the fault is a great undulating blanket of Tertiary strata.

The dominating features of the Tertiary structure are: A synclinal or monoclinal trough between the Central and Northern Mountain Ranges; an anticlinal uplift along the south side of the Central Range striking east-northeast by west-southwest, from Pointe à Pierre to Nariva Swamp; an undulating synclinal structure between San Fernando, Mayoro Point, Guayaguayare Bay, and Icacos Point with an east-west strike; the magnitude of erosion at the unconformity below the Morne l'Enfer formation. Numerous local folds, faults, kinks, anticlines, and synclines modify the broader features and are very important in the concentration of petroleum.

Occurrence of Petroleum

All the producing oil fields of Trinidad (except Tabaquite Field) are within or on the flanks of the great synclinal trough or basin of the southern part of Trinidad. Most of them are on the southwest peninsula. This undulating synclinal structure is underlain by Naparima clays, marls, and organic shales. It forms the drainage area from which petroleum has accumulated. This petroleum has concentrated in commercial quantities near anticlinal folds.

The location and richness of each productive area are modified by thé magnitude and condition of the unconformity below the Morne l'Enfer formation: by the channels of migration: by the local conditions of porosity of reservoir sands; by the lenticular condition of the oil sands; by the facility with which connate salt waters were displaced by oil. There are three principal horizons in which petroleum usually, but not always, is concentrated in commercial quantities.

The Cruse oil zone is persistent because its proximity to the organic shales permits ready saturation, has permitted much time for connate waters to be forced out, and Tertiary erosion has not attacked it as frequently as higher strata. Its thinness and high gas pressure increase operating cost. This condition applies at Parry Lands, Morne l'Enfer Forest Reserve, and Point Fortin.

The Stollmeyer oil zone overlies the organic shales and the sands are lenticular. The porosity and saturation of the oil sand varies locally.

It may or may not, locally, be conformable below the Morne l'Enfer formation or it may be entirely missing.   Where apparently conformable below the Morne l'Enfer formation, conditions are simple and anticlinal structures may prove very rich, as in the Morne l'Enfer Forest Reserve. As the unconformity increases, modifications occur.   Part of the Stoll-meyer sand may have been removed by erosion and the remainder sealed by the clayey base of the Morne l'Enfer formation.   One flank of an anticline may prove richer due to better dráinage area on that side, as may be the case at Lot One.   A flank of the anticline may be enriched but the apex barren because the sand is missing; such may be the case at Point Fortin, Barracpore, and possibly at Brighton.   Connate salt water has not been completely forced out of all the sand lenses but usually remains only in the lowest lenses.

The Morne l'Enfer formation is enriched by oil migrating from the underlying organic shales.   Where the organic shales lie close below as a result of Tertiary erosion and the Morne l'Enfer sands are not too thick or too clayey at the base, saturation may be sufficient for commercial production; such may be the condition in fields near Pitch Lake.   Where the sand is too thick and petroleum has migrated slowly, saturation may not be sufficient for commercial production; such may be the condition of pitch sands in the Forest Reserve.

Near Tabaquite, petroleum has concentrated in sands closely associated with organic shales but too distant from other fields for correlation.

TECHNOLOGY

Drilling

The rotary system of drilling has proved most successful in the productive fields.   Cable tools are usually confined to some, but not all, isolated test wells, to special work, and to repairing damaged wells; but in the early days many wells were drilled and finished with them.   Portable drilling machines have been successful for shallow wells in the central and extreme southern portions of Trinidad.   Some wells have been drilled with Canadian and Galacian outfits.

Some difficulty is encountered in penetrating pitch strata.   If sandy, they are hard and wear off rotary bits.   If clayey, they are plastic and squeeze slowly but persistently into the hole and grip the drill pipe above the bit; this has been overcome by using hot water circulation and driving casing through the pitch.

For wells expected to be over 1000 ft. (305 m.) deep, it is common practice to drill with rotary and set 15½ in. 70-lb., 13-in. 54-lb., or 12½-in. 50-lb. screw casing as the outside string.   Either this or the succeeding one is used to shut off water preferably, but not always, by cementing. Wells are usually drilled into the oil sand using 6-in. (15.24-cm.) or 8-in.

perforated drill pipe equipped with a blow-out preventer on an outer string. With all in readiness to receive a big flow of oil, drilling proceeds until the oil sand is drilled through or the flow of oil and gas prevents farther progress. Then the drill pipe is left as it is and the wash pipe recovered when convenient. In shallow fields, a common practice is to set about 100 ft. (30.5 m.) of 12½-in. (31.75-cm.) casing as a conductor and then to drill through the oil zone. Perforated casing is substituted for drill pipe and the well tubed to pump or flow as the case may be.

Casing is not perforated in the well if it can be avoided; the usual practice is to set shop perforated casing. Screen casing has not been successful because of clogging with clay. Explosives are never used to increase production and rarely to break up junk.

For a well 1500 ft. (457.2 m.) deep, 60 days is a fair average time from first actual drilling until production begins. This includes usual delays, casing setting, changing crews, waiting, etc. The actual number of days in which hole is dug may be as low as fourteen. In 1918, $15,000 was a fair average cost to the depth of 1500 feet.

Production

Wells in the thin deep sands usually begin production with a large initial flow or gust under great gas pressure, yielding up to 100,000 bbl. in the first few days and later choking with sand or shale. During the first year, the production is dependent largely on spasmodic flows aided by bailing or tubing agitation, but after the first year few wells yield over 100 bbl. daily. The shallower wells with thicker oil sands begin production sometimes as pumpers and sometimes by flowing. The initial flow averages much less than for the deeper wells, but is less spasmodic and less costly to control. Few wells flow for over a year.

After wells cease flowing they are usually pumped by the walking beam. Sand and mud must be cleaned out frequently for two years or more. None of the southwest fields have been successful in pumping from a central power or jack. Few wells have produced over eight years and many cease producing in the second or third year. The production of individual wells is greatly influenced by the local porosity of the oil sand and the size of individual oil-sand lenses.

Character of Petroleum

Trinidad petroleum varies greatly in specific gravity, not only in different fields, but also within the same field. It is (with one exception) of asphaltic base. Oil from the Trinidad Central Oil Fields, Ltd., near Tabaquite has little asphalt but some paraffine, and yields much gasoline and kerosene by distillation. The average specific gravities for

Geologic Column of Trinidad

Age	Name of Formation	Thickness, Feet	Lithology Petroleum Evidence Folding	Miscellaneous Remarks
Recent	Alluvium	40	Principally soft clay, silt, vegetable remains Less sand. Rarely conglomeritic. Asphalt cones and seepages and mud volcanoes occur by breaking through from underlying formations. Never tilted	Consists of stream alluvium and swamp deposits.
			Unconformity	
Pleistocene (?)	Llanos Formation	100	Ferruginous sands, clays and conglomerates Evidences of asphalt occur by breaking through from underlying formations Usually nearly flat; rarely tilted to 5°.	The Llanos formation consists of material deposited in the basin of which the present Orinoco Valley was a portion Large areas occur in Venezuela, particularly in the Llanos, or plains, of the Orinoco River Valley, but in Trinidad where the formation appears thinner, erosion has dissected it until only hill-top remnants and a few larger areas remain When seen from the Gulf of Paria, the topography of southern Trinidad has the appearance of a former flat surface, such as a sea bottom, uplifted to a plateau 100 to 300 ft. above sea level through which "islands" or peaks of older resistant rocks project (Erin Peak, Morne l'Enfer, Soldado Rock, Naparima Hill for example.) The present drainage system has dissected this plateau into a low, but steep topography gentler than canyon topography.
			Unconformity	
Pliocene (?)	Upper Tertiary Formation	400	Porcellanite, lignitic clay, lignite, partly altered wood, shale, clay, and sandstone exhibiting great lateral variation in character. Conglomerates not known. Rarely contains asphalt and has no commercial oil horizons. Usually found tilted but rarely over 35°.	Usually occurs within synclines flanked by the l'Enfer formation. It may be of fresh-water origin of material derived from the older tertiary rocks. In troughs, or synclines, deposition may have been uninterrupted between this and the Llanos formation. This formation corresponds to the upper tertiary strata in reports of E. H. Cunningham-Craig. Porcellanite has not been proved to exist in other formation in Trinidad.
			Unconformity (locally)	
Oligocene (?) or Miocene? or both	Morne l'Enfer Formation	2500	Sandstones of uniform small quartz grains separated by bands of clay shale and rarely by lignite. No conglomerate known. The lowest sands are commonly saturated with asphalt. Near Morne l'Enfer 300 ft. of "tar sand" has been observed in the lowest 700 ft., some of which was very rich. Some of the oil fields nearest the Pitch Lake may derive production from sands of this formation. Tilting is commonly over 20° but rarely as much as 90°.	The following thicknesses have been measured: 2500 ft. at Erin Bay, 1200 ft. at Guapo Bay, 900 ft. at Vessigny Bay, 800 ft. at Morne l'Enfer. Fossils of doubtful Oligocene age have been found near this formation. In the Central Range mountains, Miocene fossils occur in what may be the equivalent formation Because of the great unconformity below this formation, the author prefers to regard it as Miocene. The name of this formation is selected because of its occurrence in the Morne l'Enfer Forest Reserve.
Eocene or Oligocene	Forest Clay	500	Blue and gray clay often very sticky.	This forms the impervious cover over the Stollmeyer oil zone. The author is convinced that there is a great unconformity below the Morne l'Enfer formation, but owing to the clayey non-resistant nature of the

Age	Name of Formation	Thickness Feet	Lithology Petroleum Evidence Folding	Miscellaneous Remarks
Eocene or Oligocene	Forest Clay			strata the exact horizon is difficult to identify. It probably occurs in these clays, below the lowest Morne l'Enfer sand. This condition was observed by the author on a much smaller scale at a small island which rose overnight from the sea near Trinidad in 1911. A few weeks later waves had eroded it completely and deposited tne material on similar adjacent clayey material.
			Unconformity	
Eocene or Oligocene	Stollmeyer Oil Zone	500	Overlapping pancake-shaped lenses of sand and shale alternating. The sands contain oil and salt water, the best saturation of oil being in the upper part of tne zone and not far from an anticlinal axis. Salt water is usually confined to the lower lenses, but has been found at the top of the zone.	This is the most profitable oil formation on Trinidad. It is difficult to correlate individual lenses from well to well but the group or zone can easily be traced through a field.
Eocene or Oligocene	Stollmeyer Cruse Shale	600	Principally clay shales with occasional lenses of sand. Foraminifera occur in the lower part of these shales. Some of the sand lenses are highly saturated with petroleum and gas under great pressure. Lenses occasionally contain salt water.	Several oil wells yield production from restricted sand lenses in this formation.
Eocene or Oligocene	Cruse Oil Zone	40	Sand. Often saturated with petroleum and gas under great pressure. Salt water may occur.	This is the most persistent oil horizon on Trinidad, but its thinness, depth, and violent gas pressure increases the cost of exploitation. It is identified over a large area in the northern portion of the Morne l'Enfer Forest Reserve where it occurs 1000 to 1200 ft. below the top of the Stollmeyer oil zone. Many of the gas-mud volcanoes of Trinidad may occur near the outcrops of this horizon.
Eocene or Oligocene	Naparima Clay	4000	Clay, shale, and marl containing marine organic matter. Outcrops often with a perceptible odor of kerosene and where an irridescent film of oil covers pools of water. Manjak veins occur near San Fernando. Commonly tilted to vertical with abrupt changes and overturns.	Large areas outcrop near San Fernando. Folding is so complex and abrupt that it is difficult to obtain a reliable measurement of thickness. This formation may be the "mother rock" from which the petroleum of Trinidad is derived Some of the light oil from Trinidad may come from wells in this formation.
Eocene			Clay and shale and hard gritty sandstone.	Eocene fossils occur in or below the Naparima clay The author has not made extensive studies of the Tertiary strata below the Naparima clay.
			Unconformity	
Cretaceous			Dark, black or brown shale and limestone.	Cretaceous strata have been reported in limited areas in the Central Range of Trinidad and doubtfully farther south. Large mountainous areas of Cretaceous occur in Venezuela.
			Unconformity	
Pre-Cretaceous	Metamorphics		Schist, gniess (Pre-Cretaceous volcanics near Toco).	The Northern Range of Trinidad consists of a metamorphosed complex bounded on the south by an eastwest fault passing near Port of Spain and Natura Bay, and extending into the Atlantic Ocean and Venezuela.

different fields are: 0.9524, 0.9722, 0.9589, 0.9459, 0.9333, 0.9211, and 0.8092; or, 17°, 14°, 16°, 18°, 20°, 22°, and 43° Baumé,.

Transportation and Utilization

The Trinidad Lake Petroleum Co., Ltd., and the Petroleum Development Co., Ltd., together operate a 6-mi. (9.66 km.) pipe line from the Morne l'Enfer Forest Reserve to a tank farm at Brighton near Pitch Lake beginning as 4 in. (10.16 cm.) and increasing to 10 in. (25.4 cm.). At Brighton pier are facilities for docking and loading steamers up to 35,-000 bbl. in 24 hr. Much of this oil has been exported to the United States for industries using asphalt and its products.

The Trinidad Leaseholds, Ltd., operates approximately 28 mi. (45 km.) of 6-in. (15.24 cm.) pipe line from the Morne l'Enfer Forest Reserve to Pointe à Pierre, with a short side branch from Barracpore. At Pointe à Pierre is a tank farm and pipe trestles to a loading station 1 mi. (1.6 km.) from shore where full-size tank steamers can be loaded. Most of this oil has been taken by the British Admiralty, although considerable has been disposed of as bunker fuel to steamships and some has been refined at Pointe à Pierre.

The United British Oilfields of Trinidad, Ltd., operates a 6-in. (15.24 cm.) pipe line 6 mi. (9.66 km.) in length from the Morne l'Enfer Forest Reserve to Point Fortin, with an additional branch contemplated. At Point Fortin, oil is loaded in barges and towed to tankers anchored in the Gulf of Paria. Loading a tanker requires several days. A refinery at Point Fortin produces "navy fuel." Most of this oil has been taken by the British Admiralty, but some of it has been disposed of as bunker fuel oil to steamships and some early shipments went to various places.

The Trinidad Central Oilfields, Ltd., operates a 3-in. (7.62 cm.) pipe line from the Tabaquite oil field to a loading pier at Claxtons Bay. This oil is very high in gasoline and is nearly all refined for petrol, kerosene, and fuel residue.

Stollmeyer, Ltd., operates a 2-in. pipe line 2 mi. (3.22 km.) in length from near the Morne l'Enfer Forest Reserve to Guapo Bay where sail lighters can be loaded.

FUTURE POSSIBILITIES

The future of the petroleum industry of Trinidad depends on the discovery of new oil fields or units as much as on complete exploitation of the known fields. The most obvious oil fields are already in exploitation. The writer is confident that a thorough search will result in the discovery of other oil fields which will compare favorably with the known fields.

The discovery of new oil fields necessitates the drilling of isolated test wells of which most will be barren. Exploratory drilling should be guided by a thorough geological study of a broad area with special attention to: The magnitude and trend of the unconformity below the Morne l'Enfer formation, character of strata below this unconformity, and geologic folding. Such geological study will reduce the number of barren wells, which is the greatest expense of exploration. In the known fields a continuous drilling program will be necessary to maintain the production with declining wells.

DISCUSSION

RALPH ARNOLD, Los Angeles, Calif.—The Trinidad field has been the graveyard of the reputation of many drillers and production men. Apparently the effort to hold back this clay and sand by the use of strainers is unsuccessful because the well will gradually plug up to such a point that every known method will fail to loosen the pores and allow the oil to come in. As wells put down near old producers will show large initial production, the ultimate yield of oil will be increased by putting down secondary wells.

In one field, a perfect dome, the sand is in lenticular form. At first the wells showed considerable water but now the oil pumped is free from water.

E. DeGOLYER, New York, N. Y.—I have understood that the chief difficulty in Trinidad operations was to find any strainer that would hold back the sand, which is of uniformly fine grain. The ordinary sand is composed of grains of assorted sizes. The strainer lets the fine sand pass through and holds a sponge of the larger grains outside so that after a well starts producing, this coat of larger grains on the outside does as much straining as the strainer itself.

R. VAN A. MILLS, Washington, D. C.—It seems probable that several factors enter into the sanding up of wells. Underground changes in the gravities and viscosities of the oils incident to the operation of wells may play a part in this trouble. In California there are instances of the Baumé gravities of oils issuing from wells in new fields undergoing reductions of 7° in the first months of production. Under these conditions the deposition of residual matter from the oils would influence the sanding up of the wells.

A more important point is the deposition of inorganic matter (mineral salts) together with silt in the sands. This induced effect is accomplished through the agency of the waters accompanying the oils —concentration and chemical reactions being responsible for the deposition of the salts.

Water interferes with the movements of the oils to the wells especially where the oils are of high viscosity. The shutting off of the oils through the agency of waters is probably the worst of these underground troubles with which we have to deal. I believe that by reducing the rapid flows of oil and gas we can largely eliminate these troubles.

R. A. CONKLING,* St. Louis, Mo.—Mr. Macready has not made any mention of the Tabagie field, which has a very light oil, 35° to 40°, that comes from Cretaceous and other sands much higher in the Tertiary.

RALPH ARNOLD.—In the principal producing area, there is enormous production during the first three or four days and very light production thereafter. Many of the wells have given as high as 15,000 to 20,000 bbl. per 24 hr. for the first three or four days, and but a mediocre production after that.

ARTHUR KNAPP, Shreveport, La.—One other place where the same thing occurs is Louisiana. The trouble is not sand but squeezing clay. The clays in Trinidad are contaminated with oil and pass through the perforated casing. It is useless to place a screen for the clay squeezes through and appears in the overflow in the form of paper-thin sheets.

E. DEGOLYER.—I have wondered if sanding-up is not often a case of the pinching together of top and bottom clays rather than any blocking of the well sand or something of that sort. These wells, when they come in as gushers, produce large amounts of sand, so that if all the sand is blown out, there is nothing to hold up the overlying clay or mud. There must be some considerable tendency for them to close together and, where the sand had been imperfectly exhausted, a small production would continue.

RALPH ARNOLD.—We operated on that theory at one time and tried to control the flow of the wells at the start, and by holding back the sand allowed the production to be slower, but I think the records show that the wells that ran wild at the start gave the greatest ultimate production.

---

* Head Geologist, Roxana Petroleum Co.

(St. Louis Meeting, September, 1920)

IN VIEW of the frequent occurrence of petroleum in other parts of the world, it seems odd that so large an area as is contained within the borders of Brazil should be without this product. This apparent deficiency may be due, however, to our ignorance of the regions in which it may exist. In some places, indications point to the probable existence of petroleum in ages gone by; and while the presence of petroleum pools may be problematic, in several regions conditions not unfavorable to their occurrence exist.

Yet, Brazil has enormous oil resources in the rich oil-shales in different parts of the country. Many of these shales are very rich and only suitable processes for the extraction of the oil are lacking. At the present time, only a few small experimental plants are producing oil by distillation from these shales. These plants have been the subject of recent investigations by the Geological and Mineralogical Survey, a preliminary report of which is in the hands of the printer. The only regions where studies have not been made are the upper Amazon, the Acre, the Rio Negro, and the Peruvian frontier, which really seems to be the most promising field for explorations of any in the country.

In the vast plateau region of the interior north of the 20th parallel of latitude, granites, gneiss, mica schists, and very old metamorphosed sedimentaries predominate. Later, sedimentaries occupy a minor position and, where found in the interior, represent a thin veneering resting on the older rocks. At several places near the coast they have a greater development and contain considerable beds of oil-shale and may, in some cases, offer conditions favorable for the occurrence of oil. Such deposits.are found in the Permian rocks of central and southern Maranhão; in the Tertiary and Cretaceous beds along the coast of Alagôas Sergipe, Bahia, and perhaps farther south in Espirito Santo; and in the Parahyba embayment north of Cape Frio and in the interior Tertiary basin of eastern Sao Paulo.

---

* Geologist, Brazilian Geological and Mineralogical Survey. '

MARANHÃO

The information at hand as to the detailed structure and distribution of the rock formations of this state is very meager. It is derived principally from the paper by Dr. Miguel Arrojado R. Lisbôa[1] on the Permian rocks of Maranhão and from unpublished notes on these rocks in Piauhy and Maranhão by Dr. Gonzaga de Campos, Director of the Brazilian Geological Survey. The Permian beds are exposed along the Rio Parahyba for over 1000 km. and, generally, over the southern and eastern half of the state. These beds are covered largely by the thinner Trias and Cretaceous formations. In the extreme northwestern part of the state, granite appears near the coast. While the Permian rocks have suffered considerable folding in a minor way, the material in hand seems to indicate a general synclinal structure across the state with the main axis bearing northeast-southwest.

On the middle reaches of the Itapicurú and Mearim Rivers, bituminous shales are found together with calcareous sandy and marly beds associated with limestones. Occurrences are also met with near Codó on the Itapicurú, on the Rio do Inferno, at Fazenda da União on the Igarapé Sant'Anna, on the Codósinho, and on the Rio Mearim at the city of Barra da Corda.

At the occurrence on Rio do Inferno, the beds strike east and west with a dip of 30° south. The lowermost bed consists of a bog-head coal, somewhat similar to the Marahú beds of Bahia, overlying a thick bed of well-laminated bituminous shales. In the bed of the Rio Mearim, the bituminous shales are covered by a limestone with siliceous and gypsiferous intercallations. These beds have a southerly dip and are covered by over 50 m. of flaggy sandstones. At Grajahú, farther southwest, the same gypsiferous limestone occurs but without the bituminous shales, which, if present, are below the water level of the river. The limestone dips northeast with the strike **N**. 60° W. It is covered by a red conglomeratic sandstone. Similar beds are found in the extreme southwest of the state and in northern Goyaz on the Rio Tocantins.

The plains and lowlands of central Maranhão are so covered by the lateritic formation that observations on the underlying rocks are difficult especially as regards character and structure. Field work in this region is practicably limited to the dry season, from May to November. Samples of the oil-shale from this region gave the following results on analysis: bitumen, 36.5 per cent.; clays, 22.6 per cent.; soluble carbonates, 40.8 per cent.; and on slow distillation 450 l. (about 100 gal.) per ton. This appears to have been a very rich sample.

---

[1] Permian Geology of Northern Brazil. *Am. Jnl. Sci.* (1914) **37**, 425.

Some prospecting work has been done in this region, a drill having been mounted near Codó, but it seems that the attempt was discontinued after considerable depth was attained. Judging from the registered dips and strikes observed by different parties, the region has suffered considerable folding. For this reason special work should be done in determining the structural and stratigraphic features before any extensive drilling operations are undertaken.

ALAGÔAS

Knowledge of this region is obtained chiefly from the paper by Dr. John C. Branner[2] on the oil-shales of Alagôas. This has been supplemented somewhat by recent work on these shales by the Servico Geologico.[3] Shales rich in oil are found at several places along this coast. The series of rocks to which the oil shales belong are found along the coast about Cape S. Agostinho, Rio Formosa, Tamandaré, Abreu da Una, etc., but at these places the unweathered shale does not appear. Farther along the beach, in latitude 9° 3', at Maragogy, the oil-bearing shales appear at and a little above tide level. At this place they show a wrinkled synclinal structure and outcrop frequently from this point south, as at São Bento, Camaxó, Japaratubá, and in front of Pitinguy, in latitude 9° 7', where they are exposed at low tide. Dips are generally to landward.

At Barreira do Boqueirão, north of the Porto das Pedras, the shale exposed has a thickness of 2 m., with a probable thickness of 3 or 4 m. in all. At Camaragibe, the shales form a wave-cut terrace about 150 m. wide; the dips observed were from 5° to 10°. At this place several pits were put down many years ago. Samples from these pits examined by Boverton Redwood[4] showed the following composition:

	PER CENT.	PER CENT.	PER CENT.	PER CENT.	PER CENT.
Volatile	30.6	24.8	27.1	25.5	7.8
Non-volatile combustible	9.5	4.3	2.2	2.2	2.9
Ash	60.0	70.9	80.7	72.3	89.3

The shales are exposed at Barra do Santo Antonio and at Riacho Doce in latitude 9° 36'. The exposure at Riacho Doce is quite similar to those already mentioned. Several pits were sunk and the shales were found to be richer than those at Camaragibe. The composition was as follows:

	PER CENT.	PER CENT.	PER CENT.	PER CENT.	PER CENT.
Volatile	34.9	46.3	26.9	32.8	25.4
Non-volatile combustible	1.1	19.5	8.1	14.6	10.5
Ash	64.0	34.2	65.0	52.6	64 1

[2] Oil-bearing Shales of the Coast of Brazil. *Trans.* (1900) **30**, 537.

[3] Gonzaga de Campos: "Informações sobre a Industria Siderurgica." Rio de Janeiro, 1916.

[4] Boverton Redwood and William Topley: "Report on the Riacho Doce and Camaragibe Shale Deposits on the Coast of Brazil near Maceió." London, 1891.

A further examination of the second sample showed 4.7 per cent. sulfur, and upon distillation it yielded 44.73 gal. of oil and 19.58 gal. of ammoniacal water per ton. Exposures of these rocks are met with 10 or 15 km. inland in some of the river valleys and along the railway. ·

Redwood says of these shales: "The presence of sulfur would not, however, be a serious objection, if the crude oil were used as a liquid fuel or as a source of gas for illuminating purposes. One ton of such oil would, if properly burned, afford rather more heat than two tons of good steam coal, and from each gallon of oil about 90 cu. ft. of 60 candlepower gas could be produced. Results obtained on the laboratory scale of working are less satisfactory than those obtained when the shale is distilled on the manufacturing scale in retorts of suitable construction. The difference is far greater in the case of the ammoniacal liquor, and a yield of probably four times the quantity of sulfate of ammonia may be expected."

BAHIA

The better known occurrences of bituminous rocks in this state are those found in the vicinity of Marahú and southwards along the coast. The best study of the Marahú deposits is to be found in the paper by Dr. Gonzaga de Campos,[5] who examined the region in the year 1902. These rocks occur along the coast in the flat region for considerable distances and widths. The Marahú deposits are about 30 km. long by about 15 km. wide. The beds contain both fresh-water and marine fossils. Resting up against the old crystalline rocks is a fresh-water series of rocks containing plant remains, which is largely impregnated with bituminous matter. This is characteristic of the western inland part of the basin. Farther east, resting on these beds, are found limestones containing marine fossils and also with impregnations and masses of asphalt; these beds are of Cretaceous age.

The appearance of the bituminous and carbonaceous material everywhere is notable; these materials occur in the most varied forms. In these beds are found large solid impregnations having the appearance of asphalt; at some points the bitumen is viscous like pitch. At Taipúmirim, cavities a meter or more in diameter and quite deep are filled with black bituminous matter. An analysis of this material gave Dr. Gonzaga de Campos the following: volatile matter, 30.0 per cent.; non-volatile combustible matter 14.0 per cent.; ash, 56.0 per cent. The material contains much pyrites. Alcohol dissolves little of it; on evaporation, it gives a brown rosin. Ether dissolves most of the material and benzol dissolves it almost completely.

Resting on these Cretaceous beds, a clayey lignite is found in the lower beds of the Tertiary bluff formation of this coast. In the lowermost

---

[5] "Reconhecimento Geologico na Bacia do Rio Merahú, Bahia." São Paulo, 1902.

beds, almost at tide level, the boghead coal known as the "Turfa de Marahú" is found. This is a most peculiar material, being quite different from other known bitumens. It is light yellow in color with brown and gray veins, which appear as stratification planes. The rock separates along these planes and frequently plant leaves and other fossils of vegetable origin are found. To the touch, it is rather rough with a felty texture. It floats and does not absorb water readily. After many days immersion, it gave a density of 0.925 (mean) with variations between 0.850 and 1.200. On boiling in water, it becomes somewhat elastic to compression. It is easily cut with the knife and is elastic to a blow from the hammer, but is readily reduced to a fine light powder. Neither alcohol nor ether dissolves the material but it is highly bituminous. It takes fire readily from the lighted match and burns with a yellow smoky flame. An analysis gave the following: Water (at 110°), 2.75 per cent.; volatile matter, 71.65 per cent.; non-volatile combustible matter, 9.75 per cent.; mineral residue, 15.85 per cent. The residue consists principally of silica, much alumina, lime, and grains of quartz. Beds of this material are exposed for a depth of 3 to 4 m. at the mouth of the Rio Arimembeca and are said to continue in depth for over 15 m. These beds are horizontal.

On slow distillation, this material yielded 430 l. of crude oil to the ton; the density of this oil varies between 0.870 and 0.880. Neither in color nor aspect does the rock have any resemblance to coal, but the composition and the products are those of the bituminous coals. It is not a bituminous schist because the organic material greatly predominates over the mineral. The great mass of the rock is composed of yellowish brownish humic material. By fractional distillation, the material gave the following:

	PER CENT.
Below 150°, water strongly charged with pyrolignic acid	10.00
Yellow wine colored oil (sp. gr. 0.812)	9.74
Below 270°, dark brown greenish oil (sp. gr. 0.840)	21.84
Below 350°, dark oil (sp. gr. 0.884)	5.74
Residue—coke, porous, brilliant, weak	37.00
Loss	15.68

Farther south from Marahú, in the vicinity of Ilhéos, oil-shales similar to those of Alagôas are found in several places. These are small exposures of beds that appear along the coast between the granite points, which hereabouts frequently extend down to the ocean. The area of these beds seems to be relatively small; and while they are rich in oil content, their value remains to be determined. They are under investigation at the present time. At one of these exposures, near Ilhéos, a small still was erected, toward the end of the war, for the extraction of oil.

A region that may prove of more importance and worth while exploring lies farther south to beyond Caravellas in southern Bahia and northern Espirito Santo. In this region, over 100 km. long by 50 to 80 km. wide, sedimentary rocks occur and while no oil-shale is reported, the general geology would indicate that it is probably underlain by the same shale horizon as that just described. The deposits referred to, all along the coast, outcrop at or near tide level; in this region they may be slightly lower and so below that level and not exposed at low water, for which reason they have never been observed.

The shales about Ilhéos, as also those of Alagôas, on exposure after quarrying become warped and separate out into thin paper-like sheets. These sheets burn readily and frequently contain beautifully preserved fossil fish. Where more massive and clayey, the shales break into blocks and are not utilized.

SÃO PAULO

The Tertiary basin in eastern São Paulo, on the upper reaches of the Rio Parahyba, is perhaps 150 km. long by 15 to 20 km. wide. Over a considerable part of this basin, oil-shales have been found. These shales outcrop 10 to 15 m. above the Parahyba near Tremembé and Pindamonhangaba, where they are being mined. Quantities of these shales have been used at the gas works in Rio and in São Paulo at various times, especially during the war, on account of the shortage of coal. There exists at Taubaté, a small plant for the distillation of oil from these rocks. These shales also separate into thin paper-like sheets on exposure and take fire readily from the lighted match. The richer beds contain quantities of beautifully preserved fossil fish. An analysis gave the following composition: crude oil, 13.08 per cent.; water, 23.36 per cent.; gas and loss, 4.02 per cent.; mineral residue, 58.64 per cent. On slow distillation these shales yielded 27 gal. of crude oil per ton. However, at the plant that existed at Taubaté many years ago, only about 17 gal. were extracted.

SOUTHERN BRAZIL

Extending through São Paulo, Paraná, Santa Catharina, Rio Grande do Sul, and into Uruguay[6] is a very persistent bed of black petroliferous shale in the upper Permian series of rocks. This bed of shale was named the *Iraty Black Shale* by Dr. I. C. White[7] from its occurrence near the station of Iraty on the São Paulo Rio Grande railway. A freshly broken specimen of this shale generally gives off a strong odor of petroleum.

---

[6] E. P. de Oliveira: Regiões Carboniferas dos Estados do Sul. Servico Geologico, Rio de Janeiro, 1918.

[7] "Final Report of the Brazilian Coal Commission." Rio de Janeiro, 1908; The Coals of Brazil. Second Pan-American Congress, 1916.

At places, the petroleum of these shales has been oxidized into albertite or other substance resembling coal, as about Piracicaba and Rio Claro in the state of São Paulo. Material rich in oil is found between São Pedro and Piracicaba in beds of considerable thickness. A company has been formed recently in the city of São Paulo for explorations in this region. Near Rio Claro, several miles farther north, some drilling has been done during the last few years.

At about the same geological horizon as the above outcrops, but at a much lower level, deposits of asphalt occur along the Rio Tieté near Porto Martins. Farther south, in the foot hills of the Serra de Luiz Maximo between Tatuhy and Botucatú, a heavy bed of bituminous sandstone is found some distance above the black shale. An analysis of this rock showed 15 per cent. bituminous matter. This sandstone seems to represent the oxidized and eroded remains of a former pool. Small deposits of asphalt occur at different places in Paraná and Santa Catharina. Recently a plant has been installed near São Gabriel, in Rio Grande do Sul, for the distillation of oil from these shales.

Analyses of the black petroliferous shale and the albertite, as given in Doctor White's report, are as follows:

	Petroliferous Shale, Per Cent.	Albertite, Per Cent.
Moisture	3.95	0.35
Volatile matter	19.40	41.15
Fixed carbon	8.91	57.33
Ash	67.74	1.17
	100.00	100.00
Petroline	7.30	4.44
Asphaltine	0.00	6.80
Non-bituminous organic matter	24.95	87.59
Ash	67.74	1.17

These black shales outcrop in the plains region and among the foot hills along the east scarp of the great interior plateau. Farther south, this scarp gradually approaches the coast in Santa Catharina and then swings back west and southwest across Rio Grande do Sul. The rocks generally have a low westerly dip but the whole region has been somewhat faulted and folded and is cut by dikes of eruptives from which extruded the great flows of trap covering large parts of the interior.

The region west of the interior scarp has been indifferently mapped and almost no work has been done in studying the geological structure of the underlying rocks. The region merits study. It seems clear that no pools are to be expected east of the mountain scarp (the strata in which they might have occurred having been removed by erosion, as near the Serra de Luiz Maximo above noted) but conditions may exist farther west somewhere in this vast region favorable for the accumulation of such pools.

While extensive faulting and fissuring of the strata of this region may have allowed the escape of contained petroleum in their vicinity, these are neither so numerous nor so wide spread as to preclude its existence in other places. If one may judge by many examples known today, important deposits may still be present in the strata even in the vicinity of eruptive dikes. Be this as it may with regard to petroleum, the fact is abundantly demonstrated that, in these shales, Brazil has an inexhaustible supply which .only requires suitable processing to become available.

DISCUSSION

RALPH ARNOLD, Los Angeles, Calif.—The newspapers state that the Brazilian government is contemplating putting into effect rules and regulations for the oil business. I think this is a pretty good sign that there is oil in Brazil.

DAVID WHITE, Washington, D. C.—Oil-shales of the Tertiary age have long been known in the Province of Bahia. Bog heads extremely rich and comparable in constitution to the "kerosene shale" of New South Wales are reported to have been found in the coal measures of Santa Catarina and Rio Grande do Sul. Such Permian bog heads, which I examined at the time Dr. I. C. White was investigating the Brazilian coal fields, were, in fact, found to be so far identical paleontologically with the Australian rock as to arouse suspicion as to the genuineness of the Brazilian source of the material, as was noted in the report. Richly bituminous shales are, however, credibly reported to be present in great thickness in a formation of Triassic age in Brazil.

J. ELMER THOMAS, San Antonio, Tex.—1 saw recently a private report on an oil-shale occurrence in Santa Catarina. While not made by a recognized expert, it was an extremely detailed and careful report and called attention to one large deposit within 100 miles of the coast and midway between Buenos Ayres and Rio de Janeiro. Oil at this point is worth about $8 per barrel. The deposit was believed to be extensive and outcrops in a cliff to a thickness of 70 ft. Samples of the oil had been distilled in this country and showed a good gasoline content as well as kerosene, lubricating oil, and gas oil. The estimated yield was high, from 35 to 40 gal. per ton. It seems probable that an occurrence of this nature will be developed soon, as its economic importance is considerable.

JOHN C. BRANNER,* Stanford University, Calif. (written discussion).— The theory of the possible existence of oil-bearing formations in the upper

---

* President Emeritus.

Amazon region is a perfectly legitimate inference based on the known oil-bearing horizon in regions farther north, but it lacks the support of all the necessary facts. That region, however, is covered by dense tropical forests and is difficult of access on account of its great distance from the coast and the difficult navigation of the upper reaches of the Amazon River. Also, white races cannot remain in it long with impunity. Only a company with unlimited means could undertake the exploration and exploitation of such an area. The population is sparse and confined to the small towns along the larger streams.

In addition, the mining laws of Brazil do not encourage the development of these regions. In Decree No. 2933, January, 1915, article 42, paragraph 1 says that a mining claim shall contain 5 hectares (12.3 acres) and that the greatest number of claims that may be conceded to a single individual or organization for petroleum is 20 claims but "for the purposes of mining operations the limits shall be 40 claims" for petroleum. I have been informed that efforts are being made to revise the mining laws for the purpose of encouraging the development of that country's mineral resources but I do not know the result of these efforts. A proposed new mining law was published at Rio de Janeiro, Dec. 8, 1917, but I do not know if it was passed. This law provided that the unit of a claim shall be a hectare but that the number of hectares "that may be granted for each type of mineral deposit shall be established by the regulations for the enforcement of this law." These regulations are not published with the proposed law.

No one acquainted with the peculiarities of petroleum deposits of other parts of the world would venture the large capital necessary in a new and untried field for the sake of what he could reasonably expect to obtain from 100 acres of land.

Since the above was written I have received from one of the best posted legislators in Brazil the following information:

The laws now in force are those of Decree No. 2933 of Jan. 6, 1915. The provision of Art. 42, par. 1, of that decree relating to mining claims (*lote de lavra*) refer only to lands controlled by the Federal Government; and inasmuch as the Federal Government controls only limited areas, this provision is of little importance in its bearing on petroleum lands. When considerable areas are required for the development of petroleum fields, they may be obtained from the landowners by lease or purchase very much as such lands are secured in the United States.

International Aspects of the Petroleum Industry

BY VAN H. MANNING,* WASHINGTON, D. C.

(New York Meeting, February, 1920)

IN SUBSTANCE, the international aspects of the petroleum industry, as these relate to the United States, are as follows: The domestic production is not keeping pace with the domestic demands; our best engineering talent warns us of the imminence of a decreased production by our oil wells, although more oil is needed; and the only practical source whence this increasing demand can be supplied for some time to come will be the foreign fields. Other nations have given thought to the future and, in recent years, have shown a tendency to adopt strong nationalistic policies regarding their petroleum resources, policies that hinder or prevent the exploitation of these resources by other nationals. In consequence, we find that, facing a probable shortage of the domestic supply, our nationals are excluded from foreign fields; and this in spite of the fact that foreign nationals have been permitted to enter into and exploit our own oil resources on an equality with American citizens and without hindrance or restrictions. This country has supplied the larger part of the petroleum consumed by the world and yet, with a failing supply imminent, it finds that those countries that have been drawing upon our resources to supply their needs are showing a tendency to exclude us from their resources. In this way we shall be transferred from a position of dominence to one of dependence; and only by sufferance of those countries that are now seeking financial or political control of petroleum supplies, shall we be able to obtain the oil we will need.

IMPORTANCE OF PETROLEUM

Petroleum has become, during recent years, one of the essentials of our social and industrial life. All civilized countries recognize that the world is dependent on petroleum as on nothing else except textiles, foodstuffs, coal, and iron. Today, the tendency is toward an ever-increasing consumption of petroleum and its products as new and more efficient uses are found for them. The utilization of petroleum is extending more and more into the structure of our civilization. Consequently,

---

* Director, U. S. Bureau of Mines.

it becomes a matter of the gravest concern whether we can go on building up an industrial and social structure dependent on petroleum unless we make provision for obtaining the necessary supplies. Unlike foodstuffs and textiles, the world's supply of petroleum is definitely limited; moreover, it is, like coal and iron resources, a wasting asset. But petroleum is a liquid, is by nature migratory, can be quickly extracted, and an oil field is readily exhausted; whereas coal and iron are extracted more slowly and, by prospecting, reserves can be blocked out for the years ahead. Oil fields once discovered are developed almost immediately; within a short time the peak of production is passed and decline sets in. We are constantly relying upon the discovery of new fields, at the moment unknown, to make up for the decline and depletion of those that are proved. Thus, we are living a hand-to-mouth existence and although during the past decades we have been very fortunate in making opportune discoveries—first Cushing, then Kansas, and then northern Texas— each of which has made up for a threatened deficit, the time must inevitably come when fortune will forsake us and the needed new production will not materialize. Then we may find ourselves suddenly thrown upon the mercy of the nations that control foreign sources of supply.

Few of us realize in how many ways petroleum products serve our daily needs. Petroleum in one form or another is used in every household; gasoline for the motor car, lubricating oils for bearings, kerosene lamps or paraffin candles for illumination. Not one of us can sit back and say that an adequate supply of petroleum is not a personal concern. Perhaps a recent statement appearing from enemy sources may convey most convincingly the importance of petroleum in modern life. Ludendorff, in his book on the late war, in speaking of the Rumanian campaign, says, "As I now see clearly, we should not have been able to exist, much less carry on the war, without Rumania's corn and oil, even though we had saved the Galician oil fields from the Russians."

IMPORTANCE OF INDUSTRY

During the world war, the Navy demonstrated the value of petroleum as marine fuel. Having a higher heating value than coal, a given tonnage assures a ship a much wider cruising range before refueling. In the mercantile marine the smaller bulk of fuel provides larger cargo space in the hull. Cleanliness and less labor for loading and burning are two other important features. In consequence, new ships are being built to burn oil and old vessels are being changed from coal to oil burners. Our greatest maritime rivals, the British, are rapidly equipping their merchant marine to burn oil, so that it has become obligatory upon the United States Shipping Board to do likewise in order that our vessels may be able to compete on an equal basis, as regards fuel, with foreign-owned bottoms.

The production, refining, and distribution of petroleum and petroleum products is one of our greatest industries; it provides a livelihood for many thousands of families. Although it. has offered a big field for the engineer and chemist, in my opinion it has been comparatively unexploited by the mining engineer and is capable of absorbing hundreds, if not thousands, of properly trained and experienced engineers.

The oil industry also provides a wonderful field for our chemical engineers. Petroleum can be considered as a crude chemical, like coal tar, and the fuel value of all its products and the most efficient methods of utilizing them have not been discovered by any means.

Not only has petroleum furnished useful and essential products, but industries based upon these products rank among the major activities of the nation. Of such dependent industries, the greatest is the automotive industry. The automobile, the truck, the tractor and the airplane enter into our daily life. Today more than 6,000,000 automobiles are in use in the United States alone.

The three most important utilizations of petroleum are as fuel, as an illuminant, and as a lubricant. Petroleum fuels may be classified as light and heavy. The light fuels are gasoline, naphtha, and kerosene, which can be vaporized and used in the internal-combustion engine of the automobile or tractor. Heavy fuels are those that are burned directly for steam raising or for heating purposes, or can be used in internal-combustion engines of the Diesel type. About 57 per cent. of our output of crude petroleum is oil fuel of the heavy type, only a small proportion of which is used in internal-combustion engines; the other uses are relatively inefficient and for such uses petroleum is replaceable by coal. A larger use of this heavy fuel in the internal-combustion engine is hopefully expected, but with this development, the dependence of the world on petroleum will be increased still further.

This country is not as dependent upon petroleum illuminants as it was, although kerosene still is used in large quantities in districts not served by gas or electricity, and is an article of great importance in our trade with foreign countries.

Petroleum lubricants, although less in amount than the other products, are more generally used and are really more essential. They lubricate practically all bearings or moving parts. Quantitatively, there are no satisfactory substitutes and when one starts to replace, on a large scale, mineral lubricants by animal or vegetable oils of satisfactory quality, the dependence of our industrial life on petroleum lubricants becomes evident.

When we realize what petroleum, directly and indirectly, has done for our country and when we try to see what improvements in our ways of living the future holds for us, the significance of the international aspects of the petroleum industry becomes clearly evident. When we consider the number of automobiles turned out yearly, the airplanes that will play an important part in commerce, the trucks that will supplement present transportation facilities, the agricultural machinery needed to meet the lack of man power on our farms, and the relation of our merchant marine program to oil, we can understand how vitally necessary an adequate supply of petroleum will be to us.

OUR PETROLEUM RESOURCES

The United States was the first country to produce oil in large quantity by the modern system of drilling wells and, except during a few years, has led all the countries of the world in the quantity of its production. In 1914, when the World War began, the United States was in first place and produced approximately 266,000,000 bbl. of oil, or about 66 per cent. of the total output of the world. Russia was second, with an approximate production of 67,000,000 bbl., or about 17 per cent. of the world's total. Mexico came third with about 21,000,000 bbl., or a little over 5 per cent. of the world's production. Rumania, the Dutch East Indies, India, Galicia, Japan produced comparatively small amounts of oil, totaling approximately 12 per cent.

In 1914, therefore, the United States was far ahead of any other nation as a producer of oil. It was also far ahead of any other in the development of its oil fields and in the utilization of oil products. The vital importance of petroleum had not been fully recognized by the leading countries of the world, so the United States occupied a unique position, practically without competitors. Foreign countries had not begun to consider seriously future supply and there was less rivalry in gaining control of possible oil fields. Yet signs of an awakening interest were evident. Great Britain, because of having adopted fuel oil in the Navy, had begun taking steps to assure, through British nationals, an adequate supply of oil from Mexico and to encourage development in British domains. The British Government had also entered into partnership with the Anglo-Persian Oil Co. to exploit a huge concession in Persia.

The point of real importance, however, is the relative position of the United States as a consumer rather than a producer of oil. To produce the bulk of the world's production is of small consequence in comparison with producing enough to meet our present and probable future needs. In 1918, the output of crude oil in the United States was 356,000,000 bbl. Mexico had taken second place with 63,000,000 bbl. The production of the United States for the past several years has been approximately 65 per cent. of the world's total. The approximate consumption of the United States for the year 1918 was 418,000,000 bbl., or more than 80 per cent. of the world's production. This figure of consumption, however, includes the oil that was refined or partly refined in the United States and exported for consumption abroad. The exports of petroleum products approximated the imports of crude petroleum from Mexico and other foreign countries. But in addition, some 20,000,000 bbl. of oil were withdrawn from domestic storage. In substance, therefore, the United States, in 1918, was living beyond its means. The year 1919, because of the present flush production from Texas fields and the increased imports from Mexico, finds the United States in a somewhat more favorable condition, not having to draw on stocks; yet it must be remembered that the stocks have not only decreased actually but have decreased in proportion to our production and consumption. Thus, in 1915, there was over six months' supply of oil in storage, whereas at the end of 1918 stocks had been reduced to less than four months' supply.

The U. S. Geological Survey has given the following figures of the marketed production and consumption in the United States. The figures for marketed production approximate, but are not the same as, actual production:

	MARKETED PRO- DUCTION, BARRELS	CONSUMPTION, BARRELS
1916	301,000,000	329,000,000
1917	335,000,000	384,000,000
1918	356,000,000	418,000,000

Evidently the production of the United States, in spite of its having risen steadily during recent years, is not rising as rapidly as it should and is not keeping pace with the increase in consumption. The sources from which we can draw for our future needs of petroleum and its products are: Our own oil fields, foreign oil fields, oil shales, and substitutes for petroleum products. Engineers and geologists who have investigated the possible oil underground in our developed and undeveloped oil fields agree in making pessimistic reports. This is particularly true of the U. S. Geological Survey, the organization that has given most attention to our petroleum resources and has the most facts. The U. S. Geological Survey estimates our unproduced but recoverable oil in January, 1919, at 6,740,000,000 bbl. This, could it be produced as needed, would not continue our present production of oil for more than 20 years.

Many persons, especially non-technical oil men, are inclined to question these estimates and call them too pessimistic, saying that whenever in the past more oil was needed new discoveries were made and unexpected fields brought forth new supplies. However, our best-informed engineers have given this estimate, and their belief should outweigh the vague optimism of those who question it. Of course, in view of the fallibility of estimates, the figures may prove to be too pessimistic. Even if the estimates of the supply of unrecovered petroleum were 50 to 100 per cent. too low, the situation would still not be satisfactory. And the fact remains that no matter how much oil there may still be in the ground, we have not been and are not getting it to the surface as fast as it is now needed.

Clearly, we must seek other sources of supply to make up the balance between domestic production and domestic needs. Enormous deposits of oil-bearing shales occur in the western states, in the Cretaceous formations of the Rocky Mountain region. The U. S. Geological Survey estimates that the shales in the states of Colorado, Wyoming, and Utah alone contain many times the recoverable oil present in our oil fields before well drilling began. But the oil in these shales is not immediately available. The extraction of oil from the shales on a commercial scale under existing conditions in the United States is still in an experimental stage. We do not know, as yet, whether these shales can be developed profitably under present conditions, nor under what conditions they can be developed. Furthermore, it will take many years, even under favorable conditions, to obtain from these shales enough oil to replace a considerable part of that now obtained from wells.

I do not wish these statements to be interpreted as reflecting on the prospects of the shale industry, but simply wish you to realize that the production of oil in the quantities demanded by present-time needs would require development on a tremenduous scale and would require the mining of hundreds of millions of tons of shale each year, the annual amount being more than half the annual tonnage of coal now mined. There is no evidence that shale oil can be produced on such a scale at present prices and, therefore, to satisfy our petroleum needs by oils from shales involves higher prices for petroleum products. Moreover, our oil shales occur in sparsely populated regions, remote from centers of large consumption. Oil shales constitute a reserve that, fortunately, seems to provide ample protection against an ultimate future but they cannot be used to meet the present situation.

Substitutes

The products from the destructive distillation of coal can be used, in so far as they are available, to replace gasoline; but quantitatively it seems out of the question to expect more than a minor alleviation from them. Coal can largely replace fuel oils. Alcohol can replace gasoline and has the advantage that it can be made from replaceable material— that is, from plants, but because of its cost, it cannot compete in a large way with gasoline at present. Moreover, the difficulty and expense of replacing any considerable part of the gasoline supply by alcohol is not generally appreciated. Finally, no substitutes are now known that will satisfactorily replace mineral lubricants in the quantities needed.

Thus, the facts indicate that we must inevitably seek foreign supplies in order to meet our needs and to compete in the world's markets without too great a handicap. However, we should not rely upon any one solution of the problem, but should seek to put into effect every feasible means that promises to help, and should strive to anticipate our future needs rather than to go along blindly with the inevitable result of sud- denly being confronted at some future date with a shortage of oil. Steps should be taken to conserve our developed supply. This supply is tangi- ble; we already have it, and common sense dictates that we take the best possible care of it. By conservation I do not mean the tying up of re- sources, but a wise utilization, the working out of methods that will yield us the greatest quantity of oil at the least cost and will enable us to refine and use the oil with the highest efficiency. This phase of the question is peculiarly a part of the work of technical men, and I believe that this Institute should seriously endeavor to further, in every possible manner, the application of engineering methods to the oil business, for the oil industry is probably more backward in applying engineering knowledge than any other mineral industry. This statement is not a criticism of the oil industry for being backward in taking up engineering, any more than it is a criticism of the engineer in being backward in taking up the oil industry. Until recently there were few engineers who were qualified, by actual experience in the oil fields as well as by engineering training, to be of real assistance to the industry. Happily, this condition is rapidly improving. Yet there is today an under supply of competent petroleum engineers equipped to deal with practical problems.

In addition, we should further the oil-shale industry and, regardless of our individual opinions, should endeavor to determine as soon as pos- sible under what conditions the oil-shale industry is commercially feasible, and thus be prepared for a future emergency.

In the same way, petroleum substitutes should not be neglected These lie mainly in the field of the chemical engineer rather than of that the mining engineer.

FOREIGN SOURCES OF SUPPLY

Recently, the U. S. Geological Survey has shown a particular interest in questions of foreign supply, and has rendered a splendid public service by collecting all possible information on the subject. This information has been placed at the disposal of the government and also of those in- dividuals who contemplate entering foreign fields.

In the opinion of the U. S. Geological Survey, enormous resources await development in various parts of the world; but these resources have not been developed as intensively as those of the United States. The premier position of the United States to the present time has been due, perhaps, more to an intensive development of resources than to any supremacy in the resources themselves. Enough information is available about foreign countries to know that oil occurs in many places, and that there are partly developed fields of high promise.   It may well be that in vast areas which have not been studied by the geologist or tested by the prospector there are undiscovered fields of great magnitude.   For these reasons I believe that there is not nearly as much danger of a world shortage as there is of a domestic shortage.   Fortunately, the situation requires nothing more than the developing of foreign fields as supplies are needed and the accessibility of those fields to our nationals.   The problem that presents itself, therefore, is whether the United States can obtain an adequate share of oil from the known and potential fields of the world, or whether it is going to be excluded by the political and economic policies of other nations and thus find itself, so far as petroleum is concerned, at the mercy of those nations.

The key to the future is access to the sources of supply.   The strong financial position of the petroleum industry, in this country, the refining and marketing facilities of the strongest American companies will not, by themselves, suffice if we are at the mercy of the citizens of other nations for our crude supplies.

Strong Nationalistic Tendency of Foreign Countries to Exclude Other Nationals

One result of the war has been an accentuation of nationalistic spirit; the nations that were combatants and those that were neutral have shown increasingly a tendency to exclude other nationals from their domains and to develop their own resources by their own interests.   This tendency is a natural result of an awakened knowledge of the need of self-protection and of a desire to conserve for themselves the materials now essential to the world's civilization.

The United States is not an imperialistic nation, and, exclusive of Alaska, its foreign possessions are with small potential resources.   Thus we find no political control of consequence over other than the domestic sources of supply within the United States proper.

When we turn to the developed or prospective oil fields in other parts of the world, we find that their political control may be grouped under two heads: colonies and domains of such nations as England and France, and domains of smaller nations, such as the Latin-American countries, China and Persia; under present chaotic conditions perhaps Russia could be included.   The most promising oil districts now known outside of the United States are in Mexico, in the South American countries bordering the Caribbean, in Equador, Peru, Bolivia, Argentina, northern Africa, Egypt, Persia, Mesopotamia, Palestine, Russia, India, East Indies, and China.   There are other localities of smaller promise or about which less is known, and doubtless some of these will develop fields of the first magnitude when explored and prospected.

When one reviews these potential oil fields, one is struck with the fact that Latin America, Great Britain, France, and the Netherlands, apparently control the main potential sources of supply, and particularly those that are of the most concern to the United States. Thus, the policies of these countries are of the greatest interest to America. We find England and France adopting policies, already in part incorporated into laws or 'regulations, that now virtually exclude other than their own nationals from developing the resources within their own realms. Of course I do not mean to insinuate that the policies of these countries are aimed directly at Americans; the policy of each country is to look after its own citizens; hence it is directed against the citizens of all other countries, and thus affects Americans. For a detailed statement regarding the policies of these countries I refer to a memorandum by myself to the President, which was disclosed to the United States Senate by Senator Phelan of California. Copies of this document appear in the Congressional Record of July 29, 1919. Those interested in the various political phases of the situation can obtain information there, or from the American Petroleum Institute.

The members of this Institute are well informed as to the situation in Mexico. Mexico is considering stringent regulations as to oil concessions which, if enacted into law will be very detrimental to the just interests of nationals other than Mexicans, including ourselves. The policy of Argentina has been, practically, the nationalization of its petroleum resources. Other Latin-American countries have shown some uncertainty as to what their policies are to be. Japan has adopted a policy that practically excludes other nationals from its own fields in Japan, Formosa, the Island of Sakhalin, and from the fields of China so far as its control extends. The Netherlands Government has also adopted a policy of exclusion that practically restricts developments within its domains to its own nationals. France has adopted policies that are not so evident on the surface, but in effect, these policies are proving restrictive, and are seemingly intended to exclude other nationals.

RECIPROCAL PRIVILEGES SHOULD BE GIVEN TO AMERICAN NATIONALS

A review of the foreign situation, therefore, discloses the fact that whereas other nationals can enter our oil fields, acquire properties there, and work these properties on an equality with ourselves, our nationals are not receiving reciprocal privileges from many foreign governments now controlling the most important oil regions of the world, and thus in time we are likely to be largely dependent on those governments for our domestic needs. Moreover, conditions in the Latin-American countries are not as satisfactory as they might be. The question comes, therefore, as to what should be done toward removing discrimination under which Americans are practically excluded from foreign oil fields. It is not for me to discuss here such a question in detail, but it is perfectly obvious that in all fairness our nationals should be accorded the same privileges that we accord other nationals. It has not been the policy of the United States to exclude foreign corporations or individuals; in fact, they have been welcomed, as it has been recognized that the capital brought in has been, in a large way, helpful to the United States even though the profits went mostly to the benefit of other nationals. It would be, in my opinion, a mistake to forsake this policy, just as I believe it is a mistake on the part of other nationals to have put into effect such policies. It would be desirable if all countries adopted the same open policy as that which has prevailed in the United States.

In regard to individual Americans, and particularly to the members of this Institute, it seems to me that it is the duty of all to interest themselves in the situation and to do what they can to educate the people of this country and their representatives as to the situation, and to urge such wise and necessary steps as would best relieve it.

Another help that the members of this Institute can render is to transmit to the government such information as it acquires on the foreign situation, including information on the possibilities of oil fields, on laws, regulations, and policies that tend to discriminate against American nationals entering foreign fields, and on actual cases of discrimination. This information built up from many sources will prove invaluable to the government, and thus to yourselves and those interested in the foreign oil fields. I do not know whether the furtherance of such work could be made properly a part of this Institute collectively, but I see no reason why the members of this Institute should not render this service to their government.

I may also urge the opportunities and national importance of American concerns entering foreign oil fields. Evidently this country is going to need foreign sources of supply, and it will be to its great advantage to obtain these through its own nationals. Heretofore, American methods, American machinery, American brains have been employed by foreign capital to develop foreign resources. It will be more desirable if our brains and abilities are employed under our own nationals. It is desirable that every engineer realize before accepting employment with any foreign corporations competing against ours, just what this means. I believe it should be made a policy of the members of this Institute to see that the younger engineers and those unacquainted with foreign conditions, are informed on this matter.

DISCUSSION

LEONARD WALDO, New York, N. Y.—In Mexico, there are huge oil resources, but the only means of transmitting that oil to the United States is by ship, and ships seem to be forgotten on all occasions. Those we had at the beginning of the war for carrying oil were almost all under foreign charters, which were soon recalled and the ships used for transporting oil from Mexico and other points to Europe. Consequently, now we have a scarcity of ships for carrying oil; that is the most important defect in fueling the Atlantic seaboard. Every effort should be made to bring the shipping interests into line, including the government shipping. Oil is the one way of fueling the Atlantic seaboard and taking care of our steel plants, our boiler plants, our heavy industries that take oil, and ships must be used to relieve the pressure from the oil lines, which are only capable of supplying the higher uses of oil at 20 or 30 cents a gallon. For fuel, the marketable value of oil should be about 2 cents a gallon; before the war, large contracts were made at 1.8 cents per gallon for Mexican fuel oil delivered at the docks for the steel works to use.

A Foreign Oil Supply for the United States

BY GEORGE OTIS SMITH,* PH. D., WASHINGTON, D. C.

(New York Meeting, February, 1920)

TWELVE years ago, the Director of the United States Geological Survey addressed to the Secretary of the Interior a letter calling attention to the government's need for liquid fuel for naval use and pointing out that the rate of increase in demand was more rapid than the increase in production.[1] This letter, in a way, inaugurated the policy of public oil-land withdrawals, which was well founded in its primary purpose of protecting the oil industry and highly desirable in its immediate effect of checking the over-development of that day in California. Unfortunately, however, through delays in legislation, this policy may be regarded now as having outlived both its intent and its usefulness. In 1908, the country's production of oil was 178,500,000 bbl., and there was a surplus above consumption of more than 20,000,000 bbl. available to go into storage. In 1918, 10 years later, the oil wells of the United States yielded 356,000,000 bbl.—nearly twice the yield of 1908—but to meet the demands of the increased consumption more than 24,000,000 bbl. had to be drawn from storage.

Nor is this all of the brief comparison. In 1918, our excess of imports over exports of crude petroleum was nearly 33,000,000 bbl. whereas in 1908 we exported 3,500,000 bbl., which was net, as we had not begun to import Mexican oil. In this period, the annual fuel-oil consumption of the railroads alone has increased from 16,871,000 to 36,714,000 bbl.; the annual gasoline production from 540,000,000 gal. to 3,500,000,000 gal. This record may be taken not only as justifying the earlier appeal for Federal action, but as warranting deliberate attention to the oil problem of today.

NEED OF FUTURE SUPPLY

The position of the United States in regard to oil can best be characterized as precarious. Using more than one-third of a billion barrels a year, we are drawing not only from the underground pools but also from storage, and both of these supplies are limited. In 1918, the contribu-

---

* Director, U. S. Geol. Survey.

[1] This letter, drafted by Dr. Ralph Arnold and concurred in by Dr. C. W. Hayes and Dr. D. T. Day, is quoted in *Bull.* 623, U. S. Geol. Survey, 104.

tion direct from our wells was 356,000,000 bbl., or more than one-twentieth of the amount estimated by the Survey geologists as the content of our underground reserve; we also drew from storage 24,000,000 bbl., or nearly one-fifth of what remains above ground. Even if there be no further increase in output due to increased demand, is not this a pace that will kill the industry? Even though we glory in the fact that we contributed 80 per cent. of the great quantity needed to meet the requirements of the Allies during the war, is not our world leadership more spectacular than safe? And even though the United States may today be the largest oil producer and though it consumes nearly 75 per cent. of the world's output of oil, it is not a minute too early to take counsel with ourselves and call the attention of the American geologists, engineers, capitalists and legislators to the need of an oil supply for the future.

This appeal to American brains and American dollars to provide for the future needs only the backing of a brief recital of the facts of known present needs and of well-justified expectations for the future. In a single decade, then, the consumption of fuel oil by railroads has more than doubled; the consumption of gasoline has increased sevenfold. With the rapidly mounting cost of coal, the competitive field of fuel oil for steam use is expanding. But not only is the use of oil, both under boilers and in internal-combustion engines, thus increasing, there is an even more widespread use of a petroleum product, which was brought to the President's attention over 10 years ago.[2] Every new installation of machinery, whether the 60,000-kw. generator in the Government nitrate plant at Sheffield, Ala., or the 20-hp. motor in the small automobile, adds to the country's demand for lubricating oil, which is an essential in every phase of modern civilization. We may lessen the increase in coal or oil consumption for generating power by harnessing the water powers of the country; but these prime movers, whether driven by steam or water, require lubrication. With the rapidly increasing use of machinery to make labor more productive, with the almost universal use of the automobile, hardly foreseen a decade ago, and with the expected increase in railroad and steamship traffic, who can venture an estimate of our petroleum requirements, 10 years hence, in terms of lubricatin oil alone?

A most serious aspect of our oil problem presents itself when we consider the entry of the United States as a real factor in the shipping of the world—when we picture the return of the American flag to the seven seas. Any nation which today aspires to a large part in world commerce imposes upon itself an oil problem, for the future freedom of both the sea and the air will be defined in terms of oil supply. The new demand of our shipping program alone involves fuel oil in quantities equivalent

[2] Letter quoted in full in *Bull.* 623, U. S. Geol. Survey, 134.

to nearly one-half of the present domestic output, and, unless there is some corresponding decrease in other demands, this new requirement must be met with an increase in production of crude oil of nearly 200,000-000 bbl. How can such quantities of oil be supplied? Mr. Requa's earlier estimate of 52,C00,000 bbl. as the annual gain in output needed to meet the ordinary increase in consumption and to offset the expected decline in old wells would involve a task laid upon our oil companies, in their exploration and development activity, of bringing in a million-barrel new production each week. How can the oil fields of the United States maintain such a curve of new production?

Fuel oil, gasoline, lubricating oil—for these three essentials are there no practical substitutes or other adequate sources? The obvious answer is in terms of cost; the real answer is in terms of man power. On land and on sea, fuel oil is preferred to coal because it requires fewer firemen; and back of that, in the man power required in its mining, preparation, and transportation, the advantage on the side of oil is even greater. So too, the substitute for gasoline in internal-combustion engines, whether alcohol or benzol, means higher cost and larger expenditure of labor in its production. While we have great reserves of oil shales as an independent source of fuel oil, gasoline, and lubricating oil, it is necessary to consider the practical contingency suggested by Mr. Requa, that to develop this supply on a scale comparable in output with our present oil supply "would require an industrial organization greater than our entire coal-mining organization." Plainly our country cannot afford to support another such army of workers until we reach another stage in our industrial development.

A country-wide thrift campaign needs to be waged looking to the saving for this essential resource. Man power and oil ought to be conserved all along the line of production and consumption by better methods in the discovery, drilling, recovery, transportation, refining, and use of petroleum and its products. Unwarranted optimism, which seems indigenous in most parts of the United States, has led both the oil industry and the public to waste this best of fuels; the program of wastage begins with leakage below ground and above ground and continues to the indiscriminate burning of fuel oil under boilers, with regard for convenience rather than for efficiency.

The estimate by the United States Geological Survey of the oil remaining in the ground is of necessity subject to criticism as speculative—it must contain errors in the allowances made for isolated and undeveloped fields—yet the excesses of unexpected yield in one region will largely be balanced by deficiencies in another. Indeed, as has been suggested by the Chief Geologist of the Survey, if happily the estimate of reserve proves too low, this unpredicted abundance would surely raise the consumption rate. On the whole, he believes it fair to consider the official estimate of

6,500,000,000 bbl. as conservative and 8,000,000,000 as an improbable maximum. The difference between these two estimates of reserves represents only four years' supply, even at the present rate of consumption.

It seems almost as if divine providence, by the Cushing and Healdton "strikes," replenished our supply of oil "in storage" just in time to enable us to export oil and gasoline in quantities sufficient to justify Earl Curzon's statement that the "Allied fleets floated to victory on a sea of oil;" and the Ranger discovery was equally providential; yet the motto inscribed on our silver coins should hardly be made our national policy in providing a future oil supply.

It cannot be pointed out too often that while in the last 100 years the unprecedented growth in the industrial and transportation demands of our country has resulted only in the exhaustion of less than 1 per cent. of its coal resources, in the 60 years since the Drake well began our production apparently 40 per cent. of the available oil has been brought to the surface and consumed; and the rate of America's development is still an accelerating rate. American interests, commercial and industrial, thus require a future supply of crude oil outside the United States. Indeed, we have been draining our own oil pools in part to supply the needs of the rest of the world, but have made little effort to render the rest of the world self-supporting in oil production. Whether such a national policy is to be characterized as that of a spendthrift or that of an altruist, it is a short-sighted policy. With our oil reserves so plainly inadequate, it is not too much to treat our own country under a kind of favored-nation policy. Surely the United States can rightfully safeguard American interests at home and abroad, with the spirit of reciprocity in trade relations.

OBTAINING A FUTURE SUPPLY

Two methods of handling the problem of a future oil supply suggest themselves: either reserve the domestic oil fields for American development and thus prevent foreign acquisition of what is needed at home· or, encourage our capital to enter foreign fields to assist in their development, thus insuring an additional supply of oil for our needs. The one method harks back to the "Chinese wall" period, the other expresses the "open door" policy. At present the United States Government follows neither method; the British Government has adopted both.

The British Admiralty led the way in its appreciation of the advantages of fuel oil, and the British Government has led the way in assuring to its nationals control of oil resources wherever found on British territory. Advantages that American capital may once have held in Trinidad and elsewhere in the British Empire are not now enjoyed and British enterprise is narrowing the field of opportunity in Mexico, South America, Mesopotamia, and Africa. Be it said, moreover, to the credit of British efficiency and foresight, that British capital has made generous use of American brains in discovering and developing its oil properties. American geologists, American engineers, American drillers, and American rigs and supplies have been utilized in British oil exploration and we may well reciprocate by adopting the British policy of encouraging the acquisition by its nationals·of petroleum supplies in foreign fields. American capital as well as American engineering should be encouraged to help develop the new fields and so do its part in insuring the continuance of this source of power for future generations at home and abroad.

The part of the Government is to give moral support to every effort of American business to expand its circle of activity in oil production so that it will be coextensive with the new field of American shipping. This may mean world-wide exploration, development, and producing companies, financed by United States capital, guided by American engineering, and safeguarded in policy because protected by the United States Government.[3] Thus only can our general welfare be promoted and the future supply of oil be assured for the United States.

DISCUSSION

M. L. REQUA, New York, N. Y. (written discussion).—This paper calls attention to what is perhaps our most critical raw-material problem. I have spoken and written so vigorously and frequently upon this subject that it seems almost useless repetition to refer again to the subject; but, in the face of everything that has been said and done by various individuals alive to the situation, we are utterly without any national policy as related to foreign sources of petroleum supply. We build up a great mercantile marine and predicate its success upon fuel oil, but we make no really constructive effort to assure the source of supply for that material. We construct warships made to burn nothing but fuel oil, and we face a lack of preparedness and appreciation of the gravity of the situation on the part of the directing head of the Navy that would be grotesque were it not for the tragedy involved.

---

[3] In his annual report to the President, the Secretary of the Interior states (pp. 18–20) that the present situation "calls for a policy prompt, determined, and looking many years ahead." The supplemental supply needed "may be secured," he says, "through American enterprise if we do these things: (1) Assure American capital that if it goes into a foreign country and secures the right to drill for oil on a legal and fair basis (all of which must be shown to the State Department) that it will be protected against confiscation or discrimination. This should be a known published policy. (2) Require every American corporation producing oil in a foreign country to take out a Federal charter for such enterprise under which whatever oil it produces should be subject to a preferential right on the part of this Government to take all of its supply or a percentage thereof at any time on payment of the market price. (3) Sell no oil to a vessel carrying a charter from any foreign government either at an American port or at any American bunker when that government does not sell oil at a non-discriminatory price to our vessels at its bunkers or ports."

The day of reckoning must come, of course, in all things; and our Government officials have before them a very unpleasant experience when they have to explain the lack of foresight as regards petroleum. The documents on file in various governmental departments in Washington calling attention to this situation would make, if assembled as a whole, extremely interesting reading—in the light of events. To my knowledge, the Director of the Geological Survey has for at least five years been urging proper consideration of the subject.

CHESTER W. WASHBURNE, New York, N. Y. (written discussion).— It is a delightful surprise to read Doctor Smith's statement that the policy of withdrawing government oil lands ''may be regarded now as having outlived both its intent and its usefulness.'' It has indeed become a nuisance and an injustice to anyone who discovers oil on the public domain, only to have the benefits of his intelligence and daring snatched away by Presidential decree. Everyone, even the old-timers who thought differently, now recognize that the oil supplies of this country are wholly inadequate for our own future needs. The passage of a good leasing bill will help a little; the development of foreign oil fields is the great necessity.

Foreign development of any consequence requires two things. First, there should be greater backbone in the American State Department and President in protecting and helping American capital in foreign fields. The recommendations of the Secretary of the Interior outlined by Doctor Smith would help, if adopted. Secondly, there should be greater and more persistent efforts of American capital in hazardous foreign undertakings. The second element already shows manifestations of serious importance. In the first we are outclassed by the well-knit organization of the British Foreign Office and the harmoniously working British Consular Service.

In view of the great risks involved and the high expenditures, it would be wise for American companies to combine in some way for foreign work. The principal American companies working in any one foreign country should be able to pool their interests in some way, to avoid bidding against each other for concessions, etc. One way in which this could be done is exemplified by the British companies in Venezuela, which operate independently in the field, each in its own area, but which are in close, though more or less secret, association in London. They practically have cornered the best part of Venezuela, while the American companies remained uninterested. We soon will regret this oversight. As an example of the way American companies work, I will cite one experience. Several years ago, I had a distinguished oil geologist examine certain properties in Colombia. He condemned them. Since then four American companies successively have sent expeditions to examine the same properties, and since no one has taken them I presume all geologists have condemned them. The mouth of each geologist is sealed by professional ethics, but the heads of the companies in question might have been friendly enough to prevent this foolish reduplication. I believe these properties will remain on the market for future examination, until some company happens to send a poor geologist who will allow his employers to pay a big bonus and drill some dry holes. Meanwhile British capital has been strengthening its position in the more desirable territory to the east, Venezuela.

Two American companies have been trying to get another Colombian property that looks rather attractive, but their efforts have resulted in boosting the price to a foolish figure. Development is delayed until the owners will listen to reason. Cooperation would have saved this situation. Cooperation would have resulted in definite development in other foreign affairs where Americans have been competing with each other.

R. H. JOHNSTON,* Washington, D. C.—The remarks just made emphasize the fact that Great Britain enjoys a much more vigorous foreign policy than does this country. During the discussion regarding the Persian oil fields,[4] the suggestion was made that American companies should take part in the development of these fields. On Aug. 9, 1919, Great Britain signed a treaty with Persia, from which I will read three paragraphs:

(2) The British Government will supply, at the cost of the Persian Government, the services of whatever expert advisers may, after consultation between the two governments, be considered necessary for the several departments of the Persian administration. These advisers shall be engaged on contracts and endowed with adequate powers, the nature of which shall be the matter of agreement between the Persian Government and the advisers.

(3) The British Government will supply, at the cost of the Persian Government, such officers and such munitions and equipment of modern type as may be adjudged necessary by a joint commission of military experts, British and Persian, which shall assemble forthwith for the purpose of estimating the needs of Persia in respect of the formation of a uniform force which the Persian Government proposes to create for the establishment and preservation of order in the country and on its frontiers.

(4) For the purpose of financing the reforms indicated in clauses two and three of this agreement, the British Government offers to provide or arrange a substantial loan for the Persian Government, for which adequate security shall be sought by the two governments in consultation in the revenues of the customs or other sources of income at the disposal of the Persian Government. Pending the completion of negotiations for such a loan, the British Government will supply on account of it such funds as may be necessary for initiating the said reforms.

---

* Vice President, The White Co.   [4] See p. 16.

So you see that the Persian oil fields are pretty well controlled by Great Britain. Not only is Great Britain pursuing its historic policy of making the lives and investments of British subjects safe in every part of the world, but this treaty practically puts the administration of Persia's affairs into the hands of British advisers. It would not be desirable for the United States to make treaties of this kind, but the successful development of foreign fields by American capital hinges entirely upon our having a vigorous foreign policy.

Petroleum Resources of Kansas

By RAYMOND C. MOORE,* Ph. D., LAWRENCE, KANS.

(New York Meeting, February, 1920)

THE oil-producing districts of Kansas comprise the northern portion of the so-called Mid-Continent field. As shown in the accompanying map, these districts are located chiefly in the southeastern and south central parts of the state. A considerable area in southeastern Kansas, extending northward nearly to Kansas City, has long been known as oil territory, the productive wells being distributed in patches or spots of irregular size and shape, the location of which is controlled by conditions of rock structure, and by the texture and porosity of the "sands" beneath the surface. In south central Kansas, there are a number of producing fields, the location of which appears to be controlled chiefly by well-defined structure. The most important districts are those in Butler County, especially that in the vicinity of El Dorado, which was for a time the most productive district in the entire Mid-Continent field. Recently new production of importance has been brought in the vicinity of Peabody and present development is active to the north across Marion County. Tests in the western parts of Kansas have not been successful in finding new petroleum fields.

HISTORY

The first well drilled for petroleum, in Kansas, was near the town of Paola, Miami Co., about 40 mi. southwest of Kansas City, in the summer of 1860, only a few months after the completion of the famous "Colonel" Drake discovery well in Pennsylvania. Kansas appears to be the second state to engage in a serious attempt to find oil by drilling. The Civil War caused the temporary abandonment of attempts at oil development in the state.

It was in the vicinity of Paola, where numerous oil seepages had been observed, that the first well producing oil in commercial quantities was drilled,[1] where also gas was first piped to the city for commercial use. Prospecting spread southward into Linn County and northward into

---

* State Geologist of Kansas.
[1] Raymond C. Moore and Winthrop P. Haynes: Oil and Gas Resources of Kansas. Kans. Geol. Survey *Bull.* 3a (1917) 20.

Johnson and Wyandotte Counties. a number of small gas wells being obtained.   Later development extended southward toward Iola, Chanute, Neodesha and Coffeyville, reaching to the boundary of the Indian Terri-

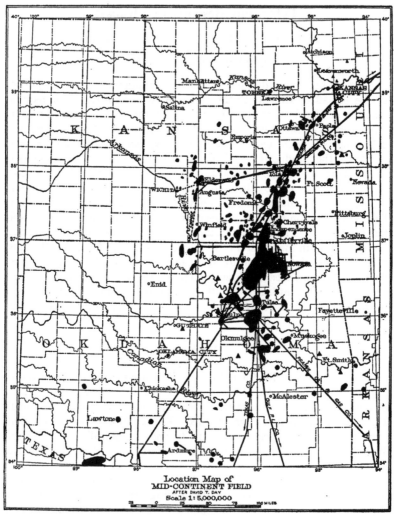

Location Map of
MID-CONTINENT FIELD
AFTER DAVID T. DAY
Scale 1: 5,000,000

FIG. 1.

tory, now Oklahoma.   From 1891 until 1894, prospectors covered the entire southeastern part of Kansas along the Neosho and Verdigris rivers. Many oil wells, though none with very large individual production, were brought in, particularly in Allen, Neosho, Montgomery, and Wilson Counties.

The production of petroleum in Kansas amounted to relatively little until tests completed in the then Indian Territory showed that beneath the Mid-Continent plains lay really important deposits of oil. The great impetus then given to drilling in Kansas resulted in a very rapid increase in the volume of production. Although in 1900 less than 75,000 bbl. of oil were obtained in the entire state, the production in 1904 amounted to 4,250,779 bbl. Due to the decline in price, drilling fell off and so large an annual production was not again reached until 1916 when, with a considerably increased market price and the recent discovery of the rich Butler County fields, the production of the state was brought to nearly 9,000,000 barrels.

The larger place which Kansas has occupied in recent years as a producer of petroleum is almost wholly due to the discovery in June, 1914, of commercial quantities of oil in Butler County, south central Kansas. It has been known for a number of years that gas was available in this part of the state. One of the wells in the Augusta gas district was drilled into an oil sand at a depth of about 2500 ft. (761 m.) and before the end of 1914 five oil wells had been drilled in the heart of the gas field. By the close of 1915, the number of oil wells was increased to twelve, one of which is reported to have had an initial natural flow of 1500 bbl. Meanwhile, geological examination of the country to the north revealed a very promising structure in the vicinity of El Dorado. In the latter part of 1915 the Continental Oil & Gas Co., now the Empire Gas & Fuel Co., brought in a 100-bbl. well on the Stapleton farm, section 29, township 25 south, range 5 east, about 15 mi. northwest of Augusta. The discovery was in a sand penetrated at a depth of about 660 ft. (198 m.) Offset wells confirmed the importance of the shallow sand but in the first well the sand was cased off and the drilling continued. A lower productive sand was encountered at a reported depth of 2460 ft., the well being completed with an initial production of 120 bbl. a day from this horizon. Succeeding wells were, for a time, drilled into the shallow sand only. Later the deeper sands were developed, culminating in the discovery and exploitation of the 2500-ft. sand in the Towanda district in the spring and summer of 1917. Some of the wells in this district are reported to have had an initial daily production of more than 25,000 barrels.

In the latter part of 1918, oil was discovered in the extreme northwestern part of Butler County east of Elbing. The wells were not important, but the drilling in the early part of 1919 on a favorable structure south of Peabody, Marion Co., was marked by large production. The present activity in development is in this region and northward across Marion County into Dickinson County.

STRATIGRAPHY

In general, the geology of Kansas is almost ideally simple. The state is a typical part of the Great Plains region and has the uniformly gentle slope and simplicity of geologic structure which characterize the plains. The surface of Kansas has a general inclination from west to east amounting to about 10 ft. (3 m.) per mile, the elevation of the western state boundary being about 3500 to 4000 ft., that of the eastern boundary from 750 to 1000 ft. The rock formations of which this sloping plain is built lie almost flat and are exposed in broad north-and-south bands across the state. They sag slightly in central Kansas, the rock slope, or dip, being toward the west in the eastern counties and to the east in the western part of the state. The oldest beds appear at the surface in the east and dip beneath the younger overlying formations, which appear in succession as the state is crossed to the west.

The rocks in the general region of the Mid-Continent field range in geologic age from almost the oldest known to the youngest. The oldest rocks are granites and other crystalline rocks of pre-Cambrian age, which are exposed in the southeastern part of Missouri, in the Arbuckle and Wichita Mountains of Oklahoma, in the Rocky Mountains of Colorado, and at points north of Kansas farther distant. The pre-Cambrian nowhere appears at the surface in Kansas, but recent exploration for oil and gas in the central part of the state suggests that it approaches the surface much more closely than was supposed. Sufficient tests have been made to indicate quite clearly the presence of a buried ridge or mountain range of granite, which appears to trend in a direction slightly east of north from Butler County to the northern limits of the state. No evidence of metamorphism of the sedimentary rocks immediately overlying the granite has been found, and it is probable that the ridge represents a part of the pre-Cambrian floor.

Table 1 presents the chief stratigraphic divisions of the rocks of Kansas.

Strata which belong to the Cambrian and Ordovician, consisting of dolomites, limestones, shales, and sandstones, and aggregating about 2000 ft. (609 m.) in thickness, underlie eastern Kansas and perhaps other parts of the state. They have been penetrated in a number of wells but in no place found to contain commercial quantities of petroleum or natural gas. Upon the eroded surface of the rocks of the older Paleozoic, in the Great Plains country, is found the Mississippian system, or, as it is called by drillers, the "Mississippi lime." The Mississippian is a clearly defined, readily traceable, stratigraphic unit, consisting chiefly of crystalline limestones containing a rather unusual amount of hard flinty chert.

TABLE 1.—*Geologic Section of the Kansas Region*

	System	Groups	Formation	Character of Rocks
Cenozoic	Quaternary	Recent		Alluvium, dune sands
		Pleistocene	Wisconsin stage Kansas stage	Glacial deposits
	Unconformity			
	Tertiary	Pliocene Miocene	Ogalalla	Gravel, sand, clay
	Unconformity			
Mesozoic	Cretaceous	Montana	Pierre	Shale
		Colorado	Niobrara Benton	Limestone, chalk, shale
		Dakota	sandstone	Sandstone, shale
	Unconformity			
	Comanchean	Washita	Kiowa Cheyenne	Sandstone, shale
	Unconformity			
	Permian	Cimarron	Greer Woodward Cave Creek Enid	"Red beds," sandstone, shale, dolomite, gyp- sum, salt
		Big Blue	Wellington Marion Chase Council Grove	Shale, limestone
Paleozoic	Pennsylvanian	Missouri	Wabaunsee Shawnee Douglas Lansing Kansas City	Limestone, shale, sand- stone
		Des Moines	Marmaton Cherokee	Limestone, shale, sand- stone
	Unconformity			
	Mississippian	Chester *Unconformity*		
		Osage	Warsaw Keokuk Burlington Pierson	Limestone
		Kinderhook		Limestone, shale
	Unconformity			
	Ordovician		Joachim Jefferson City Roubidoux	Dolomite, sandstone, shale
	Unconformity			
	Cambrian		Gasconade Proctor Eminence Potosi	
	Unconformity			
	Pre-Cambrian			

Not exposed in Kansas

TABLE 2.—*Divisions of Pennsylvanian Rocks of Kansas*

Group	Formation	Member	Thickness, Feet
Missouri	Wabaunsee formation	Eskridge shale	30–40
		Neva limestone	3–5
		Elmdale shale	120–140
		Americus limestone	6–10
		Admire shale[1]	275–325
		Emporia limestone	5–10
		Willard shale	45–55
		Burlingame limestone	7–12
	Shawnee formation	Scranton shale	160–200
		Howard limestone	3–7
		Severy shale	40–60
		Topeka limestone	20–25
		Calhoun shale	0–50
		Deer Creek limestone	20–30
		Tecumseh shale	40–70
		Lecompton limestone	15–30
		Kanwaka shale	50–100
	Douglas formation	Oread limestone	50–70
		Lawrence shale[2]	150–300
		Iatan limestone	3–15
		Weston shale	60–100
	Lansing formation	Stanton limestone	20–40
		Vilas shale	5–125
		Plattsburg limestone	5–80
		Lane shale	50-150
	Kansas City formation	Iola limestone	2–40
		Chanute shale	25–100
		Drum limestone	0–80
		Cherryvale shale[3]	25–125
		Winterset limestone	30–40
		Galesburg shale	10–60
		Bethany Falls limestone	4–25
		Ladore shale	3–50
		Hertha limestone	10–20
Des Moines	Marmaton formation	Pleasanton shale	100–150
		Coffeyville limestone	8–10
		—— shale	60–80
		Altamont limestone	3–10
		Bandera shale	60–120
		Pawnee limestone	40–50
		Labette shale	0–60
		Fort Scott limestone	20–40
	Cherokee shale[4]	Undifferentiated	400–500

[1] Possibly contains shallow oil sand at El Dorado.
[2] Includes Chautauqua sandstone member; probably 1500-ft. sand at Augusta and El Dorado.
[3] Possibly horizon of oil sand at 2400 ft. at Augusta and El Dorado.
[4] Includes the main oil sand outside Augusta and El Dorado and Peru; contains Bartlesville and Burgess sands.

In Oklahoma and northern Arkansas, it includes important beds of shale and some sandstone; but where encountered by the drill in Kansas and throughout most of Missouri, it is essentially a limestone series. An exception, apparently, is found in central Kansas, according to recent information from well records, which indicate a disappearance locally of the limestone and a replacement by clastic material. The thickness of the system in the south central part of the Mississippi basin is more than 2000 ft., but in Kansas it is not more than 300 or 350 ft.

The oil and gas deposits of the Mid-Continent field are confined almost wholly to rocks of the Pennsylvanian system, which outcrop in a broad belt across eastern Kansas and Oklahoma. The rocks of this system consist of a thick series of alternating shale and limestone formations, with irregular beds of sandstone and some beds of coal. Though not great in thickness, many of the beds are surprisingly persistent horizontally, having been traced in most cases some hundreds of miles along the outcrop. They have a total thickness of nearly 3500 ft. in the southern part of the state and a slightly smaller amount to the north. A total thickness of about 3000 ft. has been measured along Kansas River.

Table 2 shows the stratigraphic divisions of the Pennsylvanian that have been recognized in Kansas, with their approximate thicknesses.

Permian rocks are found in a north- and- south band across central Kansas. The zone of outcrop is narrow at the north, where it is overlapped from the west by the much younger beds of Cretaceous age, and reaches its maximum width near the southern border of the state. The lower Permian beds are marine and overlie the upper Pennsylvanian strata without unconformity or other prominent mark of stratigraphic division. The upper Permian, which is confined to the southwestern part of the Permian area in the state, consists chiefly of red beds. The subdivisions which have been made are listed in Table 3, with approximate thickness.

The remainder of the surface in Kansas is occupied by rocks of Cretaceous and Tertiary age. The former consists of an important basal division of sandstone, the Dakota, and of middle and upper divisions of chalky limestone and shale. The total thickness is approximately 1300 ft. Seepages of oil have been reported in the Cretaceous area and there are some excellent structures, but no commercial production of oil has been obtained. from. these rocks or in the part of the state in which they outcrop.

In common with the Mississippian and older systems that underlie it, the Pennsylvanian strata have a gentle inclination outward from the Ozark highland. In northeastern Kansas, they dip toward the northwest; in central eastern Kansas, almost due west; and in the southern counties, slightly southwest. If the Pennsylvanian is continuous beneath the thick overlying formations of Permian, Cretaceous, and Tertiary age in the western part of Kansas, the system is a part of the broad

TABLE 3.—*Subdivisions of Permian System in Kansas*

Group	Formation	Member	Thickness, Feet
Cimmaron	Greer	Big Basin sandstone	12
		——shale	20
	Woodward	Day Creek dolomite	1–5
		Whitehorse sandstone	175–200
		Dog Creek shale	30
	Cave Creek	Shimer gypsum	4–25
		Jenkins shale	5–50
		Medicine Lodge gypsum	2–30
	Enid	Flowerpot shale	150
		Cedar Hills sandstone	50–60
		Salt Plain shale	155
		Harper sandstone	350
Big Blue	Wellington	Undifferentiated	.500–800
	Marion	Abilene limestone	4–8
		Pearl shale	70
		Herington limestone	12–15
		Enterprise shale	35–44
		Luta limestone	30
	Chase	Winfield limestone	20–25
		Doyle shale	60
		Fort Riley limestone	40–45
		Florence flint	20
		Matfield shale	60–70
		Wreford limestone	35–50
	Council Grove	Garrison shale and limestone	135–150
		Cottonwood limestone	6

shallow sag, or syncline, that characterizes the general structure of the state. However, when examined in detail it is seen that there are many irregularities in the structure of the Pennsylvanian rocks. In many places in eastern Kansas, the rock strata are absolutely horizontal, and in a number of places, they are inclined to the east for short distances. These irregularities are minor waves on the major structure of the Pennsylvanian but are, in most instances, the controlling feature in the accumulation of commercial deposits of oil. Most of the minor structures are of the unsymmetrical dome type, the rocks dipping away in all directions. Others are merely terraces, or "noses," where the western dip is diminished sufficiently to permit local accumulation of petroleum. None of the structures are very prominent, the vertical distance from the top of one of the best defined anticlines to the upper part of the adjacent saddle, that is the closure, being only 160 feet.

. The texture of the "sand" is a controlling factor in the production of areas in southeastern Kansas. Oil and gas wells with an important production are located in many instances without relation to structure, the supply of oil and gas being controlled by the lenticular character or the "patchy" texture of the sands.

TECHNOLOGY

Two types of drilling are employed in the Kansas fields, the standard, or cable drilling, which is used in all the deeper wells, and the Star, or Parkersburg type, which is commonly used in the shallow fields of the eastern part of the state.

On account of water conditions in certain parts of the Kansas fields, especially in the El Dorado and Augusta districts, the depth to which the oil-producing sand is penetrated and the casing of the well are important considerations. If the well is drilled too deep, there is danger of drowning within a comparatively short time. In most cases, only the ordinary requirements of casing are met. The use of cement and the mud-laden fluid has been successful, where employed in the Butler County wells, but there has been no uniformity of practice, due to varying conditions in the field and to lack of state supervision.

PRODUCTION STATISTICS

Most of the wells in the Kansas fields are not large producers, the average yield amounting to but a few barrels a day. The largest production from individual wells has been found in the El Dorado-Towanda district, where at least one well is reported to have flowed more than 25,000 bbl. a day. The initial production of many wells in this part of the state has exceeded 2000 bbl. a day. Table 4 shows the average production of oil wells in Kansas from 1910 to 1918 based on available data.

TABLE 4.—*Average Production of Oil Wells in Kansas, 1910–1918*

Year	Total Production, Barrels	Total Number, Oil Wells	Dry Holes	Average Annual Production per Well, Barrels	Average Daily Production per Well, Barrels
1910	1,128,668	1,831	25	616	1.7
1911	1,278,819	1,787	25	715	1.9
1912	1,592,796	1,757	41	906	2.5
1913	2,375,029	1,812	87	1,310	3.5
1914	3,103,585	3,054	156	1,016	2.8
1915	2,823,487	3,460	158	810	2.2
1916	8,738,077	3,673	360	2,379	6.5
1917	36,536,125	5,843	420	6,253	17.1
1918	43,253,470	8,950	925	4,833	13.2

The crude petroleum has a specific gravity ranging from about 20° Baumé, for some of the southeastern oils, to 40° or slightly higher for some of the oils in the Butler County district. The specific gravity of the oil in the vicinity of Chanute, Coffeyville, and Independence is about 30° to 32°. This heavy oil has considerably less gasoline than the oils with higher specific gravity. On account of this there has been a tendency for some of the refineries to move to points from which a larger supply of higher grade oil could readily be obtained.

The oil is gathered by pipe lines from the producing fields in the Butler County district and from other important areas, pipe lines converging toward the northeast in the vicinity of Kansas City. A considerable number of tank cars are used both in the transportation of crude oil from some of the fields and in the distribution of the refined product.

According to the best available records, Kansas had produced to the end of the year 1918 a grand total of 119,898,233 bbl. of crude oil. The character of the Kansas oil fields is in part indicated by the statistics of wells drilled. Throughout the larger part of the producing area, especially that located in the southeastern counties of the state, the wells are numerous, but none have a large output. The average yield for each producing well is from 1 or 2 to 25 bbl. a day. In the Butler County fields, some of the wells were credited with a very large individual daily output. In general, the interest in development and activity in the fields is also shown by the number of new wells drilled. Field operations follow more or less closely the fluctuation of the market, periods of greatest activity accompanying times of highest crude-oil prices. In the years 1912 to 1918, inclusive, 13,649 wells were drilled, of which 10,979 were producing and 2670 dry. Of the 4671 wells put down in 1918, 2549 were oil producing and 272 gas wells.

FUTURE POSSIBILITIES

At the present writing, the rich Butler County fields are past the zenith of their production, the climax having been reached with the development of the Towanda district, which reached its peak in 1918. The discovery of new fields, east of Elbing and extending toward Peabody, in Marion County, has given new impetus to development in this part of the state. Tests south of the El Dorado and Augusta fields, toward the Blackwell area, have thus far given little encouragement, but satisfactory showings in structures located in Marion County and northward into Dickinson County are attracting considerable attention.

Southeastern Kansas fields have been thoroughly tested and, with the exception of new wells in porous sands that have not yet been drained, there is little additional production to be expected. It is possible that new pools will be discovered in part of the state between the old oil and gas fields in the vicinity of Chanute, Iola, and Independence, and the fields farther west, Butler County and trending toward the north. Development in this area, however, cannot be foreseen.

In summary, the Kansas oil fields are, in all probability, beyond the zenith of their production. Much of central and western Kansas may yet be tested, but conditions are difficult or impossible to predict, and the result cannot be foreseen.

Rise and Decline in Production of Petroleum in Ohio and Indiana

By J. A. BOWNOCKER,* D. Sc., COLUMBUS, OHIO

(New York Meeting, February, 1920)

THE existence of petroleum in the rocks of Ohio and Indiana seems to have been first shown by wells dug for salt. The fuel, however, was objectionable owing to its odor and inflammability. Not until the Drake well was drilled in 1859 did the people appreciate the value of rock oil, and then they at once began plans to secure the coveted fuel. The first successful well in these two states was near Macksburg, in southeast Ohio, where at a depth of 59 ft. (17 m.) oil was found in commercial quantity (1860). A year later this fuel was secured on Cow Run, in the same county, and at about the same time in Noble, Morgan and perhaps other counties in that part of Ohio.

The second great step in the production of oil in Ohio and Indiana was taken in 1884 when the reservoir of natural gas in the Trenton limestone of northwest Ohio was tapped, and where a year later oil was secured in the same formation. Petroleum in this limestone was obtained in Indiana in 1889.[1]

The third step in the production of oil in Ohio and Indiana is associated with the Clinton sand of Ohio. Natural gas in large volume was discovered in that rock at Lancaster, in 1887, and the area has been extended until it has become the largest individual producer in the world. The presence of natural gas in such great volume all but demonstrated to the driller that oil lay hidden near by. Soon the search for it was started but not until 1899 was petroleum in commercial quantity found in the Clinton sand, and a large pool was not located in it until 1907.

The fourth step in the development of the industry was taken in 1913, when the pool in Sullivan County, in western Indiana, was opened. The producing rock is the Huron sandstone, which is the topmost member of the Mississippian system.[2]

---

* State Geologist and Professor of Geology, Ohio State University.
[1] Dept. of Geol. and Nat. Res. of Indiana, 28th Ann. Rep. (1903) 82.
[2] Edward Barrett: Dept. Geol. and Nat. Res. of Indiana, 38th Ann. Rep. (1913) 9-34.

TABLE 1.—*Production in Ohio and Indiana*

	OHIO, BARRELS	INDIANA, BARRELS	TOTAL BARRELS
1876........	31,763		31,763
1885................ ..	661,580		661,580
1889....................	12,471,466	33,375	12,504,841
1896....................	23,941,169ᵃ	4,680,732	28,621,901
1904....................	18,876,631	11,339,124ᵃ	30,215,755ᵇ
1914................	8,536,352	1,335,456	9,871,808
1918....................	7,285,005	877,588	8,162,563

ᵃ Maximum production for state.
ᵇ Maximum production for the two states combined.

PRODUCTION FROM THE TRENTON LIMESTONE IN OHIO AND INDIANA

When natural gas was discovered in the Trenton limestone at Findlay, Ohio, in 1884, and petroleum a year later, it marked an epoch in the geology of petroleum. Heretofore the source of these fuels had been in sandstones or conglomerates, and limestones were regarded as non-petroliferous. For this reason the new field was looked on by the practical oil man with much suspicion, which was increased when he noted the dark color and bad smell of the oil. It was soon found, too, that the methods of refining hitherto practiced would not apply, which, of course,

PRINCIPAL PRODUCING OIL ROCKS IN OHIO AND INDIANA

Pennsylvanian	Mitchell sand (Ohio)
	First Cow Run or Macksburg 140-ft. sand (Ohio)
	Macksburg 500-ft. sand (Ohio)
	Macksburg 800-ft. sand (Ohio)
Mississippian	Huron sand (Indiana)
	Keener sand (Ohio)
	Big Injun sand (Ohio)
	Berea sand (Ohio)
Devonian....................	Corniferous limestone (Indiana)
Silurian......................	Clinton sand (Ohio)
Ordovician..................	Trenton limestone (Ohio and Indiana)

was another objection to the fuel. However, the oil had a market, which insured further drilling. Wells were completed as fast as the tools could be forced through the rocks and the production of oil increased at a rapid rate. In fact, the supply grew faster than the demand so that the storage of the fuel became a serious problem. To check production, the Standard Oil Co. reduced the price time and again until, in July, 1887, it was listed at only 15 cents per barrel. In spite of this, drilling continued and the production kept apace. The development of an improved method of refining strengthened the market for the crude oil and the price advanced, with some fluctuations, until at present (October, 1919) it is $2.48 per barrel.

While the Lima-Indiana field was opened 34 years ago, drilling has been continuous and is still in progress.   In 1917, 534 wells, of which 174 were producers, were drilled in Ohio and 266, of which 174 were producers, were completed in Indiana.   The magnitude of the drilling is well shown

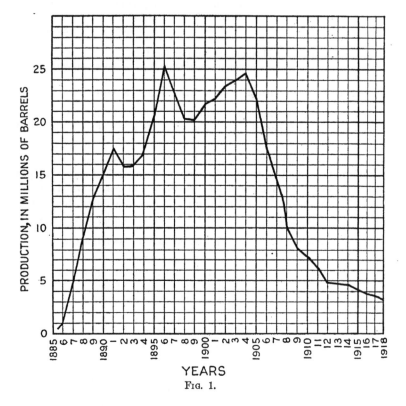

FIG. 1.

by the fact that from June, 1905, to Dec. 31, 1917, 12,514 wells were sunk in the Ohio part of the field and 15,005 on the Indiana side.[3] .Barrett estimates that 30,000 wells have been drilled to the Trenton in Indiana and the number in Ohio must have been larger because the producing area is much greater.

With Findlay as a center, drilling was carried on in all directions, but it was soon learned that the producing territory had narrow limits along an east-and-west line and that it was continuous in a

---

[3] Petroleum in 1917.   U. S. Geol. Survey *Mineral Resources* (1919) 750.

northeast-southwest direction. By 1890, the field had been pretty definitely delimited and was found to form an arc of a circle from the western end of Lake Erie southwest through Wood, Hancock, Allen, Auglaize, and Mercer Counties to the Indiana line, which it later crossed and extended slightly northwest to near Marion, Grant Co. The total length of the field is about 150 miles (241 km.) but the width varies greatly; in places it is only a fraction of a mile while elsewhere it may have a width of 20 miles.

The maximum production from the Trenton limestone of Ohio was reached in 1896; it was 20,575,138 bbl. The Indiana part of the field, as it was developed later, did not attain its maximum until 1904, when it produced 11,317,259 bbl. Fig. 1 shows the rise and decline of the entire field.

The size of the wells varied greatly but the maximum seldom reached 10,000 bbl. Initial productions of from 100 to 500 bbl. per well, however, were common, and the shallowness of the wells made them profitable though the price of oil was low. Everywhere salt water was found, and the work of pumping this has been enormous. Streams were made brackish and the water therefore unfit for use. The oil has a density of from 36° to 42° Bé., and is consequently heavier than the Pennsylvania oil. Its base is chiefly paraffin, though some asphalt is present. Sulfur is an objectionable constituent. The oil is darker than the Pennsylvania product and the odor more disagreeable.

Because of its thickness and continuity, the Trenton limestone may be thought of as the rock floor of both Ohio and Indiana. It rises to the surface in only one locality—the Ohio Valley from Coney Island (near Cincinnati) to Ripley, Brown County. From this locality it dips to the east, north, and west, but rises to the south and forms the surface rock in part of the blue-grass region of Kentucky. At Findlay, it lies 1092 ft. (332 m.) below the surface; at Columbus, 2035 ft.; and at Cleveland, 4445 ft. In Indiana similar, though not as large, variations are found. Everywhere in both states, outside of the narrow area of outcrop located above, the Trenton has always been found if the drill has penetrated to its horizon.

In thickness, the Trenton shows much variation, but using the name in its older and broader sense the rock is everywhere measured by hundreds rather than by tens or scores of feet. Thus, at Findlay its thickness is 729 ft. (222 m.)[4]; at Columbus, 475 ft.;[5] and at Waverly, in southern Ohio, 808 ft.[6]

The composition of the Trenton limestone varies both horizontally

[4] D. D. Condit: *Am. Jnl. Sci.* (1913) **36**, 125.

[5] Edward Orton: Geol. Survey of Ohio (1888) **6**, 107.

[6] J. A. Bownocker: Geol. Survey of Ohio *Bull.* 12 [4] (1910) 48.

and vertically.   Where the rock yields oil or gas in commercial quantity, it is magnesian, as is shown by the following analyses:

	CaCO$_3$	MgCO$_3$	INSOL. RESIDUE	Al$_2$O$_3$ AND Fe$_2$O
Findlay, Ohio..................	53.50	43.05	1.70	1.25
Bowling Green, Ohio............	51.78	36.80	4.89	
Lima, Ohio....................	55.90	38.85	0.75	2.94
Kokomo, Indiana...............	52.80	39.50	4.60	2.40

In places, at least, the Trenton changes rapidly with depth.   Thus, at Bowling Green an analysis of the rock from 100 ft. (30 m.) below its top showed more than 88 per cent. of calcium carbonate and less than 7 per cent. of magnesium carbonate, while the top beds gave less than 42 per cent. of calcium carbonate and nearly 37 per cent. of magnesium carbonate. Outside of the producing territory, the Trenton appears to lose its magnesian character, for the calcium carbonate rises in most places to at least 75 per cent. of the rock.[7]   Since there is an increase in porosity with magnesium carbonate, the composition of the rock may have much to do with its capacity to store the oil.

The color of the Trenton limestone is dark gray, judging from pieces thrown to the surface when the wells are torpedoed.   It is finely crystalline and contains numerous veins of dolomite.   Cavities, doubtless the work of solution, are common and greatly increase the storage capacity of the rock.

The depth of wells usually ranges from 1000 to 1500 ft. (304 to 457 m.) in Ohio; in Indiana the depth averages about 1000 ft.   The great body of the oil is in the upper 50 ft. of the reservoir rock, but a second and even a deeper pay has been found in places; these, however, have yielded but little oil.   Efforts have been made to find oil in rocks below the Trenton, but without success.   In a well drilled near Findlay, Ohio, in 1912, work did not cease until the tools had penetrated the rocks to a depth of 2980 ft.; granite or a similar rock was struck at 2770 ft.

The thickness and structure of the rocks are remarkably uniform, and while there is considerable variation in depth of wells, it results almost wholly from the anticlinal structure.   Naturally, wells located on the top of the arch are the most shallow.   Following are representative well records from Ohio and Indiana:

TABLE 2.—RECORDS OF OHIO AND INDIANA WELLS

OHIO	THICKNESS, FEET	INDIANA	THICKNESS, FEET
Drift.....................	43	Drift.....................	50
Monroe limestone..........	107	Niagara limestone..........	153
Niagara limestone..........	140	Cincinnati shales..........	751
Niagara shale..............	13	Trenton limestone at........	954
Brassfield limestone........	89		
Gray shale................	47		
Red shale.................	45		
Cincinnati shales...........	706		
Trenton limestone at........	1190		

[7] Edward Orton: Geol. Survey of Ohio (1888) 6, 103–104.

The petroleum in the Lima-Indiana field is, in places at any rate, closely related to structure. As is well known, the Cincinnati axis, a broad fold, crosses the Ohio River about 25 miles east of Cincinnati and, extending west of north for perhaps 40 miles, bifurcates, one arm running a few degrees east of north toward the western end of Lake Erie and the other arm west of north toward the southern end of Lake Michigan. In Lucas, Wood, Seneca, and Hancock Counties, Ohio, the principal oil fields are on the summit or eastern slope of this arch. Farther southwest, in Allen, Mercer, and Auglaize Counties, Ohio, the productive territory is on the west side of the arch, while in Indiana it is on the north side. The Ohio part of the field was one of the first to lend support to, if not to demonstrate, the anticlinal theory that has recently been announced by I. C. White.

The Lima-Indiana field passed its zenith nearly a quarter of a century ago. It is still an important producer but is steadily decreasing. While new wells are completed by the hundreds each year, these by no means equal the number of old wells abandoned. Manifestly this field cannot be relied on to meet the present, much less the rapidly growing, demand for petroleum.

PRODUCTION FROM THE CLINTON SAND FIELDS IN OHIO

The Clinton sand nowhere outcrops in Ohio; hence our knowledge of it has been obtained entirely from the drill. The rock was first struck at Lancaster, in 1887, and was considered limestone, but this error was soon corrected. While still called Clinton, it has been pretty definitely shown that the rock forms a part of the underlying formation, the Medina.[8] The drill has demonstrated that the Clinton sand does not underlie the western half of the state and that its place is there occupied by shales. Since the rocks in eastern Ohio dip to the southeast, the horizon of the Clinton sand is found at increasing depths as the Ohio River is approached. Its position at Wheeling is about 6560 ft. (1999 m.) below the Ohio Valley, though the drill has not penetrated to so great a depth at that place. While it is probable that the Clinton sand underlies eastern Ohio, its presence has not been demonstrated in the counties east of Tuscarawas, Muskingum, Athens, and Gallia. Natural gas is now secured in this sand in Tuscarawas County at a depth of nearly 4000 ft., which is the deepest source of either oil or gas in Ohio.

The Clinton sand is usually light colored and clean, but in places it is brick-red. The range in thickness is generally from 10 to 40 ft. (3 to 12 m.) but the maximum occasionally reaches 100 ft. Along its western edge, the sand is thinner and somewhat patchy. According to the testimony of drillers, the sand is free from water when first penetrated, making the territory unique among the oil fields of Ohio.

---

[8] J. A. Bownocker: *Econ. Geol.* (1911) 6, 37.

Petroleum is now secured in the Clinton sand of Ohio in Hocking, Perry, Fairfield, Muskingum, and Wayne Counties and, to a very small extent, in several others. The pools, however, are nearly all small, the largest being in Perry County. The oil has a density of from 35 to 46° Bé and much of it is of Pennsylvania grade. Few wells have had an initial production as large as 1000 bbl. per day; in fact, those starting at as much as 500 bbl. have been rare. The production, however, is well maintained, which helps compensate the operator for his great labor and expense. The maximum production of the sand was about 1,300,000 bbl. per year, but it is now smaller. Much time and money have been expended in an effort to extend the producing territory to the east, but the results have been unsuccessful.

The depth of wells varies with surface altitude and with dip. Near Pleasantville, Fairfield County, the depth to the producing sand is about 2325 ft., while near Crooksville, in the eastern part of Perry County, the depth is more than 3400 ft. These two wells represent very well the present extremes for oil production from this sand in Ohio.

The position of the Clinton sand is usually from 100 to 150 ft. (30 to 45 m.) below the Silurian limestone, or "Big lime" as the rock is known by the driller, and hence it is very easy for him to determine his position with reference to the desired sand. The following well records, one in southern and the other in northern Ohio, show very well the rock succession:

	Perry County		Wayne County	
	Thickness, Feet	To Bottom, Feet	Thickness, Feet	To Bottom, Feet
Mantle rock..................	55	55	57	57
Big Injun sand...............	100	235		
Berea sand...................	33	718	30	495
Bedford and Ohio shales........	1032	1750	1335	1830
Devonian and Silurian lime- stones......................	798	2548	1085	2915
Clinton sand.................	33	2708	31	3135

The Bedford and Ohio shales form a great wedge-shaped mass with the apex in central Ohio and the base near Wheeling, W. Va., where its thickness is at least 2500 ft. The Devonian and Silurian limestones increase in thickness to both the east and the north. At Columbus they measure 770 ft.; at Zanesville, 1012 ft.; and at Wheeling, 1900 ft. To the northeast, the thickness increases from 770 ft. at Columbus to 1085 ft. in Wayne County, and reaches a maximum of 1400 ft. at Cleveland. It is the increasing thicknesses of these rocks that give the underlying Clinton sand its sharp dip, hence its rapidly increasing depth.

The structure of this sand, in its broader aspect at any rate, is easily stated. It dips to the southeast, while in the longitude of Columbus it thins and is replaced by shales. It may, therefore, be compared with one arm of an anticline. Along the western, or higher, part of this arm, great volumes of natural gas have been found; while a little farther east, and hence at a lower level, reservoirs of oil have been located. How the oil got into its position is not clear, for the absence or scarcity of water deprives us of the usual agent. Neither is it plain how the oil is held in its present location, but possibly it rests in shallow basins.

PRODUCTION FROM THE CORNIFEROUS LIMESTONE OF INDIANA

The Corniferous limestone that forms the base of the Devonian in Ohio and Indiana is a source of petroleum in the latter state, but not in the former. However, even in Indiana, the reputation of this rock as a source of fuel rests on a single well, the Phoenix, which was drilled in Terre Haute in 1889 and is credited with being the best payer ever drilled in the state.[9] The limestone was struck at a depth of 1660 ft. and for at least 12 years the production averaged 1000 bbl. of oil per month. In 1908, it averaged 340 bbl. per month; few wells in this country have so large a daily yield after 30 years continuous production.[10] Later, a few small wells were secured in the same formation south and southeast of Terre Haute, but were it not for the remarkable Phoenix well, the territory would not be mentioned.

PRODUCTION FROM MISSISSIPPIAN AND PENNSYLVANIAN SANDSTONES IN OHIO AND INDIANA

Since these producing rocks have similar physical and chemical properties and representatives of both groups may be yielding oil in the same territory, they will be reviewed together. Next to the Trenton limestone, they have been the largest producers of oil in each state, and at the present time they are the largest source in Ohio.

The producing territory in Ohio is restricted to the eastern half, since that is the only part where rocks of this age are present. Trumbull and Lorain have been the northernmost counties, though neither was ever a large producer. The large sources of oil, now as in the past, are Jefferson, Harrison, Belmont, Monroe, Noble, Washington, Morgan, and Perry Counties, with Monroe and Washington far in the lead. As previously stated, drilling in this territory started in 1860 and is still in progress. In 1891, oil in the deeper sands of Monroe County was first secured and that marked the beginning of the large source of oil in eastern Ohio. Most of the pools are small, but some of these in Monroe and Washington Counties compare favorably in size with the largest of the Appa-

---

[9] W. S. Blatchley: Dept. of Geol. and Nat. Res. of Indiana, 25th Ann. Rep. (1900) 517. Also 33rd Ann. Rept. (1908) 373.
[10] This well is still producing between 3000 and 3500 bbl. per year.

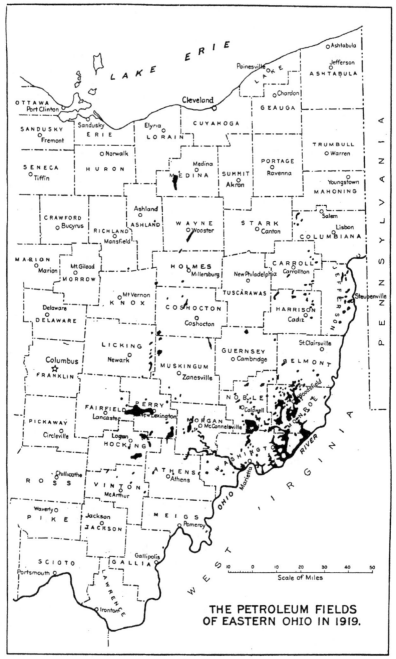

THE PETROLEUM FIELDS
OF EASTERN OHIO IN 1919.

Fig. 2.

lachian field. Wells have been sunk in great numbers and no large area remains untested. On the Woodsfield quadrangle alone about 2000 wells have been drilled, and the number is very large on numerous other quadrangles in Monroe, Washington, and Morgan Counties. The future discovery of large pools is therefore very improbable, and the production of 5,586,433 bbl. in 1903 will probaby not be equaled.

In Indiana, the producing territory in rocks of this age lies chiefly in Gibson and Sullivan Counties in the southwestern part of the state. This territory assumed commercial proportions in 1913, and in 1914 the production of the state increased 40 per cent, the increase being from these two counties. Later, oil was found in Pike and Daviess Counties, and the territory to be prospected was thus largely increased. Of the 266 wells drilled for oil in Indiana in 1917, 187 were in these four counties. Notwithstanding the yield from this territory, the production for the state has decreased, that for 1917 being smaller than for any year since 1892. While the outlook for an increased production from southwest Indiana is not so unfavorable as it is from the Trenton limestone, the prospect is not very promising.

The following composite record of two wells in Washington County shows the important producing sands in Ohio and their relative positions:

	Thickness, Feet	To Bottom, Feet
Pennsylvanian.		
Meigs Creek (Macksburg) or No. 9 coal...............	5	15
First Cow Run or Macksburg 140-ft. sand............	35	378
Macksburg 500-ft. sand............................	17	702
Macksburg 800-ft. sand............................	51	826
Salt sand...	190	1095
Mississippian.		
Mountain limestone (Big lime)......................	35	1325
Keener sand.......................................	55	1430
Big Injun and Squaw sands.........................	115	1545
Berea sand..	14	1953

Strenuous efforts have been made by drillers and producers to find below the Berea sands that are the equivalent of the deep sands of Pennsylvania, but it has been proved that when the Berea is passed, in eastern Ohio, the last hope of securing oil is gone. The Clinton sand should be present, but it lies so deep that the drill has not as yet reached it.

The persistence, texture, and thickness of the sands in eastern Ohio vary greatly. The deepest sand, the Berea, may be put in a class by itself for persistence and texture. It is extensively quarried near Cleveland and it outcrops in the middle of Ohio from Lake Erie south to the Ohio River. From its outcrop, the sand dips to the southeast and it is almost invariably present in its proper place. In thickness, the usual range is from 10 to 40 ft.; but in northern Ohio, much greater measurements have been made. Lying interbedded in shales, the Berea is easily recognized and is an important guide, or key, rock of the driller. While this sand is coarse grained, it is much less so than the higher sands of the eastern part of the state. The workmen report it much harder to drill than the overlying sands, and it therefore receives larger charges in torpedoing. The Berea is productive in spots over much of eastern Ohio. In counties north of Belmont and Guernsey, it is the only source of oil, and it is of still more importance farther south. Nevertheless the total yield from this formation is not large, and it is greatly surpassed by the Keener and Big Injun sands.

The Big Injun and Keener sands are the largest producers of oil in eastern Washington and in Monroe County where they are at their best. Where these sands outcrop in the middle of Ohio, they are pebbly or coarse grained and constitute the Black Hand and Logan formations. Under cover, the same texture is maintained and the rocks yield readily to the drill; their storage capacity is large. These sands are everywhere present where due, but they are not so sharply delimited above and below as is the Berea.

The Mountain limestone, or Big Lime, where present, is an excellent guide, or key, rock of the driller. It is well developed in the eastern part of Washington and in the southern half of Monroe County, but from there it thins rapidly to the north and west and is rarely if ever reported. Its thickness seldom reaches 100 ft. Along its outcrop in Muskingum, Perry, and other counties this rock is known as the Maxville limestone. The Big Lime is not a large source of oil, but an occasional producer is gotten in it. The term Big Lime, however, is used by the drillers for two entirely different rocks or groups of rocks in Ohio. As the following table shows, the upper one forms the top of the Mississippian system; while the lower one belongs to the Devonian and Silurian systems.

Mississippian (Lower Carboniferous)........	Maxville limestone, the Big Lime of southern Ohio.	
	Black Hand and Logan..	Keener sand Big Injun sand Squaw sand
	Berea sand	

| Devonian.......... | Olentangy and Ohio shales
Delaware limestone
Columbus limestone
Monroe limestone | |
| Silurian........ | Niagara limestone
Brassfield limestone | The Big Lime of the Clinton sand fields. |

The sands of the Pennsylvanian are much alike in properties. They are coarse grained, light colored, open textured, and easy to drill. Lack of persistence is a striking feature; this is particularly true of the First Cow Run sand. They are beach or near-shore deposits, and their variations are a result of changes in direction and in strength of movement of the water.

The sands of the Pennsylvanian are not productive except in the southeastern part of Ohio. The First Cow Run, or Macksburg 140-ft., sand is an important source of oil in Morgan, Washington, and Noble Counties. Its place is in the Conemaugh formation and about 160 ft. below the Pittsburgh coal. The Macksburg 500-ft. and the Macksburg 800-ft. sands are still more restricted in their producing area, which is limited to northern and central Washington County.

The variation in depth of wells is notable. The range is from 12 ft. (3.6 m.) to 2200 ft. (670 m.) and one of only 38 ft. is still being pumped. Compared with other important fields, the wells have been small producers. Very few have had an initial production as large as 500 bbl. per day, and the maximum was about 2500 bbl. The wells are long lived, which in a measure compensates for their small size. Thus one well, only 98 ft. deep, near Joy, Morgan County, is said to have been producing continuously since 1872. The oil is nearly all of Pennsylvanian grade and ranges in density from 42° to 50° Bé. The color is dark green, red, or black. Salt water is everywhere present.

The structural features of the oil-producing sands in eastern Ohio are not conspicuous. The rocks dip to the southeast at a varying rate, in most places from 25 to 40 ft. per mi. A few well-marked anticlines or domes are found, the most conspicuous being the Burning Springs anticline, which crosses the Ohio Valley about 12 miles east of Marietta. Another one is at Cambridge, but neither oil nor gas in large quantity has been found with it. The Cow Run pool, in Washington County, is on a well-marked dome. Smaller folds have been located in Washington, Belmont, Harrison, and other counties, and oil or gas in most cases has been found beneath them. Some of the larger fields, however, are not known to be on structures of this kind. The contour maps of the oil sands, which the Federal Survey has been issuing in recent years, show oil in almost all positions except synclines. Probably the three most important features that determine the presence of oil are structure, texture, and salt water.

Within the past few years Smith and Dunn, of Marietta, Ohio, have patented a process for increasing the production of oil wells that have been pumped. They force air into the oil sand under a pressure that varies in most places from 40 to 350 lb. to the square inch. This, it is stated, increases the production on an average from 100 to 150 per cent., with a maximum of 800 per cent.[11]

DISCUSSION

L. S. Panyity,[*] Columbus, O. (written discussion).—Under the subdivision of "Production from the Clinton sand field in Ohio," relative

---

[11] J. O. Lewis: U. S. Bureau of Mines *Bull.* 148 (1919).

[*] Chief Geologist, Ohio Fuel Supply Co.

to the depth to the sand in Tuscarawas County, the wells are in all cases at least 4500 ft. (1371 m.) in depth and the deepest well producing oil or gas from the Clinton found that sand at a depth of over 5000 feet.

In regard to "how the oil got into its position," the possibility that the main accumulations rest in shallow basins is untenable, as structure maps indicate homoclinal accumulations. The main controlling factor is lensing and differential cementation. This belief is further strengthened by the presence of scattered gas wells down dip from the oil. The present eastern edge of the oil fields in the Clinton is already at great depths, which fact has prevented extensive prospecting still farther down dip, but it is a question of time when deeper tests will be made. If the water conditions remain as they are, water thus far having been found in but a few wells, the homocline at lower structural levels promises that newer pools may be opened up. Another point in favor is that the Clinton, as well as all other formations, being deposited on the eastern flank of the Cincinnati anticline, at the same time when the folding was taking place, i.e., folding and deposition being contemporaneous, the formations can be expected to thicken away from the axis. Thus, we may expect a thicker Clinton stratum farther east and down dip. Should an abundance of water make its appearance in the formation, structural conditions must be more favorable; and as we have more pronounced structures eastward, as noted from surface outcrops, they will offer sufficient inducements for drilling.

Relative to the structure of the eastern, or shallow sand, fields, especially the effect of the Cambridge anticline upon production, the older surveys have indicated this arch to take a northeast-southwest direction. The writer's study of this structure indicates that the main fold takes a southern and a little easterly direction; commencing at Cambridge, it passes through Caldwell and Macksburg, and apparently extends farther southward in the direction of the Burning Springs anticline of West Virginia, which may prove to be a continuation of this fold. The southward plunging axis brings gas accumulations just south of Cambridge and oil pools are found all along this axis as far as the Ohio boundary line at the Ohio River; thus we have good oil pools all along it south of Caldwell, including the well-known Macksburg pool which is known to all oil men. The secondary folds, which radiate from the main fold in a northeasterly direction, also control the accumulations to the east. It is the writer's opinion that the greatest number of accumulations have been directly caused by the general main fold known as the Cambridge anticline and the secondary folds radiating from it.

It is true that there are many good pools not so situated, where a different explanation is needed. North and west of the city of Cambridge small scattered gas wells are found but not what may be considered as real pools. Here the anticline loses its prominence and the sand conditions are entirely different; this phase has been discussed by the writer[12] in a former paper.

The Scio pool is often quoted as one of the large "off-structure" accumulations. The main controlling factor here is the water level on a homocline above which the oil is found. That the sand conditions are not the main factors at Scio is brought out by the fact that the percentage of dry holes inside the producing territory is exceedingly small. We have here a very extensive formation, what may be called a sheet sand on a somewhat smaller scale than is generally understood. Oil accumulates above the water and is found in almost every well drilled above the water level, up the dip, until gas showings are encountered. Corning offers a similar case, where the water level is a factor, however; the normal dip is arrested to a considerable extent, giving a terrace structure, and several smaller pools northeastward and along the strike are claimed to be on small domes.

There have been very few quadrangles mapped by the Federal and State surveys, thus the impression gained from them should not be a criterion for the entire shallow sand production of the state. One noticeable feature of the so-called "off-structure" accumulations here is the way the pools adhere closely to the direction of the strike, and that production is found along certain structural levels, which is very evident, even though there may be considerable barren areas between pools. In that section of the state where the structural conditions are homoclinal, the prospector will do well to pay strict attention to these apparent producing levels, and also to make a careful study of lensing.

---

[12] *Trans.* (1919) **61,** 478.

Oil Fields of Kentucky and Tennessee

By L. C. GLENN,* Ph. D., NASHVILLE, TENN.

(New York Meeting, February, 1920)

IN THE preparation of this paper the writer has drawn freely upon the writings of Orton, Munn, Shaw, Mather, Miller, Hoeing, St. Clair, Jillson, and others, as well as upon his own personal knowledge of the fields of both states. It is to be regretted that certain data gathered by him and his assistants last fall are not available for publication.

OIL IN TENNESSEE

A few wells drilled for brine for salt making in Tennessee between 1820 and 1840 obtained oil, but no definite search was made for oil until just after the close of the Civil War. Active drilling was then begun in Overton and counties southwest of it on the eastern half of the Highland Rim. A number of strikes were made at shallow depths in the basal part of the Mississippian but the wells were soon exhausted and abandoned. Drilling was revived, about 1892, when the Spurrier district in Pickett County was developed and was followed by the Riverton district in the same county in 1896. A pipe line was laid from the Wayne County, Ky., fields and about 60,000 bbl. of oil were run before a very heavy slump in the production, a failure to find an extension of the field, and excessive local taxation caused the removal of the pipe line in 1906. There was then no production in Tennessee until the discovery, in 1915, of oil near Oneida, Scott Co., at about 950 ft. (289 m.) in fissures in the Newman or St. Louis limestone. This field, however, soon failed and was abandoned.

In 1916, oil was found at Glen Mary, Scott Co., in the Newman limestone at a depth of 1232 ft. (375 m.) A number of wells have since been drilled there, some of which were dry while others, close by, were producers. The largest one yielded, at first, about 340 bbl. per day and produced for several months, when it suddenly went dry. Several of the first wells began at 6 or 8 bbl. per day and are still maintaining that output. Production is from a fissured part of the limestone and varies greatly in accordance with the size and extent of the ramification of the

---

* Consulting Oil Geologist and Professor of Geology, Vanderbilt University.

fissures. In some areas, the limestone has no fissured zone and wells go through it without obtaining even a show of oil. Fissuring, when present is not always at the same horizon in the limestone and failure to obtain a well in one location does not necessarily mean that the next location may not be a successful producer. The production at present is probably not over 1000 bbl. a month. The oil is shipped to Somerset, Ky., in tank cars and there delivered to the Cumberland pipe line.

The limestone in which the oil occurs has, so far as has been ascertained, a monoclinal structure and rises gently to the west. There is only a little gas with the oil and little or no salt water is encountered. No production curves can be given since the wells vary greatly. Some decline rapidly and fail in a few months while others show scarcely any decline after several years and bid fair to have a long life as pumpers of about 6 to 8 bbl. per day. The gravity is from 36° to 38° Baumé.

There is now considerable activity in both leasing and drilling, especially in the western half of the Highland Rim, west and northwest of Nashville, although the eastern and southern parts of this Rim are also receiving some attention. The surface of the Highland Rim is almost everywhere of Mississippian age and is underlaid, at a maximum depth of not more than a few hundred feet, by the Chattanooga black shale. Oil shows are often found just above or just below this shale. Much of the activity has been stimulated by the finding of oil in Allen County, Ky., under geological conditions very similar to those that obtain in the adjacent Highland Rim section of Tennessee.

There has been occasional deep drilling in Tennessee for a number of years, especially in the Central Basin, where the surface rocks are of Ordovician age, in the hope of obtaining a deep pay usually spoken of as the Trenton. All such attempts in this part of the state have so far failed. There have been a few slight shows and a little gas has been found, but no good sand has been encountered. The only Ordovician production from Tennessee has been that in Pickett County.

A half dozen holes or more-have been bored in the last 10 years in the western part about Memphis, and to the north of it, near the axis of the great trough in which the Gulf embayment deposits have been laid down. These wells usually range from 2000 to 3000 ft. (609 to 914 m.) in depth and several of them reported shows of oil or gas in the lower part of the section. Very recently, activity in this part of the state has been revived and preparations for further deep testing of the embayment deposits in the vicinity of Reelfoot Lake in the northwestern corner of the state are now being made.

The history of attempts at oil production in Tennessee give meager data on which to base any predictions of a large future oil production. No well-defined oil sands of any considerable extent are known, although large areas of the Newman limestone exist beneath the Cumberland plateau, under conditions very similar to those at Glen Mary, and remain untested by the drill. Should portions of these be notably fissured, they might furnish an oil field of much importance. It is entirely possible that oil may be found in various parts of the Highland Rim, either in the Waverly rocks close above the Chattanooga black shale or in Onondagan or Silurian limestones close beneath it. Such rocks appear to the writer as the most promising for further drilling. Oil is much less probable in the Ordovician rocks, since sand and other conditions do not usually seem favorable there.

The surface of the Cumberland plateau consists of Pennsylvanian sandstones and shales of Pottsville age that attain a maximum thickness of 1000 ft. (304 m.) or slightly more, beneath the general plateau level. So far, there is no evidence, either from occasional wells that have gone through them or from their character as they outcrop on either side of the plateau, that the Pottsville rocks contain oil in Tennessee. Should it occur, it would most probably be found in that portion nearest the Kentucky line, as oil is obtained from several Pottsville horizons in Knox County, Ky., not far to the northeast.

The Gulf embayment sands and clays of western Tennessee attain a thickness in excess of 2500 ft. (762 m.), and may be 3500 ft. (1066 m.) thick along the axis of the trough, before the Paleozoic floor on which they rest has been reached. The lower part of these embayment rocks are of Cretaceous age and are the equivalents of the rocks that yield oil and gas in northwestern Louisiana. It is possible that they may contain oil in western Tennessee, although structural relations are so obscured by a blanket of surficial sands and by the general flatness of the region that drilling there must be largely a matter of chance and success mainly the result of luck. It is further possible that some part of the old Paleozoic floor beneath the embayment deposits may contain oil, although there is no means of determining either the lithologic character or the structure of the older rocks from surface inspection. Where they go under the embayment ·deposits near the Tennessee river, they vary in age from Silurian to Mississippian. Their surface is usually regarded as a beveled erosional one, so that it is probable that the floor of the embayment may, in the deeper parts, be composed of Ordovician rocks.

OIL IN KENTUCKY

Oil is produced in Kentucky in a large number of separate areas, most of them small. They are widely scattered through the east central, eastern, southeastern, southern, and southwestern parts of the state. Only one of these, situated in Estill and Lee Counties and generally known as the Irvine field, is of very great size. This includes a recent extension to the southeast known as the Big Sinking Creek field. The most northeasterly are the Fallsburg and Busseyville pools in Lawrence County, and the most eastern is the Beaver Creek field in Floyd County. Closely connected with the Irvine-Big Sinking pool in the central eastern part of the state are the Station Camp, Lost Creek, Campton, Stillwater, and Cannel City pools; and a short distance to the northeast is the Ragland pool. In the southeastern part of the state is the Knox County area north of Barbourville, and a number of small pools in Wayne County. In Lincoln County, there is a small area northeast of Waynesburg. In the southwest, there are the Barren County fields, a small area in the eastern edge of Warren County and a number of small detached areas in Allen County, the most important of which are grouped about Scottsville. Elsewhere, there are a few isolated wells or very small groups of wells not important enough to be given specific mention.

The first oil in Kentucky was discovered, by accident, in 1819 while drilling for a salt well near the south fork of Cumberland River in what is now McCreary County. The oil came probably from the Mississippian. The next find was made, in 1829, on Renox Creek near Burksville, Cumberland County, and was from Ordovician rocks. This well flowed for many miles down Cumberland River, caught fire, and burned for some time. Later its products were used for medicinal and other purposes, until about 1860.

Following the discovery of oil in Pennsylvania, discoveries were made in Wayne and other counties along the Cumberland river, from 1861 to 1866. Most of the oil obtained was shipped by barges down the Cumberland to Nashville, although a part was refined locally. Just after the war, there was renewed interest in the search for oil and additional discoveries were made, especially in Allen and Barren Counties, where oil was found close beneath the Devonian black shale. Interest waned between 1870 and 1880, but was revived during the last two decades of the century, when additional discoveries were made in Barren County west of Glasgow, in Lawrence County on Big Blaine creek, and in Floyd and Knott Counties on the right fork of Beaver creek; while in Wayne County renewed activity led to important discoveries in a number of localities. The most important production in Wayne was found in the Beaver sand in the lower part of the Waverly, but some oil was also obtained below the black shale. During the two decades from 1880 to 1900, the average production for the entire state was not over 5000 bbl. per year; the maximum production was in 1899 when 18,280 bbl. were produced.

The modern period in the development of oil in Kentucky may be said to date from the discovery of the Ragland field in Bath County, in 1900. In this field, oil was found in the Onondaga limestone at a depth of 300 to 380 ft. (91 to 115 m.) beneath the Licking River valley. By

1904, the field was practically drilled up and production since then has gradually declined until it is very small. The Sunnybrook pool, Wayne Co., was discovered in 1901. Oil was obtained from the Trenton, which came to be known locally as the Sunnybrook sand. There was a considerable yield, but it was short lived and within a few years the field had been abandoned. Many further attempts have since been made to obtain oil from the Trenton, or Sunnybrook, both in Kentucky and in Tennessee, but so far they have been without success. Following the Sunnybrook development, much drilling was done elsewhere in Wayne and adjoining counties, and a number of small pools were developed, chiefly in the Beaver sand.

In 1901, oil was found in the northern part of Knox County on Little Richland Creek, near Barbourville. The oil came from three sands in the Pottsville, named in descending order the Wages, Jones, and Epperson. The wells were small producers and were practically all abandoned in a few years. Recently there has been renewed activity in the Barbourville region, but nothing noteworthy has developed.

In 1903, the Campton pool was discovered and by 1909 had been drilled up. Oil was found in the Onondaga limestone. Many of the wells have since been abandoned, but others in the field are still pumping ½ bbl. or more per day. Shortly after this, wells were gotten at Stillwater on the eastward continuation of the Campton structure. They were very similar to the Campton wells and have had a similar history. The same structure yielded oil at Cannel City in 1912, and by 1913 a production of 12,000 bbl. per month had been attained. This rapidly declined, however, and the production today is merely nominal.

For many years oil has been known near Irvine, having been originally found in borings made for salt wells. Soon after the discovery of oil at Campton some shallow wells were bored at Ravenna, near Irvine, on the westward continuation of the structure on which the Campton wells were located. This structure is now generally known as the Irvine structure. The wells were very shallow, but yielded considerable oil for a number of years, until their decline led to the removal of the pipe line that had been laid in the early years of their development, and they were entirely abandoned. In 1915, a well drilled 3 mi. northeast of Irvine started the development of the present Irvine fields and ushered in the present period of intense activity of oil development in Kentucky. The producing sand is the Onondaga limestone, just beneath the black shale, and is generally known as the Irvine sand. The field was rapidly extended eastward and by 1917 had reached the Pilot section near Torrent, making the field about 12 mi. long and from 1 to 2 mi. wide. In 1918, there developed what might be called a southeastward extension of the Irvine field along Big Sinking Creek. Development in this new area was rapid and by the early part of 1919 its southern limits had been reached a mile or two northwest of Beattyville.

On Station Camp creek, some 8 mi. south of Irvine, a small pool was found in 1916, at less than 100 ft. (30 m.) beneath the valley floor. It was drilled very closely and was soon practically exhausted. In similar fashion another small, shallow pool was discovered and developed on Ross Creek. Decline has set in there also but exhaustion has not yet been reached. About these two are grouped several still smaller productive areas of like character but of still more recent development.

Meantime, in 1903, oil was discovered at Busseyville and Fallsburg, Lawrence Co., in the Berea sandstone about 1400 to 1600 ft. (426 to 487 m.) in depth. The wells are small but maintain their production for years with but slight decline.

Although oil was produced in Allen County about the close of the Civil War, it was not until about 1915 that the modern period of production there was ushered in by the drilling near Scottsville of a number of small wells 200 to 300 ft. deep. The oil came from close beneath the black shale from either Onondaga or Niagara limestone. Development has been checked until very recently by inadequate transportation facilities. Most of the development is to the south of Scottsville, but there are several small areas in the northwestern part of the county and recently an important well or two have been drilled just across the line in the eastern edge of Warren County. Wildcat wells are being drilled in numerous places in nearly all sections of the state except the central and northern part, where the surface rocks are of Ordovician age, and in the extreme western part within the area of the Mississippi embayment deposits.

Geology of Kentucky Oil Fields

A list of geological horizons designed to include all sands that have at any time furnished oil in Kentucky would be quite lengthy. A list confined to horizons producing oil today in commercial quantities follows:

PRINCIPAL PRODUCTIVE OIL SANDS IN KENTUCKY

PERIOD	EPOCH	OIL AND GAS HORIZONS		
Carboniferous	Pottsville	Beaver Horton Pike Salt } of Floyd and Knott Co.	Wages Jones Epperson } of Knox Cos.	
Mississippian	Waverly	Berea of Lawrence Co.	Stray and Beaver Creek } of Wayne and Mc-Creary Cos.	
Devonian	Onondaga (Corniferous)	Of Olympia, Ragland, Cannel City, Stillwater, Campton, Irvine, Big Sinking Creek, Ross Creek, Station Camp Creek, Lanhart, Buck Creek, Miller's Creek, Heidelberg, Barren Co., Warren Co., Allen Co., Ohio Co.		

In Floyd and Knott Counties, four sands occur in the lower part of the Pottsville; these, in descending order, are: the Beaver, Horton, Pike, and Salt sands. They are all sandstones and each varies in thickness from less than 50 ft. (15 m.) to more than 300 ft. (91 m.). The interval between them also varies from a few feet to over 100 ft., making it practically certain that the sands split and reunite in such irregular ways that correlation of them is uncertain. In Knox County, the Wages, Jones, and Epperson sands of the lower Pottsville are also sandstones and vary considerably both in thickness and in interval. Their correlation from well to well is doubtful at times and no correlation has so far been possible with the Floyd County sands. The Berea sand of Lawrence County is a medium grained sandstone that usually runs from 50 to 100 ft. in thickness and lies at the base of the Waverly.

In Wayne and McCreary Counties, the principal oil-bearing horizon is a cherty, geodal limestone known as the Beaver Creek sand. It lies just above the Chattanooga black shale and forms the basal member of the .Waverly. It varies greatly in thickness, texture, and porosity and the production of the wells in it varies accordingly. In some cases, a similar oil-bearing limestone is found near the top of the Waverly in these counties and is known as the "Stray sand." It is usually from 10 to 30 ft. (3 to 9 m.) thick.

The Onondaga, or Corniferous, limestone is by far the most important oil-bearing horizon in the state. It lies close beneath the Genesee or Chattanooga black shale. It is a soft brown, porous to cavernous, magnesian limestone which, in the Irvine fields, thickens to the east from 20 or 30 ft. (6 to 9 m.) about Irvine to from 70 to 95 ft. on Big Sinking Creek. The pay exists in from one to several streaks that have no regular distribution or position. Between the pay portions, the limestone is hard and close grained. In Allen County, the pay may extend down into fissured or porous limestone of Silurian age.

Genuine sandstones occur in Kentucky as oil-producing sands only in the Pottsville and Berea, and their aggregate production amounts to less than 2 per cent. of the total production of the state; 98 per cent. of the production comes from limestones. In a sandstone, the distribution of porosity is usually more uniform than in a limestone, where the porous, fissured, or cavernous condition is apt to be irregular in occurrence. This difference in the nature of the two rocks explains the marked differences in the amount of pay, in the yield of nearby wells, and the freakish occurrence of dry holes in the midst of production where the sand is a limestone.

If one takes the percentage of the present production from the several sands given in the preceding table, it will become evident that the producing horizons in the state vary greatly in their relative importance and that the one sand of prime importance is the Onondaga, or possibly, the

Onondaga linked with the Niagara for Ohio and parts of Barren, Warren, and Allen Counties.   The aggregate production, however, from these counties is so small, relatively, that the importance of the Onondaga as the premier oil horizon of the state is not materially diminished.

APPROXIMATE YIELD OF OIL BY GEOLOGICAL HORIZONS IN KENTUCKY

	PER CENT.
Pottsville of Knox, Floyd, and Knott Counties.............	½ to 1
Berea of Lawrence County.............................	1
Stray and Beaver Creek of Wayne and McCreary Counties..	2
Onondaga, of Allen, Barren, Warren and Ohio Counties (?)..	4
Onondaga, of Irvine-Big Sinking and other nearby areas....	92 to 92½

STRUCTURE IN RELATION TO OIL OCCURRENCE

All of the oil fields in the central eastern part of the state are on the eastern or southeastern flank of the Cincinnati anticline.   The rocks in which they occur rise gently to the west out of the great Appalachian trough, whose axis lies along the extreme eastern border of the state. Oil has migrated up the slope of these rocks to the westward until arrested by an anticline with a northeast-southwest axis, whose northwestern limb has usually been faulted in simple or compound fashion.   The most important part of the great major anticline of this region extends from near Irvine eastward to Paint Creek, though the extreme limits are more remote at either end.   Subordinate and somewhat parallel anticlines occur in the Ragland and in other minor fields near the Irvine field. There has apparently been some cross folding also that has corrugated the slope up which the oil has migrated and concentrated it in certain more favorable localities.   The Irvine field, however, presents certain anomalies worthy of mention in this connection.   The axis of the anticline pitches to the northeast at a rate more than sufficient to cause the migration of oil westward along it and without, so far as the writer knows, any cross folding sufficiently strong to check such movement; yet oil is found along this axis at intervals from Irvine eastward to Cannel City with only a few dry areas between the separate pools.   Again, the eastern end of the Irvine field proper has a broad southeastward tongue that extends a number of miles down the dip in the Big Sinking Creek area. It seems that this oil should have migrated farther up the slope to the northwest and have been found nearer the axis, since it has salt water below it to push it onward.

In the Lawrence field, the Berea sand seems to have an anticlinal structure, which combined, perhaps, with difference in porosity may explain the occurrence of the oil there.

In the Pottsville sands, in Floyd and Knott Counties, oil moving up the dip to the westward has been arrested either by slight terraces or by encountering tight places in the sand. There is no anticlinal structure. Similar structural conditions prevail, so far as the writer knows, in the Knox County wells north of Barbourville.

In the Wayne County field, the oil favors, according to Munn, the sides and bottom of synclinal troughs that slope gently eastward.

In Allen County, recently published work by Shaw and Mather show a number of small anticlines and domes with an area of 2 to 3 sq. mi. each, superimposed upon a prevalent northwestward dip of perhaps 40 ft. (12 m.) to the mile. These have a closure of 25 to 30 ft. (7 to 9 m.) or less, and their location, from a study of the surface, is often difficult or impossible because of lack of exposures. These same features and lack of exposures characterize much of Barren County and the eastern part of Warren County. In the western part of Warren County, exposures are better and pronounced doming and terracing occurs. These structures have yielded considerable shows of oil near Gasper River.

Water usually follows the oil in the Onondaga rather closely. It soon begins showing in the wells in the lowest part of the structure and, as time passes, invades the field farther and farther up the dip. Water has thus encroached upon part of the Irvine field and has appeared in the Big Sinking field. Concerted efforts should be taken by operators there to combat this invasion.

TECHNOLOGY

Drilling was formerly by standard rig; and in deep tests in wildcat territory this method is still used. Most of the known production can, however, more easily be reached by drilling machines. Wells in the Allen fields 250 to 300 ft. (76 to 91 m.) deep cost about $1000 complete. In the Big Sinking field, wells 800 to 900 ft. deep cost about $3500; while those 1000 to 1200 ft. deep cost from $5000 to $6000. Prices for drilling tend to rise in harmony with all other prices at present.

The size of Kentucky wells varies greatly both for the various pools and for adjacent locations in the same pool. This is true especially if the sand is a limestone. The rate of decrease also varies greatly. Reliable determinations of this rate are made difficult by the development of the more important pools having been so recent that their records of production extend over only a very few years. This difficulty is further increased by the fact that pipe-line facilities have until very recently been entirely inadequate to take care of the production. In the Allen County fields, transportation conditions have been especially bad, and while partly remedied are not yet entirely satisfactory.

In Lawrence County, wells in the Berea sandstone come in at from 4 to 8 or 10 bbl. and show only a very slow decline over a long period of years.

In Floyd County, where the oil is also derived from sandstone, the initial production is likewise small but is well maintained. Some of the wells drilled 10 to 20 years ago show only slight decline.

In Wayne County, where production is from a limestone, the initial yield varies greatly, though some of the largest wells produced from 100 to 500 bbl. daily for a short time. The average initial production, however, is well below 100 bbl. These wells soon settle to 20 bbl. or less per day and then usually show only a slight further decline. In some cases there has been practically no decline in 15 years; in other cases, the yield in that time has decreased to a barrel or two or even less. Many of these old wells have been overhauled recently and put on a vacuum with a gratifying increase in yield.

In the Irvine district, initial production also varies greatly. The average, given by Shaw, for successful wells drilled between October, 1915, and February, 1917, is about 39 bbl., and the producing wells were 89 per cent. of the total number drilled. Few exceeded 100 bbl. each. In Big Sinking Creek a number of wells have had an initial production of several hundred barrels and a few have probably yielded 1000 bbl. per day. The decline in the Irvine field by the end of the first year has been to about 10 per cent. of the initial yield, although some wells have held up considerably better. This rate of decline has been due to the porosity of the sand and the close spacing of the wells in many cases. In parts of the Station Camp and Ross Creek pools, wells have been spaced one to an acre or less. The well spacing in the Big Sinking field has also been entirely too close on certain properties and has been attended with a rapid decrease in production.

In the Allen County region, about 75 per cent. of the wells drilled have been successful. Initial production for the larger wells has varied from 25 to 100 bbl. per day with a few exceptional wells yielding 200 to 300 bbl. The gas pressure behind these largest wells, however, is quickly relieved and in a few days they decrease greatly. By the end of the first month, the larger ones yield from one-fourth to one-third of their initial production, while the smaller ones hold up somewhat better. These smaller wells come in at from 5 to 20 bbl. per day.

In Barren County, a well recently abandoned because of decreased flow and the eating away of the casing produced oil for over 40 years and during that period showed a remarkably low decline curve. It was probably next to the oldest well in the country at the time of abandonment.[1]

Future production curves and tables have been published by the Internal Revenue Department for Floyd County, Beaver Creek in Wayne County, Ragland and Irvine, in its "Manual for the Oil and Gas Industry."

The oil varies considerably in character. Most of it is dark green by reflected light, but dark brown when seen by transmitted light in thin films. A little amber oil has been reported from Barren, and occasion-

---

[1] A well in Wirt County, W. Va., drilled in 1860 is still producing.

ally elsewhere, but the quantity of such oil is negligible. In gravity, it ranges from 26° to 45° Bé. In the Floyd field, the average is about 40°. In Wayne county, it varies from 36° to 43°. In the Irvine field, the average range is 30° to 36°. In the Ragland field, the average is 26° or 27°. Allen county averages from 35° to 38°; and Barren about 40° to 42°. The gasoline content is usually high.

In the Lawrence, Floyd, Knox, and Wayne County fields, no abnormal values have attached to lands; but in the Irvine district values, especially in Big Sinking field, have rapidly risen until prices of $2000 to $5000 per acre have been reached with extra royalties at times. In Allen County and near the Moulder well in the eastern part of Warren County, high prices have also been given recently for acreage. Wildcat acreage has, in many places, been held at high figures when compared with equal grade acreage in many other states and much development in certain sections has been retarded by these prices.

The great bulk of the oil in Kentucky is transported by the Cumberland pipe line, which has lines serving practically all of the eastern and southeastern parts of the state. It does not, however, reach the fields of Allen and adjacent counties. Until recently its capacity was inadequate to care for the possible full production. A little oil in the eastern fields is handled by short private lines, by barges, or by tank cars. In Allen County, several small pipe lines gather the oil and deliver it to loading racks at Scottsville and Bowling Green for shipment to Nashville, Louisville or elsewhere.

FUTURE POSSIBILITIES

There is a good chance for finding a number of small pools in the Pottsville and the Berea in the eastern part of the state on small structures or under favorable conditions of the sand. Such pools may be expected to have the general character of those in Lawrence, Floyd, and Knox Counties, starting with a small production, but sustaining it well for a long period.

The Onondaga oil is seemingly confined to a narrow belt near the outcrop of these rocks in the central eastern part of the state, which has already been pretty thoroughly tested and developed. The writer looks for no large new pool from that horizon there. Where the Onondaga crosses the saddle between the Cincinnati and Nashville domes in the Barren, Warren, and Allen areas, there doubtless remain a number of new finds; but the difficulty in determining the structure because of the prevalent surface soil covering will make their discovery either a matter of slow detailed work or of chance.

There should be chances of finding oil on the sides of the basin in which the west Kentucky coal field lies where the Mississippian, Devonian, and perhaps Silurian rocks rise from that basin to the east and southeast, wherever domes, terraces, or other favorable structures can be located. The chance on the south side of this basin is less favorable because of the extensive faulting there.

Within the West Kentucky coal field, the writer believes the only favorable chance of finding oil is along the Gold Hill-Rough Creek disturbance and conditions there are complicated because of the severity of the folding and faulting. In the Gulf embayment deposits of West Kentucky, there are no known structures; and it is too soon to make prediction worth anything until the results of the testing soon to be done in the nearby Reelfoot Lake district in Tennessee are known. Much light should then be thrown on the oil possibilities of these embayment rocks in Kentucky.

Little or no oil need be expected in the Ordovician or in any older rocks and drilling in any part of the central blue grass limestone region of the state is practically money wasted.

DISCUSSION

MORTIMER A. SEARS, Huntington, W. Va. (written discussion).—In dealing with the future possibilities for oil and gas in Kentucky, I regret that Doctor Glenn has failed to mention the Paint Creek Dome, which lies in parts of four counties, viz., Johnson, Magoffin, Morgan, and Lawrence. This immense structural uplift has possibilities second to none in the state. It lies along the line of structural uplift known to extend from the Irvine field through Kentucky, and into West Virginia, where it is known as the Warfield anticline.

In an article that I wrote for the *Oil and Gas Journal* (May 21, 1917), I stated the geologic facts in connection with this field, which at that time was strictly a wildcat proposition. It is true that wells have been put down at various times since about 1860, but such operations were spasmodic and haphazard. So far as I know there had been no geologic report relating to oil and gas upon this area at the time I made my examination (February, 1917)—except in the form of a communication from Prof. J. P. Leslie to the American Philosophical Society in 1865. After February, 1917, development dragged along slowly until about a year ago, when more energetic measures were inaugurated, with the result that about 20,000,000 cu. ft. of gas per day has been developed and several oil wells having capacities of from 3 to 50 bbl. per day have been brought in.

Commercial quantities of gas occur in the Weir sand at a depth from the surface of about 850 ft. (259 m.); it varies in thickness from 20 to 40 feet. Part of the product is sold to the Central Kentucky Natural Gas Co., and part to the Louisville Gas and Electric Co. These two companies have main gas lines extending through the field about 5 mi. from the particular area in which gas has been found. Lateral lines have been laid and compressor plants are in operation.

The largest oil wells in the field find their product in the Weir sand also, although commercial quantities of oil have been found in the Berea. The Weir sand appears to be a "genuine sandstone" and seems to prove an exception to Doctor Glenn's statement that genuine sandstones occur in Kentucky as oil-producing sands only in the Pottsville and Berea. It may correspond to one of the oil sands of Wayne and McCreary Counties, but it certainly cannot be called a "geodal limestone."

The Keener, also, has produced small amounts of oil. It is from this sand that a well recently brought in produced an oil of 51° Bé. gravity. The weir oil runs about 38° Bé.   The Cumberland Pipe Line Co. expects soon to lay a line into the field.

The last well brought in showed the Weir sand to be over 60 ft. (18 m.) thick with a 16-in. (40-cm.) break.   Thus far, wells drilled to the Onondaga (Coniferous) have proved disappointing and no production has been found in the Clinton. Comparatively few wells are drilled below the Weir so that it is yet too early to condemn the lower formations.

Leases are constantly. changing hands. Very little acreage remains in the hands of the land owner. Whenever a well is brought in, leases sold on adjacent property bring from $100 to $150 per acre. With the opening of spring there is no question but that this area will be the scene of the greatest activity in the state of Kentucky.

WILBUR A. N. NELSON,* Nashville, Tenn. (written discussion).— Certain pertinent facts in regard to the oil produced in Tennessee in the past and to the extension of the different formations of Allen County, Kentucky, into Middle Tennessee are not given in this paper.

The very heavy slump in production that occurred in the old Riverton Spurrier district of Pickett, Tenn., was due to fresh-water troubles. A recent study of the water troubles of this field brings out these facts: Under the Chattanooga "black" shale occurs a practically uniform bed of Ordovician limestone, bedded or creviced so as to permit a connection between the different gas shows in the upper part of the limestone immediately under the black shale and the oil horizons in the base of the limestone, some 165 to 270 ft. below the black shale. In the old wells, the casing was set below the gas shows and just above the oil horizon. That the release of the gas pressure permitted the fresh water to flow down through the limestone joints, bedding planes, or fractures to the oil horizon and thus drown out the well, seems to have been proved by Mr. J. H. Compton, of Riverton. Several years ago he reset the casing in one well above the first gas show and, after plugging the other wells,

* State Geologist, Tennessee Geological Survey.

above the gas horizon, started pumping. After several weeks, the well again commenced to produce oil.

A structural report recently made on this area by the Tennessee Geological Survey, in coöperation with the U. S. Geological Survey, shows that the best old producing wells were located on the crest and north flank of a long narrow anticline extending in a direction of approximately north 60° east and that the oil probably occurs in pools of small extent with a radius of about $\frac{1}{2}$ mi. Several similar anticlines were mapped in this district, which are yet untested. The Cumberland Pipe Line Co. laid a 2-in. line into this field in 1902, which was removed in 1905, due to a decline in oil production but primarily to the levying of a $10,000 annual tax on the line by Pickett County. During this time 58,776 bbl. of oil were piped from this field, of which over 36,000 bbl. came from one well, known as the Bobs Bar well, which shortly went to water.

In Sumner County, Tenn., and in adjoining counties to the west and southwest on the Highland Rim, there is at present much drilling going on, but the majority of these wells have been drilled without paying any attention to structure. This was recently shown in Sumner County, which joins Allen County, Ky., on the south. A detailed structural map of part of this county made by the Tennessee Geological Survey, in coöperation with the U. S. Geological Survey, shows that of over 30 holes drilled only two were located on favorable structure. But on that particular dome, one could have little hope of finding oil, as the oil horizon had been cut through on the south flank of the dome. The structurally favorable places are still untested.

In Allen County, Ky., around Scottsville, the oil is found at three horizons below the Chattanooga black shale. These three sands are not always present at one place; and when present, as a rule only one is producing. The upper, sometimes the two upper, sands are considered of Devonian age and probably correlate, in Tennessee, with the Pegram limestone. The lower sand, which produces most of the oil to the south of Scottsville, is thought to be of Silurian age and to be Louisville limestone, as this formation outcrops in Sumner County, Tenn., just south of Allen County, Ky., at the base of the Chattanooga black shale, the Corniferous beds of Devonian age being absent.

The location of the old shore line of the Pegram limestone, as it is known in Tennessee, and of the Corniferous limestone, as it is known in Kentucky, is important. Exposures of this limestone are not known south of Petroleum, Allen County, Ky. No outcrops are known in Sumner County, Tenn., but it appears again 12 mi. (19 km.) west of Nashville, at Newsom Station, where it has a thickness of 3 ft. (0.9 m.); a few miles farther west, at Pegram, in Cheatham County, it has a thickness of 12 ft. From these exposures it would appear that this shore line would extend from Newsom Station northeastward through Cheatham and Robertson Counties, Tenn., probably passing in the vicinity of Springfield, and crossing the state line in the proximity of the northeast corner of Robertson County, near Mitchellville. All territory as far west of this area as the Tennessee River is underlain by Devonian limestones. A well was brought in, in January, 1920, in Simpson County, Ky., about 3 mi. from the northeast corner of Robertson County, Tenn. in a very peculiar sandy limestone 61 ft. below the Chattanooga black shale. The sand was penetrated to a depth of 7 ft. and may be a phase of the Harriman chert of Oriskany age, which outcrops about 50 mi. to the southwest near Cumberland City, Stewart County, Tenn.

The shore line of the other supposedly oil-bearing limestone, the Louisville limestone of Silurian age, is of interest because of the effect it would have on possible oil territory in the counties on the western Highland Rim of Tennessee. In Sumner County south of Westmoreland, it is 20 ft. thick, while about 25 mi. to the southwest near Ridgetop, in Robertson County, it only shows a thickness of 10 ft. Farther to the southwest, in southern Cheatham County around Pegram, it is very thin, having a 15 ft. exposure. On the western edge of the Highland Rim along the Tennessee River, this formation changes to a shaly phase, known as the Lobelville, which varies in thickness from 10 to 75 ft. These facts would indicate that the extent of the limestone phase of the Louisville formation would lie just to the southeast of the present edge of the Highland Rim on the Middle Basin of Tennessee, as far south as Pegram, and that at this point the line would turn to the northwest, swinging back into Kentucky. The extreme thinness of the formation, except in the northern part of Sumner County and probably in the northern part of Robertson County, would indicate that only in these two areas would it be thick enough to act as a commercial oil reservoir. The long narrow embayments in which this and the overlying formations were laid down make it probable that there are areas in the northern Highland Rim counties lying outside of these old embayment areas in which these formations were never deposited. In the more southern counties on the Highland Rim west of Nashville, overlapping formations come in between the Louisville limestone and the Chattanooga black shale, which would keep this formation from containing oil if such oil is derived from the Chattanooga black shale. That this formation is probably absent in the southwestern part of Robertson County is indicated by the fact that a recent well on Sulphur Fork, 6 mi. southwest of Cedar Hill, which went to a depth of 1015 ft. and passed through the Chattanooga black shale at 615 ft., encountered no water, oil, or gas below the Chattanooga. This hole probably passed through the rocks of Trenton age at 950 ft. The presence of blue phosphate sand in the limestone above this level is taken as evidence of the presence of the Hermitage formation of Trenton age at this depth.

In western Tennessee, two deep tests are being drilled, one in Lake County at Proctor City on the west side of Reelfoot Lake and the other in Obion County near Walnut Log on the northeast side of Reelfoot Lake. From numerous exposures of the formations just under the loess bluffs northeast of Reelfoot Lake, it is thought that there is a marked anticlinal area just to the northeast of Walnut Log and extending over the Tennessee state line into Kentucky. The oil and gas rights on Reelfoot Lake, which belongs to the state of Tennessee, have been leased by the Governor to the men who are drilling near Walnut Log. This hole is on land joining the state property. Among other things the state requires that the well be drilled to a depth of 3000 ft. The Paleozoic floor of the gulf embayment should be reached inside of that distance, while the formations producing oil in the northwest corner of Louisiana should be reached at about 2200 ft.

In Allen County, Ky., detailed structural work done to the south of Scottsville shows that in the area thus mapped the best production comes from the northwest or west side of small structural domes, with closures of about 20 ft., but where the dome has a very steep dip on the north or northwest side, with gentle dips to the south, the production is obtained on the south and southwest sides. Such production is always less than the production from the northwest sides of Allen County domes. In small wells that are shot, the production often drops off four-fifths after the first two or three days. In several cases, wells that have come in producing salt water change to oil after about two weeks pumping, and make average producers. No fresh water is encountered in the Allen County wells below the Chattanooga black shale. The average production in this section is probably not more than 5 bbl. per pumping well.

STUART ST. CLAIR, Bowling Green, Ky. (written discussion).— Doctor Glenn's paper is interesting as an historical résumé of the oil development in these states, the former of which has come into prominence during the past few years, producing, in 1919, approximately 8,000,000 bbl. of high-grade oil.

The writer had hoped that Doctor Glenn would give more detailed data on the accumulation of oil in the Onondaga limestone, as that formation furnishes about 96 per cent. of the oil production of Kentucky. If he had, in his discussion of the eastern part of the state, he would have noticed that his statement that oil has migrated westward up the slope of the rocks which rise from the great Appalachian trough until arrested by an anticline with a northeast-southeast axis, would need some modification or further explanation. Between the Appalachian trough and the Irvine District, the latter comprising the oil fields of Lee, Estill, Powell, and Wolfe Counties, there are a number of well-defined anticlinal structures that have been drilled upon with unsuccessful results. The Onondaga formation does not have a continuous bed of such porosity as would be needed for migration of oil, except within a restricted distance from its outcrop and from the Irvine fault. Therefore, migration of oil in this formation took place only over a short distance. As explained by the writer in a paper[2] on the Irvine Oil District, the greater part of the porosity in certain beds of the Onondaga from which oil is produced is caused by solution by circulating meteoric water which has entered at the outcrop and along fault planes. It is this theory that explains the position and structural relations of the prolific Big Sinking Creek pool of Lee County.

In view of what has been said, Doctor Glenn's statement that water usually follows the oil in the Onondaga rather closely may need partial revision. It is true that wherever there is oil in commercial quantities there is also water, for water has in most cases caused the porosity in the rock in which the oil has accumulated. However, there are areas where there are very small wells of doubtful commercial value, the oil having accumulated in the Onondaga where there may have been a little porosity induced by recrystallization or partial dolomitization, where there is a total absence of water. Outside of a restricted distance from the outcrop of the Onondaga or from major faults, wells drilled on anticlines or in synclines show a general absence of both oil and water.

How far the thought developed by the writer, in his paper mentioned above, showing the relation between the area affected by circulation of meteoric water and oil accumulation in the Onondaga limestone in Kentucky can be applied to other fields where the oil production is from a porous limestone, cannot be stated, but he hoped that the idea advanced might be used with additions or modifications in helping to explain accumulation problems in other limestone fields.

Two minor corrections should be made in Doctor Glenn's paper for the benefit of those unfamiliar with the Kentucky fields. First, the gravity of the oil for the Irvine field is given correctly but it should not be thought to include the adjacent Big Sinking field. In the latter the gravity is much higher, running from 38° to 42° Bé. and the gasoline content is exceptionally high. Second, the great bulk of the oil in Kentucky is not carried by the Cumberland Pipe Line Co. at the present time. In the fields of Lee and adjacent counties, the Cumberland runs but little more than half the production; the balance is handled by six other pipe line companies, chief among which are the Indian Refining, Great Northern, and National Refining. In Allen County, the Indian Refining handles nearly all the production, although recently two other pipe lines have entered the field.

The writer fully agrees with Doctor Glenn in his outline of the areas of Kentucky that contain possibilities for future production. In his

---

[2] See p. 165.

opinion, even the Allen, Barren, and Warren County areas are about outlined at the present time. In the western Kentucky coal field, development will be slow, but something of importance may be opened in the Chester or lower Mississippian sands. Kentucky cannot hope for a second Big Sinking, which is the most important field in the history of oil production in the state. The flush was taken from this pool in 1919 and the production for that year will mark the apex of the production curve for the state. The decline in the curve will not be great for 1920, but after this year the decline will be noticeable.

In Tennessee, aside from a probable few small pools along the Highland Rim in the limestone underlying the Chattanooga black shale and within a restricted distance from the outcrop of this limestone formation, or from a major fault, and a possible few small pools in the coal-measure area in the eastern part of the state, the oil possibilities, in the writer's opinion, lie in certain areas within the Gulf Embayment province west of Tennessee River.

Oil Possibilities in Northern Alabama

BY DOUGLAS R. SEMMES,[*] PH. D., UNIVERSITY, ALA.

(Lake Superior Meeting, August, 1920)

THE possible oil territory of Alabama can be readily divided into two regions, the Paleozoic area of the north, and the Coastal Plain province of Cretaceous and younger formations lying to the south. This latter area has received much attention in the last few years and has been described by a number of writers.[1] Although the possibilities of the Cretaceous series have been much emphasized by recent writers, the fact remains that the two, or possibly three, localities where oil or gas have been found in anything like paying quantities are confined to the area of Carboniferous rocks. Moreover, almost all of the oil seeps and a good percentage of the gas seeps are confined to this area.[2]

Topographically, as well as structurally, the Paleozoic area can be divided into three rather well defined provinces: (1) The broad, open Coosa Valley lying adjacent to the crystalline oldland, with comparatively little relief, except for occasional longitudinal ridges and rather intense folding; (2) the plateau region of horizontal or gently warped Pennsylvanian strata broken by occasional anticlinal valleys aligned northeast and southwest, outliers of the Coosa Valley proper, in which the older Paleozoic formations are exposed—a region of much relief (200

[*] Associate Professor of Geology, University of Alabama.

[1] Eugene A. Smith: Report on the Geology of the Coastal Plain of Alabama. Geol. Survey of Alabama (1894).

Eugene A. Smith: Concerning Oil and Gas in Alabama. *Circular* 3, Geol. Survey of Alabama (1917).

O. B. Hopkins: Oil and Gas Possibilities of the Hatchetigbee Anticline, Alabama. U. S. Geol. Survey *Bull.* 661 (1917) 281.

Dorsey Hager: Possible Oil and Gas Fields of the Cretaceous Beds of Alabama. *Trans.* (1918) **59**, 424.

[2] Among the more important references on northern Alabama are:

Henry McCalley: Report on the Valley Regions of Alabama. Part I. (Tennessee Valley.) Geol. Survey of Alabama (1896).

M. J. Munn: Reconnaissance Report on the Fayette Gas Field, Alabama. Geol. Survey of Alabama *Bull.* 10 (1911).

Eugene A. Smith: Historical Sketch of Oil and Gas Development in Alabama. *Oil Trade Jnl.* (Apr., 1918) **9**, 133.

to 300 ft.) and thorough dissection, well wooded, and of little agricultural importance; and (3) the Tennessee Valley region of horizontal or gently warped Pennsylvanian and Mississippian strata, where the relief is not so marked, the wooded area is less extensive, and the country is of more importance agriculturally.

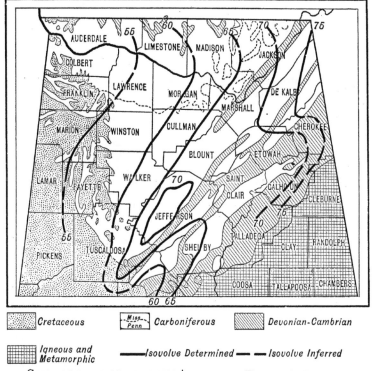

| Cretaceous | Carboniferous | Devonian-Cambrian |

Igneous and Metamorphic   ———Isovolve Determined — —Isovolve Inferred

GEOLOGICAL MAP OF NORTHERN ALABAMA, SHOWING CARBON RATIOS.

STRATIGRAPHY

The following generalized section gives an approximate idea of the thickness and lithologic character of the formations of the region as a whole. The Carboniferous series, especially, shows lateral variations of striking prominence, but certain horizons are persistent throughout the area.

Of these formations the Carboniferous cover much the larger part of the whole region, the Pennsylvanian, or Coal Measures, forming the surface throughout large portions of Cullman, Winston, Walker, Blount, Jefferson, and Tuscaloosa Counties, and the Mississip-

STRATIGRAPHIC SECTION FOR NORTHERN ALABAMA

AGE	FORMATION NAME	THICKNESS	LITHOLOGIC CHARACTER AND SUBDIVISIONS
Pleistocene Pliocene	Lafayette	0- 50	Unconsolidated and semiconsolidated gravels and sands. Red, pinkish, maroon, and whitish clays.
Cretaceous	Tuscaloosa	0–1000	Gravels, sands, and clays. Red to gray or white. Non-marine.
Pennsylvanian	Coal measures	200–2500	Shales, arenaceous shales, massive sandstones, and conglomerates near base. Basal conglomerate, or Millstone grit. Coal seams.
Upper Mississippian	Mountain limestone	400- 900	Massive bluish crinoidal limestone (Bangor limestone), overlying a series of coarse to medium-grained sandstones with alternating thinner beds of limestones and shales. Thick, locally massive, brown sandstones at base (Hartselle sandstone).
Lower Mississippian	Fort Payne chert	200- 500	Cherty limestone or limestones with thin chert seams or layers of nodules. Readily eroded, valley-making formation (Tuscumbia limestone). This overlies a series of hard cherts (Lauderdale chert), very resistant to erosion and forming prominent ridges.
Devonian	Black shale	0- 50	Black, highly bituminous shales and locally thin sandstones and bluish shales.
Silurian	Clinton	200- 400	Shales, limestones (Niagara), and ferruginous sandstones and some conglomerates. Iron ores.
	Unconformity (?)		
Ordovician	Trenton limestone	500–1000	Bluish, thin-bedded limestone (Pelham or Chickamauga) with coarse-grained siliceous layers, making excellent oil sands.
	Unconformity		
	Knox dolomite	2000–3300	Sharply folded and faulted thick-bedded crystalline dolomite with chert seams, overlying 600 ft. of thick-bedded non-cherty gray crystalline dolomite.
Cambrian	Coosa shales	1000–1500	Thin-bedded blue limestone and gray and yellow shales.

pian forming the surface in Madison, Limestone, Lawrence, Lauderdale, and Colbert Counties. To the west, the Carboniferous strata are overlapped by the Cretaceous, which in turn is covered in places by the Lafayette, but throughout a large part of this Cretaceous area the underlying Coal Measures are exposed along the courses of the principal streams. The pre-Carboniferous rocks are only exposed in small areas in the north, along the anticlinal valleys farther south, and in the Coosa Valley.

Oil and Gas Horizons

The possible oil and gas horizons throughout the section are rather numerous; many of these horizons have, locally, given very promising shows. Owing to the striking lateral variations in lithologic character, horizons that may at one point be promising have little or no possibilities at another; this is especially true in the Carboniferous series.

Pennsylvanian Horizons

In the Fayette gas field in Fayette County, the wells encountered the first shows at about 500 ft. (152 m.) below sea level, or at a depth of from 850 to 950 ft. (259 to 289 m.). This sandstone, though giving good shows of oil, proved of no value and the wells were continued 500 ft. lower and there encountered the Fayette gas sand proper, a soft, white sand of excellent quality. The best well of this group was estimated at 4,000,000 cu. ft. (112,000 cu. m.) per day. At other points, the sand was found more tightly cemented and proved less productive. About 200 ft. below the Fayette sand a thick sandstone (250 ft.) was encountered, which also gave gas shows and is locally known as the Second Gas sand. Drilling has been continued 900 ft. below the Fayette sand and another thick sandstone, containing salt water, was encountered near the bottom. This sandstone has been correlated with the Pine sandstone member of the Birmingham Folio, in which case the Fayette sand should be underlain by some 1000 ft. of shales and massive sandstones to the base of the Pine sandstone, then 500 ft. of shales and shaly sandstones, and finally 500 ft. more or less of massive sandstones with conglomerates at the base (Millstone grit). In other words, the Fayette gas sand should be about 2000 ft. above the base of the Pennsylvanian, if maximum thicknesses were represented. Deep borings in the area have shown, however, that the total thickness of the Pennsylvanian is not over 2500 ft., or only about 1200 ft. of sediments underlie the Fayette sand. It should be remembered, moreover, that in the Birmingham district the Boyles sandstone (Pine sandstone) is found lying directly on the Bangor limestones of Mississippian age, representing a hiatus of much over 2000 ft. of sediments. The two basal sandstones of the Coal Measures, the Pine sandstone and the Millstone grit, can be considered as possible oil horizons, but owing to their great thickness, thorough cementation, and massive character, as well as the fact that they have no adequate source of oil below, the writer would not consider them horizons worthy in themselves of extensive testing.

Mississippian Horizons

As early as 1865, wells were drilled in Lawrence County in the vicinity of asphaltum and maltha showings in Mississippian strata, and the majority of all such occurrences of bitumen, maltha, and asphaltum that have been reported since are in these formations. In the Bangor limestone, many such occurrences have been found and it is not unreasonable to suppose that a sandy layer in this limestone might prove a paying oil sand. Below the Bangor comes the Hartselle group of thick sandstones and interbedded limestones. In Morgan, Lawrence, and Franklin Counties, numerous seeps have been found in the Hartselle, and, in many places, this group is found saturated with residual petroleum. In many localities this sandstone is coarse-grained and friable, with a large amount of pore space, which should make it an excellent oil sand; but elsewhere it is fine-grained and highly cemented and has so little pore space that it is improbable that it would prove a pay sand. Unfortunately, no good test has been made in this horizon as none of the wells started in the Pennsylvanian has reached this depth, while those started below the Pennsylvanian have usually commenced operations on the Hartselle itself, or immediately above it. The Lower Mississippian Tuscumbia limestone has given rather promising shows in certain recent tests. In it are found sandy horizons that, according to the driller's statement, make excellent oil sands.

Trenton Horizons

The only well that has struck oil in commercial quantities in Alabama found it in the Trenton (Pelham) limestone. This horizon is, in the opinion of the writer, the most favorable for commercial oil and gas to be found in the northern part of the state. The Trenton series is composed of thin-bedded bluish and shaly limestones, throughout which there are horizons of coarse-grained, sandy limestones making good oil sands. The well mentioned (Goyer No. 1), drilled in 1891, is located in the southwest quarter of the southeast quarter of sec. 29, T. 7N, R 6W, Lawrence County, and was not a geological location. A log of this well[3] is as follows:

		FEET
36	Soil	10
35	Limestones; Bangor	290
34	Sandstones; first gas, in the upper part	35
33	Shales; a dark blue color	110
32	Limestone; of a pearly white, sulfuretted hydrogen gas was struck in this rock at 55 ft. below its top and salt water at 53 ft. from its top; the salt water on evaporation gave a good flavored salt	80
31	Limestone; of a light drab color	320

[3] McCalley's Tennessee Valley Report, 239.

FEET

30 Limestone; impure, coming out as a coarse powder like corn meal and hence
    called "corn-meal sand".. 28
29 Shales; Devonian, black.. 32
28 Limestones; shaly 17
27 Limestones; blue... 2
26 Shales; sandy and of a mottled (red and white) color.................... 9
25 Limestone; it carries some little oil................................. 422
24 A gritty calcareous sand, likely from an impure limestone................ 100
23 Limestone; blue... 45
22 Limestone; coarse grained, the lower 5 ft. is an oil sand though it carries
    no oil.. 9
21 Limestone; coarse grained, impure and siliceous, a good oil sand........... 20
20 Limestone; blue.. 261
19 Limestone; white.. 32
18 Limestone; blue with greenish specks................................. 6
17 Limestone; white or cream colored................................... 6
16 Limestone; blue.. 63
15 Limestone; bluish with a slight reddish tinge......................... 26
14 Limestone; white.. 27
13 Limestone; gray with a slight reddish tinge........................... 4
12 Limestone; white... 4
11 Limestone; gray with a few reddish specks............................. 3
10 Limestone; of a light gray color..................................... 49
 9 Limestone; white... 5
 8 Limestone; of light gray and reddish specks........................... 2
 7 Limestone; of a brownish gray color................................... 5
 6 Limestone; white... 4
 5 Limestone; of a grayish color with white and blue specks................ 7
 4 Limestone; with large white specks that resemble pieces of fossils.......... 4
 3 A dark grayish powder with blue and white specks; it may be shale........ 5
 2 Limestone; with white and light gray colors with reddish specks........... 22
 1 Limestone; white... 3

Of this above log, 35 is in the Bangor limestone, from 34 down into 32 is the Hartselle sandstone group; from 32 to 30 is the Tuscumbia limestone and Lauderdale chert; 29 is the Devonian; and the rest is Trenton.

The pay sand in this well was found 625 ft. (190 m.) below the Devonian. The well was estimated at 25 bbl., but owing to the collapse of the casing and the letting in of salt water the well was lost. The oil was of a light greenish color and had a specific gravity of 38.7° Bé.

Knox Horizon

The Cambro-Ordovician Knox dolomite has occasionally given small shows of gas; and since dolomitization with its attendant shrinkage should indicate increase in pore space and in capacity as a reservoir, this formation has been tested several times in the hope of its proving productive. But, as the Knox lies unconformably below the Trenton and where exposed shows a high degree of deformation, its possibilities are so slight as to be unworthy of serious consideration.

STRUCTURAL FEATURES

The Coosa Valley region and the adjacent outlying valleys are characterized by a rather intense type of parallel folds, trending northeast and southwest, with which are occasionally associated faults showing displacements as great as 3000 ft. (914 m.). In many of these folds, the Ordovician formations are exposed and, consequently, have no possibilities as oil traps. In addition to the northeastern series of prominent Appalachian-type folds, there is a series of undulations running northwest and southeast. These folds, or "waves" as they have been termed by the earlier writers, are sometimes quite pronounced and show a reversal of 100 ft. or more. At the intersection of these undulations quaquaversal structures are formed, such as the Blount Springs dome, which should make excellent oil traps if the oil horizons themselves are not exposed. The intensity of the deformation of this area, however, is considered an unfavorable factor in the development of valuable accumulations of oil.

In the Plateau region, running parallel to the Appalachian folds farther east and south, there is a series of subsidiary undulations, which can be traced out in the beds of the Coal Measures. In this area are likewise developed the series of northwest-southeast waves, and at the intersection of these two series more or less perfectly developed domes are occasionally established. Unfortunately, the most favorable portion of the Pennsylvanian area, the western part, is rather extensively covered by the Lafayette gravels and the Cretaceous series, or by a mantle of residual soil, which makes the locating of favorable structures a difficult task; and after evidence of folding has been found, it is often impossible to map the structures in detail.

In the Tennessee Valley region, especially in the area underlain by the Mountain Limestone group (Morgan, Lawrence, Franklin, and Colbert Counties), the two series of undulations can be readily detected; and owing to the character of the surface rock the structure can be accurately mapped. Since the most favorable oil horizon, the Trenton, occurs in this area at a depth of 1000 to 2000 ft., the area is considered very favorable for future prospecting.

SIGNIFICANCE OF CARBON RATIOS OF COALS OF AREA

A recent paper by Fuller,[4] discussing the relation of the carbon ratios of the Pennsylvanian coals to the oil fields of northern Texas, has interested the writer in collecting and plotting the fixed carbon percentages of the Pennsylvanian coals of northern Alabama.[5] In certain localities, a

---

[4] Myron L. Fuller: Relation of Oil to Carbon Ratios of Pennsylvanian Coals in North Texas. *Econ. Geol.* (1919) **14**, 536.

[5] For these analyses the writer is indebted to the publications of the U. S. Bureau of Mines, to the Geological Survey of Alabama, and to R. S. Hodges, chemist of the Geological Survey.

sufficient number of analyses were obtainable to locate the isocarbs[6] definitely; in other areas, their location was largely inferred. On the accompanying map, the direct relationship between the fixed carbon and the amount of deformation is very apparent. Even such outlying folds as the Sequatchie anticline have their definite effect upon the percentage of fixed carbon in the coals mined along their flanks. The degree of metamorphism attending this deformation has been shown by David White[7] to be definitely related to the distribution and composition of the oils found in the formations affected. When metamorphism has reached such an extent that the fixed carbon in the coals, considered on a basis of pure coal, has reached a percentage of 70, it is very improbable that oil pools of commercial importance will be found. In the case of northern Alabama, the writer believes that all areas where the fixed carbon runs as high as 65 per cent. may be considered as unfavorable territory and unworthy of extensive tests, until at least the areas of lower percentages have been tested and oil found in paying quantities. An examination of the accompanying map shows that the 65-per cent. isocarb becomes definitely fixed near Huntsville and swings southward and westward across Madison, Marshall, Cullman, and Blount Counties, and along the northwestern line of Jefferson County, thence across Tuscaloosa County, swinging around the southwestern end of Jones' Valley anticline into Shelby County, thence southward again following the crystalline area. Farther east, the 70-per cent. and the 75-per cent. isocarbs swing around the areas of more intense folding and attendant metamorphism. To the north and west of the 65-per cent. isocarb, the 60-per cent. isocarb is less definitely located; and owing to the scarcity of analyses of the coals of Marion and Franklin Counties, the location of the 55-per cent. isocarb is in part inferred.

Favorable Areas

If we are to consider the relation between metamorphism and oil distribution and composition as established, we are led to the conclusion that all the Coosa Valley region and much of the Plateau Region is unfavorable territory, and would expect in this area only small accumulations of light oil. West and north of the 65-per cent. isocarb there should be better chances for larger accumulations, provided we have the other necessary requirements of favorable section and structure.

The Pennsylvanian series shows a fairly favorable section, several good sands, and good structure where it can be worked out. Most of the

---

[6] The term "isocarb" has been adopted by David White, to signify a line drawn through points of equal carbon ratio.

[7] David White: Some Relations in Origin between Coal and Petroleum. *Jnl.* Wash. Acad. Sci. (Mar. 16, 1915) 6, 189.

David White: Late Theories Regarding the Origin of Oil. *Bull.* Geol. Soc. Amer. (1917) 28, 727.

area underlain by the Coal Measures lies within the 65-per cent. isocarb; to the west, the Coal Measures are covered by the Lafayette and Cretaceous formations. In the Fayette district, which is near the 55-per cent. isocarb, only gas was found in paying quantities, which would indicate that the degree of metamorphism was still too great for accumulation of oil. Considering, therefore, all evidence obtained so far, the Pennsylvanian area is not considered favorable territory except toward the western line of the state. In this area a hole drilled to 2500 or 3000 ft. would test the Hartselle sandstone as well as the Pennsylvanian horizons.

To the north, the Pennsylvanian formations break off forming a pronounced scarp facing the north. Passing down over this scarp, one comes upon a fairly level plain underlain by the lower members of the Bangor limestone. The upper, more massive members of this limestone form the base of the scarp. Farther north, beyond the plain underlain by the lower Bangor limestone, another scarp is formed by the Hartselle sandstone, locally known as Little Mountain. Between these two scarps there might be found favorable structure, where the Hartselle is sufficiently covered to prove productive; this area is rather limited, however,. for which reason practically no tests have been made of the Mississippian horizons.

By drilling at any point north of the Pennsylvanian scarp, the upper Trenton horizons (Goyer horizon) would be encountered not more than 2000 ft. (609 m.) in depth. This would give a large territory comprising most of Morgan, Lawrence, Colbert, and parts of Franklin and Lauderdale Counties, in which the type of structures commonly considered as oil traps are fairly abundant, the degree of metamorphism is comparatively low, and the section is decidedly favorable. It is in this area, and especially in Franklin, Colbert, and Lawrence Counties, that the writer would suggest that future tests be made. Numerous anticlinal folds can be found in the Hartselle sandstones and in places definite closure can be worked out.

Past and Present Development

The extent and results of past development in northern Alabama have been fully described by Doctor Smith in his historical sketch of developments already cited. The more important of these tests, with their dates and producing horizons, are as follows:

1865. Watson wells, southeastern Lawrence County; good shows in two wells; Trenton horizon.
1890. Newmarket well, Madison County; strong petroleum odor, but no sand; Trenton horizon.
1891. Goyer wells, southeastern Lawrence County; one estimated at 25 bbl. a day; Trenton horizon.

1893. Allen wells, Florence, Lauderdale County; one dry, the other showed small quantities of very light oil and some gas, well spoiled by shooting; Trenton horizon.

1904–5. Huntsville and Hazel Green, Madison County; gas shows at both localities; Trenton horizon.

1909. Fayette wells, Fayette County; one estimated at 4,000,000 cu. ft. of gas; Pennsylvanian horizon.

1910–11. Shannon wells, Jasper, Walker County; 50,000 cu. ft. of gas, oil show; Pennsylvanian horizon.

1911–12. Woodward Iron Co., Russellville, Franklin County; small gas show in Knox dolomite (?).

1912. Bryan, Jefferson County; oil and gas show; Pennsylvanian horizon.

1916. Cordova, Walker County; good show, black residual oil; Pennsylvanian horizon.

1917–18. Atwood well, Franklin County; gas indications; Pennsylvanian horizon.

1917–18. Aldrich Dome wells, 6 mi. southeast of Birmingham, Jefferson County; gas shows; Pennsylvanian horizon.

1918. Guin well, Lamar County; oil shows in two sands; Pennsylvanian horizon.

1919. Hobson well, Frankford, Franklin County; drilling at 1765 ft. (Dec.), small oil and gas shows; Trenton horizon.

The evidence of the above tests strongly supports the carbon ratio hypothesis, as all localities near the 65-per cent. isocarb showed gas and only small shows of oil. The heavy oil found at Cordova (26.5° Bé) is an exception, but this was undoubtedly a pocket of residual oil, the lighter volatile constituents of which had been driven off.

Future Prospecting

The area the writer considers most favorable for future testing is the northwestern portion of the state, where the Trenton limestone would be the producing horizon. Even this area is not without its disadvantages. The degree of metamorphism increases not only near areas of deformation but in depth in any locality. Therefore the degree of metamorphism of the Ordovician formations, once covered to great depth by the Pennsylvanian series, may be much greater than is indicated by the Pennsylvanian coals, in which case commercial accumulations would be improbable. Moreover, there is a possibility of an unconformity below the Silurian. Nevertheless, considering the structure, the lithologic character of the section, and the evidence of the carbon ratios of the overlying Pennsylvanian coals, the area is undoubtedly worthy of further tests, provided they be well located on carefully determined structure. In addition to this area, the Coal Measures, where exposed in Winston, Marion, and Fayette Counties, should be well worth testing, especially where drilling is continued to sufficient depths to test the Hartselle and the Trenton horizons.

DISCUSSION

DAVID WHITE, Washington, D. C.—Pessimism regarding the capacity of the Hartselle sandstone of Alabama should be discouraged. The outcropping sandstone south of Tuscombia carries asphalt seepages that are still fresh. In fact, the sandstone was once drilled near this outcrop. The Hartselle is remarkably persistent throughout a great area, and, in regions where the carbonization of the organic debris has not progressed too far, this sandstone offers oil possibilities in favorable structures.

It is possible that the Carboniferous of western Alabama, beyond the zone of too great carbonization, may contain oil deposits as important as any to be found in the arches of the Coastal Plain formations.

MOWRY BATES, Tulsa, Okla.—Last spring I examined diamond-drill cores of Hartselle sandstone from holes drilled in nearly every section of Alabama. Every core showed oil but it was thick and the sand was so tight that it was impossible to move the oil. Under the microscope no pore spaces could be found. None of these wells have shown oil in appreciable amounts.

Résumé of Pennsylvania-New York Oil Field

By Roswell H. Johnson, M. S., and Stirling Huntley, Pittsburgh, Pa.

(New York Meeting, February, 1920)

Pennsylvania will be remembered, as long as oil is produced, as the cradle of the industry of petroleum in North America. It was on Oil Creek, near Titusville, Venango Co., that Col. Edwin L. Drake, superintendent for the Seneca Oil Co., brought in the first commercial oil well on Aug. 28, 1859. Great difficulty was experienced in getting the well down to the producing depth of 69 ft. (21 m.) with the spring-pole system then in vogue for punching shallow water wells, so the novel expedient of driving an iron tube through the surface clays and quicksand was finally resorted to. The well had an initial yield of 25 bbl. a day on the pump, but soon went off, though 2000 bbl. were produced by the end of the year.

With the Drake well a success, a young industry sprang into being, the rapid growth of which has been second to none in the country and the value of whose product is only surpassed by that of coal. For years the only producing territory, the Pennsylvania-New York field attained its greatest production in 1891, when the bringing in of the Mc-Donald pool, between Pittsburgh and the West Virginia line, gave a total of over 33,000,000 bbl. At present the field, combined with West Virginia and southeast Ohio, gives 25,000,000 barrels.

Geology and Stratigraphy

In general, the Appalachian field is a huge geosyncline, the axis of which runs roughly northeast-southwest, from north of the New York state line south through Brookeville, Kittanning, Pittsburgh, Washington, through the southwest corner of the state of Pennsylvania into West Virginia. Minor folding has accompanied or followed the dominant fold of the field, and it is from these structures near the Pennsylvania-West Virginia line and their influence on the accumulation of oil and gas that I. C. White, state geologist of West Virginia, obtained his evidence for anticlinal guidance of prospecting.

Many horizons in the geologic column of the field serve as reservoirs, and there are several pools that have wells producing side by side from different and widely vertically separated strata. The strata show a promising succession of porous sand and conglomerate horizons alternating with numerous gray and dark brown or black shales, admittedly the ideal section. In southwestern Pennsylvania and northern West Virginia, for the last few years, deep wells have been drilled in the hope of revealing new deep gas reservoirs. It is interesting to note that the last two wells have established deep drilling records for the world, the last, the Hope Natural Gas Co., on the Lake farm, 12 mi. east of Fairmont, W. Va., having reached a depth of 7579 ft. (2311 m.)

As a rule, in the Pennsylvania-New York field, the dips in the producing oil fields are very gentle with the exception of the Gaines pool, producing from a fissured shale horizon, which has dips ranging up to 30°. The sand bodies, though locally lenticular, are on the whole fairly persistent. Indeed, with so many sand horizons, an operator usually considers that he has a good chance in deeper drilling, even though his principal sand is poor or absent.

Pennsylvania production runs from the top of the Conemaugh to the base of the Chemung. The Murphy, Cow Run, and Dunkard sands are in the Conemaugh; the Maxon appears in the Mauch Chunk shale; and in the Pocono sandstone, below the Greenbrier limestone, are found the Big Injun, Squaw, Papoose, Butler, Berea, Gantz, Fifty-foot, and Hundred-foot sands. In the Catskill occur the Ninevah, Snee, Gordon, Fourth, Fifth, and Sixth sands; and in the upper Chemung are the Elizabeth, Warren, Speechley, Tiona, Bradford, Elk, and Kane sands.

In the latter part of 1919, a gas pool was developed south of McKeesport, Pa. where the dominant structure is the Murraysville anticline, with a northeast-southwest axis. The Foster-Brendel No. 1 had an extraordinarily high initial flow and unusually favorable marketing conditions enabled it to yield over 50,000,000 cu. ft. the first day it was controlled, which was about a week after its completion. The production is from the Speechley sand at a little below 3000 ft. Lithology here seems to play a more important part than structure. Dry holes drilled along the axis of the structure revealed a dry tightly cemented sand, which until recently condemned the territory.

The lateral limits of large production seem to be fairly well established; and, due to the small area and the close spacing of the wells, it is expected that the pool will have a short life. The rock pressure has already been lowered from an estimated pressure of 1450 lb. to 350 lb. and is declining at about $3\frac{1}{2}$ lb. a day.

The excitement over the one really large well has led to unjustifiable claims that this is the world's largest gas field; it has also led to an orgy of promotion and speculation. Except the Foster-Brendel lease, the pool will show a net loss to the producers.

GRADE OF OIL

Nearly all the oil of the field is listed as the Pennsylvania grade and is taken the world over as a criterion of high-grade crude oil. It is a light, greenish-colored oil with paraffine but no asphaltum. It varies around 44° Bé. and has a high gasoline content. It has always commanded a premium over other grades, and its present price of $5.50 will keep alive many old wells longer than seemed probable a few years ago, and also noticeably encourage new production. Little difficulty is experienced in marketing the oil and gas produced. A number of pipe lines collect the runs of the field and carry them either to one of the several refineries along the Allegheny and Ohio Rivers, or to one of the large pipe lines, such as the Tidewater and the National Transit, which run down to the huge refineries of the Atlantic seaboard.

TABLE 1.—*Natural Gas Production of Pennsylvania in* 1916–1917

Year	Volume in 1000 Cu. Ft.	Average Price in Cents per 1000 Cu. Ft.	Value
1916...................	130,483,705	18.78	$24,513,119
1917...................	133,397,206	21.53	28,716,492

TABLE 2.—*Natural Gas-gasoline Production of Pennsylvania in* 1916–1917

Year	Number of Operators	Plants		Gasoline Produced			Estimated Volume of Gas Treated, 1000 Cu. Ft.	Average Yield of Gasoline per 1000 Cu. Ft.
		Number	Daily Capacity, Gallons	Quantity Gallons	Value	Price per Gallon Cents		
1916	167	195	46,487	9,714,926	$1,726,173	17.77	38,490,621	0.252
1917	287	251	59,164	13,826,250	2,778,098	20.01	49,487,056	0.279

Great numbers of gas-gasoline plants have sprung up and are realizing handsome returns from the utilization of casing-head gas as a source of gasoline, before turning over the dry gas to be used as a fuel. The residual gas is taken up by public-utility corporations and marketed in the nearby industrial centers both for manufacturing and domestic purposes. Pennsylvania gas seldom has nitrogen in important amounts and so gives an average heating value of about 1000 B.t.u.

The increasing scarcity of gas in this field has been a source of considerable worry both to householders and to industries dependent on it as a fuel. The scarcity has resulted in the gradual increase in price to consumers and a careful redevelopment of old pools and a utilization of former leakages and wastes. The recent deep drilling was the direct outcome of this search for deeper gas to replace the gas from present reservoirs, which are, of course, gradually becoming exhausted.

COSTS AND DRILLING

The cable-tool system was developed in this field to present standards of efficiency. The ranks of drillers in Kansas and Oklahoma are composed, to a great extent, of men whose apprenticeship was served in this field. The cost of drilling has risen rapidly in this field, as in all others, though perhaps not in so great a measure because of the proximity to the iron and steel supply centers, and the comparatively greater supply of labor at hand. The Drake well was 69 ft. deep; the well that has recently established a new world's record is 7579 ft. About 2000 ft. may be taken as a fair average for the depth of present drilling. The average cost of a producing well is about $16,000, although the variation is great. The percentage of dry holes in 1918 has been estimated to be 22 per cent.

FUTURE POSSIBILITIES

In view of the long period of testing nearly all parts of the field since its inception, there is little possibility of many new pools of considerable extent or production being brought in from present producing horizons. Max W. Ball estimates the present per cent. of exhaustion of the field at 69.5 per cent. One encouraging feature of the field, however, is the remarkable evidence given by the decline curve of the Appalachian field. The longevity is good, which is to be attributed mainly to the high price which keeps the well alive for a long period after the rate of decline has naturally become slow. Here, as elsewhere, close drilling gives the usual sharp decline.

It is a remarkable fact that the land near the Drake discovery well near Titusville, which was drilled in soon after the date of that well, is still producing. The great richness of the casing-head gas permits some leases in this field to be operated when it no longer pays to pump the wells. We should get a much higher extraction. There is still the possibility of the deep reservoirs of oil and gas, which was the goal of the recent deep drilling. The disappointing results to date should not be taken too seriously, in view of the fact that these wells were for the most part not on the strongly marked domes, which should be chosen for such tests.

DISCUSSION

G. H. ASHLEY, Harrisburg, Pa.—There are two or three things regarding Pennsylvania that are of interest. Mr. Johnson spoke of the Gaines oil field which lies far east of the main oil belt. Another oil field, very small, occurs near Latrobe, well east of the main belt, and over in Somerset County there is a well that is reported to have yielded some oil. These suggest the possibility of oil over all of the gas, or eastern, side of the field. Again, in the southeast corner of this state, during the last year or two, some oil has been found in seeps, which has raised the question whether there may not be commercial oil in that section. The matter is one we are still studying. We are not quite ready to say that the oil is actually coming from its apparent source, that is, from the pre-Cambrian rocks.

Some question has arisen as to the eastward extent of the gas fields of the state. On the flank of the Chestnut Ridge anticline, there is a little bench with a gas pool, and a little gas has been found in Cambria County just east of the anticline. These facts would lead to the supposition that there might be gas on that anticline, but all efforts so far have failed to show any.[1] The only explanation we can give for the failure to find gas in that region is that the rock there is not favorable. There are other places in the state where the structure and other conditions seem to favor the presence of gas, but drilling finds none.

---

[1] Since this was written, a drilling of the Peoples Gas Co., in the center of the anticline where cut by Loyalhanna Creek, has struck 300,000 ft. of gas at 6822 ft.

Geology of Cement Oil Field

BY FREDERICK G. CLAPP, NEW YORK, N. Y.

(New York Meeting, February, 1920)

ALTHOUGH many oil fields have been, and still are being, discovered in Oklahoma, the geology and structure of most of them have not become familiar to the general public because of the delay in securing government geological surveys and the reluctance of oil companies and other interested parties to give out their "inside information" Therefore, until official surveys are available, it behooves us to publish geological results as soon as possible. Fortunately the writer has been authorized by the Cement Field Oil Co. to publish his data on the Cement field, Caddo County, at this time.

LOCALITY AND DESCRIPTION

The Cement field is situated in the part of Oklahoma known generally until recently, as "Healdton fields," and lies 60 mi. (96 km.) northwest of the Healdton field proper. Like the Healdton field, it forms an approximate ellipse, trending northwest and southeast through the village of Cement on the St. Louis-San Francisco Railroad, on which it is reached in 2½ hours from Oklahoma City.

In its geological structure, the field constitutes an anticline over 13 mi. (21 km.) long and from 1 to 3 mi. wide; the point of greatest width being not far from its intersection by the above-named railroad. The major axis trends north 75° west from the village of Cement; but eastward appears deflected (if field interpretations are correct) to about south 45° east. About 5 mi. south lies an approximately parallel syncline, which may be conveniently called the Cyril syncline, the south and west boundaries of which may be distant many miles, but forming a closed basin south of the Cement anticline. The position of the offsetting synclinal axis, which is believed to lie north and east of Cement, has not been discovered.

TOPOGRAPHY

In the southwest part of the state, nearly all maps of Oklahoma show two mountain areas—Arbuckle and Wichita—which are conspicuous geological and topographic landmarks. In addition, some maps show a third, and smaller, range, named the Keechi Hills, in the vicinity of Cement. These hills also form a conspicuous feature in the landscape; but for some reason they have been neglected by geologists, and the merest references to them have appeared in state and private reports. In particular, they have been neglected until recently by oil geologists.

So prominent are Keechi Hills that they can be seen many miles away, rising above the generally rolling agricultural surface in the form of mesa-like and conical treeless masses 100 to 400 ft. (30 to 122 m.) above the neighboring valleys. Perhaps their dwarfing by Wichita Mountains, visible in the distance, is what has prevented their being studied and tested for oil years ago; or, perhaps it is the presence of the sometimes pure gypsum rock which caps the isolated mesas, and in one place caps the main mass of Keechi Hills. The relief of the land surface in Keechi Hills is about 500 ft., ranging from about 1150 ft. above sea-level near the Little Washita River on the east, to 1630 ft. on the crest of Keechi Hills.

HISTORY OF DEVELOPMENTS IN CEMENT FIELD

The first development took place, many years ago, in the village of Cement, where a hole was sunk only a few hundred feet in depth and abandoned. About 1916 a well was started on the Funk farm in section 6, township 5 north, range 8 west, 3 mi. (4.8 km.) east of Cement, and at 1415 ft. it discovered 500,000 cu. ft. (14,000 cu. m.) of gas and a showing of oil; its total depth is 1685 ft. (513 m.) Shallow tests were also drilled years ago in township 6 north, range 9 west, 3 mi. northeast of Cement; and in section 21, township 5 north, range 8 west, 5 mi. southeast of Cement. The first real excitement, however, was caused, about 1917, by the drilling of a well by the Oklahoma Star Oil Co., on the Kunzmiller farm in the southwest quarter of section 32, township 6 north, range 9 west, 2 mi. northwest of Cement. At a depth of about 1700 ft., an unknown quantity of oil was found which flowed into the tank. The production is reported to have been 10 to 25 bbl. per day; but we have no definite information on the subject, except that it still flowed slightly when first visited by the writer in the fall of 1917. The main point of interest is that the oil was found in a comparatively shallow sand of Permian age, 800 ft. or more above the Fortuna, or next important group of sands.

In September, 1917, Fortuna Oil Co. completed a gas well at a depth of 2340 ft. and an initial production of 35,000,000 cu. ft. of gas per day, on the Thomas farm in the southwest corner of section 31, township 6 north, range 9 west, 3 mi. west of Cement and 1¼ mi. west of the Oklahoma Star well. In 1918, the first well of Prosperity Oil & Gas Co., in the southeast quarter of section 5, township 5 north, range 9 west, was drilled into the same sand at a depth of 2345 ft. and obtained a flow of oil, which has been variously estimated from 50 to 150 bbl. per day; but the well was badly handled and was thereafter continued in an effort to reach the deeper sands. The first well of Gorton Oil & Refining Co., in section 2,

township 5 north, range 9 west was completed in 1918, having an estimated capacity of 15,000,000 cu. ft. of gas per day. The second well of
the Gorton Oil & Refining Co., known as the "Betty G," was completed
later the same year in the northwest quarter of section 32, township 6
north, range 9 west; and while it has flowed oil, its production is not known,
because it has not been thoroughly cleaned out, but it was reported to be
good oil well. Fortuna No. 2 well, situated in the northwest quarter
of section 6, township 5 north, range 9 west was completed in December,
1918, with an initial production reported at 150 bbl. per day. The oil
is from the same sand as the gas in Fortuna No. 1, 1 mi. to the north. A
few days later the first well of Gladstone Oil & Refining Co. in southeast
quarter of section 31, township 6 north, range 9 west was finished, with
a reported initial production of 90 bbl. per day. Since that time three
other oil wells have been drilled along the north flank of the anticline, and
two near its center. Three wells are now being drilled or about to be
drilled to deeper sand. About sixty derricks stand in the field at the
time this paper is printed.

The well of the Cement Field Oil Co. on the site of the old Oklahoma
Star well, bought out by the aforesaid company, was only a small gas
well, as was the well of Hill Petroleum Corpn. in the southeast corner
of section 33. Fortuna No. 3, in section 35, township 6 north, range 10
west, missed the sand but is drilling deeper; while Fortuna No. 4, in
the center of section 6, township 5 north, range 9 west, only had a showing
and is likewise preparing to drill deeper. These facts indicate considerable irregularity in the group of sands. Several wells are now being
drilled.

STRATIGRAPHY

The formations at the surface appear to be entirely of Permian age,
being designated technically as the Enid, Blaine and Woodward formations. The vertical section of the outcropping beds covers a stratigraphic
range of about 300 ft. (91 m.). In this section, two members demand
principal consideration; the Whitehorse sandstone and the Cyril
gypsum bed.

CYRIL GYPSUM BED

The most persistent formation in the field is the Cyril gypsum, which
ranges from 20 to 80 ft. (6 to 24 m.) in thickness. It is believed to underlie the Whitehorse sandstone of northern Oklahoma and here overlies
a great mass of generally gray sandstones that might be supposed to be
Pennsylvanian, but which are nevertheless Permian in age. In southern
Caddo County, this gypsum bed is thought to be practically synony-
mous with the Blaine formation. There is no sign of division into three
gypsum beds as in central Oklahoma.

FORMATIONS OVERLYING THE CYRIL GYPSUM

The strata directly overlying the principal gypsum bed are classified as Whitehorse sandstone of the Woodward formation, the intermediate Dog Creek shales of the Oklahoma Geological Survey being generally absent. Quite outside the limits of the field, however, are great masses of red shale which may belong in the overlying Greer formation.

The best sections of the uppermost strata are visible south of Cyril and on the north slopes of Keechi Hills, where deeply cut ravines intersect the surface and expose the beds throughout a thickness of more than 100 ft. These are found to be mainly red sandstones and red sandy shales, regularly or irregularly stratified, that in some places north of Keechi Hills are so confused with recent dune sands as to raise the question whether they too may not have been wind-deposited.

In the vicinity of Wichita Mountains, strong winds are almost constant and sometimes fill the air with such clouds of dust and sand as to simulate a desert sand storm. In some sections these winds have piled the sand into considerable hills; for instance, over considerable areas 5 to 10 mi. southeast of Cement and also 1 to 5 mi. north of that town, forming a belt parallel with and north of Keechi Hills. In this belt these generally prevalent southwest winds have piled up the sands so that few rock exposures are now visible; and beyond the impression of a synclinal or homoclinal slope, little can be learned. Just where the Permian sands end and where the recent dune sands and sandstones begin in these areas is hard to determine in most cases.

Overlying the red sandstones are great thicknesses of red shales and shaley sandstones, such as constitute most of the Permian series of western Oklahoma.

FORMATIONS UNDERLYING THE CYRIL GYPSUM

The Enid formation ·is a name applied by the Oklahoma Geological Survey to the lowermost 1500 ft. (457 m.) of Permian red beds up to the base of the lowest heavy gypsum; therefore, in the Cement field, it includes all strata up to the base of the Cyril gypsum. So far as the Cement field proper is concerned, the Enid consists of massive gray sandstones of great hardness and persistence with overlying red sandstones; but few shales have been found. The base of the Permian series is thought to lie about 2700 ft. from the surface but some geologists place it at 1700 ft. Beyond the east end of the field, a gypsum bed, generally only about 2 ft. thick, outcrops in a few places; and gypsiferous sandstones occur in the Enid formation in many localities.

GEOLOGICAL STRUCTURE

The geological structure of the Cement field is better known and more easily determinable than that of any other known dome in the Permian series. Only at its east end is there any considerable difference of opinion as to details. It is an excellent anticline, or elongated double dome, on which the zone of closure appears to be approximately 13 mi. (21 km.) long and from 1½ to 3 mi. wide. The Cyril gypsum bed, on the basis of which the structure contour lines of the accompanying map are drawn, rises from the center of the Cyril syncline, 1 mi. south of Cyril, at an elevation of less than 1350 ft., to an eroded position of more than 1650 ft. above the east end of the main Keechi Hills, 3¼ mi. north of Cyril.

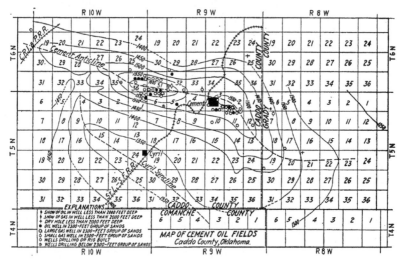

Fig. 1.—Geological structure and distribution of wells in Cement field. Contour lines are on the basis of the Cyril gypsum bed. Contour interval 50 feet.

Descending the north side of the anticline, the gypsum drops below 1400 ft. (426 m.) in a distance of 2 mi. (3 km.). West the axis plunges to about 1450 ft., 6 mi. north-northwest of the apex, and to 1300 ft. on its eastern end; the amplitude of closure is apparently about 200 feet. Although the main apex of the anticline is situated 3½ mi. west of Cement, a subsidiary dome with an apex above 1550 ft. centers 1 mi. east of that village.

In the eastern end of the anticline, the horizon of the main gypsum bed appears to descend to below 1300 ft., and northeast it drops below 1200 ft. between Ninnekah and Chickasha, and is believed to go much deeper; but the exact correlations and amount of descent are disputed by some geologists.

COMPARISON WITH OTHER FIELDS

When comparing the geological structure of the Cement field with that of other fields in southern Oklahoma, we must acknowledge that its prospects appear excellent. So far as known it appears to be more symmetrical than the Healdton field; but a buried mountain range may just as naturally exist beneath the Cement field as at Healdton. The general trend of the Cement field corresponds with that of the Healdton, Burkburnett, Fox, Two-Four, Velma, Loco, and other less thoroughly developed fields. While the general geological structure and attitude of the formations are rather similar to those in the Kilgore field, which is being developed in the extreme southeastern corner of Grady County and in the adjacent edge of Stephens County, the trend of the Kilgore field is, however, north and south, in contradistinction to that of the Cement field, which is nearly east and west.

Correlations with the Duncan, Healdton, Loco, Wheeler, and other southern Oklahoma and northern Texas fields are difficult; but we have some data, and estimate that the Fortuna—or principal producing group of sands—may be identical with the principal gas sand of the Graham-Fox field. On this basis we might expect the deep new sands of that field at about 3400 ft.; but in the only deep well in the Cement field it has not been struck at that depth.

ATTEMPTED PREDICTIONS RELATIVE TO POSITIONS OF DEEP SANDS

In connection with studies made in Carter, Stephens, and adjoining counties, the writer has had occasion to collect, compile, and plot many well logs. Since these logs, when compared carefully with those in the Cement, Kilgore, Fox, Graham, and Walters fields, give certain light on geological conditions previously unknown and since this information may be of value in the Cement field, it is given herewith.

The wells referred to as being the deepest or nearly deepest in their respective fields are Prosperity No. 1 in Cement field, Magnolia No. 1 in Walters field, a well of the Kirk Oil Co. (which produced 36,000,000 cu. ft. per day of gas) in Graham field, and Pierce No. 2 of the Oklahoma-Fox Oil Co. in the so-called Oklahoma-Fox field. These wells will be referred to here as the Prosperity, Magnolia, Graham and Oklahoma-Fox wells, respectively. Many logs of the Cement Oklahoma-Fox fields and of the Santa Fe No. 1 of the Kilgore field have also been studied in attempting correlations.

It must be acknowledged that the results are far from satisfactory, on account of the variable nature of the red-bed formations, the unconformity at the top of the Pennsylvanian series, and the personal equation in the case of records kept by different drillers with differing degrees of care. There are, however, several sandstones, limestones, and shale beds of enough persistence and definite characteristics that some sort of correlations have been arrived at which, although not positive, are definite enough to give certain ideas in the nature of predictions. The information given should be accepted in this spirit, rather than as an absolutely certain exposition of what will be found by deeper drilling.

First, it is barely possible that the Prosperity and Fortuna No. 5 wells of Cement field passed through the horizon of the Magnolia 2150-ft. sand of the Duncan field at about 2550 ft. without finding it. It is much more probable, however, that the horizon of the Magnolia sand was passed in the Prosperity well at about 2700 ft.

In the early days of the development of Cement field, we had no basis by which to predict the position of the top of the Pennsylvanian series of rocks or the position of the Healdton group of sands, because the records of wells in the Healdton field were too discordant and the unconformities of the buried "Healdton Hills" are too enormous to allow of deep-lying correlations. Now, however, we have the log of Pierce Nos. 1 and 2 of Oklahoma-Fox Oil Co. in section 7, township 2 south, range 2 west in northern Carter County. These wells have gone to a greater depth than others in the region, the producing sands being apparently about 1000 ft. below the producing sands in the Graham and Fox gas fields. One of these deep sands has produced over 100 bbl. per day of oil into the pipe line. The oil is of low grade but will presumably be lighter in fields farther from the Arbuckle Mountains. The sand is a thick one and is considered as about the stratigraphic position of the best oil sands in that part of Oklahoma. At least three possibilities exist:

1. The Fortuna sand of the Cement field may have been penetrated at about 600 ft. (182 m.) in the Graham field and missed entirely in the Oklahoma-Fox field. In this case the Magnolia sand exists at about 1000 ft. at Graham. Then the horizon of the Graham gas sand (1480 ft.) should lie at about 3200 ft. in the Cement field, and the Oklahoma-Fox sands at about 4200 ft.

2. The Fortuna sand may lie at 1000 ft. (304 m.) at Graham, and have been missed in the Oklahoma-Fox wells. In this case the Magnolia sand is the same as the Graham gas sand (1480 ft.) and probably the same as the 1540-ft. sand in the Oklahoma-Fox wells. Then the Oklahoma-Fox deep sands will be found presumably at about 3200 ft. in the Magnolia well and about 3700 ft. in the Cement field.

3. The Fortuna sand may be equivalent to the big gas sand in the Graham and Fox fields (1480 ft.) and to the 1540-ft. sand in the Oklahoma-Fox wells. In this case, the Magnolia sand was missed around 1900 ft. in the last-mentioned wells, and the horizon of the Oklahoma-Fox deep sands may be expected at about 3400 ft. in the Cement field. This depth has now been passed by Fortuna No. 3 well in shale.

Whichever hypothesis is correct, there seems no possibility of finding the Healdton or Oklahoma-Fox sands at Cement at less than 3400 ft., and they may be as deep as 4200 ft. The Oklahoma-Fox sands are considered as constituting a part of the "Healdton group," which are often referred to informally but do not correlate satisfactorily in records of wells in the Healdton field, on account of considerable unconformities existing there.

These attempted correlations open a wide field for thought and consideration. Our next starting point must be that the Magnolia sand of the Duncan field appears to be at about the top of the Pennsylvania series (bottom of the Permian "Red Beds") above which it is not usual to expect oil in large quantity. In the great fields of Wichita County, Tex., and to some extent in the Healdton field, large producers were found in sands above this horizon; but these were, and are generally, believed to be seepages from lower sands. Because the red beds are not capable of having originated oil in themselves, geologists generally agree that oil contained in them has risen from formations of the Pennsylvanian series. This is the main reason for confidence that the main sands at Cement lie below anything yet encountered.

Operators in Cement field should not allow themselves to become discouraged over the outlook; they should bear in mind the following facts:

1. Since large gas wells and excellent showings of oil have been found both at Cement and Kilgore in sands of the Permian series, which are not normally oil bearing, we may expect something better in deeper sands lying in the Pennsylvanian series.

2. It will not be necessary to go to the full 3200 or 4200 ft. throughout the Cement field, as several sands are generally present in the Permian and Pennsylvanian and some of these are productive at shallower depths.

3. It is decidedly possible that the Fortuna group of sands may be more productive than usual somewhere in the field, as is the case with Permian sands in the Healdton and Wichita County fields.

4. The finding of still shallower sands in the so-called Two-Four field in western Carter County and southeastern Stephens County proves that the sands of the Permian formations are very irregular, and some of these lenticular sands may hold oil somewhere in the Cement field.

5. It is possible, and even probable, that the Keechi Hills, which are coincident with the Cement field, overlie a buried mountain range, as the Healdton field overlies the buried "Healdton Hills." In such case the conditions at a depth generally become irregular and unusual and numerous new sands occur, expanding the field laterally in these "stray" sands.

6. Deep drilling will not be a permanent obstacle to the development of the deep sands, as wells have been drilled economically in this field by the rotary process. The cost of 4000-ft. wells in the future will be less than the cost of existing wells. New wells started in the field should be drilled with a rotary prepared to go to 4500 ft. if necessary.

The one fact that appears undoubted is that a considerable series of oil sands should be expected below the Cement, Kilgore, and other fields and some of these may be expected to produce oil at Cement.

Most of the statements here made apply also to the Kilgore field. Although correlations and predictions are only of relative value, it is believed that since the sands in these fields are deep and that both of them are surrounded by deep synclines, the chances are good for large productions as soon as the difficulties in deep drilling have been overcome.

CONCLUDING STATEMENTS

The foregoing is an exposition of conditions, developments, and probabilities in Cement field according to the information and belief of the writer. As in all fields in the course of development, it is not practicable for any person other than a resident geologist to have all the facts at hand, therefore some details may be in error. Especially is it thought some geologists may have further light on the probabilities and predicted depths of deeper sands.

Irvine Oil District, Kentucky

By Stuart St. Clair,* M. S., E. M., Chicago, Ill.

(Chicago Meeting, September, 1919)

In view of the great interest shown in the oil possibilities of Kentucky, one is impressed with the paucity of reliable literature on the oil fields of the state. A few brief reports by the Federal and State Geological Surveys are about the only reliable data available. When the estimated production figures, for 1918, are published by the U. S. Geological Survey, they will show a revival of the oil industry in the Blue Grass State during the past half decade. There will also be an increase in the production for 1919 and 1920, at least. Although as an oil-producing state Kentucky is small, compared with some of the other oil states, the present production and the area of undrilled proved territory is large enough to classify it as one of the important oil states of the Union. This paper will be confined to the Irvine District and the immediately adjoining areas which have been prospected with varying success. The Corniferous limestone or Irvine sand is the oil-producing formation in the area discussed.

In my divisional nomenclature, the Irvine District includes the Irvine field, which extends from the town of Irvine eastward toward Campton; the Big Sinking area, which joins and lies to the south of the eastern part of the Irvine field; the Beattyville area, which lies to the north and northeast of the town of that name and joins the Big Sinking area; and the Ross Creek pool, which lies to the southwest of the big production and across the Kentucky River. Except for the Ross Creek pool, the main producing area is bounded on the east by the L. & E. R. R. and on the west and south by the Irvine Branch of the L. & N. R. R. Winchester and Lexington form the gateways and Torrent on the east, Irvine on the west, and Beattyville on the south are the principle entrances to the main fields. Evelyn, on the L. & N. R. R., south of Irvine, is the point of entrance to the Ross Creek area.

Geology

The geology of the Irvine District is very simple. The rock formations with which the oil man should acquaint himself lie between the lower measures of the Pennsylvanian sandstones and shales and the Devonian or Corniferous limestone, or Irvine formation. Only a very brief description of these formations will be given, as they have been described fully by E. W. Shaw.[1]

---

[1] U. S. Geol. Survey *Bull.* 661d.

Capping the hills, and forming a rim-rock over the eastern part of the Irvine field, the Big Sinking, and Ross Creek areas, is a cliff-forming sandstone above which are yellowish-gray and dark colored shales with irregular sandstone and conglomerate members, and also some valuable coal beds.   Below are dark colored shales, with irregular thin sandstones, with a thickness up to about 50 ft. (15.24 m.).   Underlying these is the Big Lime of the driller, or Newman or St. Louis limestone.   It is typically exposed along the L. & N. R. R. from a point south of Irvine nearly to Heidelberg and in this district varies in thickness between 100 ft. (30.48 m.) and 125 ft. (38.1 m.).   The underlying Waverly formation is composed chiefly of a bluish-green shale and has an average thickness of about 450 ft. (137.16 m.).   The Waverly and the underlying Black Shale formation increase in thickness eastward and southeastward from the Irvine field.   The Berea sand is found in the lower part of the Waverly formation in the eastern part of Kentucky and in adjoining oil sections of West Virginia and Ohio.   The Devonian, or Chattanooga, black shale varies in thickness between 120 ft. (36.57 m.) and 170 ft. (51.81 m.) in the Irvine District.   The base of the formation in most places is a white shale, or fireclay as it is locally called, which varies up to 20 ft. (6.09 m.) in thickness.   In some localities, brown shale from a few feet to 25 ft. (7.6 m.) in thickness underlies this and directly overlies the Corniferous limestone, which is the Irvine sand of Kentucky. This formation is a dolomitic limestone, sandy at a few irregular strata, and contains chert in varying amounts.   Porous beds, irregular in their continuity, are the oil sands of the formation.   The outcrop at Irvine is brown in color and about 8 ft. (2.43 m.) thick, but eastward the formation thickens rapidly, attaining about 100 ft. (30.48 m.) at the eastern and southern edges of the main producing field.   In Wolfe County, near the Breathitt county line, the formation is approximately 175 ft. (53.34 m.) thick.   Underlying the Devonian are the Silurian shales and interbedded limestones and the Ordovician limestones, one of which is, perhaps, in part the equivalent of the Trenton.

 The Irvine District is affected by the Cincinnati geanticline, which extends in a general north-and-south direction, and the Chestnut Ridge uplift, the axis of which crosses Kentucky in an east-and-west direction. All beds dip away from the former structure, thereby making the general dip in the Irvine District southeast.   The Chestnut Ridge uplift has been described, by Gardner, as extending from Pennsylvania through West Virginia, Kentucky, and southern Illinois.   This is a general disturbance which has been of paramount importance in the formation of some of the principal oil structures in this district.   The northern boundary of the Irvine field is marked by the Irvine fault, which is part of the Chestnut Ridge disturbance.   The general position of this faulting is shown by Shaw.   Most of the anticlines of the district closely parallel the direction of this faulting. An eastward extension (including the Ashley pool, the production near Zachariah, and to the east of Torrent) is similar structurally to the Irvine field. The Campton extension was reported on by Munn.[2]

From his observations, the writer is led to believe that there are anticlinal structures of two ages in this part of Kentucky and very probably in other parts also. The main structures apparently are the younger, have a general east-and-west trend, and were formed by the Chestnut Ridge uplift, which was probably completed in the Tertiary period. The general trend of the older structures is north-and-south, but the writer hesitates to venture a statement regarding their age as he has not had the opportunity of studying all the facts that may bear upon this problem. Some evidently antedate the last period of general base-levelling of the region and are probably associated with the Cincinnati uplift. These conclusions have been reached from a study of the relation between present topographic features and the structure. The localization of some of the large producing areas may be due, in part, to the intersection of the younger with the older structures.

Occurrence of Oil

Oil men who have had experience in a field where the oil is found in a limestone will appreciate the eccentricities of sand condition encountered in development work in Kentucky. In localities where the limestone is hard and tightly cemented, the whole formation may be barren. Tight sand conditions may be encountered in the center of an area with producing wells on all sides and but a few hundred feet distant. Such a condition is most unfortunate when encountered in a prospect well in undeveloped territory, as the operator may prematurely abandon the area. Drilling has shown parts of the Corniferous limestone to be generally porous over the area described here. Usually a porous bed of variable thickness, from a few feet to 10 ft. (3.04 m.) or more, occurs directly under a hard limestone cap-rock. Under favorable structural conditions, this porous bed may be entirely filled with oil, and under less favorable structure part or all of it may contain salt water. Over part of the Irvine District only one pay sand is found but in some localities two or more with a little, though in many wells without any, intervening salt water. In some areas abundant salt water may be encountered under the cap-rock, where the first pay sand should be found, and must be cased off before lower pay sands can be drilled into. In fact, each locality presents varying conditions, so an intimate knowledge of the field should be acquired in order that the operator may know how to handle the problems confronting him in the development of his property.

[2] M. J. Munn: U. S. Geol Survey *Bull.* 471a.

In general, the oil accumulation has taken place under anticlinal conditions. In the producing areas, all folds and well-defined terrace structures have been productive and salt-water wells have been the result of drilling in the less favorable places. Exception may be taken to this statement by some geologists when applied to the Big Sinking area, but when other factors governing oil accumulation in this pool, which factors will be mentioned later, are considered in association with the anticlinal theory, this general statement will be found to cover the case.

Only a generalized description of the Big Sinking area will be given here, as a complete report on the oil pools of Lee County is being contemplated. From certain deductions that will be given regarding this field, the reader may appreciate the writer's reasons for having strongly recommended the Big Sinking Creek area as early as the spring of 1917, when the nearest drilling was some miles distant. At that time the writer outlined on the map of a prominent Kentucky oil producer the probable western boundary of the Sinking Creek area as being the divide between Little Sinking Creek and Billy's Fork of Millers Creek, and the probable eastern limit as the divide between Hell Creek and Walkers Creek. So far, nothing of importance has been developed beyond these bounds close enough to be classified as part of the Sinking Creek area.

From what has been said regarding the anticlinal occurrence of oil in the Irvine District, the reader has probably formed the opinion that drilling on such structure outside of the present proved area should also result in favorable strikes. However, experience has shown the opposite results. Wells located near Beattyville, farther south in Owsley County, in southern Lee County, in eastern Lee County north of Tallega, in Breathitt, Wolfe, and Elliott counties, all located on well-defined anticlinal structures, have struck only shows of oil. In the wells closest to the producing fields, salt water was encountered; in those farther away, the Corniferous limestone was hard and, in most cases, practically dry.

BIG SINKING POOL

The structure of the Big Sinking area is not complicated. The regional monoclinal dip, which is southeast, is crossed by several very low folds, the axes of which are in a general east-and-west direction. The resultant would be plunging anticlines with a general southeast-by-east trend. Minor irregularities have produced terraces. In conjunction with this type of structure, a broad low fold extends in a general north-and-south direction and definitely outlines the western and eastern limits of the Big Sinking pool. The axis of this fold roughly follows Sinking Creek from Bald Rock Fork northward and the more productive part of the pool is along the crest and on the southeast flank of the fold with the most productive points determined by the east-and-west folds.

The number of pay sands and their total thickness is variable in the Big Sinking pool. As many as three sands, with a thickness of 40 ft. (12.19 m.) have been reported but in some places 5 ft. (1.5 m.) will cover all the actual pay sand encountered. Probably 15 to 20 ft. (4.57 to 6.09 m.) is an average for the more productive parts of the pool. In general, the pay sand is very porous, although in short distances it may tighten up materially. The best sand condition for quick recovery apparently lies along the crest and the southeast flank of the north-and-south fold. Westward from this fold, the sand changes rapidly. Eastward, the change is more gradual, but with the increasing depth, due to the regional dip, the sand becomes tighter and the pay is not so thick nor uniform. However, wells in the tighter sand, although the production is smaller per day, will be longer lived than the wells in the porous sand. The writer has estimated that from the very porous sand as much as 1000 bbl., and in a few selected spots 1200 bbl., to the acre-foot of actual pay sand will be produced. This amount will decrease to 500 bbl. and probably as low as 200 bbl. to the acre-foot as the tighter sand areas are approached.

The thickness of pay sand reported by various owners and lease men is often misleading or in error. In most cases, the thickness of the true pay sand is much less than it is thought to be when the well is being drilled. Therefore, in making computations, the thickness of actual pay sand should be carefully determined or the error in the calculated production per acre will be large. This rough method should be used only when no production data on the property or adjoining properties are available from which decline and future-production curves can be constructed.

The writer has been frequently asked why the oil is found, throughout the Big Sinking area, on the higher and in the lower structural positions. It is quite evident that this condition is due to water pressure, chiefly from the south and southeast, which is behind a sufficient body of oil to cause the sand to be filled with oil over the entire area, thereby forming one large pool in which nearly every location is proved. Salt water will encroach upon the field from the south and the wells in that part of the field and in the lower structural positions will be affected first.

EXTENSION OF EASTERN FIELDS

The writer's personal experience in the Kentucky oil fields has led to some deductions that may throw some light on the conditions described, and which may be of value in prospecting the Corniferous limestone or

Irvine sand in the eastern part of the state. The area under which the Irvine sand will be productive of oil in commercial quantities is dependent on three primary conditions, which are listed in the order of their importance: (1) Distance from the outcrop of the Irvine sand and from the major faults. (2) Porous or non-porous character of the Corniferous limestone. (3) Geologic structure. The writer will probably be criticized for making the second condition less important than the first, but in this field the porous or non-porous character of the oil formation is dependent almost entirely on the distance from the outcrop, and especially from the major faults along which meteoric water is able to reach the Corniferous limestone. It will be remembered that such a system of faulting extends along the northern boundary of the Irvine field and on the eastward occupying a similar relation to the Campton and Cannel City fields. Water circulating through certain beds of the Corniferous limestone has dissolved some of the mineral matter and left the rock porous. The irregularity in the thicknesses of these porous beds indicates that solution has taken place under non-uniform conditions of underground circulation. The distance to which this circulation of meteoric water has taken place from the outcrop or fault will mark the area of continuous porous formation, which may contain either oil or salt water according to structural conditions, and will mark the limit of the area that has been unaffected by water circulation and in which the Corniferous limestone will generally be tight and hard. The water in the Irvine sand, being heavier than the oil, will occupy a position farthest down the regional dip and will be dammed back by increasingly less porous beds. This would form what we may term a monoclinal trough produced by differential cementation of the limestone. Obviously then, a well drilled on the crest of an anticline near the lower limit of this water area would encounter water and not oil in the Irvine sand. The position of the oil on the regional monocline would depend on the amount of water in the trough. As stated before, it is the writer's belief that this pressure of salt water to the south of the Big Sinking area, together with very porous rock conditions underlying the Big Sinking area, is the explanation for the unusually large concentration of oil in this field and may explain the presence of oil-saturated sand strata in both higher and lower structural positions in the Big Sinking pool. The pressure that causes flowing wells is water pressure and gas pressure combined. Some water has been encountered at the southern edge of this pool although none has yet been found in the main part of the pool.

Outside of the large area that has been affected by circulating waters, there may be small areas where the Corniferous limestone has sufficient porosity, either primary or through recrystallization or dolomitization, or where minor faults have allowed some meteoric water to reach the Corniferous limestone and dissolve some of it, thus producing porosity in

the rock, to allow small accumulations of oil. Chance drilling may
strike small pools of this class, which may be found on anticlines or in
lower structural positions. Tests, however, should be made on the most
favorable structures that can be found. No wells in this category of any
importance have yet been struck unless the Little Frozen Creek production is extended beyond its present limits.

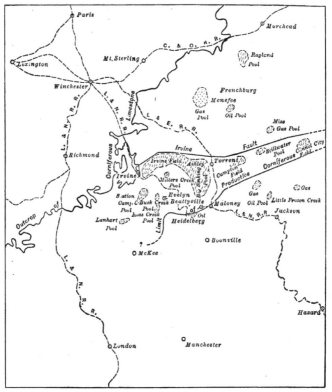

FIG. 1 —OIL POOLS PRODUCING FROM THE IRVINE SAND.

Roughly, the probable line of separation of the possible productive
area and the non-productive area for the Corniferous limestone may be
drawn. Inside this line proper structural conditions are necessary for
commercial accumulations of oil. The line should run along the east
side of Ross Creek near the Lee County line to Kentucky River; then
eastward through Heidelberg, Beattyville, and Maloney; then northeastward, probably passing close to Holly Creek, Wolfe County; then
paralleling and closely following the Campton-Cannel City anticline.
This line is shown in Fig. 1 as the limit of the area of productive Corni-

ferous limestone. The approximate position of the Irvine fault and the outcrop of the Corniferous or Irvine sand are also shown together with the oil and gas pools producing from the Irvine sand.

If the theory that the distance from the outcrop is of primary importance in determining whether the Corniferous limestone will be porous enough for oil to accumulate in is correct, in areas where this formation is present and of similar character to what it is in the Irvine District, and if the overlying petroliferous black shale is present, under favorable structural conditions there should be some accumulation of oil or gas; provided the structure is far enough away from a fault of any magnitude, which is on the down-dip side of the structure, to allow a sufficient area from which accumulation may come. Reference to Fig. 1 will show that the Menefee gas field, the Ragland oil field, and the Station Camp, Ross Creek, Buck Creek, and Lanhart oil pools are not far from the outcrop of the Irvine sand. These fields are located on structures and, therefore, other small fields may be opened, under favorable structural conditions, within a restricted distance from the outcrop of the Irvine sand.[3]

ECONOMIC CONDITIONS

For the past two years, Kentucky has enjoyed a great oil boom and prices for acreage have gone sky-rocketing in certain parts of the state, following each strike of importance. Probably no area has had such a quick advance in speculative prices in such a short time as the Big Sinking area in Lee County. Two years ago, the Irvine field extending northeastward from the town of Irvine to Pilot was receiving nearly all the attention of the oil men. Acreage along Big Sinking Creek was selling for a very small bonus. With the opening of the Ashley pool, Pilot district, the price of acreage began to advance and by the fall of 1917, acreage in the Big Sinking area was selling from $50 to $100 per acre. During the year 1918, prices advanced to $2000 and $3000 per acre, with extra royalties attached, in some cases. On such high-priced acreage, flowing wells are sometimes drilled, but the average initial production will probably be between 100 and 300 bbl. per day. The writer knows of a few wells that produced close to 1000 bbl. the first day and another that flowed 24,000 bbl. in a little over four months. Although these are the exceptions, there are many wells that are producing far above the average.

The practice of additional royalty was started when there was no valid reason for such action. It has kept many strong, conservative companies out of the field and has brought in many operators who had

---

[3] Since this article was written, several oil wells have been drilled in southeastern Menefee County on a well-defined structure.

the oil business to learn and some stock companies who had to get production.

The Big Sinking area is a remarkable oil pool. The depths of the wells vary from 800 to 1200 ft. (243.8 to 365.7 m.), dependent on the topography. It has two, and in some places three, pay sands with an aggregate thickness of 10 to 30 ft. (3.04 to 9.14 m.). Much of the pay sand is very porous and the oil, in parts of the field, is under considerable pressure. The gravity of the oil is around 40° Baumé and the gasoline content above 30 per cent.[4] The cost of drilling and equipping a well varies from $3000 to $5000. These advantages, however, do not warrant the payment of such high prices for acreage and additional royalty when the average size of the initial production is considered. On account of the porosity of the pay sands, the production will decline rapidly and the average life of the wells in the Big Sinking field will not be long.

The Big Sinking pool is so young and in the older Irvine pool such poor records of production were kept that it is difficult to get many accurate figures from which depletion can be computed. Some leases that were located on favorable geologic structure and had average sand conditions were producing about 10 per cent. as much oil at the end of the first year as they were producing at the point of maximum flush. Others have declined considerably more where the wells were put too close together. However, a few of the properties in the older Irvine field have held up remarkably well, a fact that may be due to a thicker pay and tighter sand, or inability of the pipe line to take all the oil. The oil sand in the best parts of the Big Sinking pool is so porous that wells should not be too close to one another. One well to five acres should be sufficient; but on many leases the wells are 300 ft. (91.4 m.) apart and not infrequently there is one well to the acre. Under such conditions the wells must decline rapidly, unless there is an unusually thick pay sand.

The writer has estimated that on properties where wells are properly spaced and where there is an average thickness of pay sand, a well that is continuously pumped would be producing about 10 per cent. of its initial capacity at the end of a year. Where the pipe line does not take all the oil produced, or where protracted shut-downs are experienced, these estimates must be made proportional to the lengths of the unpumped periods. Further, the average well, which will have an average thickness of pay sand, in the very porous sand areas of the Big Sinking pool, that is pumped regularly to its capacity will probably produce as much in the first six months as during the remainder of its life. In tighter sand areas it will probably take a year to produce half the oil under regular pumping conditions, and to get the maximum recovery the wells should be spaced

---

[4] The writer has been informed that some tests have shown a gravity of 42° Bé. and a gasoline content of 45 per cent.

closer than in the porous-sand areas. What has been said of the individual well may be applied collectively to a lease. Where new wells are being drilled and production is being added all the time, the decline in the producing wells is not as noticeable as when the property is fully developed. Production and decline records of wells and leases should be carefully kept so that the owner may profit by the depletion allowance for which he can claim exemption from taxation.

A company purchased a small lease that was producing and only partly drilled up, paying at a specified rate per barrel of production that amounted to approximately $30,000 per acre. With a very porous sand and 20 ft. (6.09 m.) of pay sand, probably 15,000 to 20,000 bbl. per acre should be recovered, if the pay is as thick as claimed. At the end of about five months of regular pumping 7000 bbl. per acre was removed, which is between one-half and one-third the calculated recoverable oil. At the end of two or three years, the property will be practically exhausted. A small profit will be shown on the investment if depletion is charged off each year at a rate commensurate with the decline of the property. The importance of correct estimates of depletion of properties in the Big Sinking pool and in other pools of similar character in Kentucky cannot be too forcibly impressed upon the operators.

The producing area north and northeast of Beattyville is but a southern extension of the Big Sinking field and the remarks made are applicable to it. Drilling is a little deeper and the wells are smaller, but there is room for expansion of area. Care, however, should be taken in drilling to avoid the salt water. The Ross Creek pool is the Big Sinking on a small scale. It has been greatly overdrilled and must, necessarily, have a short life. The area is rather limited but there are possibilities for a small extension southward. Prospects for a few smaller areas of production in this general region are good.[5] The Irvine field has been quite fully exploited and only drilling up of proved territory remains. This field has now been extended eastward several miles beyond Torrent, the eastern entrance to the producing areas. Efforts to carry production north of the fault in this field have failed.

Kentucky was a poor-man's field during the early stages of its recent oil development and many have made comfortable fortunes. To get acreage anywhere near the proved fields today requires capital. Wildcat acreage can be gotten cheap but the chances of success are commensurately lessened. There are opportunities for consolidation today that did not exist a short time ago. Many local organizations could be handled much more economically and efficiently if they were under larger

---

[5] Since this article was written, in the fall of 1918, a pool has been opened on Buck Creek, Estill Co., about 4 mi. (6.4 km.) north of Ross Creek. The area is one of the prospects referred to and promises to furnish a number of small producing wells.

managements. Conservation is a pertinent question; mismanagement and waste can be seen in many places and must cause injury to the field. Considerable gas, which should be rich in gasoline, is allowed to go to waste by the thousands of cubic feet per day. However, the field is young and such conditions are remedied with time.

Conclusion

The object of this paper is to call attention to the probability that the area of production from the Irvine sand in eastern Kentucky is a function of the distance from the outcrop of the oil formation and from the major faults. The results of many wells drilled to the south, southeast, and east of the producing Irvine sand pools uphold this theory. However, the writer does not wish to discourage prospecting in the deeper areas as there are possibilities of opening up small, isolated pools. Prospecting should also be carried on with a view to opening up production in sands higher than the Corniferous limestone. Southeast and east of the Irvine District, the Berea, the Big Injun, and sands higher than the Big Lime may be found to contain oil in commercial quantities. In Knox, Floyd, Magoffin, and Lawrence counties, there has been small production from these higher sands for many years. A number of wells have been drilled below the Corniferous limestone in the Irvine District, but no oil formations were found. Sands are reported but oil shales apparently are absent. The possibility of finding a deeper pay sand is not very promising.

Genetic Problems Affecting Search for New Oil Regions

By David White,* Washington, D. C.

(New York Meeting, February, 1920)

In these days, when detailed investigations of stratigraphy, structure, and sand conditions so frequently result in the discovery of new oil fields, and applause from oil companies and the public, geologists do well to walk humbly, and punctiliously to admit that the geologic principles controlling the distribution of oil and gas have as yet been discovered only in part, and that what remains yet to be learned is probably vastly more than what is already known.   The few experiments already attempted have been fragmentary, and somewhat desultory, and however positive each of us may be with respect to certain theoretical conclusions, many of the fundamental questions as to the origin and mode of occurrence of petroleum are subject to radical disagreement.   Of the chemical changes attending the generation of petroleum from organic matter, little is actually known.   Most of the postulated formulas are liable to be misleading, through ignorance of essential factors.   Open-mindedness is therefore a prime essential at the present stage of our science. Nevertheless, adopting the hypothesis that oil originates in some manner fundamentally connected with the organic theory, and in possible departure from such open-mindedness, the writer will pay no attention to the so-called inorganic theory, since every attempt to apply this theory to the study of old oil fields, or to the discovery of new ones, affords cumulative evidence of its inadequacy.

In this paper, some of the factors affecting the occurrence of petroleum that the writer believes worthy of consideration by the prospector for oil in any new region will be discussed.   Some of these, which are less generally understood, will be considered somewhat in detail.   Other points, the significance of which cannot now be determined, require more field study, and for that reason are here brought to the attention of the field geologist.   On the other hand, certain theoretical points which do not bear especially on the oil possibilities of a new region will be given little or no attention.   The main topics that will be discussed are: (1) sufficiency of carbonaceous detritus and residues in the oil-forming rocks; (2) stage of carbonization of the organic matter in the oil-bearing formations; (3) folding of the strata; (4) thickness of sedimentary formations; (5) conditions of deposition.

---

* Chief Geologist, U. S. Geological Survey.

Sufficiency of Carbonaceous Debris and Residues in the Oil-forming Rocks

Most oil and gas geologists agree that in those formations in which oil is found there must be sufficient organic matter genetically to account as mother substance for the oil, which is believed to have escaped from its mother rock into some suitable and accessible reservoir rock where it is confined beneath impervious strata. However, very little seems to be known as to the requisite quantity of mother substance or as to the maximum distance at which this substance may be situated from the reservoir.

Most geologists assume that this mother substance is carbonaceous, but others hold that recognizable carbonaceous debris or visible residues are not necessarily present. "Bituminous" or other carbonaceous shales and limestones are almost invariably searched for because, seemingly with good reason, such deposits are regarded as the principal materials from which petroleum may be generated; certainly they are the rocks from which oils nearest to typical petroleum may be artificially produced by distillation. As shown by Orton and others, similar carbonaceous matter adequate for supplying oil and gas may be found in most regions disseminated through the rock or concentrated in certain layers; it is present in ample amounts even in less distinctly carbonaceous shales and limestones, and in some sandstones, and there seems no room for doubt that oil in commercial amounts has been derived from such deposits. Most dark limestones, sandstones, and shales, as well as ordinary black shales, owe their dark tones to the presence of carbonaceous residues, which are easily recognized under the microscope. Yet it remains to be seen how much of such organic matter is requisite, as a minimum— probably, in reality, an average minimum. Circumstantial evidence— the conditions actually presented in certain oil fields—seems to indicate that the carbonaceous matter need compose but a very small percentage of a supposed mother formation, if the matter is of the right sort, and if other requisite conditions are fulfilled, and that a very great thickness of the mother formation is not indispensable. In general, however, our most productive oil deposits are found in districts containing formations in which there is evidence of abundant life, with ample vegetal matter. That only smaller productions are found in districts containing little carbonaceous matter may prove to be a rule with numerous exceptions.

In the search for oil in regions containing thick series of strata so barren of carbonaceous matter as the "Red Beds" of New Mexico, Arizona, and the northern Rocky Mountain States, or as the Jurassic of Utah, southwestern Colorado, northern Arizona, and northwestern New Mexico, or as the Newark formation of the Connecticut Valley and Pennsylvania, the question as to the quantity of organic matter appears, at the present moment, to be somewhat insistent.   As bearing in a practical way on this problem, the demonstrated occurrence of oil in the Conemaugh of the Appalachian Basin, in the Embar and associated beds of Wyoming, in the "Red Beds" of north Texas and Oklahoma, and in small amounts in red beds near Roswell, New Mex., may be cited.   It must be admitted, however, that the Conemaugh carries thin coals and carbonaceous shales; that the Permian reds of Oklahoma and Texas contain rare beds of coal and carbonaceous shales, usually of limited horizontal extent; and that disseminated carbonaceous matter, in aggregate amounts larger than at first thought, may be present in portions of the Embar and in intercalated shales or sandstones in the "Red Beds" of New Mexico. On the other hand, it is a question whether, in at least some of these cases, the oil has not migrated upward from more carbonaceous beds in relatively remote, underlying formations; or even whether the oil has not migrated downward.   The presence of other oil sands lying in more richly carbonaceous formations, at different and sometimes great depths beneath the "Red Beds" sands in the Appalachian Basin and in the Mid-Continent field, lends weight to the supposition that in some cases the oil has migrated upward instead of originating in the "Red Beds" themselves. If the oil in the latter regions has ascended into the "Red Beds," deeper sands should be tested in the possibly less forbidding shales beneath the Embar of Wyoming and the Abo of New Mexico, and beds to the base of the Percha will be explored in southern New Mexico, if the Percha is present and not too greatly altered.

A. W. McCoy, who has had most excellent opportunities for studying the composition of the beds penetrated by the drill in the Mid-Continent field, points out[1] the presence of ample carbonaceous material, including oil shale, intimately associated with the Bartlesville sand in northern Oklahoma, and suggests that closer inspection will reveal the presence of sufficient mother substance in close proximity to the oil sands in other regions.   The discovery, somewhere, of oil in a series of distinctly non-carbonaceous "Red Beds," directly underlain by metamorphic rocks or igneous masses, with no possible source in nearby unaltered sediments, would have an important bearing on this problem, and should be recorded; drilling under such conditions, however, will probably be done with great hesitation.   The argument that oil in the above-mentioned "Red Beds" has migrated downward suggests the inquiry whether the associated gas also gravitated.   Certainly, if oil has not been generated in beds which, on casual view, appear to contain very little organic matter, the petroleum in some of our sands must have migrated across many hundred feet of strata before finding hospitable storage in its present reservoirs.

---

[1] *Jnl. Geol.* (1919) **27**, 252.

The term "organic matter" should be restricted to carbonaceous debris and residues, as distinguished from non-carbonaceous mineral deposits of organic origin, such as shells, diatoms, etc., which may now be devoid of any associated hydrocarbons. Such mineral deposits do not, I believe, serve as mother substance of oil, although, when porous, they may offer excellent storage. In many, perhaps most cases, however, limestones contain some matter that is strictly organic and may have been mother substance. Impure, especially argillaceous, bituminous limestones should well serve the purpose of mother rock, and have undoubtedly done so.

The question as to whether oil may not have been generated in the biochemical stage at the time of the decay and deposition of the organisms, such as mollusca or diatoms, as believed by Stuart and many other oil geologists, is a debatable point germane to this subject, but will be considered in connection with the influence of diastrophic movements. The discovery of oil pools in a great thickness of strata actually barren of carbonaceous or so-called bituminous matter, but containing limestones largely of "organic" origin, and underlain by metamorphic or igneous basements, would give force to this theory.

STAGE OF CARBONIZATION OF THE ORGANIC MATTER IN THE OIL-BEARING FORMATIONS

A study of the incipient regional metamorphism of carbonaceous deposits in the coal and oil fields of the United States and other countries shows that no commercially important oil fields have yet been discovered in any area where the fuel ratios of the coals, occurring in the formations in which oil is sought or in overlying formations, exceeds 2.3. The progressive devolatilization by which the coals in any region or formation have been transformed from peats to lignites, bituminous coals, etc., and finally to graphite, is the first indication of incipient metamorphism[2] of the rocks of the area. The proximate analysis of the coal or coaly deposits, as the writer has shown,[3] is a sort of "ultra-violet" method of observing this initial stage of regional metamorphism of the ordinary type. Other attending criteria include the stages of dehydration, consolidation or lithification, development of jointing and cleavage, and, in due time, schistosity and mineralization.

More observations and tests are necessary to fix more exactly the stage of regional alteration beyond which commercial oil pools, though

---

[2] In this transformation the mass of organic debris (coal or coaly matter) is altered both in chemical composition and physical characters. In other words, technically, it is genuine metamorphism.

[3] *Jnl.* Wash. Acad. Sci. (1915) **6**; *Bull.* Geol. Soc. Amer. (1917) **28**, 727-734.

formerly present, will not have survived, but it is probable that the limit falls, in general, slightly lower than the point at which coals of the ordinary bituminous type show a fuel ratio of 2.2, or 68 per cent. of fixed carbon in the pure coal; it may approach nearer the ratio of 2.0, or 66 per cent. fixed carbon. Coals verging toward the sapropolic type, such as are believed by many to approach more closely the typical mother substance of oil, are more fatty and accordingly richer in hydrogen and lower in fixed carbon (pure coal basis) than the other types, until, in the course of alteration by geologic processes, they approach the above limit, when the volatile matter seems to disappear rapidly. At the semi-bituminous stage (fuel ratio 3.0, fixed carbon 75 per cent.), their carbonization is approximately on a parity with typical bituminous coal.

It is important that, in a new region under consideration as to oil possibilities, every precaution be taken to ascertain whether the alteration of the rocks, as indicated by the stage of carbonization of the carbonaceous deposits, has not gone so far as to preclude the survival of oil in commercial amounts. As I have shown in the papers already cited, drilling in regions of greater metamorphism will find only gas or mere showings of "white oil"—approximately kerosene—generally little more than samples, and nowhere in commercial amount. This principle appears to be proved by thousands of tests in the Appalachian field, in the Mid-Continent region, and in other parts of the world.

Oil in commercial amounts should not be expected in the Devonian of east-central and southeastern New York and eastern Pennsylvania; in the Paleozoic regions of Georgia; in the Arkansas coal field;[4] nor in those areas of northeastern Kentucky, of eastern Tennessee, of Alabama, of the Paleozoic region in southeastern Oklahoma, and portions of New Mexico, Colorado, Montana, Utah, and Washington, as well as of Pennsylvania, Maryland, Virginia and West Virginia, where the regional carbonization has passed the stated limit. The Utica, Genessee, Hudson, Ohio, Chattanooga, and Woodford shales are splendid depositories of mother substance, but it is futile to search for oil in the associated "sands" in regions where the organic matter of these shales is too far altered.

It is unfortunate that so little attention has been given to this factor of control of the distribution of oil, and so little systematic effort has been made to secure such evidence as might have been gained. Data are needed, for example, as to the carbonization of the organic matter in the

----

[4] Over 300 holes have been drilled in the Arkansas coal field with but a showing of "white oil" in a single instance, although, as in Pennsylvania, West Virginia and other areas, gas may be present in commercial amounts in anticlines, where the carbonization has progressed too far for the survival of oil pools. An asphaltic dyke at Mena, in the altered region of Arkansas, has been anthracitized.

Percha (Devonian) shale and the Magdalena limestone and Sandia formations in portions of New Mexico,[5] for the information concerns not only the probability of finding oil pools in or adjacent to these formations, but also the problem as to the source of oil that may be found in the overlying Red Beds. It is known that the coals of the Mesa Verde, in portions of the Trinidad, Crested Buttes, and Durango coal fields, approach, if they do not pass, the fuel-ratio limit, but the boundaries of the areas in which oil should not be expected in this or the underlying formations have not been determined for lack of sufficient and properly distributed coal analyses. The high probability that the abundant organic debris in formations like the Mancos and Graneros, and the still older formations beneath the Mesa Verde, have been still more altered must not be overlooked in any search for oil deep below the coals in these regions of relatively high carbonization.

Also, in the lower Saline River Valley, in southeastern Illinois, where the carbonization advancing toward the south approaches the oil limit, some uncertainty will attend exploration for oil in anticlines of the Mississippian, Devonian, and Trenton, which furnish oil in other parts of the state. The degree of carbonization of organic deposits in the exposed beds, and the probably greater alteration of the underlying beds, deserve further consideration, wherever the data are procurable, in the regions of some of the anticlines located in the direction (southwest) of advancing carbonization in Montana; and it should not be ignored in the vicinities of the coal fields near Sunnyside, Utah, and in Washington and Oregon.

Disregarding contact metamorphism, which from the present standpoint is unimportant, it is probable that regional alteration in much of the Newark formation of the Atlantic States has advanced too far to encourage the driller, even where the series has great thickness, contains ample carbonaceous matter, and is not too closely folded. If found to be not too far altered, it should, where sufficiently thick, be reviewed by the oil geologist. Reliable information, if it can practicably be obtained, is to be desired, as to the stage of alteration of the Upper Paleozoic in portions of Montana, Utah, and Arizona, though it is possible that in some areas inferences based only on cleavage, induration, incipient schistosity, and mineralization (not contact alteration), can be drawn. In many instances, valuable deductions may be based on distillation tests of oil shales or other richly bituminous shales which, if far devolatilized, will yield little oil, though containing much carbon.

In passing, it should be noted that: (a) local, slight variations of carbonization are not to be ignored, for they are to be expected, especially

---

[5] The coal of the lower Pennsylvanian appears to be too far altered in the Pecos Valley, about 10 miles above New Pecos, and samples from Bernalillo County cast suspicion on the same formation in that county.

in closely folded and faulted areas; (b) in general, carbonization advances downward, according to the law of Hilt,[6] so that the fuel ratios of coals in underlying formations will, in most cases, be higher than in the exposed formations, thus offering no hope of getting oil at greater depths where the regional alteration of the exposed rocks is too great; (c) the carbonization rule applies only to areas in which the alteration is regional, not contact metamorphism; (d) the fuel ratios are typically based on coals or coaly deposits of the so-called bituminous group, and may be satisfactorily determined in coaly streaks, in very earthy and bony coals, and in shales containing great amounts of organic matter, though it is not yet proved that they can be determined in shales carrying but small percentages of carbonaceous matter. Attempts to determine the percentage of fixed carbon in the organic matter of ordinary carbonaceous shales have not yet been wholly successful, but experiments are now in progress with the object of learning the minimum of carbonaceous matter that may be reliably subjected to proximate analysis in the average carbonaceous shale. If methods can be devised for successfully ascertaining the fuel ratio in the organic matter of even moderately carbonaceous shales, criteria of the greatest value will be available to the driller.

As bearing upon the grade of oil that may be expected in a new region, attention may again be called to the observation that, in general, the oils in regions of relatively high, but not too high, carbonization are characteristically of the highest grade, that is, of low gravity; while in regions of less carbonization the oils average higher in gravity. Going still further, as the writer has elsewhere pointed out, the oils found in regions of low-rank coals, such as lignites (brown coals), are also characteristically, though not without exception, lowest in rank, notwithstanding the lack of satisfactory explanation of the fact, on what may at the present moment be considered a reasonable chemical basis. The true explanation may come from the thorough application of experimental physics and physical chemistry to the oil problem.

The causes of carbonization (alteration) of the organic debris and residues in sedimentary formations have been more fully discussed in my previous papers, but will be briefly reviewed in the following section.

FOLDING OF STRATA

Folding of the strata, or the development of structure, is almost universally regarded as an essential feature of any oil region. The migration and "gravitational" segregation of oil, gas, and water are commonly supposed to be connected with the existence, if not indeed with the origin, of folds, and in particular with minor local

---

[6] U. S. Bureau of Mines *Bull.* 38 (1913), 125.

anticlines and domes. Therefore, folding is always looked for and analyzed in detail.

However, to what extent and through what processes folding operates as a cause, or a means, or, on the other hand, whether it is to be regarded only as an effect or a mere indication, is yet to be shown. Most of us hold that folds facilitate the segregation and localize the distribution of oil and gas pools,[7] and are therefore of great consequence in the search for new oil fields; contrasted with this view, folding seems to be regarded by some geologists mainly as an effect of questionable importance.

One of the most thoughtful advocates of the operation of folding as a cause of the migration of oil and gas is Marcel R. Daly,[8] who starts with the assumption that the oil already exists, presumably from the date of deposition of the terrane, disseminated in the clays, sands, etc. in the form of minute spherical globules between the mineral particles. Under increasing loading by deposition of superincumbent strata, the argillaceous and organic deposits are compressed and the water, oil, and gas are gradually squeezed out of the compacting deposits, moving in the direction of least resistance into the less compressible sandy beds and sandstones. Coalescence of the globules and concentration of the liquids proceed en route. In the sandstone, separation of the water, oil, and gas tend to go forward according to the size of the pore spaces, the water, with its greater capillary tension, tending to occupy the fine-grained portions and forcing the oil and gas into the larger voids. Horizontal stresses of diastrophism, causing new and greater differential compression of the beds, produce waves of unequal compression and, overcoming friction, drive the water and hydrocarbons into the zones of less pressure, some of which are the forerunners of anticlines as buckling proceeds, the tops of the anticlines offering zones of least compression, while the bottoms of the synclines are most squeezed.

The important point of Daly's presentation is the function of loading and thrust pressure in causing the escape of the water and oil from its matrix into the sands, and in overcoming capillary resistance to further migration into reservoirs and anticlines. It is hoped that this paper will bring partial support to some of Mr. Daly's conclusions.

In previous discussions[9] of the features common to the genesis of coal and of oil, the writer has insistently pointed out that the evolution of each is brought about through the common agency of dynamic forces—mainly horizontal stresses—acting on loaded strata and causing the progressive

---

[7] Preliminary compilations by K. C. Heald indicate that over 88 per cent. of the anticlines and domes in the Osage Nation are oil-bearing, as compared with about 15 per cent. of the synclines.

[8] *Trans.* (1916) **56**, 733–753.

[9] *Jnl.* Wash. Acad. Sci. (1915) **6**; U. S. Bureau of Mines *Bull.* 38 (1913) 91; *Bull.* Geol. Soc. Amer. (1917) **28**, 727.

devolatilization of the organic debris and residual products buried in the sedimentary deposits. Both coal and oil are products of alteration, by geologic processes, of organic matter not only similar, but, at least, in part, identical in composition. Coal consists of the mass or stratum of relatively pure organic debris, including the residual solid hydrocarbons left in the process of transformation from peat, or its genetic equivalent (deposited under different conditions), to graphite. Oil, on the other hand, is a volatile product of this natural "distillation" by the same agencies, of the organic debris and residues buried in the sedimentary deposits.

The transformations or geochemical changes are intimately associated with, if they are not actually caused by mainly horizontal stresses, under loading, with consequent molecular displacement, and some incidentally generated heat. The temperature developed during the process was probably very moderate, and almost certainly was not great enough to distil the organic matter at slight pressures.[10] There is generally but little trace of alteration of the rock except progressive dehydration, compression, and lithification in the earlier stages, with some sericitization; the latter can, however, hardly be attributed to hydrothermal action,[11] since there is no evidence of the percolation of magmatic waters. Deformation of crystals has not yet been observed, except in sands of regions where the carbonization is approaching anthracitization, in which case a change to quartzite, and some deformation of quartz grains, may be noted, as well as occasional thin platy cleavage, probably representing incipient schistosity. As the regional alteration approaches the graphitic stage, mineralization and considerable deformation of the rock grains, including pebbles, may locally be noted. In short, the transformation of the organic debris and the concomitant changes in the surrounding rock are such as are characteristic of the earliest phase of normal regional metamorphism. The chemical reactions in the organic matter are not yet convincingly explained. The processes are now in operation, though they are more energetic and efficient in regions and during periods of diastrophic movement.

Experimental evidence strongly, but not conclusively, supporting

---

[10] Observations by C. E. Van Orstrand, of the U. S. Geol. Survey, of temperatures in several deep wells in the Northern Appalachian region indicate temperatures, at the present time, of approximately 170° F. (77° C.) at depths of 7500 ft. (2286 m.), the increase averaging 1 to 50 ft. in depth. [See *Ohio Gas & Oil Men's Jnl.* (Sept., 1919) 1, 22.] The temperature gradient is found to be steeper in other regions. ·Thus, at a depth of 3000 ft. (914 m.) at Newkirk, Okla., it is 128.1° F. (53.5° C.) or 1° F. per 46 ft., while in the Ranger, Tex., field, at 3000 ft. the temperature was 134.9° F. (57.1° C.), the rate of increase being 1° to 45 ft. No doubt very much steeper gradients will be found in regions of more recent movement.

[11] Studies of the petrology of oil sands are now in progress by M I. Goldman, of the U. S. Geol. Survey.

the pressure theory of the origin of oil has recently been adduced[12] by Alex. W. McCoy, geologist of the Empire Gas and Fuel Co. By means of pressure on the ends of a cylinder of oil shale enclosed in a tube, the walls of which were thinner in the central zone than at the ends, so as to allow bulging, Mr. McCoy was able to induce flowage in the oil shale, and, without causing an appreciable amount of heat, developed small globules of oil in the shale which were visible with a hand lens. The material used in the experiment was typical oil shale, capable of yielding 25 gal. (94.6 l.) of oil to the ton, and having a crushing strength of about 3000 lb. per sq. in. (211 kg. per sq. cm.). No oil could be removed by solvents prior to the experiment. From this and other experiments, Mr. McCoy concludes: (1) that the solid bituminous material in the rocks is only changed to petroleum by pressure in local areas of differential movement; (2) that "the accumulation of oil into commercial pools is accomplished by capillary water; and this interchange only takes place in local areas where the oil-soaked shale is in direct contact with the water of the reservoir rock," such conditions being explainable either by joints or faults; (3) that "some adjustment takes place" until the oil in the sand has found the larger openings, where it remains indefinitely; (4) that "the amount of oil in any field could have been derived from normal bituminous shales in close proximity to the pay horizon;" (5) that areas of maximum differential movement are in accord with anticlinal structures, that the maximum sub-surface faulting is on the flanks and sides of the anticlines, and that the best production runs in trends parallel to the faulted zone. The most important part of Mr. McCoy's experiments, as it seems to me, is the production of petroleum by pressure alone acting on unaltered shale.

As noted in my discussions of coal, regional carbonization results from the progressive devolatilization of carbonaceous matter in the strata on a regional scale under dynamic stresses, dominantly horizontal thrusts, probably with development of moderate temperatures. It is most advanced in the regions of apparently greatest thrust compressions and hence of greatest molecular displacement; and in any region it is seen to be greatest on the side of greatest cumulative and sustained horizontal stresses.

With reference to both carbonization and folding, it is important for a field geologist, prospecting for oil in new regions, to remember that folds are likely to mark lines of pre-existent weakness resulting from former anticlinal buckling or faulting in the deeper strata, or that they may occur in zones of less competence, such, for example, as along zones of marked or abrupt unconformities; also that buckling and, in particular, overthrusts are the structural changes (really strains) that compensate and relieve

---

[12] Notes on Principles of Oil Accumulation: *Jnl. Geol.* (1919) **27**, 252–262.

the pressure stresses and tend to neutralize them through a relatively easy and quick shortening of the arc which would otherwise take place through compression only. The buckling may occur at an early stage of the thrust, giving comparative relief through the remainder of the movement and even through the periods of greatest intensity of movement. Accordingly, a buttress of horizontal competent strata under adequate loading may endure, and undoubtedly has in many cases, more vigorous and long continued differential stresses, and has sustained greater molecular displacement and compacting of the rock, incident to actual compression, than a folded series, even though the thrust may actually have been stronger and covered a greater distance in the latter region. The study of the carbonization in a number of coal fields shows this to be true.

It appears probable that, in regions where the thrusts have been sufficient to cause well loaded strata to form anticlines, the stresses have been great enough to cause the generation of petroleum. If these deductions are well founded, the earlier and minor stresses are connected with the production of the heavier oils, anomalous or even inexplicable as this may seem from the chemical standpoint, while the highest grade oils are usually found where the carbonization, resulting from more intense stresses, has approached the limit of oil production.

According to these observations, and contrary to the views of most geologists and chemists, it would appear that the heavy oils, occurring in regions of less thrust and alteration, are the first products of oil generation, while the light oils, occurring in the regions of greater thrust, are the more refined products. Whether the latter are to be regarded as the direct result of the greater compression of organic matter or, perhaps more likely, as oils that have undergone subsequent migration, probably with fractionation by geologic processes, remains to be proved. In this connection, it is to be borne in mind that the solid residues of heavy hydrocarbons, devolatilized in the shales and other strata during the destructive stages beyond the oil limits, are now in evidence as particles of carbon, causing the blackness of slates, some of which were once richly carbonaceous shales, and undoubtedly productive deposits of oil mother substance.

On the other hand, it would appear probable that, in general, oil either is not present or is not segregated in series of sedimentary formations that have never been thrust sufficiently to cause some buckling or undulation under favorable conditions, with the requisite amount of loading. If not sufficiently loaded, they are likely to remain unconsolidated though they may have been folded. In the Coastal Plain formations of the Atlantic States, which appear to be but slightly warped and possibly lack good anticlines and domes, as though the region had been lifted bodily, without local disturbance, on the back of the metamorphic basement complex, the apparent absence of oil pools is attributed by some geologists to the lack of folds; this explanation is more likely to be correct than the view that it is due to the absence of sufficient organic matter in the formations. But it is also probable that, over much of the area, the unaltered sedimentary strata have not been thick enough to assure the requisite loading had moderate folding taken place.

In the genesis of an oil pool not only is the organic debris altered and devolatilized, with the generation of petroleum and natural gases, as the result of dynamic thrust stresses attending diastrophic movements, but the migration and segregation of these hydrocarbons, disseminated in their place of origin in the mother rock, are promoted, if not caused, by the molecular rearrangement and the movement of rock grains consequent to these stresses. Most, by far, of the oil and gas is generated under the influence of differential stresses in "impervious" beds, the larger part being formed in the midst of typically impervious deposits, mainly organic muds, carbonaceous clays, fine-grained shales, and dense organic strata, such as oil shales, than which few unaltered sediments can be more impervious. The molecular displacement and the readjustments of the particles of the rock are essential to the migration of the newly formed oil and gas, and of the water, in the directions of least resistance, which, other things being equal, will be toward those beds, or regions of beds, most resistant to pressure and within the pore spaces of which the pressure will be relatively less. Sandy strata, sandstones with grains varying in size and shape, porous limestones, lavas, and, finally, coarse sandstones composed of round grains of uniform size, display varying resistances to compression, with corresponding variation of pore-space pressure. Coalescence of the infinitesimal globules of oil will take place enroute from the yielding to the resistant strata; and as the porous resistant beds with stable grains are traversed, concentration and segregation of the oil, gas, and water will ensue, the water driving the oil and gas into the larger voids by reason of its greater capillary tension, whereby it tends to seize and hold the smaller ones.

The extent to which argillaceous and organic sediments are reduced in volume under pressure is better realized when one recalls that the subsurface layer of a peat bog contains from 80 to 90 per cent. of water, and sub-surface slimes and muds carry nearly as much. At the lignitic (brown coal) stage, the average water contents of the coal bed approaches 38 per cent.; the proportions of water in sub-bituminous, low-rank bituminous, high-rank bituminous, and semi-bituminous coals average 23, 15, 6, and 3 per cent. respectively.[18] To the water losses, a part of which may be attributed to mere loading soon after deposition, are to be added the progressive losses of organic volatile matter, including

[18] G. H. Ashley: *Trans.* (1920) **63**, 782.

petroleum. The necessity for readjustments of the rock material, as the process goes forward under heavy loading and powerful lateral thrusts, is obvious.[14] Rearrangement of the grains in an impure sandstone, or in one composed of grains varying in size and irregular in shape, will permit less compression than purer coarse sandstone; while a coarse, porous sandstone composed of round grains, if not too rigidly cemented, may even change its shape under lateral thrust without change of volume, until the stresses become so great as to deform the grains, at which stage carbonization will have passed the oil limit.   All these conditions tend to drive the oil into the sand having the largest, roundest, and most uniform sized grains.

It may not be out of place here to note that diastrophic movement is not simple or cataclysmic. It is always in progress in one region or another, though its magnitude and vigor are specially noticeable in periods of most marked isostatic adjustment. These periods, though for the most part relatively short, geologically speaking, doubtless span thousands or perhaps hundreds of thousands of years.   The complex movement of a lateral thrust may be considered as the product of a cycle, or perhaps a series of cycles of complex differential stresses, possibly cumulative for a period, then decreasing in force, probably to be renewed again and again in greater power, until compression, buckling, or displacement have so far relieved the stresses that they are no longer able to overcome the rigidity and friction of the strata.   There is an obvious contrast between those strata which relieve the intensity and continuity of a thrust by buckling, folding, or faulting, and those more competent strata which, though enduring even more intense stresses, are able to relieve them only by horizontal compression.

Plainly, then, during these periods of horizontal diastrophic stresses, the opportunities for progressive readjustment of the particles may have been almost without number. It is reasonable to conclude also that molecular rearrangements have attended these stresses, since the chemical composition of the organic debris and residues has from time to time certainly been altered, with the generation and expulsion of volatile matter, including oil.   A study of coals shows an apparently uninterrupted gradation from lignite to anthracite and graphite. It would appear, therefore, that during a period of diastrophic stresses, the conditions have repeatedly been favorable for the evolution of the oil, the displacement and rearrangement of the organic particles and rock grains, the coincident

---

[14] Lateral transfer or flowage, under differential pressures, of the more plastic argillaceous and organic strata in a series of beds varying in composition and thickness is most natural, and is illustrated by the "horses," "squeezes," "veining," and "pocketing" of coal and clays, so familiar to the miner in the bituminous, semibituminous and anthracite fields. Such local flowage may cause thin included sandstones or even environing shales to bend in accommodation, thus producing small local anticlines, some of which may be misinterpreted as depositional.

rupture, enlargement, decrease or rearrangement of the pore spaces and capillaries, the development of zones of varying pressure, the overcoming of friction, and the disorganization of capillary resistance. In short, the conditions must have been most favorable (a) for squeezing oil, gas, and water out of their impervious source, through the intervening, impermeable, organic and argillaceous deposits, into the less compressed regions of the sandy rocks, sandstones, and porous limestones; (b) for their migration in spite of capillary resistance; and (c) for their eventual escape into the most porous, coarse-grained reservoir available, where, under a relative stability of the rock material, segregation and gravitation may be assumed to have taken place, subject to the effect of capillary tension. In some respects, the effects of diastrophic stresses in compressible sedimentary strata may be likened to a jigging of rock particles and mineral grains, in which process existing capillaries may become unstable and disrupted, pore spaces reorganized as to number, form and size, and friction repeatedly overcome; thus the escape, migration, concentration, and segregation of water, oil and gas, into less compressible sandstone and limestone reservoirs were promoted.

Consistent with this interpretation, it would appear that:

1. Oil will be generated only at depths sufficient to assure the necessary loading, which may vary somewhat with the composition and rigidity of the strata and, to some extent, with the intensity of the thrust.

2. In oil fields where the stress has been slight and probably confined to a single period, carbonization (alteration) being in the early stage, the oil is not likely to be found far, stratigraphically, from the carbonaceous sediments. If the thrusts have not been sufficient to drive the water, oil, and gas to a suitable storage "sand," the disseminated oil may not be recoverable. Water, with its stronger capillary tension, will tend to drive the oil into the largest pores available. Accordingly, a lenticular body of open-pored coarse sand may be filled with oil under heavy pressure, independently of anticlinal structure, or even in a shallow structural depression.

3. The largest oil pools normally occur where ample suitable storage is convenient to abundant organic mud or mother substance, unless the thrusts have been too great and carbonization has gone too far. Insufficient storage in very thin or fine-grained sands may be found in extensive carbonaceous formations; for example, the thin sands of the Graneros in the Thornton field, Wyoming, and the fine-grained sands in the Mancos shale in northwestern Colorado and in the Chattanooga shale in Barren County, Ky.

4. The stresses of a diastrophic movement may be sufficient to generate only a part of the oil and gas derivable from the organic mother substance, leaving some to be evolved under later stresses, until oil is no

longer produced, though gases may continue to be eliminated until the organic substance is wholly devolatilized, leaving only the "fixed carbon."[15] From field observations on the progress of devolatilization of organic matter, it is concluded, as already noted, that the first oils are generally heavy, usually with considerable asphalt; the later products, generated in areas of more advanced alteration, are lighter; while the oils from formations and regions where the carbonization limit has been approached are characteristically of the highest grade. This is the reverse of the order in which fractions are obtained by heat distillation.

5. In the course of successive periods of lateral diastrophic stresses, the water, oil, and gas, under cumulative pressures, may be carried through relatively impervious rocks for long distances in the direction of least resistance, if the thrusts and consequent pressures are sufficiently energetic, capillary tension and friction being to some extent counteracted by the forces which cause rearrangement of the rock particles. For this reason, several sands may yield oil generated from a single deposit. Enormous pressures should develop in the lower sands. In fields containing many oil sands, the oil is more likely to be of deep origin.

6. Oil pools generated and localized during one period of stress may be, and probably have been, carried on to new reservoirs, possibly at different horizons, at a later period of greater stress. This may be termed secondary migration and secondary storage. It seems within the limit of probability that some of the oil found in sands stratigraphically remote from recognizably carbonaceous beds may have come from the latter by secondary if not by primary migration. Given sufficient stresses in a great thickness of compressible strata, or pressures sufficient to compress the interlaminated somewhat arenaceous beds, it would seem possible that some of the water, oil, and gas may be forced comparatively near the surface before they are trapped in a sandstone beneath impervious cap-rock; finally, if these sandstones lie sufficiently near the surface to crack, fracture, or buckle under thrust displacement, the oil and gas may even escape from the strata. Consideration must be given to the probable depth of erosion that has occurred in a field, where sands now near the surface are productive.

Whether there is a sort of natural fractionation when the oil pool, at a later period of stresses, is forced into new and perhaps stratigraphically higher reservoirs, cannot now be answered definitely. The facts that (a) the oil disappears eventually in a process of advanced carbonization, leaving only its solid residues as dark carbonaceous matter in the rocks,

---

[15] This is indicated by the artificial distillation of oil from oil shales in regions which have undergone varying degrees of carbonization, up to the oil limit; only small amounts of oil can be obtained from shales which have been carbonized beyond this limit.

and that (b) a thin film of oil, including some of the heavier hydrocarbons, is left on the grains around which oil has stood, point toward the improvement of the product with each such transfer. This might account for the progressive refinement of the oils in the course of recurrent periods of thrusting, as mentioned under paragraph 4. The possible depreciation of the oil by percolating surface waters, especially those carrying sulfates, or by escape of the lighter matter to the surface, must not be overlooked.

The problems of secondary migration of oil may be as important as they are interesting, and require further study in both field and laboratory.

The disappearance of oil pools in areas of too advanced carbonization may be due to leakage when jointing and cleavage become more highly developed; or the oil may have been driven to the surface up the dip of the sands; or, as I am inclined to believe, it may have been volatilized, the solid residues remaining in the rock.

Whether it is possible for oil and gas to pass through impermeable clay shales or other cap-rocks except at times of diastrophic readjustment may well be doubted. The extent to which such readjustments are essential to the migration of oil, gas, and water along a stratum so composed, as to size of grains and porosity, as to comprise an oil sand, remains to be experimentally proved; but I am disposed to believe that their migration through "tight" sands and other so-called impervious beds takes place under dynamic stresses of diastrophism.

C. E. Van Orstrand suggests that the geologists of the country may not have given due consideration to the possible influence of osmotic pressure in moving the oil from deeper and warmer strata, in which it originates, to the overlying cooler strata or up the dip into the zones of lower temperature at the apex of an anticline or dome. This subject has been mentioned by Mr. Van Orstrand in the record of his temperature observations in several deep wells of West Virginia and southwestern Pennsylvania.[16] In this connection, attention is called to a brief discussion, by H. B. Gillette,[17] of the influence of osmotic pressure in transferring rock solutions from warmer to cooler zones, as relating to the deposition of orebodies.

The oil in the salt domes of the Gulf Coastal Plain may have originated in the strata in which the salt plug is found, or it may have ascended more or less of the distance traversed by the salt. The pressure theory as to the origin of the salt plugs, which seems to demand acceptance,

[16] Discussion of the records of some very deep wells in the Appalachian oil fields of Pennsylvania and West Virginia, by I. C. White, State Geologist, with temperature measurements by C. E. Van Orstrand.

[17] *Trans.* (1903) **34,** 710.

premises local pressures in the surrounding rocks which might be sufficient to cause the generation of such low-grade oils as are usually found in these domes. Oil of higher grade would be expected, in general, at great depth. On the other hand, it has not been proved that the oil did not ascend with the salt, which seems possible. In the latter case the domes may deserve testing to a maximum depth. Proof that the oil was disseminated in the strata, in readiness to migrate horizontally into the monoclines about a dome when the latter was formed, would support the theory that the oil was biochemically formed at the time of deposition of the strata, as suggested by Murray Stuart,[18] separated by pressure, and segregated gravitationally in the upturned beds surrounding the plug.

THICKNESS OF SEDIMENTARY FORMATIONS

Whether or not the geologist follows my conclusion as to carbonization, and its use as an index of incipient regional alteration, the degree of which approximately determines the limit beyond which productive oil fields will not be found, he must in any case take into account not only the alteration of the sedimentary formations, according to his own conception of the metamorphic limits, but also the thickness of sediments that, according to his judgment, are not too altered, and hence must furnish the oil. However, this subject has awakened less discussion, and therefore less systematic observation, than its very great importance demands.

It will probably be generally agreed that the requisite thickness of sedimentary strata in any oil basin depends on the character, composition, and competence of the strata; the position of the sands and the caprock; the distribution of the mother substance; the structure, the jointing, faulting, erosion, conditions of deposition, etc. Mother substance, reservoir sands, or cap-rock may, of course, be lacking, but for purposes of discussion, it must be assumed that they are all present and favorable, i.e., the organic matter is near but not at the bottom, and the reservoir and cap-rock are next above it. Further, a consideration of the requisite thickness must take into account the probable depth of strata eroded since the oil was generated and brought to its present storage. In other words, the original thickness at the time of deformation by horizontal stresses is to be regarded, rather than the present thickness in the producing basins, for the original thickness is what determined the amount of load on the organic beds when dynamic action occurred.

A review of the field evidence circumstantially presented by oil fields possessing relatively thin and not too altered strata, lying on a crystalline or thoroughly metamorphosed basement complex, would be both

---

[18] *Records* Geol. Survey India (1910) **40**, 320–333.

interesting and valuable, but the data seem insufficient for definite conclusions. I do not recall any oil field, meeting the conditions above mentioned, in which the non-metamorphosed sediments originally aggregated less, in round numbers, than 2000 ft. (600 m.); in most cases the thickness is over 2500 ft. (760 m.). ⸱ Exceptions should be made of series marginally overlapping on metamorphics, like the Pennsylvanian and Permian on the buried igneous and metamorphic rocks in portions of the Mid-Continent region, where the hydrocarbons may have migrated diagonally through the littoral sands of the relatively steeply transgressive formations.

The question of thickness is possibly of great importance in regions like the Atlantic Coastal Plain, where unaltered and largely unconsolidated sediments lie on pre-Cambrian complexes; also in portions of the Atlantic Trias. As to the Atlantic Coastal Plain, over the greater portion of which the thickness of the Coastal Plain deposits is almost certainly less than 2200 ft. (670 m.), while throughout large areas it is less than 1200 ft. (366 m.), it may be questioned whether, if thrusts sufficient for the generation and migration of oil into coincidentally induced folds had been exerted, the sediments were sufficiently thick to provide enough loading to favor the generation, segregation, and retention of the oil and gas. Almost surely the thickness has been too little, also, over considerable areas in those marginal zones of the Coastal Plain formations in the Gulf embayment, where the Cretaceous and Tertiary sediments lie on metamorphic or crystalline series.

On the other hand, the presence of oil in relatively thin sediments overlying other sedimentary formations, in which the carbonization has not gone far beyond the limit cited above, may not be precluded, in accordance with the suggested migration of hydrocarbons during recurrent periods of thrust stresses. The Madill, Okla., field seems to offer an illustration. The stresses inducing incipient metamorphism in an unconformably overlying formation must further alter the lower formation, which may already have nearly reached the carbonization limit. However, the presence of oil pools in thin unaltered sediments, where the alteration of the carbonaceous debris in underlying formations has progressed considerably past the carbonization limit—say, into the semi-bituminous (fuel ratio 3.0 or more) or the semi-anthracite rank—would be very interesting and worthy of record. In such an occurrence, the questions will be: Did the oil (a) originate in the lower series and pass into the younger by primary or secondary migration, as seems most probable; (b) condense from vapors generated in the lower series during the progressive alteration after the upper sediments were laid down; or (c) originate in the upper series?

The evidence bearing on these questions is not sufficiently complete and coördinated to encourage a satisfactory discussion at this time, due largely to the lack of observation on carbonization and other indices of incipient metamorphism of the sediments, including the carbonaceous deposits. Factors to be considered in this connection include unconformities at erosional intervals, as affecting the escape or deterioration of oils near the old erosion surfaces; migration of oil and gas up the dips to the margins of transgressing formations or through littoral zones to higher formations; thicknesses of rock eroded from producing formations; and the effects of sealing on oil pools in the sands. The application of the problem to many regions is obvious.

CONDITIONS OF DEPOSITION

Aside from such facts as the presence of adequate organic matter, of sands suitably composed and situated for service as oil reservoirs, of cap-rocks properly located, etc., some of which have already been mentioned, a question which should not be ignored in the search for oil in a new region, such, for example, as the Tertiary freshwater basins of eastern Washington and Oregon, or in the Great Basin region, is whether or not the beds concerned in the generation and storage of the oil are strictly of freshwater origin, and particularly whether the oil-bearing series was laid down in an exclusively non-marine basin. Inseparably connected with this question is the related one as to the importance of the association of salt water or gypsum in the oil-producing formations, as is so insistently urged by some geologists, with citation of circumstantial evidence. On these matters opinion differs widely, possibly without succinct data sufficient for a final decision.

As criteria to be considered in the answering of this problem the following may be noted:

(1) Ample organic matter undoubtedly suitable for the generation of oil and gas was deposited with the sediments in many of the freshwater basins. These deposits contain oil shale of high quality, which, on distillation, yields oil essentially like and possibly indistinguishable from that obtained from oil shales of marine origin. Many of the organic products are common to both habitats.

(2) The mechanical constitution of the deposit in both marine and freshwater formations is essentially the same.

(3) Important oil-bearing sands and organic remains were deposited during intervals, sometimes of considerable length, during which only freshwater sediments were laid down, these deposits being intercalated in brackish water or marine sediments.

(4) Oil-bearing sands and organic deposits were laid down in waters, but slightly saline, in the younger formations of the Appalachian trough.

(5) Some salts are present in freshwater deposits.

(6) Natural gas is present in freshwater basins, and has been developed at considerable depth in such basins.

(7) While it may be true that, in the geologic processes of oil generation, salt in amounts premising marine or brackish water deposition may
be essential as a catalyzer or otherwise, the fact remains to be demonstrated, possibly in the laboratory. The absence of salt does not appear
to affect the artificial production of oil by distillation of shale.

It is possible that in some of our oil fields, salt water may have found
its way downward through joints or along the dip into fresh-water beds
subsequently submerged beneath the sea, somewhat like the invasion
of fresh water down the dips of some of the marine oil sands in California.
On the other hand, fresh-water basins in which the requisite conditions
as to depth, organic matter, sands, cap-rocks, thrusting, and incipient
alteration favorable for oil pools are fulfilled, and which have never subsequently been submerged beneath the sea, have been too little tested
to justify conclusions as to their possibilities for oil production. Such
basins, if actually closed and without outlet, will be more or less distinctly alkaline. At the present stage of our knowledge, fresh-water
basins appearing otherwise to meet the requirements should be wildcatted
without prejudice.

DISCUSSION

R. H. JOHNSON, Pittsburgh, Pa.—It seems to me that the case for
the fresh-water origin of natural gas must be accepted, since the coal
progressively loses methane. We know that much natural gas must
have arisen in that way. My reserve in connection with petroleum in
contrast with natural gas comes from the fact that if the fresh-water
deposits have been as productive of petroleum as the marine, the field
evidence ought to show us more petroleum in close proximity to the coal;
it is this that leaves me skeptical as to very much fresh-water petroleum,
although we must admit a great deal of fresh-water natural gas.

H. W. HIXON, New York, N. Y.—In the Appalachian field, how
much oil and gas do you find above the coal? I am not well informed
on that subject, but I have not heard of a case where oil and gas occurred
above the coal in Pennsylvania or West Virginia, except possibly that
which had migrated there along a fault. There is another thing, you
cannot saponify petroleum and you cannot make glycerine out of petroleum. If you could do that, why should we have paid such high prices
for glycerine during the war? You can saponify organic oil and you can
make glycerine out of animal fats, so that the petroleum and animal
fats differ entirely. Also, petroleum has no food value.

We differ fundamentally as to our ideas about the origin of petroleum and natural gas, and I think we will have to let it go at that. As
regards the origin of the force, the dynamic force, that causes these
deformations, the elevation and folding, I consider that these gentlemen

are laboring under the impression that it is contraction. I have studied that question for a number of years and find that the best authorities on the subject state that the total amount of contraction of the interior of the earth, due to loss of temperature, in 100,000,000 years would prove a circumferential contraction of about 7 miles, and that if all of the folding and faulting and overthrust of the various mountain ranges of the earth were ironed out, they would amount to something like 150 to 200 miles. There is a decided difference that has to be accounted for. I account for it in an entirely different way; that the gases which cause elevation by reduction of density, as the surface of a loaf of bread is raised by reduction of density, tend to accumulate and migrate toward the axis of elevation. They carry the crust in two directions and, because of the reduced density, creep toward the axis of elevation or toward the center of the dome. The hydrocarbons are in the gaseous interior of the earth for exactly the same reasons as all the other gases—because of the diffusion of gases in the original gaseous planet in that gaseous core, and they have remained there ever since, being held by the power of diffusion. The change from the gaseous to the solid condition is by loss of temperature. It then becomes lighter at the same time, because the gaseous core must be denser than the solids that lie upon it.

My contention is that petroleum, natural gas, and the helium with them are of volcanic origin. The origin of oil and gas is connected with the whole theory of earth physics, which is entirely different from the old contraction hypothesis, on which most of the geologists base their theory of mountain formation.

The authorities on the subject state that the earth's crust, considered as a dome, is not capable of supporting one five-hundredth part of its own weight. If it is not capable of supporting any more than that portion of its own weight, it must be supported by the material below it at all times, whether it is above sea level or below sea level. We find rocks of marine origin in the highest mountains. They did not get there by accident; and if they were supported by the material below them at all times, and they are above sea level at one time and below sea level at a previous time, the only possible solution of that problem is that the matter below the zone of fracture has varied in density between those two periods. So that you come to a question of accounting for that variation in density between two geological periods; that, I maintain, is due to the accumulation of magmatic gases derived from the gaseous core denser than the solids which will form out of it when cold.

DAVID REGER,* Morgantown, W. Va.—Regarding the statement just made that oil has not been found above the coals in the Appalachian Basin, I would like to say that one of the first wells in the West Virginia

---

* Assistant Geologist, West Virginia Geol. Survey.

fields was about 300 ft. deep, at Burning Springs on the Little Kanawha River in 1860. When I visited the well in 1909, 49 years after its completion, it was still producing 30 bbl. of oil a month and 10 years later, in 1919, was still producing. Only the owner knows how many thousands of barrels the well has actually produced in that time. The producing formation there was the Cow Run sand, which is right in the middle of the Coal Measures. This well is on the Burning Springs which extends half way across the state. The great producing sands have been the two Cow Runs, which have coal measures above and below them. It is my opinion that while the first oil in the state was found in the coal measures, it is entirely possible that the last drilling and the last production of oil to be eventually found in the state will be in the same coal measures.

H. W. HIXON.—I believe I stated that if the oil or gas migrated to that particular place along a fault it might be found there, but it all comes from below the coal.

DAVID REGER.—There is no fault there.

J. F. DUCE, Denver, Colo. (written discussion).—The question of the presence of carbonaceous material in the "Red Beds," is an old one. The Triassic "Red Beds" of New Mexico are certainly carbonaceous and at times bituminous. Certain of the sandstones are crowded with fossil wood, while directly above these will be found bituminous sands. This is true also in the Dockum of Texas. Of the underlying strata in northern New Mexico we cannot be so certain, as they are largely arkoses. The basal members of the Manzano group contain some lime-stones, but M. G. Girty states that the fossils are not well preserved, which suggests erosion on the sea floor before entombment.

In the southern part of New Mexico, the occurrence of the limestones and bituminous shales of the Guadalupe Group at the base of the Triassic and the great thickness of limestones in the Manzano (Hueco) suggests the presence of petroleum; these same rocks are probably the source of the oil in the artesian wells of the Roswell area.

It is perhaps well to bear in mind that the criterion that White suggests concerning the state of metamorphism of the coals in an area is applicable but locally in the Rocky Mountains. We are confronted there with exceedingly rapid structural changes, and it is along the axes of these changes that the coal fields White has mentioned occur. Recent investigations by Richardson, Lee, and Ziegler have changed our conception of Rocky Mountain structure. The steep monoclines that form the mountain front die out abruptly both east and west, and seem in some cases to have been accompanied by strike faults. The zone of intense folding is therefore narrow and is frequently associated with volcanic activity.   Along the flanks of these sharp folds most of the producing Rocky Mountain coal fields are grouped; here, too, there is the maximum metamorphic effect, so that the coals are of high grade.   As we pass from the folding, the coal becomes poorer and poorer.   The coals in the Trinidad field are associated with the intrusives of the Spanish Peaks group, the coals of the Durango field with the intrusives of the San Juan group, the coals of Crested Butte with the Crested Butte intrusives, and those of the Anthracite Range with the Elkhead Mountain intrusives.   This connection is surely not accidental.   In one case, however, at New Castle, the high-grade coals are not associated with eruptives but with sharp folding alone.   (Basalt flows are present in the near vicinity but I am speaking here of intrusives).   If now we pass from the focus of folding and igneous activity but a short distance, the grade of the coal changes markedly, and in accordance with White's theory.   We must, therefore, restrict this criterion of the fuel ratio of the associated coals to the locality in which the coals occur and cannot extend it generally to formations beyond the field in which the high fuel ratio coals occur.   Further than this, if petroleum migrates up the flank of Rocky Mountain monocline, we would expect even within the metamorphic areas petroleum that had migrated from farther down the slope where unmetamorphosed sediments occur, unless the metamorphism has reached a point where the porosity of the strata through which it must migrate has been affected.   Lighter oils would, however, be expected, as the long journey would result in the fractionation of the original oil.

In connection with the origin of petroleum, it is interesting to note that almost all the Cretaceous oil of Wyoming is produced from the lower Colorado group, and that the oil sands are directly associated with the bituminous shales of the Mowry and equivalent formations.

Petroliferous Provinces *

By E. G. Woodruff,† Tulsa, Okla.

(Chicago Meeting, September, 1919)

The earlier struggles in petroleum geology were directed to solving the origin and method of accumulation of petroleum. We are now fairly well agreed on those subjects. Most of us think that the great mass of petroleum commercially produced comes from plants or animals, or possibly from both. We are confident that the oil was not produced where it is now found but has accumulated in reservoirs of various kinds. The types of reservoirs are certainly variable but they just as certainly follow definite geologic laws. Some of these types of reservoirs can be determined from surface study; others cannot. We know, too, that these types of reservoirs (largely structures such as anticlines, domes, and terraces) are much more widespread than the oil pools. In other words, there are many places where good sands and good structures exist but where oil is not found. It is the purpose of this paper, therefore, to attempt to analyze, from a regional standpoint, some of the conditions that control the presence or absence of oil pools and to group them in a regional way, hence the term "Petroliferous Provinces." The paper lays no claim to presenting new facts but attempts to group and classify the information that so many have expressed again and again.

The essential factors for an oil field are petroleum, a reservoir material, and conditions under which the petroleum can enter the reservoir but cannot escape except through the drill holes. The paper will first discuss the source of petroleum as it occurs in definite regions, then the regional arrángement of reservoir strata, and finally the areal arrangement of structures.

To have petroleum, there must be a source. Since living matter is considered the source of the petroleum, geological conditions must have been such that living organisms were abundant. Arid regions on the earth's surface have not given rise to living things in sufficient abundance to produce oil; similarly, too cold regions and saline inland lakes. The converse of this is that warm moist conditions must prevail to produce an abundance of vegetable matter. Before an area can be

---

* Paper prepared for meeting of Tulsa Section, Feb. 25 and 26, 1919.
† Chief Geologist, Oklahoma Producing and Refining Co.

considered a petroliferous province, it must have had an abundance of living things from which the oil could have been derived. On this basis, certain classes of petroliferous provinces may be distinguished.

Igneous Rocks.—It is evident at once that petroleum cannot come from provinces in which there are nothing but igneous rocks. One does not expect petroleum in the granite regions of the Rocky Mountains or the Hudson Bay, large areas in western Georgia, North and South Carolina, central Maryland, southeastern Pennsylvania, and northeastern New York. If life was ever abundant in these provinces, the remains have been eroded away. They are certainly non-petroliferous areas.

Metamorphic Rocks.—Organisms may have been abundant in the rocks from which the metamorphics came but the geologic processes are such that the petroleum must have been driven from the rocks if there was ever any in them. Geologists exclude these areas of metamorphic rock from the petroliferous provinces, because there can be no source for the oil in them.

Sedimentary Strata.—Almost any sedimentary rock may be a source of petroleum but to the commercial geologist some are impossible of petroleum production, whereas others are improbable and others probable.

Lower Paleozoic Strata.—The very old sedimentaries, pre-Cambrian, Cambrian, and Ordovician, have not been productive of petroleum to any considerable extent. It is probably because, during those ages of the earth, life was not sufficiently abundant to accumulate in quantities large enough to produce petroleum in commercial quantities. Possibly, too, the geological forces have been operative so long and locally so intensively that the petroleum has been driven from the strata if any ever existed in them. On this basis, we look with doubt on a large part of Arkansas, part of Missouri, certain areas in Ohio, Tennessee, Kentucky, most of Minnesota and Wisconsin, northern Illinois, the belt of closely folded strata from northeastern Alabama to New York, and practically all of New England. By this process of elimination, possible petroliferous provinces are greatly restricted.

Middle Paleozoic.—The Middle Paleozoic strata have been productive but have produced only locally. It is probable that, by that geological time, the animal life had become sufficiently abundant locally but only in the most favorable localities to be a source for the oil; therefore, if the province under consideration has only Silurian or Devonian strata it should be considered and probably classed from the standpoint of the life condition that prevailed during the period of deposition. If the paleogeographic conditions were such that life was abundant, the province may be petroliferous; but if life could not abound, then the province must be non-petroliferous.

Carboniferous and Younger Strata.—The upper Paleozoic and all younger strata must be classed as possibly petroliferous. But in classify-ing provinces embracing these strata, a criterion that should be applied is the presence or absence of such paleogeographic conditions as supported life in abundance or suppressed it. Largely on this basis the writer has tentatively classed areas in Iowa, Nebraska, Kansas, and Arkansas as non-petroliferous. He is fully aware that the facts are as yet meager and incompletely studied and that petroleum may be produced in some of them, but certainly they must be considered doubtful. Areas classed as promising on this basis are in Pennsylvania, West Virginia, Kentucky, Tennessee, Alabama, Ohio, Indiana, Illinois, Kansas, Oklahoma, and Texas. Even some of these must be classed as non-petroliferous on the basis considered later in this paper.

Our second broad division in classifying any province as petroliferous or non-petroliferous is the character of the reservoir stratum. As we know, the most reliable stratum is a sandstone, its continuity of porosity and the resistance to closing of pores under compression render it the most reliable; next to the sandstone is porous limestone or dolomite; and, finally, shale. Other classes of reservoirs are practically negligible because the amount of oil reservoired in them is very small.

The ability of a sandstone to reservoir petroleum depends on its freedom from material that will fill the pore spaces. This may be diffi-cult to prophesy in advance of actual drilling, but we can achieve a considerable degree of success by studying the conditions under which the material was deposited. Sandstone composed of quartz derived from the granites must be open, if deposited in fresh, or comparatively fresh, water not far from the source. On the other hand, if the sand has been transported a long distance from the source or deposited in land-locked or very saline basins, its pore spaces are filled and it cannot become a reser-voir stratum. As a concrete example, the sandstones now forming along the rivers debouching from the front range of the Rocky Mountains are almost universally porous but none of us will expect the sandstones form-ing in Great Salt Lake to be porous. My associates who studied the petroleum conditions of Cuba found the sand there to be derived largely of fragments of gabbro from the great gabbro masses nearby. Appar-ently, these fragments were fresh when deposited as sand but after deposi-tion they disintegrated sufficiently to allow enough clay to fill the pore spaces and compact the whole mass, thus closing the porosity of the sand and preventing it from becoming a reservoir stratum. Some sand reservoirs are derived by the disintegration of previously deposited sandstones, such as the Tertiary sands along the Texas Gulf Coast. They follow the same laws as the sands deposited primarily from the granites.

Thus, the basis on which to classify sand reservoirs must rest on paleogeography or on the character of the sand and the relation of the

present position of the sand to the source.   Let us look at it in another
way.   Take the map of the oil fields of the United States, with the
possible exception of the Gulf Coast; in the fields in which sandstones
are productive, the sandstone beds were laid down just off the flanks of

FIG. 1.—MAP OF NORTH AMERICA SHOWING PETROLIFEROUS PROVINCES.

paleozoic mountains.   The converse of this is that sands far from their
source are not productive because the pore space is closed by clay or salts.
The writer feels that even the Gulf Coast fields are an apparent exception
only because those sands are secondary and derived from the breaking up

of strata not far away.   On the basis of sand study, the petroliferous provinces outlined on the accompanying map are presented.

With our present very limited knowledge of the condition of limestones underground, no reliable classification of provinces can be made.   Some geologists have presented data to show that the limestones are cavernous or porous, whereas others have shown that they are creviced only.   At present we must consider all limestone possibly capable of reservoiring petroleum until proved otherwise, but the writer clings to the idea that the time will come when certain provinces will be delimited in which the limestones will be known to be non-petroliferous because the limestones are non-porous or non-creviced.

The writer is beginning to feel that possibly one distinction may be made, based on the purity of the limestone, which, of course, again depends on the paleogeographic conditions that prevailed when it was deposited.   The limited number of petroliferous limestone cuttings that have come under the writer's observation are very siliceous (generally cherty).   The writer is inclined to believe, though he is not ready to apply it as a criterion, that only cherty limestone beds produce petroleum in commercial quantities.   The crevices in shale offer a very limited reservoir space, so limited in fact that shale beds as such must always be considered as having doubtful commercial petroleum value.

If the third, but probably the most important, set of criteria to be applied in delimiting petroliferous provinces is structure, the types of structure necessary for the accumulation of petroleum have been so thoroughly discussed that a repetition is unnecessary.   These structures are the results of tectonic forces and are, therefore, grouped according to certain laws.   Again we are without sufficient data on which to base a grouping of these structures.   Certainly structures are most numerous on the periphery of the great structural basins.   They are not too close to the mountains surrounding the basins but certainly not far away. They seem to bear a certain zonal arrangement.

To apply these criteria a set of maps may be constructed: first, to show the petroliferous provinces on the basis of geologic age; then to restrict the petroliferous provinces thus outlined by striking from them the overlapping parts of the non-petroliferous provinces on the basis of reservoir material; and, finally, to restrict on the basis of structural groupings.   On these bases the writer presents the accompanying map of North America.   He recognizes that it is imperfect but hopes that it may form a basis on which an accurate map may be constructed.

This map shows petroliferous provinces as follows:

1. In Western Alaska.   Oil seeps are known in this province; there has been some drilling but as yet no considerable production.

2. In Western Canada from the Arctic Ocean to and including

Canada. Some commercial production may be found in the northern part of this province; the southern part is of doubtful value.

3. Along the Pacific Coast in California, Oregon, and Washington. Only the southern part of this province seems important.

4. In Wyoming, Colorado, and a part of New Mexico. Only small areas in this province will be productive.

5. In Oklahoma, Texas, and Louisiana. Considerable areas in this province are productive and others probably will be found.

6. From Pennsylvania to and including Illinois and extending southward into northern Alabama and Mississippi.

7. In Lower California. This province seems of doubtful value but may be productive.

8. On the eastern coast of Mexico.

These are the broader subdivisions of North America. On the same basis and by the same methods each province may be subdivided in areas in which petroleum may be found and thus a set of maps built up that will limit the areas in which the geologist and prospector may hope for success. Then, as our knowledge is perfected, the principles may be applied to South America, Europe, Asia, Africa, and Australia, thus greatly aiding pioneer work in those countries and rendering the fuller application of geology immensely valuable in the ultimate development of the world's petroleum resources.

DISCUSSION

CHARLES SCHUCHERT,* New Haven, Conn. (written discussion).— I embrace the opportunity to take part in a discussion of Mr. Woodruff's paper because a successful discerning of what actually constitutes petroliferous areas from the geologists' standpoint is worthy of our endeavors, not only from the intellectual side, but also because it may lead, as Mr. Woodruff hopes, to the more certain establishment of principles that can be applied to other continents in exploiting them for petroleum. This discussion will also embrace the results of two other recent papers, one by Alexander W. McCoy and one by Maurice G. Mehl.[1]

Sources of Petroleum.—Mr. Woodruff is agreed that petroleum comes from plants and animals, or possibly from both, and that it has accumulated by migration into reservoir rocks. These reservoir rocks must of course be porous to become catch basins for the oil and gas, and then, too, their present structures are variable, as they occur in anticlines, domes, terraces, etc. The structures, he states, are more

---

* Curator, Geological Dept., Peabody Museum of Natural History.

[1] A. W. McCoy: Notes on Principles of Oil Accumulation. *Jnl. Geol.* (1919) **27**, 252–262.

M. G. Mehl: Some Factors in the Geographic Distribution of Petroleum. *Bull. Sci. Lab., Denison Univ.* (1919) **19**, 55–63.

widespread than are the oil pools, and the same is true for the reservoir rocks. Accordingly, there must be many good sands and structures that have no petroleum. On the other hand, there are conditions in the making of the hydrocarbons that are not formulated by Woodruff or are not clearly in mind. These are: Petroleum is not formed in sufficient quantities to be commercially available in the fresh-water or subaerial deposits of the lands, the continental deposits. For practical purposes all such should therefore be excluded, at least for the time being from further consideration. Moreover, land climates have but little direct bearing on the temperature necessary for life in the seas where the petroleums are formed, because there is an abundance of life in all shallow, marine waters of whatever clime. Again, there has been abundant life in the seas of all times, not only since the Cambrian, but ever since the Archeozoic. The proof of this is seen in the high state of organic evolution attested by the earliest Paleozoic fossils, and in the nature of the marine formations of the Proterozoic and Archeozoic strata, with their high carbon content. All of these differences between us will be discussed later.

Areas With and Without Petroleum.—Mr. Woodruff is seeking for the regional conditions that originally controlled the formation of the hydrocarbons and their later storage into oil reservoirs. In this way he is led to point out the petroliferous provinces. The conditions that make for oil provinces he holds to be three:

1. The source of petroleum lies in the end-results brought about through the decay of organisms, and the preservation of the residues is limited to certain environmental conditions. There are great areas that have always been devoid of the required life conditions, and others where the entombed organic residues have been dissipated by the deformational processes.

2. Petroliferous areas are limited by more or less definite characters in the oil-preserving and oil-storing strata.

3. Petroliferous strata have more or less definite deformational structures.

The ideas which, in our opinion, lead to the ascertaining of the petroliferous and non-petroliferous rocks of North America are:

1. The impossible areas for petroliferous rocks.
   (a) The more extensive areas of igneous rocks and especially those of the ancient shields; exception, the smaller dikes.
   (b) All pre-Cambrian strata.
   (c) All decidedly folded mountainous tracts older than the Cretaceous; exceptions, domed and block-faulted mountains.
   (d) All regionally metamorphosed strata.
   (e) Practically all continental or fresh-water deposits; relic seas, so long as they are partly salty, and saline lakes are excluded from this classification.

(*f*) Practically all marine formations that are thick and uniform in rock character and that are devoid of interbedded dark shales, thin-bedded dark impure limestones, dark marls, or thin-bedded limy and fossiliferous sandstones.

(*g*) Practically all oceanic abyssal deposits; these, however, are but rarely present on the continents.

2. Possible petroliferous areas.

(*a*) Highly folded marine and brackish water strata younger than the Jurassic, but more especially those of Cenozoic time.

(*b*) Cambrian and Ordovician unfolded strata.

(*c*) Lake deposits formed under arid climates that cause the waters to become saline; it appears that only in salty waters (not over 4 per cent.?) are the bituminous materials made and preserved in the form of kerogen, the source of petroleum; some of the Green River (Eocene) continental deposits (the oil shales of Utah and Colorado) may be of saline lakes.

3. Petroliferous areas.

(*a*) All marine and brackish water strata younger than the Ordovician and but slightly warped, faulted, or folded; here are included also the marine and brackish deposits of relic seas like the Caspian, formed during the later Cenozoic. The more certain oil-bearing strata are the porous thin-bedded sandstones, limestones, and dolomites that are interbedded with black, brown, blue, or green shales. Coal-bearing strata of fresh-water origin are excluded. Series of strata with disconformities may also be petroliferous, because beneath former erosional surfaces the top strata have induced porosity and therefore are possible reservoir rocks.

(*b*) All marine strata that are, roughly, within 100 miles of former lands; here are more apt to occur the alternating series of thin- and thick-bedded sandstones and limestones interbedded with shale zones.

Experience has shown that commercial quantities of petroleum do not occur in areas of igneous rocks, nor in regions of highly folded, mashed, and decidedly metamorphosed strata that as a rule are older than the Tertiary. Nevertheless, it will not do to say, because strata are decidedly folded and faulted, that in the areas of mountains there can be no commercial quantities of oil, for we know that the petroleum fields of the Coast ranges of California and those of the trans-Caspian countries have yielded vast quantities. Here, however, the oil-yielding strata are essentially of Cenozoic age. It appears that the main regions for oil production in North America will be the more or less flat-lying sedimentary formations —the vast geologically neutral area—to the east of the Rockies and to the west of the Appalachians. Also, in a broad and general way, the older the geologic formations, the more devoid they are apt to be of petroleum; and the more often a given area has been subjected either to mountain folding or to broadly warping movements, the more certain it is that all or most of the volatile hydrocarbons have been dissipated. Such places are apt to have the hydrocarbons only in fixed form and not as kerogen. In strata older than the Cretaceous, the deformed geologic structures of varying sorts should be rather of minor than of major strength as an essential to oil accumulation in commercial quantities.

Original Oil Strata.—It appears that zones of petroleum, in general, do not occur in thick deposits that are continuously of the same kind of material, as sandstones, limestones, or shales, but in or near sandstones and thin-bedded porous limestones that are interbedded with bituminous shales. McCoy says that in the mid-continent field the petroliferous shales "are generally dark colored, often black, and carry bands of highly bituminous material." Such bands "are often described by the drillers as coal, asphalt, or black lime, according to the hardness and appearance of the material. The shales are typical oil shales, quite similar in character to those (of the Cenozoic) of Colorado and Utah."

Petroleum of Organic Origin.—The hydrocarbons are the chemical end-results of organisms and, in the main, are the fatty substances derived through bacterial decomposition from the plants and animals once living in the sea waters. This is a conclusion not always clearly in the minds of petroleum geologists.

One is led, Dalton[2] states, "to regard the great majority of oils as derived from the decomposition during long ages at comparatively low temperatures of the fatty matters of plants and animals, the nitrogenous portions of both being eliminated by bacterial action soon after the death of the organism. The fats and oils from terrestrial fauna and flora may have taken part in petroleum formation, but the principal role must, from the nature of most petroliferous deposits, have been played by marine life."

The decomposition bacteria attack the cellulose of the plants and the nitrogenous tissues of animals, leaving untouched the fatty materials. The reason why the fats remain untouched is probably because the feeding of the bacteria is stopped by sedimentation, which buries and kills the decomposing organisms living beneath the surface of the sea bottom. Dalton further states that "Peckham's view, that asphaltic oils are mainly of animal origin, while paraffin is largely derived from vegetables, is worthy of acceptance on general chemical as well as geological grounds, since Krämer and Spilker, and others, have shown that vegetable fats produce paraffin either with or without artificial distillation, and the limestone oils, which on geological grounds are generally held to be mainly

[2] Leonard V. Dalton: On the Origin of Petroleum. *Econ. Geol.* (1909) 4, 603-631.

of animal origin, are notably asphaltic." In general, the Palæozoic petroleums have paraffin bases, and it seems probable that all those derived from black petroliferous shales are largely, if not wholly, of marine algal origin. Usually we do not realize the extraordinary importance and abundance of plant life, but when we think that all animals are in the ultimate dependent for their existence upon plants, we begin to perceive the truth of the following forceful statement by the English botanist, F. F. Blackman,[3] who recently said that "Botany, as the science of plants, claims dominion over some 99 per cent. of the living matter on the surface of the earth and over most of the fossil remains under the surface."

Petroleum Essentially of Marine Origin.—It is, however, plain to all who have looked into the matter that petroleums cannot accumulate upon the dry land in deserts, grassy plains, or forests, for here the oxidizing influences are so active that all the volatile parts must be taken away or completely changed. In lakes, organic decay is, as a rule, so rapid that limy marls are deposited, and it seems to be exceptional that black petroliferous muds are of fresh-water origin. The extensive oil-shale deposits of the Green River series of Utah and Colorado are certainly not of marine origin, as they are devoid of marine organisms and are underlain and overlain by river flood-plain deposits of early Eocene age, as is shown by their contained land animals and plants. The hydrocarbons appear to be of drifted plant origin, according to Charles A. Davis, and as kerogen does not form in large quantities, the evidence appears to indicate that the water in which the Green River shales were deposited was slightly saline. Therefore the chemical end-result of organic decay, the kerogen, cannot accumulate in commercial quantities except beneath a sheet of salt water, and these sheets of water probably are in the main within the limits of a few hundred feet of depth; the deeper the water basins, the more certain the amount of oil accumulation, under these given conditions. Salt water and organisms are the first requisites for kerogen making in nature and, accordingly, the hydrocarbons are stored almost always in marine sediments; these are chiefly the black and brown shales and the impure dark thin-bedded limestones. All rock formations accumulated directly beneath the atmosphere, as the pure continental deposits, must therefore be devoid of commercial quantities of petroleum. Then, too, all deposits, either of the fresh waters or of the seas, which are periodically subjected to atmospheric weathering during their time of accumulation, are also lacking in oil in paying quantities. Hence we may further conclude that all red or reddish, yellowish or white, rain-pitted or sun-cracked deposit, either of conti-

---

[3] *New Phytologist* (1919) **18**, 58.

nental, fresh-water, or semi-marine origin, are lacking in petroleum in large amounts.

McCoy informs me that an average oil shale yields, at temperatures between 500° and 1200° F. (260° and 648° C.) about 20 gal. (75 l.) of oil, and from 15 to 18 lb. (6.8 to 8.1 kg.) of ammonium sulfate per ton of shale. In the spent shale there still remains from 15 to 20 per cent. of fixed carbon, but no ammonium sulfate. The bituminous material in unspent shales, he states, "occurs in solid form, as none of the ordinary solvents show coloration after solution tests. Upon distillation, such shales have given off petroleum." This "solid organic gum called kerogen" can be changed in the laboratory to liquid hydrocarbons by heat. In nature, this may be brought about possibly by intense friction developing heat, but more probably only in deep-seated water-bearing strata—accordingly, in formations that are under greater pressure. However, "pressure alone can cause no change in this material when the included water is not allowed to escape." On the other hand, "the maximum static pressure available in any porous zone is a function of the size of the openings around that stratum. The determining factor is the capillary resistance of the water in the adjoining small openings." In other words, the solid kerogen "is only changed to petroleum in local areas of differential movement. . . . After such a change is made, the accumulation of oil into commercial pools is accomplished by capillary water; and the interchange only takes place in local areas where the oil-soaked shale is in direct contact with the water of the reservoir rock. Such conditions are explainable either by joints or faults." A. B. Thompson, however, in "Oil Field Development," states that, according to the observations of C. W. Washburne, "since water has a surface tension of about three times that of crude oil, capillary attraction exerts about three times the force on water that it does on oil. As the force of capillarity varies inversely as the diameter of pore, it is contended that this force tends to draw water into the finest tube in preference to oil and displaces contained oil and gas: the result being that oil would be expelled from fine-grained material like clays into coarse-grained beds like sand."

How thick must a petroliferous shale be to furnish the necessary amount of oil for a productive field? McCoy states that "the amount of oil in any producing field could have been derived entirely from shales immediately surrounding the oil sand. A series of shales aggregating 10 ft. (3 m.) of bituminous sediment, yielding 25 gal. (94 l.) to the ton, would furnish 17,000 bbl. of oil per acre. Assuming a 25 per cent. extraction, the acre yield would be over 4000 bbl. The average acre yield in Oklahoma and Kansas ranges from 2500 to 3000 barrels."

Petroleum is probably forming today in many marine waters. Dalton says it is present "in the mud of the Mediterranean sea-floor between Cyprus and Syria. . . . It was also found in the Gulf of Suez, and in each case ammonia and iron sulfide or sulfur occur with the oil." Potonié showed its presence in the Gulf of Stettin, Germany, and Fritsch showed that humus is forming rapidly in the salt marsh in the Bouche d'Erquy, Brittany.   In all these cases, the muds are of the black, putrid type that Potonié calls sapropel.   Why, then, does petroleum not occur more uniformly in the geologic deposits?   Because the hydrocarbons universally tend to escape into the air or water from which they were originally taken by the living entities.   Muddy waters with the finest of silts and not too much agitated by currents or winds are the places where the hydrocarbons naturally may accumulate, because here the organic fats and oils have great affinity for, and unite with, the minute clay flakes, and are thus held in more or less solid form and deposited as kerogen with the shale formations.   Evidently, the hydrocarbons can accumulate and be preserved in large quantities only in areas of argilaceous sedimentation.   Therefore, in order to accumulate petroliferous deposits, the waters must have life in them; and the freer they are from oxygen, the more certain will be such accumulations.   On the other hand, almost all life fails to exist where there is no oxygen, because oxygen is the first essential of nearly all life, and where the petroliferous materials are gathering in greatest quantities, there the waters are free of this gas and the bottoms are black and foul—putrid muds reeking with odors.   Where, then, does the life come from in these places of hydrocarbon-gathering?   It develops in great abundance in the sunlit, agitated, and oxygenated surficial areas of the water basins, and after death the organisms rain into the deeps, where they very slowly decompose, due to peculiar forms of bacteria existing in the stagnant waters that are depleted of oxygen.   Are the surficial waters the only source for the life that is gathered into the oil shales?   No.   The life may develop hundreds of miles away from the place of accumulation and be drifted by winds, or by tidal or even oceanic currents into bays, cul-de-sacs of the seas, and into the shallow but extensive depressions on the sea bottoms. The petroliferous deposits are accumulating today in greatest amount in the shallow waters bordering the lands rather than in the greater depths.

However, not all shales are oil shales.   As all geologists know, about 80 per cent. of the sedimentaries are mudstones, and yet petroliferous shale formations are not common.   If forced to guess what percentage of shales are decidedly petroliferous, I should reluctantly say probably not more than 10 to 15 per cent.   The combination of conditions necessary to deposit an oil shale is present in but few bays or other deeper, stagnant areas where clay muds are collecting.   Therefore the importance to all petroleum geologists of knowing the nature of the sedimentary formations of the areas they seek to exploit.

The rich oil shales of Utah and Colorado appear to be of fresh-water origin—shallow lakes that existed in Eocene time.   We are told that they yield on distillation up to 90 gal.(340 l.) of oil, about 18 lb. (8 kg.) of ammonium sulfate, and up to 4500 cu. ft. (126 cu. m.) of gas per ton of shale.[4] This is the only striking occurrence known to me of fresh-water deposits in North America with an abundance of hydrocarbons. The organic materials are, in the main, plants and their present condition suggests peat deposits. But we must again point out that the age of the rocks is comparatively recent (Green River = early Eocene), and that they have undergone but one slight deformation. Therefore the kerogen still remains.

Abundance of Life Necessary to Petroleum-Gathering.—The petroleum geologist thinks that there must have been a vast abundance of life to make such great storages of oil as are now still present in the shales. In this he is undoubtedly correct, but what he does not keep in mind is the long time it has taken to accumulate the black shales. Accordingly, the quantity of life necessary to oil accumulation need not be so vast at a given time as he thinks. On the other hand, he holds that life did not become abundant enough to result in petroliferous deposits until Middle Paleozoic time. In this connection it should be said that paleontologists have long been familiar with an abundance of macroscopic fossils in rocks dating from the very beginning of Cambrian time and hence from the beginning of the Paleozoic. The seas ever since that period have been filled to their limit with life, microscopic and macroscopic, and in constantly increasing variety. What the geologist sees and gets are the larger fossils; but for every one of these individuals there certainly existed hundreds of thousands and probably millions of invisible plants and animals. It is this minute life, and especially the plants, that is so important in the life cycle, for these microscopic organisms make alive in their bodies the inorganic materials on which they feed. The micro-plants are the basis not only of the subsistence of all the animals of the seas and oceans but, what is equally as important, the accumulation of the hydrocarbons. In this connection we may also add that the almost pure chemically precipitated limestones are due to the metabolic processes of minute plants, the denitrifying bacteria. Accordingly, it is the invisible, and not necessarily the visible, fossils that have gone to the making of the petroliferous deposits of the geologic ages. Most of these forms are short-lived and propagate quickly and in prodigious quantities; the great majority pass through the life cycle in from a few hours to a few days or at most a few months. In this way they make up in quantity what they lose in individual size.

We know of some animal fossils in the late Proterozoic, and even though they are as yet few in number, their high organization teaches unmistakably that there was a host of greatly varying organisms. Of

[4] Dean E. Winchester: U. S. Geol. Survey, *Bull.* 641-F (1916).

lime-secreting algal plants in the Proterozoic, we know vastly more; and from the course of all organic evolution as revealed by the living world, supported by the chronogenesis of the geologic past, we can safely state that at all times, even as far back as the beginning of the Archeozoic as now known in the oldest of geologic deposits, there must have been an abundance of life in the waters of the earth. Hence the abundance of graphite in the Archeozoic and the vast amounts of dark carbonaceous strata in the Proterozoic. Even so, it is hardly probable that commercial quantities of petroleum will be found in the rocks of the Proterozoic, and certainly none at all in those of the Archeozoic, because these very ancient deposits have either been subjected to frequent deformation, or because, due to their great antiquity, the volatile hydrocarbons have long since been liberated into the atmosphere.

The Climatic Factor in Petroleum Making.—The question of land climates probably does not enter at all into the matter of petroleum accumulation, because it is not in the land deposits that the commercial quantities of oil are found. As has been said before, the oils occur nearly everywhere in marine deposits and only rarely in fresh-water ones. This being so, and as the marine shallow waters of today abound in life, whether in the warm, cool, or coldest areas, it follows that we may look for petroliferous formations in almost all continents where the ancient oceans have spilled over them; and this without paying much attention to the changing climates of geologic time. On the other hand, as the greatest amounts of carbon and carbonaceous deposits occur in the north temperate belt, we should seek here in the main for the petroliferous strata. This does not mean that petroleum is absent in tropical lands—far from it. It only points out that the greater quantities will **not** be found in the deposits of former tropical seas, and for reasons to be setf orth.

Since the previous paragraph was written, there has appeared the suggestive paper by Mehl, already cited, in which he points out that all of the major oil fields of the world are situated between 20° and 50° north latitude. Further, that there are no major oil areas within the tropics or in the southern hemisphere. As the known major oil fields lie between the present isotherms of 40° and 70° F., he thinks this distribution "does suggest a distinctly zonal distribution of petroleum in which temperature may have been an important factor." The question that here arises is, Is this suggestion of present climatic conditions also true for the times when the oil was deposited in the strata in which it is now found, remembering that the oil fields were not made recently but are the accumulations of hydrocarbons of the seas of the geologic ages? The answer is not at all in harmony with Mehl's suggestion, for we are living in an exceptional time of stressed climates and marked zonal conditions, while the mean temperature conditions during the geologic ages were warm and equable

throughout most of the world.   And this is even more true of the temperature of the oceans than of the lands.   This being so, much of the
value of Mehl's surmise falls away.   On the other hand, it is undoubtedly
true that high temperatures in clear waters and well oxygenated seas
make, as a rule, for complete destruction of the volatile hydrocarbons,
while those of temperate waters in currentless and muddy areas tend to
preserve them.   The temperature factor, when high, appears to be destructive of volatile hydrocarbon preservation, but in this connection it
should not be forgotten that the seas are far more equable in temperature
than are the lands, and that during most of geologic time the seas were far
more equable in heat content than they are today.   This is thought to
mean that the ancient tropical seas were somewhat less warm than they
are now, while those of the polar areas were no colder than the present
temperate shallow-water areas.   Corals were common in Alaska in
Silurian and Devonian times, corals and warm-water fusulinids lived in
the Carboniferous in Spitzbergen, and there were magnolias and bread-
fruit trees in Greenland during the middle Tertiary.   The writer also
knows that hydrocarbons have accumulated in large amounts in seas
within the tropics, yet seemingly the amount is far the greatest in what
is now the north temperate zone.   That this zone has the greatest amount
of petroleum is apparently due wholly to the greater land masses here,
along with the necessary storage strata accompanied by the proper
amount of deformation.

Even if Mehl's suggestion were correct, and we should accordingly
think of next exploiting the temperate region of the southern hemisphere,
we must not overlook the fact that the northern hemisphere is the land
hemisphere, while the southern one is the water hemisphere, and therefore has greatly reduced continents.   Therefore between latitudes 20° and
50° south we have only the attenuated southern half of South America, the
southern tip of Africa, the southern half of Australia, and New Zealand.
Southern Africa and most of Australia are, furthermore, continental
nuclei or "shields" and therefore have hardly at any time been under the
sea, but in regard to South America the story of marine submergence is
very different.   Even now petroleum fields are known in Peru ("one
of the finest oil fields in the world," according to Thompson), Bolivia,
and Argentina.   Then, too, the fact should be emphasized that "shields"
are largely made up of pre-Cambrian rocks and therefore are barren of
petroleum.

In regard to Mehl's other suggestion of a "barren equatorial belt,"
I am inclined to believe that he is correct in the main; not, however, on
the ground of temperature and climate, but on that of the tectonic geologic and physiographic conditions of the continents.   On the other hand,
attention should be directed to the fact that productive petroleum
fields occur in the Tertiary strata of the tropical zone in the Lake Mara-

caibo area of Venezuela and in the Caribbean Piedmont of Colombia, Trinidad, and Ecuador. Further, highly productive fields are those of the Indo-Malay region, in Java, Borneo, Ceram, and New Guinea. We know that Africa is a continent that was more continuously high above the strand-line than any other, and is loaded with continental deposits, while South America and more especially Australia are not especially rich in marine sediments, and when these are present they have been subjected to mountain-making to such an extent that all of the volatile hydrocarbons have long since vanished into the air or been transformed into fixed carbon. In the northern hemisphere, most of Asia east of the Caspian has also been too much the seat of crustal movements to have much petroleum accumulation in the Mesozoic and Paleozoic formations. From these observations, it appears that the northern hemisphere will always remain the greater for favorable petroleum possibilities.

In all that has so far been said, the statements relate in the main to folded continental masses, but as some most wonderful oil fields, like that of the Baku area in Trans-Caucasia, are of very small extent, it follows that many restricted and highly productive fields are possible even in areas of decided crustal movements, but I should look for such places only in regions of Cenozoic marine formations, and mainly in Asia.

Paleogeography as an Aid in Locating Oil Areas.—The importance of paleogeography in petroleum geology is as yet but little appreciated. Foul sea bottoms, where the hydrocarbons accumulate, and sandstones, in which they are stored, are usually connected with nearness to land. Their physical characters have to do with shallow seas and more especially with headlands, off-shore spits and bars, barrier beaches and river mouths, which divert and from time to time change the currents of the sea. On the other hand, the open seashores, with their more or less long "fetch of the winds," are the washeries of the land-derived detritus. Here the cliff-derived materials are broken up by the waves of the seas in their grinding mills, and the finer erosion materials of the weathering lands, brought by the rivers, are assorted and reassorted many times according to specific gravity and size of grain. The coarsest material lies on the strand and near the shore, and seaward the material becomes, broadly stated, finer and finer of grain. All of this assorting and sea-transporting depends on the size of the waves "kicked up" by the winds, and the shallowness of the waterways. It makes no difference whether it is long or short rivers that deliver the unassorted muds and sands to unagitated and stormless seas, the deposits will be neither petroliferous-making areas nor good rock reservoirs for oil. If, however, such materials are delivered into the open and stormswept seas, there will be assorting according to size of grain, and the sandbars will make headlands behind which current-less waters will accumulate the hydrocarbons. In all this we see that as the places of natural hydrocarbon manufacture and its future storage are conditioned by the nearness of the shore and the depth of water, it behooves petroleum geologists to pay close attention to the discerning of the myriads of constantly changing geographies of the geologic past.

Petroliferous Provinces.—We have now defined the essential principles that underlie, in nature, the gathering of petroleum in commercial quantities and can next consider the question, What constitutes a petroliferous province? Clearly it cannot be merely an area that produces oil, because the word province is significant of embracing things more or less of a kind. Shall the criterion be whether the area has solid or fluid-gaseous hydrocarbons? Or whether the strata are dry or wet? Probably neither. Shall it be the nature of the oil, whether it is light or heavy? Probably not. Seemingly it should rather be the age and time of deformation of the strata having oil, combined with their governing structures. In other words, the classification should express the chrono-orogenetic origin of the oils. For instance, in the Ohio Basin province, a subprovince would be the oil fields in the vanishing Appalachian folds along the western sde of the Allegheny area; another, the eastern Ohio oil fields; and a third, the Trenton area of Ohio and Indiana. A beginning in such mapping has been made by Johnson and Huntley in their "Oil and Gas Production," plates 91 and 92. However, in the course of time we shall here, as in other studies, undergo an evolution in our classifications.

In general, Mr. Woodruff's map and plate 92 of Johnson and Huntley bring out the areas of worth-while exploiting, those of improbable value, and the regions that can have no petroleum. However, these maps are so small that other and even more essential features cannot be depicted; these are the structural trend lines, the periodically rising areas or "crustal highs," the long-enduring ancient lands and their shore-lines, and whether the region has strata of more than one era. Of course, all of these things cannot be plotted on a single map, however large, but until this is done on a series of maps, we cannot define what are the genetic characteristics of each petroliferous province and the proper guidance to its exploitation.

The most important of all geologic problems connected with oil exploitation, the geologic structures, will not be discussed here. Among the most important maps necessary for the broad guidance of petroleum geologists is one to show the "highs" or positive areas and the deformational structure lines, drawn in symbols according to geologic age, *i.e.*, to show the trends of the mountain folds, the many low axes, like the Cincinnati axis, and the greater fault lines. Such a map, of even a limited area, would be a prophetic guide to oil exploitation in the region so mapped.

Can such highly desirable maps be made quickly? Naturally no one geologist can alone make such maps of the North American continent,

or even of the United States. They can be made only through coöperation. . A special commisson for this work should be organized by the
larger oil companies and a philosophical study made of all of the geologic
problems involved in petroleum discovery. For this we have an example
in the study of the principles underlying copper genesis made by the cop-
per-producing companies of the United States, at a cost of about $50,000.
A similar contribution by the oil companies would go far and might,
even in a few years, make all of the required generalizing maps. But will
the companies believe in these possible solutions, and that they will
undoubtedly lead to a more certain and a more constantly successful
exploitation of petroleum in North America? We have faith in our
prophecy, but will the operators have faith in the prophets?

IRVING PERRINE, Hutchinson, Kans.—I think in reading this paper
one should bear in mind its relation to Dr. David White's paper on "Some
Relations in Origin between Coal and Petroleum."[5] In that paper he
discusses the relationship between the percentages of fixed carbon in the
coals, the gravities of the oils, and commercial gas possibilities. His
paper has a map showing certain areas which Doctor White believes to be
hopeless as far as oil and gas possibilities are concerned.

THE CHAIRMAN (C. W. WASHBURNE, New York, N. Y.).—I would
like to emphasize one point brought out by Professor Schuchert. The
southern hemisphere has had an exceedingly monotonous geological history, except the northern border of Africa, the eastern border of Australia, and the western and northern borders of South America. In
other parts of these continents there has been little deposition of marine
sediments and very little deformation since Paleozoic time. Therefore
they are not attractive places to the prospector for oil.
.There is probably truth in Schuchert's idea that the composition
of the sea water may have had something to do with the preservation of
organic matter. I followed the outcrop of an oil sand about 700 kilo-
meters along the western coast of Africa. The fossils in it are exceedingly
minute, showing that the condition of the sea water was not suitable
for vigorous life, the oysters are not much larger than the head of a lead
pencil, and nearly all forms are dwarfs. In Madagascar there is the
same formation with similar faunal conditions. If the water in semi-
enclosed basins is very salty water, bacteria cannot thrive in it much
better than the molluscan forms of life. This is probably an indication
that the composition of the sea water in enclosed basins may have something to do with the preservation of fats and waxes in the sediments of
certain areas. ' ..

. 5 *Jnl.* Washington Acad. Sciences (Mar. 19, 1915).

Nature of Coal

By J. E. Hackford, London, Eng.

(St. Louis Meeting, September, 1920)

In some research work carried out by the writer, certain results have been obtained which bear on the fundamental nature and origin of coal and the relationship between coal and petroleum. Without entering into a discussion of the details of the experiments, which were conducted on petroleum and derived bitumens, there are given here, by way of definition, some of the relations that the writer has established between certain classes of bitumens of petroliferous origin.

Bitumen.—A natural organic substance, gaseous, liquid, or solid, consisting of hydrocarbons and the oxy- or thionic derivatives of the same, or of a mixture of all three.

Diasphaltenes.—Those portions of bitumens that are soluble in ether or carbon disulfide, but are insoluble in a mixture of equal parts of ether and alcohol. Diasphaltenes are produced by the oxidation or thionization of petroleum oils; they have, as the name indicates, twice the molecular weight of asphaltenes, into which they are converted when subjected to moderate temperature. For example, an artificially produced diasphaltene, which was readily soluble in pentane and ether, was quite insoluble in either of these solvents after heating for three weeks at a temperature of 100° C., and was converted into an insoluble asphaltene.

Asphaltenes.—Those portions of bitumen that are insoluble in ether or ether alcohol but are soluble in carbon disulfide.

Asphaltites.—Those solid or semisolid natural bitumens that are composed, for the most part, of asphaltenes or diasphaltenes. A pure asphaltite[1] would be composed wholly of asphaltenes and diasphaltenes, but most asphaltites contain small percentages of oil and wax, which have not yet been converted into asphaltenes; they may also contain a small percentage of kerotenes, which represent the next stage of the metamorphosis of asphaltenes. Among the naturally occurring oxyasphaltites may be mentioned grahamite; and among the thioasphaltites, gilsonite.

---

[1] The term asphaltite, as recommended by Eldridge (*22nd Annual Report*, U. S. Geol. Survey, 1901) is preferable to Dana's term "asphaltum" ("Descriptive Mineralogy," 6th edition, 1906, 1017), for the reason that the naturally occurring representatives have the generic ending "-ite," *e.g.*, gilsonite, grahamite, etc.

Kerotenes.—Those portions of bitumen that are insoluble in carbon disulfide. They are produced, by gentle heat, from asphaltenes. It can be demonstrated experimentally that artificially produced thioasphaltenes and oxyasphaltenes, when kept at a temperature of 100° C. for three months, are converted, with but slight gaseous losses and without change in sulfur content, into kerotenes.[2] Most of the kerotenes produced by gentle heating from asphaltenes in this manner were entirely insoluble in any known solvent, including pyridine, chloroform, and quinoline.

Kerols.—Those portions of kerotenes that are soluble in chloroform as well as in pyridine.

Keroles.—Those portions of kerotenes that are soluble in pyridine but insoluble in chloroform.

Kerites.—Natural solid bitumens composed, for the most part, of kerotenes. A pure kerite would be composed wholly of kerotenes, but the natural kerites generally contain small percentages of one or more of the following: asphaltenes, diasphaltenes, wax, and oil, whose conversion to kerotenes has not been completed. Of the natural examples, wurtzilite may be mentioned as a thiokerite and albertite as an oxykerite.

It has been demonstrated, in the course of these experiments, that either sulfur or oxygen can play a predominating role in the formation of these classes of bitumens. If a straight Pennsylvania lubricating oil with a negligible sulfur content is digested at a temperature of 100°, with either sulfur or oxygen, a darkening in color first takes place (owing to the formation of thio- and oxydiasphaltenes); this discoloration gradually increases to black with the formation and precipitation of asphaltenes, which constantly increase until the whole, except for gaseous losses, is converted into kerotenes. Similar results have been obtained from sulfur-free paraffine wax[3] and from natural petroleum oils of all characters; that is, by oxidation or thionization, accompanied by gentle heat, any natural petroleum oil may be converted first into oxy- or thioasphaltenes then into kerotenes. Certain kerotenes are wholly insoluble in any of

---

[2] This term is derived arbitrarily from the word "kerogen," the term introduced by Crum Brown (Oil Shales of the Lothians, Geol. Sur. of Scotland, 1912, 43) to denote the organic matter present in oil shales, in ordinary solvents, and from which hydrocarbons are obtained by dry distillation. It was at first proposed to use the term kerogen, which would be entirely appropriate in this general sense, but it was felt that some confusion might arise because the word kerogen has become associated with the bitumen of the oil shales alone.

[3] Allen (*Pet. Rev.*, Apr. 26, 1913) and Redwood ("Treatise on Petroleum" 1, 275) consider the black precipitate formed in paraffine wax, when heated with sulfur, to be carbon, but the writer has demonstrated that this precipitate dissolves entirely when heated with benzol; it therefore cannot be carbon. The addition of an excess of ether or pentane to this benzol solution throws down a black precipitate, which is simply a thioasphaltene.

the known solvents, including chloroform, pyridine, and quinoline. As these experiments progressed, it became evident that bodies closely analogous to coal were being produced from petroleum in the laboratory by oxidation, thionization, and gentle heat; this gave rise to certain inferences, which it is the purpose of this paper to state.

RESULTS OF PREVIOUS INVESTIGATORS

The elucidation of the nature of a body, like coal, that is only sparingly soluble in solvents and cannot be made to yield crystalline derivatives without previous violent manipulation has naturally presented no little difficulty. During the past five years a large amount of work has been accomplished respecting the nature of coal by numerous investigators.[4]

These investigations have been mainly along two lines: one was the examination of solvent extracts, and the other was the study of the products of low-temperature distillation. The results are scattered and the inter-relationships have not been fully pointed out. Briefly stated, the studies of these investigators have shown:

1. That by low-temperature distillation work and by the examination of solvent extracts, paraffine, olefines, and naphthenes have been isolated and identified.

2. That the tar distilled from coal at high temperatures is a decomposition product of coal tars previously formed at low temperatures.

3. That the cellulosic compounds present in coal result in the formation of phenols upon dry distillation.

4. That the temperature at which coal was formed cannot have approached 300° C.

RELATION OF SOLUBLE PORTIONS OF COALS AND KERITES

In 1913, Messrs. Clark and Wheeler[5] described experiments in which a soft bituminous coal, upon extraction with pyridine, yielded a substance representing by weight a percentage of the original sample, which upon subsequent low-temperature distillation yielded a mixture of paraffine-hydrocarbons and hydrogen. In view of his research, the writer suspected that the portions of coal extracted in this manner by pyridine consisted, largely, of asphaltites and the soluble kerites; accordingly the following experiment was carried out:

---

[4] D. T. Jones: *Jnl. Soc. Chem. Ind.* (1917) **36**, 3–7; Jones and Wheeler: *Chem. Soc. Trans.* (1916) **109**, 707, 714; Burgess and Wheeler: *Chem. Soc. Trans.* (1910) **97**, 1917–1935; (1911) **99**, 649, 667; (1914) **105**, 131–140; Clark and Wheeler: *Chem. Soc. Trans.*, **103**, 1704–1713; R. Maclaurin: *Jnl. Soc. Chem. Ind.* (June, 1917); Pictet and Bouvier: *Compt. Rend.* (1913) **157**, 779–781; Pictet, Ramseyer and Kaiser: *Compt. Rend.* (1916) **163**, 358–361; Fischer and Glund: *Berichte* (1916) **49**, 1469–1471; and Fraser and Hoffman: *Tech. Paper 5*, U. S. Bureau of Mines.

[5] *Chem. Soc. Trans.* (1913) **113**, 1704–1713.

A sample of 250 gr. of Yorkshire coal was extracted with pyridine. The bulk of the pyridine was then distilled off under reduced pressure and a. large excess of ether added. A voluminous black precipitate was thrown down, which was pumped, washed with ether, and weighed. By weight, it represented 5.1 per cent. yield. This black powder was found to be 15 per cent. asphaltenes and 84.9 per cent. kerotenes. The 84.9 per cent. of kerotenes was found to be a combination of 17.9 per cent. of kerols and 67 per cent. of keroles. We thus succeeded in splitting up this black precipitate in a similar manner and in similar fractions to those obtained when working upon natural kerites, as, for example, albertite and wurtzilite, which gave the following results:

	OXYKERITE (ALBERTITE) PER CENT.	THIOKERITE (WURTZILITE) PER CENT.
Asphaltenes	9.0	12.8
Kerotenes	89.03	81.37
Sulfur	Trace	5.83
Oxygen	6.97	0.00

The similarity, however, does not end here, for many of the fractions upon heating melted with decomposition, evolving oil containing (in the case of albertite) quantities of paraffine wax; while the asphaltenes and kerols evolved sulfuretted hydrogen. The most sparingly soluble fraction, keroles, do not intumesce to any extent upon heating, as do the asphaltenes. The solubilities of these substances are exactly the same as those similar fractions derived from natural kerites, e.g. the asphaltenes

TABLE 1.—*Analysis of Unfractionated Precipitate*

	Asphaltenes and Kerotenes from Coal Per Cent.	Natural Kerite, e.g. Albertite from New Brunswick Per Cent.	Kerite in a Transformer Sludge[a] Naturally Produced by Oxidation of Transformer Oil Per Cent.	Synthetic Oxykerite Prepared by Passing Oxygen Through Lubricating Oil Per Cent.
Asphaltenes and kerotenes	100	98.03	100	100
Carbon	73.64		76.0	74.0
Hydrogen	4.87		7.1	6.2
Sulfur	1.07	trace	?	1.58
Nitrogen	2.83	1.4	?	
Oxygen	16.67	6.97	16.97	18.22

[a] Dr. A. C. Michie [*Jnl.* Inst. Elec. Engrs. (1913) **51**, 213] gives an analysis of a sludge deposited by a transformer oil when used in an auto-starter for a considerable period. The writer has carried out detailed experiments on a similar sludge. The original oil in this case was known to be a straight cut oil. The sludge was found to consist of 10.1 per cent. of oxykerotenes and 79.9 per cent. of oxyasphaltenes. The oxyasphaltenes, after gentle heating for a month, were converted into oxykerotenes, portions of which were insoluble.

both from the coal and from a sample of a natural kerite were soluble in carbon disulfide, benzene, phenol, nitrobenzene, chloroform pyridine, etc. but were insoluble in petroleum ether, ethyl ether, ethyl alcohol.

The analysis of the whole unfractionated precipitate is given in Table 1, and, for the sake of clearness, is contrasted with a natural kerite, a naturally produced kerite, and a synthetic kerite.

RELATIONS OF INSOLUBLE PORTIONS OF COALS AND KERITES

It has been found that, upon prolonged heating, a portion of the kerotenes becomes insoluble in pyridine or any known solvents; by inference it is believed that most of the insoluble portion of coal consists of a true bitumen that has been transformed by gentle heating into an insoluble kerotene, and that a small portion is due to the decomposition products of cellulose, as shown by the formation of phenol upon dry distillation.[6] The writer has proved that the kerites experimentally produced from petroleum yield, at both low- and high-temperature distillation, exactly the same products as are obtained under the same temperature conditions from the kerites of coal.

THEORY OF FORMATION AND NATURE OF COAL

The following theory is put forward as to the mode of formation and nature of coal, comparing it at the same time, for the sake of clearness, with the mode of formation of oil.

First, consider a stratum containing a deposit of either animal remains or marine vegetation. These substances, on decomposition, form oil and gas which, if contained in a sandy bed, are swept away from their source by either gravity or water as rapidly as formed, since neither the animal remains nor the marine vegetation contain cellulosic material capable of forming a spongelike mass, which would hold the oil *in situ* during the decomposition stage.

Second, assume a buried deposit of terrestrial vegetation. Decomposition takes place, resulting in the formation of oil and gas, as in the case of the marine vegetation. However, owing to the cellulosic nature of the material and its porous spongy nature, the oil is kept *in situ* while decomposition proceeds. Accompanying this decomposition, there is probably a rise in temperature, which even if not above 100° C. is quite sufficient, as we have proved in the laboratory, to convert into kerotenes the oxy- or thioasphaltenes that are simultaneously formed with the oil. As the process goes on, the kerotenes become more and more insoluble until they are insoluble in pyridine and quinoline and so remain as a solid in the spongelike mass afforded by the cellulosic structure of the terrestrial vegetation.

[6] Jones and Wheeler: Chem. Soc. *Trans.* (1916) 109, 707-714.

It has been recorded by Hodgland and Lief[7] that the algae on which they made tests contained from 5 to 13 per cent. of sulfur. It therefore follows that in those coals that contain algal ingredients in quantity, some undoubted cases of which White[8] puts on record, a larger amount of thio-bitumens should be present with a corresponding reduction in the oxy-bitumens and the cellulosic residues.

According to this theory, the amount of soluble bitumens should be greatest in peat and should decrease through lignite, sub-bituminous, bituminous, and semibituminous coals to anthracite, which indeed is the case. It is interesting to note that where pure kerite deposits have been found, they have nearly always been mistaken for coal. It took ten years' litigation to decide whether the New Brunswick oxykerite was coal or bitumen. Similar instances are given by L. L. Hutchison[9] in the case of the Jackfork Valley, the Impson Valley, etc. A similar case of a thio-kerite is a deposit in Nova Zembla, where coal suitable for metal smelting was reported to be situated near an ore deposit. Samples of this deposit were forwarded to the writer and yielded on analysis: ash, 0.72 per cent.; sulfur, 15.54 per cent.; nitrogen, 0.76 per cent. The sample possessed a bright luster and had the appearance of a bright soft coal. It was, however, totally insoluble in solvents and on heating gave off little gas. No oil whatever was evolved; in fact, the sample behaved in nearly every respect like anthracite. The volatile matter was only 1.8 per cent. However, from a comparison with certain experiments then in progress, it was decided that the material was a kerite. A subsequent geological examination showed the deposits to occur in small lenses in a metamorphosed deep-sea limestone, which contained none of the depositional associate of coal and, in fact, confirmed the oil origin of the deposit. This is regarded as a pure sample of a thiokerite. It is probably true that certain so-called coals from Colombia that have a sulfur content of 13 per cent. are simply thiokerites.

The main differences between these so-called coals and true coal rests in the fact that they possess no cellulosic residue, which upon distillation can produce phenols, as is the case in true coals. It is conceivable that a kerite produced from microscopic vegetal remains containing some cellulose—but not in sufficient quantities to act as a sponge—would yield phenols on dry distillation; this would be but another connecting link between coal and petroleum.

Petroleum oils, such as occur in nature, are clearly not derived from coal; but given a quantity of vegetal material, petroleum may be produced under a given set of circumstances if no cellulose is present and coal will be formed if the vegetal matter contains sufficient cellulose to form a sponge.

---

[7] *Jnl. Biol. Chem.* (1915) **23**, 287–297.

[8] David White: U. S. Geol. Survey *Bull.* 29, 48 *et. seq.*

[9] Oklahoma Geol. Survey *Bull.* 2, 81–89.

DISCUSSION

W. E. Pratt,* Houston, Tex.—Mr. A. W. McCoy some time ago, after pressing or squeezing, extracted oil with ether from oil-shales which before squeezing yielded no oil upon extraction. Mr. C. W. Washburne, in discussing McCoy's results, attributes the formation of oil in the shale to heat induced by pressure rather than to pressure directly. This seems to be McCoy's idea also; that is, the ether-soluble content increased upon the application of heat (through pressure). Mr. Hackford finds that similar materials which have a certain ether-soluble content suffer a decrease in ether-soluble content through the direct application of heat. There is an apparent contradiction in this situation which may be explained, perhaps, by assuming that heating "cracked off" new ether-soluble combinations in each set of experiments, but that these new compounds were allowed to escape in Mr. Hackford's work, leaving the residual material less ether-soluble, whereas Mr. McCoy retained the cracked products in the original material until he extracted them with ether.

David White, Washington, D. C.—The theory that beds of coal are bituminized from outside sources is, I believe, to be regarded with great skepticism. That the bitumens, so called, are generated in the process of the evolution of the coal bed itself appears more tenable, and will, I anticipate, be ultimately proved.

The distinction between the origin of the normal series of coals, namely from terrestrial or vascular vegetation, and of the oil-shales, from aquatic and largely cellular plant debris, is emphasized very properly by Mr. Hackford. Putting the distinction in terms related to the chemical distinctions, coals may be said to be characterized by ingredient carbohydrates, while oil-shales embrace waxy, resinous, gelatinous, and other plant products.

E. DeGolyer, New York, N. Y.—Since it had always been held that the oil found in the coal mines of England was distilled from the coal, it became extremely important to prove whether or not it was a coal-tar distillate or true petroleum. This was Mr. Hackford's contribution to that work.

He has given also some interesting suggestions as to the origin of Mexican oils and the Gulf Coast oils. His theory provides for the sulfur content of the oils of coastal Texas and Louisiana, the Isthmus of Tehuantepec District and the Tampico District. The Tampico area is not a salt-dome region, but its oils have a high sulfur content.

I have not paid much attention to oil-shales, but I have observed that the English chemists and geologists, who are best acquainted with oil-

---

* Chief Geologist, Humble Oil & Refin. Co

shales and not so well acquainted with petroleum as their American colleagues, think that the oil-shales are derived from petroleum; the petroleum came first and the oil-shales as some sort of secondary product. American geologists and chemists seem to argue in the other direction. We are better acquainted with petroleum than with the oil-shales and there is a marked tendency at present to regard petroleum as resulting from the natural distillation of oil-shales. Both groups are trying to explain the known by the unknown.

REINHARDT THIESSEN,* Pittsburgh, Pa. (written discussion†).—The writer agrees with three of the conclusions drawn from investigations of coal by means of solvent extracts and low-temperature distillation, but does not fully agree with the conclusion that the cellulosic compounds in coal result in the formation of phenols upon dry distillation. He believes that there is not enough proof to warrant so definite a conclusion. Experimental proof does not indicate that phenols result entirely or exclusively from the cellulosic derivatives of coal. Only relatively small amounts of tar are formed in the dry distillation of cotton; according to Cross and Bevan,[10] this tar is composed of water, furfurol, phenols, liquid and solid hydrocarbons; according to Tollens[11] it also contains allyl alcohol and creosote. Schwalbe[12] questions whether any of these products are formed from pure cellulose; he believes that they are distillation products of the substances associated with the cellulose. Why should the cellulosic derivatives in coal be considered the source of the phenols when the plants, as a whole, contain so much and so many phenols?

The writer has given considerable time to the study of the origin and constitution of coal and allied substances by examining thin sections under the microscope which has given abundant proof that the important coal beds have been formed from woody plants, trees, and shrubs, rather than from herbs, grasses, mosses, and algæ or similar organisms.

The mode of deposition and formation of the peat bogs that formed the present coal seams may be studied by examining the various types of existing peat deposits. It is probable that each kind of coal seam has its analogous deposit in present deposits of peat. For example, the ordinary bituminous and subbituminous coals and lignites in the arboreal-peat swamps; the cannel and boghead coals in the quaking bog or marsh; and the bituminous or oil-shales, in the open bog.

A study of arboreal-peat deposits, such as the Dismal Swamp and

---

* Research Chemist, U. S. Bureau of Mines.
† Published by permission of the Director, U. S. Bureau of Mines.
[10] C. F. Cross, E. J. Bevan, C. Beadle: "Cellulose," 69, 1895.
[11] B. Tollens: "Handbuch der Kohlenhydrate," 1, 233, 1891.
[12] Carl G. Schwalbe: "Die Chemie der Cellulose," 33, 1911.

those found abundantly in Wisconsin and Michigan, shows that they are composed of semi-decayed logs, branches, twigs, and stems. These are embedded in a general debris consisting of semi-decayed chips or fragments of wood and bark, leaves, cuticles, rootlets, small twigs, mosses, lichens, in all degrees of fragmentation, and of spores, pollens and resins which, in turn, are embedded in an attritus derived from all kinds of plant parts and the whole mass has been transformed into peat by means of putrefying organisms. The resinous contents of the woody parts are still in place.

There is only a relatively short step from peats to the lignites. A description of the composition and the constituents of peat will do equally well for that of lignite, except that in the transformation from peat into lignite, coalification process has taken place and the mass has been greatly compressed and hardened.

The subbituminous coals are formed from the same or similar kind of plants and plant products laid down under the same or similar conditions, and often during the same time as the lignites. In certain cases a lignite bed and a subbituminous coal bed are parts of the same deposit, but the transformation, or coalification, has gone further and the mass has been further condensed and compressed in the one part than in the other. The ordinary bituminous coals form but another step in the chain of the transformation of peat into coal. The chemical nature and structure must necessarily differ widely from those of peat or lignite for the coalification process has been carried on for a longer period.

By far the largest part of coal is derived from logs, stems, branches, twigs, and roots, which are represented in the coal by the black glistening band of varying thicknesses and widths. The thicker and wider bands represent logs and stems, while the thinner delicate black glistening bands represent smaller chips or fragments. The duller bands between these represent the general debris, which consists of coalified fragments of all kinds of woody plant parts and smaller fragments of wood, bark, leaves, petioles, cuticles, and macrospores, and an attritus. The attritus consists of finely macerated coalified plant degradation matter, spores, pollens, resins, and cuticles.

The bituminous coals are generally of the Paleozoic age, when the plants were chiefly Calamites, plants belonging to our modern horsetails; Lepidodendrons, and Sigillarias, plants belonging to our modern club mosses or lycopods; and secondarily of Cycadophytes, plants belonging to the modern Cycads; Cordaites, trees belonging to the recent conifers and ferns. The lignite-forming plants consisted chiefly of conifers. This difference in the kind of plants does not necessarily account for the chemical differences since the chemistry of plants is, in general, quite the same.

The cannel and boghead coals are composed largely of attritus, which consists chiefly of spore matter, some resinous matter, and finely divided plant degradation matter, of which the spore matter usually forms by far the largest part; they usually contain a large amount of inorganic matter. Anthraxylon is but sparingly present in the boghead and cannel coals. Beds of ordinary coals often include layers that are in every respect like cannel coal. When such layers are thick enough to be easily noticed, they are called bone or cannel coal.

The oil-shales are in many respects similar to the cannel coals; the chief difference is in the higher mineral contents of the oil-shales. Torbanite is called both a boghead coal and a rich oil-shale; before petroleum was discovered it was extensively distilled for oil.

Oil-shale, examined with the microscope, is seen to contain the same kind of objects as cannel coal but present in different proportions. In many shales, spore-exines form the largest part of the organic matter; in others, spore-exines and plant degradation matter are present in about equal proportions; in some, few or none are recognizable.

As deposits that contain constituents very similar to those contained in oil-shales are being laid down at the present time, much may be learned through their study. Such deposits are being laid down in depressions without proper drainage contàining a rather shallow body of water. The water is not deep enough to prevent vegetable growth, but it is too deep for a woody plant growth, consequently a luxuriant aquatic plant and animal life is sustained. As long as this area is maintained, the dead plant and animal matter largely decays and disintegrates. Certain parts of the plants and certain plant products, though, resist decay; pollen grains, spores, resinous matter, waxes, cuticles, certain woody parts, etc., are among these. But even in the decay of the more delicate parts a resistant degradation matter is left. All of these, together with the mineral matter of the plants and that blown into it as dust and washed into it by streams, form a slimy ooze at the bottom, which on drying has much the appearance and consistency of art gum. In many respects it is similar to peat. The constituents are of the same kind as those of the oil-shales.

Plants consist mostly of cellulose and its modified form known as lignocellulose; unfortunately, too little of the chemistry of this substance formed through decay is known. After wood has partly decayed, as the wood in peat, it is no longer cellulose nor lignocellulose; nobody knows what it is. In addition to lignocellulose, plants contain resins, gums, waxes, fats, oils, tannin, proteins, chloroplasts, various kinds of alcohols, ketones, aldehydes, acids of the aliphatic series; and phenols, quinones, alcohols, aldehydes, ketones, acids, turpenes, camphors, glucosides, tannins, alkaloids, and others of the aromatic series. Many of these are stable compounds and resist decay; others have resistant radicles and, after the end products or side chains have been torn away through putrefying and coal-forming agencies, leave a resistant substance. Particularly significant in this respect are the heterocyclic and the cyclic plant compounds and their derivatives. All organic plant substances are organic, or carbon, compounds either in a straight carbon chain, a carbon ring or rings with side chains or end groups, and all are capable of losing their side chains or end groups and leaving a hydrocarbon compound. During the transformation of plant substances into coal, cannel coal, or oil-shale, there is a reduction reaction and the organic compounds tend to form hydrocarbons; geologists term this deoxygenation. This process also constitutes what is generally termed bituminization. But we have no clear idea of what bituminization is nor what constitutes a bitumen.

We have some knowledge as to what is going on in the transformation of the plant substance into peat. Many of the organisms bringing about fermentation and putrefaction have been isolated and their activities studied and their products have been analyzed and are known. But after the deposits have been covered for years, and the activities of the organisms have ceased, changes continue. What these changes are and how they are brought about should be a fruitful field for research.

C. E. WATERS,* Washington, D. C. (written discussion).—The paper is of interest because of its bearing on the behavior of petroleum oils when used as lubricants in internal-combustion engines. Up to 90° or 100° C., petroleum oxidizes very slowly in the dark; at 200° C. and above, the oxidation may be very rapid, with the formation of compounds that are precipitable by the addition of petroleum ether. "Sludging" tests for transformer oils, Kissling's tar- and coke-forming tests, and the writer's "carbonization" test are based on this fact.

The rate of oxidation is accelerated by increasing temperature and by the presence of alkalies, iron oxide, sulfur and sulfur compounds, and the oxidation products. Filtration through bone black or fuller's earth largely removes these oxidation products, which are in solution, or perhaps more correctly in colloidal suspension, in the oil.

The precipitates thrown down by petroleum ether are almost completely soluble in benzene. They are usually dark brown and fine grained, so that they form porous lumps after the oil is washed out and they are dried in an air bath. Some of the precipitates are granular and some, after drying, are jet black with a coaly luster and look as if they had been fused, or at least sintered together, during the drying.

Some chemists reject the idea that the carbon deposits in an engine can be formed by partial oxidation of the oil, with formation of asphaltic matter. They regard cracking and incomplete combustion, both of which reactions deposit carbon, as the causes. But these deposits, after the removal of the adhering oil by extraction with petroleum ether, contain

---

* Chemist, Bureau of Standards.

much soluble matter. Benzene extracts several per cent. Following this pyridine gives a dark-brown solution that filters easily. Five per cent. caustic soda yields a dark-brown solution that is difficult to filter, evidently on account of its colloidal nature. When the residue on the filter is washed, enough runs through to render the filtrate turbid. The addition of sodium chloride makes caustic-soda solution easier to filter because the colloids are partly precipitated, as is shown by the lighter color of the solution.

Value of American Oil-shales*

By Charles Baskerville,† Ph. D., F. C. S., New York, N. Y.

(Chicago Meeting, September, 1919)

Shales containing "kerogen," or bituminous matter, which on destructive distillation yield oily and tarry matters resembling petroleum are here designated as *oil-shales*. They differ from *oil-bearing* shales from which petroleum may be obtained by so-called mechanical means. The educts obtained by the destructive distillation resemble some or all the varieties of petroleum, depending on the character of the shale and the mode of treatment. Some shale oils have a paraffin base, some an asphaltic base, or a combination; some run high in sulfur compounds. The methods of refining and cracking, therefore, are essentially the same as are used in refining petroleums.

In 1860, in this country, over fifty companies were successfully distilling various natural bituminous materials for the production of "coal oil," used for illuminating purposes. The discovery of petroleum and the failure of these companies to save and utilize the valuable byproduct, ammonia, brought about their inevitable doom. Prior to that time, more or less successful efforts were made to produce from the shales of Scotland oils for illuminating and heating purposes. Competition of native petroleum from the United States early eliminated some of these companies and with the entrance of oil from the Russian and other fields into the world's markets, the Scottish oil-shale industry underwent serious and trying experiences until, in 1916, only four (Scottish) were paying concerns. These survived only through energy and the application of skill in saving valuable byproducts.

A few companies have successfully operated in France and New Zealand. The Canadian Government showed active interest in the New Brunswick shales, which exist in quantity and are more valuable than the Scottish shales. The retarded development of that valuable asset of the Province of New Brunswick was most unfortunate, especially when the product was so much needed in the prosecution of the war.

The economic success of a shale-oil industry depends on the following factors:

---

* This paper was presented by request at the Denver meeting of the Institute. Delay in its publication gave the author an opportunity to revise that part which had to do with the prosecution of the war. However, the fundamental features concerning the economic development of an important natural resource are given as indicated in the original communication.

† Professor of Chemistry and Director of the Chemical Laboratories, College of the City of New York.

1. The shale, on distillation, must yield an oil simulating petroleum in character and composition. The distillation is carried on in retorts variously designed, preferably to make the process continuous. Normally the shale, in pieces of suitable size, is fed into a retort near the top of which the shale is subjected to a fairly low lateral heat. The products of distillation thus produced are swept out by a current of gas produced below. As the shale passes through the retort it is subjected to a more intense heat, which brings about the distillation of the heavier products. The carbonized residuum then comes into contact with regulated blasts of steam (and air), which generate water (or producer) gas. This gas passes through the cooler parts of the retort and assists in sweeping out the products evolved at the lower temperatures. The entire gaseous product passes through suitable condensers to remove the oils, paraffin, tar, etc., and through scrubbers to remove the ammonia; and the residual gas is then burned in annular chambers to provide the lateral heat referred to. The ash, often more than 50 per cent. of the original shale, is automatically removed from the other end of the retort by various mechanical devices, somewhat similar to the Mond-Lymn sealed gas producer. The Scottish practice involves four retorts in a unit, which units are multiplied into banks. A unit, four retorts, handles about 10 tons of shale per day of 24 hr. The condensers and scrubbers resemble those of ordinary gas (coal and water) works. In other words, there is no great necromancy in distilling oil-shales and refining them, as some might have one suspect or believe.

2. The shale must yield oil in such abundance as to pay the costs of mining and treatment, or the character of the oil must be such that it possesses unusual value; for example, a high percentage of paraffin, or a notable amount of ichthyol.

3. Since the last-mentioned conditions are comparatively rare in the oil-shale industry, a valuable byproduct is essential to carry the burden of mining and treatment. The combined nitrogen, which is largely converted into ammonia in the distillation, has been the salvation of the few surviving Scottish companies and must be an important consideration in any shale-oil industry anywhere.

4. Assuming adequate oil educts (30 to 60 gal. per ton of shale distilled) and a supporting ammonia output, the oil shale must be in ample quantity and so situated as to be mined in the cheapest manner. Adequate water supply is essential for condensing and other purposes for the crude-oil works.

5. An adequate supply of sulfuric acid for the absorption of the ammonia is essential. If 30 lb. of ammonium sulfate were obtained per ton of shale, it would call for some 25 lb. of sulfuric acid per ton of shale treated (round figures are used), or 12,500 tons of 92 per cent. sulfuric acid for every million tons of shale treated. A 50,000,000 bbl.

annual production would thus call for 625,000 tons of sulfuric acid, which is no mean quantity. An annual increase of over 800,000 tons of ammonium sulfate from such an operation would materially affect the market for that substance. However, the product has a variety of valuable uses, not only in agriculture and chemical manufacturing, but in refrigeration.

6. As observed, the character of the shale and the mode of distillation determine the quality of oil obtained. Although the process and its products are simple in outline much unknown along these lines awaits investigation. It is known, however, that different modes of treatment yield crude oils of entirely different composition. Furthermore, the field-test methods practiced, while giving valuable empirical information as to the character of the shale under a uniform system, fail utterly to tell the proper procedure to be followed to secure the best values. Laboratory methods come nearer the truth, but the only truly accurate way is by commercial tests in full-sized units.

The character of the shale, whether caking or non-caking, is important in determining the proper mode of treatment. For the present we may dismiss consideration of the "caking" shales, which really involve methods for treating cannel coals, and consider only the non-caking; that is, the "curly" (massive) and "paper" shales. Curly and slickensided shales are characteristic in Scotland; these and paper shales are found in Canada. The paper shale appears to predominate in certain parts of the United States.

Much discussion has arisen as to the best method of treating the shales found in very large quantities in Colorado, Nevada, Utah, and Wyoming.[1] It has been claimed that the Scottish practice is not the best for our American shales. To be sure, a successful industry in one environment may fail when transplanted, but experience has led me in a new field to adopt the best practice of a given environment and then allow it to evolve with the changed conditions. There is reason to believe that this procedure will be pursued by the Bureau of Mines, which is expected shortly to erect a commercial experimental plant in the field. Initiative has already been shown by some companies, whose engineers, as a result of research, have erected small experimental plants.

Several processes have been devised to strip the oil of its gasoline as fast as it is produced. Some attempt to fractionate even further (light oils, fuel oils, and residuum) during production; this line of attack does not commend itself to me. The crude oil, stripped of gasoline, will have an inferior value and will still require refining, as will also the gasoline thus stripped. One of the most noteworthy processes is based on a circulation of gas, which, after scrubbing, passes back through the distilling mass, thus taking advantage of the vapor pressure of the dis-

---

[1] See the reports of the Bureau of Mines and the Geological Survey of the United States, especially Bull. 641–F by Winchester (1916).

tillation.   The distillation is thus accomplished at a much lower temperature, with a saving of fuel and a larger yield of valuable products.

Whatever process may be proved to be most suitable, and no doubt several may be shown to possess distinct advantages, it must be remembered that the production of shale-oil in the West is not so much a problem of mining as of manufacturing.   Indications point to the easy application of the simplest mining methods to this field.   The mining question has been dealt with in reports by Winchester,[2] Hoskins,[3] and others, especially George,[4] whose advice in regard to oil-shales in Colorado in particular should be sought.

The production of petroleum in the United States is not keeping pace with consumption.   This condition, while it was accentuated by the war, is not an actual outgrowth of it.   The extension in the use of the gas engine and the development of oil-power energy producers have caused notable increases in the consumption of liquid fuel.   The rich Mexican fields may supply the deficit in production within the United States and the untapped oil reservoirs of South America may yet flow to our refineries, but the difficulties of transportation and the establishment of satisfactory trade relations, which are not unsurmountable, impress one with the importance of self-containedness, especially in connection with a raw material on which so much of our national industry depends.

The annual production of crude petroleum within the United States for 1918 is estimated at 300,000,000 bbl.   This will require a material addition to keep the 477 refineries in operation up to their capacity of 490,000,000 bbl. annually.   Hence new oil fields or new sources of crude oil, or both, must be developed.   Rumors of prospecting in some new fields and of active attempts to open up new pools in old oil regions are current.   War demands, which obtained and are likely to continue for some time, and the lack of a universal carburetter inhibit the use of such substitutes as benzene and ethyl (grain) and methyl (wood) alcohols for the time being.   To meet the deficiency, within recent months attention has been directed acutely to the enormous latent fuel-oil resources dormant in American oil-shales.

Recently my attention has been drawn to a variety of flamboyant advertisements in connection with the shale-oil industry, which were so misleading that I hope the Institute will take adequate steps to safeguard, as well as foster, a promising industry.   It is no business for an individual who expects quick returns.   Too much stress cannot be laid upon the fact that it is a manufacturing industry requiring ample capital for large operations with the very best of technical skill.   With these and with patience, the enormous resources now dormant in American oil-shales may be developed into a great and profitable industry.

[2] *Op. cit.*   [4] State Geologist of Colorado.
[3] *Min. & Sci. Pr.* (Apr. 13, 1918).

DISCUSSION

ARTHUR L. PEARSE, London, Eng. (written discussion).—In the last paragraph Professor Baskerville correctly sums up an important position. The paper was probably written some months ago, as is indicated; if it were written today he would have further emphasized these conclusions. The oil-shale is a great industry, has been for many years, and bids fair to become one of the most important. This industry and its twin—the carbonization of coal—are the most important unorganized industries in the world today.

We are not precise enough when we talk about the Scotch shale-oil practice. If reference is made to the system of retorting that reached an assumed standard some 10 yr. ago, I would say that no one would build such retorts today; but they are good enough to wear out and there are more of them in Scotland than there is shale to keep them going. If the reference is to the Scotch system of treating the oil, evolved out of much experience and generally adopted as standard 6 yr. ago, I would say that this method has been replaced by fractional distillation and cracking plants. The old Scotch retort is not the best to use on either American or any other shale. The adoption in the latest English plant, of which the first unit is 1000 tons daily, of an entirely different retort proves this.

Principally owing to better practice, evolved out of work on the carbonization of volatile coals and other hydrocarbons, to say nothing of shale, we have learned a great deal. With the exception of the cases when the carbonized residue is required in such shape as metallurgical coke, for instance, and for which the coal or material is primarily treated, all the older methods of carbonization in ovens, intermittent or continuous verticals, etc., and where mass carbonization is adopted, are obsolete. By mass carbonization is understood the heating of a body of material, the particles of which are in close contact with each other, in contradistinction to a condition in which each particle is unconfined. Mass carbonization involves the passage of the heat units from the wall of the retort into the center or through the charge; as this action proceeds, it sets up the best heat screen with the corresponding costly results. This is why the consumption of heat is so great in coke ovens or vertical retorts. The act of carbonization under proper conditions is almost instantaneous. The aim of modern designers is to approximate this condition. It has been proved that, provided the gases are properly taken care of, the product is better and there is more of it. Besides, if gasoline, or motor spirit, is a desideratum, the faster the carbonization, the better the spirit, for the destruction of olefines is less, especially at low and similar temperatures.

It must not be forgotten that the whole tendency of destructive distillation, or as an authority has recently named it, "constructive" distillation, is toward lower temperatures. In the United States 700° F. is used by one plant as its standard; while in England 600° F. is used with the best of results; but these temperatures necessitate other considerations if a reasonable recovery of ammonia is required.

The adoption of the principles mentioned have resulted in low first cost per ton-day for retorts because the "through put" is greater owing to better heat application. The amount of heat used is one-third less and the quality of the product is better, for the gases are withdrawn nearly as and when evolved.

While the retort has been the most serious question to many, the disposal of the gases has also been troublesome, especially where there is a shortage of water. The ponderous and, usually, leaky air and water condensers formerly so universal have been replaced, even in Scotland, by systems of fractional condensation, whereby the products are taken down in nearly the fraction or fractions desired. The cost of this section of a plant is practically cut in half and so is the trouble and expense of running.

A big through put, or divisor, is essential to the best plants; the necessary capital involved, even for a Scotch plant, was enormous, and the plant was very complicated. Today the cost of a modern plant can be reduced to 70 per cent. of what it would have been 2 yr. ago and at least the same reduction can be made at the operating end.

Although a great deal has been done toward cheapening and simplifying the process of carbonization, Professor Baskerville is right when he says that it is an industry requiring capital and skill. There are many angles and many economic conditions to be considered; not the least of which is "distribution." Notwithstanding all these, it may now be safely assumed that it is quite as easy to distil oil from shale as to drill for and distil oil for its products, and on the whole it will be quite as profitable commercially.

E. A. TRAGER, Bartlesville, Okla.—I have distilled something over 800 or 900 samples of western oil shales and find that it is possible to get different products by different types of distillation. I have also found that by the same method of treatment the shales are divided into different groups. One type of shale tends to yield gas almost entirely; the majority of them yield mostly oil; while there are some that give a good yield of both gas and oil. This summer I found a type that by dry distillation will yield B. S. almost entirely.

THE CHAIRMAN (C. W. WASHBURNE, New York, N. Y.).—What conditions do you find give the best results in distillation?

A. W. AMBROSE, Washington, D. C.—The matter of heat control is perhaps one of the biggest factors in determining the quality of the different byproducts.

E. A. TRAGER.—You can produce all gas and no oil from any shale by heating too rapidly, but as near as it is possible to tell, by a uniform method of distillation the different shales will divide themselves into different groups, this division being based on the resultant products.

A. W. AMBROSE.—Did you try any experiments by grinding shales to different sizes?

E. A. TRAGER.—Yes; but the size does not seem to affect the product. We tried everything from $\frac{1}{2}$ in. to $\frac{1}{200}$ in. mesh and the product is very much the same. The method of heating is the important factor.

CHAIRMAN WASHBURNE.—It is evident that this matter of distillation of oil shales is something for our grandchildren, possibly our great-grandchildren, but let us hope that scientists will begin to study the problem so that the next generation may have some good out of it. I believe that there has never been any gasoline or kerosene of good commercial quality produced from our Western oil shales in any quantity. The best American shale, with the best method we have, would take too much sulfuric acid in treatment. What little first-class oil would be left after the treatment would not pay for the cost of the operations.

E. A. TRAGER.—I found some oil shales that yield from 30 gal. to 60 gal. per ton, which on distillation will yield about 23 per cent. gasoline and 33 per cent. kerosene; this was treated with H_2SO_4 and the loss wasn't very great. The samples of shale which contain only a small amount of oil yield a low grade of oil; while at the same time, the better shales will yield more oil and contain a larger percentage of light constituents. The best shales which have been found to date come from Colorado. The gasoline is apparently of very good grade but the great objection is the offensive odor—it is very undesirable—just what it is, I don't know.

CHAIRMAN WASHBURNE.—Does that last remark apply to most oil shales in Colorado or to just a few samples?

E. A. TRAGER.—It applies to all Colorado shales. We have studied quite a number of samples and in every case the shale that yields a low amount of oil will yield a heavy gravity oil. Some of the crude shale oil is quite light; the first of the yield looks somewhat like the old fashioned kerosene. It is only the odor you will have to contend with.

R. A. SMITH, Lansing, Mich.—Mr. H. A. Buehler recently told me that a new type of retort for coke manufacture has been developed by G. W. Wallace of the St. Clair County Gas Co. of East St. Louis, Ill. This retort has been found to be especially adapted to oil shales. It is entirely different from the standard types in use at the present time. Coke is produced in 4 hr. and the treatment of oil shales is completed in about the same time.

Industrial Representation in the Standard Oil Co. (N. J.)

CLARENCE J. HICKS,* NEW YORK, N. Y.

(Lake Superior Meeting, August, 1920)

THE labor policy of the Standard Oil Co. (New Jersey) is founded first of all on paying at least the prevailing scale of wages for similar work in the community; on the eight-hour day at the refinery, with time and one-half for overtime; one day's rest in seven; sanitary and up-to-date working conditions; just treatment assured each employee; opportunity for training and advancement; payment of accident benefits beyond the amount prescribed by the State compensation law; health supervision by a competent medical staff; payment of sickness benefits after one year's service; coöperation with employees in promoting thrift and better social and housing conditions; and assurance of a generous annuity at the age of 65, guaranteed for life after 20 years of service. Most of these features have been a part of the company's policy for many years, but it is only during the past two years that the coöperation of employees in determining these matters has been definitely assured through industrial representation.

Industrial representation, in the Standard Oil Co. (N. J.), is a principle rather than a procedure. It is built upon the belief that personal association of those interested in any problem leads to a mutual understanding and a fair decision as to what is right. Fully believing in this principle, representatives of employees and representatives of management evolved a simple plan, the basis of which is that it gives every individual employee representation at joint conferences on problems and fundamental principles affecting all those interested in the industry. It is based on coöperation, not antagonism; its operation makes perfectly clear both to management and to employees that their interests are identical, and not at variance with the interests of the stockholders; and that mutual understanding and coöperation insure progress and success for all. Furthermore, experience has definitely shown that representatives of the employees are not only alert for the employees' interests but are as keen as the representatives of the management in determining and insisting upon fairness to the employer.

Though the plan has been in operation nearly two years, it is an experiment, in that, being based on a principle rather than on cut-and-

---

* Executive Assistant to the President, Standard Oil Co. (N. J.).

dried formulas of procedure, it is still subject to change and improvement It has proved to be equally applicable in a refinery, where thousands of men are assembled, and in the sales department and the producing field, where men are scattered in small groups over a wide territory. It is also in operation in several subsidiary companies. This adjustment to diverse conditions is possible because hard-and-fast rules were avoided, in the belief that the human element must play an important part. Therefore the plan, to a large extent, has been permitted to build itself through experience, and trial.

The plan was brought into operation by an invitation to employees to coöperate in maintaining the company's established policy for fair treatment in matters pertaining to wages and working conditions. This invitation outlined a simple method by which the employees, by secret ballot, might elect from their own number men in whom they had confidence to represent them in conference with representatives of the management. At the first joint conference a brief plan or agree-ment was evolved, which provided that adjustment of wages, including matters affecting working hours and working conditions, shall be made in joint conference between the employees' elected representatives in, the division affected and the representatives of the company. From the beginning, the plan stipulated that no discrimination shall be made by the company or its employees against any employee on account of membership or non-membership in any church, society, fraternity, or union. Agreement was made as to offences for which employees may be dismissed without notice and also as to the offences for which an employee should be warned or suspended. Further, each employee was guaranteed recourse against unjust treatment or unfair conditions by means of a definitely prescribed method through which he, personally, or his repre-sentative, may appeal his case to joint conferences of employees' and management's representatives and, if necessary, up to the highest officers of the company.

The joint works (or plant) conferences are held at regular intervals to consider all questions relating to wages, hours of employment, work-ing conditions, and any other matters of mutual interest that have not been satisfactorily settled in the joint division conferences. These joint division conferences meet whenever needed to discuss and adjust matters within the smaller confines of a division. Many problems never go beyond the joint conference, unless the problem develops into one that concerns other divisions. In case any matters were to come up on which the joint works conference could not agree, they would be referred to the Board of Directors for final decision. But as yet not a single case has been referred in this manner. The decisions of the joint works conference, when they involve serious matters, such as a general increase in wages, are subject to the approval of the Board of Directors.

At the inception of the plan, a basis of representation was determined' upon that would allow one employees' representative to be elected by approximately 150 employees, with provision for a minimum of two employees' representatives from each division. In extending the plan to other departments of the company, such as the producing field and a refinery where fewer employees are required, this basis was amended to meet the conditions obtaining in that field. On this point two essentials have been borne in mind: First, that an elected representative must not have more constituents than he can easily keep in touch with; second, that the joint conferences must not be so large as to be unwieldy at times when important discussion and decisions must be had. Experience has shown that there are many advantages to be gained by personal contact of employees' representatives and managements' representatives, and therefore full joint conferences are preferable' to numerous smaller subcommittees.

Entirely apart from the industrial representation plan, but equally established as a policy in the Standard Oil Co. (N. J.), is a method of protection for employees and their families. This is attained in several ways: Group life insurance covering, at the company's expense, every employee after one year's service, affords some financial resources to dependents in case of death of an employee—a provision that was greatly appreciated during the influenza epidemic of 1918–19. There is a fully equipped and competently manned medical department to look after the health of all employees; and there is provision for half pay during a period of sickness. An annuity plan provides for employees who retire after 20 years of service or who are incapacitated after even shorter service. These forms of financial security are considered by the company as being good business and therefore are maintained solely by the company's funds, not by either voluntary or involuntary assessments on the employees.

The company is committed to a policy of training for employees as a means of assuring, to each one who desires, an opportunity for fair advancement to greater responsibilities. The administration of training is coördinated with other personnel functions, such as selection of new employees, transfers and promotions. Thus each employee not only has the feeling of security in his position and his earnings but also knows that the company is ready to help fit him for advancement to any position within his capacity.

DISCUSSION

R. A. CONKLING,* St. Louis, Mo.—After three months' service with our company, on the recommendation of the head of the department, any employee can join the provident fund; then a fixed percentage of

---

* Head Geologist, Roxana Petroleum Corpn.

his salary is retained and deposited with the parent company, with a record of the fund. The maximum is 10 per cent. and the minimum is 5 per cent. The company sets aside an equal amount, which is invested at about 5 per cent. If the subsidiary has been successful during the year, the company declares a bonus which is deposited in this fund. For the last three years the bonus was 15 per cent., this year it was 20 per cent. That makes 40 per cent. of our salaries going into a fund into which the employee only pays 10 per cent. This money can only be drawn out after three years, if the employee leaves the company. If he leaves the company before that time, he gets only his 10 per cent.

W. E. PRATT,* Houston, Tex.—A point worthy of note is that this company has undertaken to insure its men against sickness, to pay reasonable insurance policies in case of death, and to retire the men after various periods of service on livable wages. A few years ago, although we heard a great deal about sick benefits and annuities, most of the plans called upon the employees to contribute something from their salaries, but at the present time, as exemplified by the policy of this corporation, general opinion seems to hold it to be fairer practice, as well, perhaps, as better business, to supply these benefits without placing any share of the burden involved on the employee.

---

* Chief Geologist, Humble Oil & Refin. Co.

Petroliferous Rocks in Serra da Baliza

BY EUZEBIO P. DE OLIVEIRA,* RIO DE JANERIO, BRAZIL

(Wilkes-Barre Meeting, September, 1921)

ONE of a recent batch of samples from the Serra da Baliza, in the state of Paraná, Brazil, contained asphalt and a dark heavy oil; and workmen on the railway from Porto União to Uruguay discovered asphalt coming from eruptives that outcrop along the Rio de Peixe. The occurrence of asphalt in the triassic eruptives of southern Brazil, however, has been known a long time, according to Dr. Gonzaga de Campos.

It is generally believed that the Botucatú sandstone is always a hard vitrified rock, from the metamorphic action of the overflowing eruptive contacts. In this region, however, the contact metamorphism is almost nil; the sandstone is slightly hardened in a narrow zone about 20 to 30 cm. wide. In many places, the sandstone is so friable as to be easily reduced to sand, which is used in mortar for building in Guarapuava and Palmas. South and west from Porto União, this bench of sandstone is about 50 m. thick, and is capped by a heavy bed of basic eruptives, many of which are amygdaloids.

NATURE OF ROCKS

Dr. Geo. P. Merrill, after studying the triassic eruptives collected by the Coal Commission, reached the conclusion that "All these rocks are of typical basalt-diabases, not in any essential different other than in structure. An interesting mineralogical phase is its paucity in olivine, which in many cases is completely lacking."

Professor Hussak, who carefully studied these rocks and their accessory minerals, decided that in the dikes they are granular (diabase) and that in the lava sheets, porphyritic (augite-porphyrite or melaphyre), and that the latter pass evidently to normal diabases and are always typical of effusive rocks. The examination of many slides from dikes and sheets leads us to adopt the opinion of Professor Hussak; the rocks of the dikes are of ophitic structure, while that of the sheets show a great variety of structure and may vary from almost granular to basaltic. The great paucity in olivine had been noted by Hussak, who classes as melaphyres, the porphyritic triassic rocks in Brazil which contain olivine.

---

* Geologist, Serviço Geologico e Mineralogico do Brazil.

Between Porto União and the Serra da Baliza, all the rocks, apparently, are in sheets, *i.e.*, they are all porphyritic, containing plagioclase, augite, iron, etc. in a variable proportion as well as the many decomposition products. The predominant rock is black, or greenish black, and so fine grained that even with a lens, none of the essential constituents can be made out; it contains, however, cavities or amygdules, empty or full of various accessory minerals, products of its decomposition. Another rock type is chocolate colored, of a cavernous structure, containing many geodes full of accessory minerals. This rock is intercallated in a black porphyry, in a cut in the Serra da Baliza. In a cut made below the humus and surface-earth, rounded blocks of decomposed eruptives and blocks of metamorphosed sandstone were found, and, below this, the more or less decomposed eruptive porphyry, *in situ.*

The Serra da Baliza (1040 m.) is a ridge resulting from erosion and lies between the Rios Jangada and Iratim and the creeks Antas and Jangadinha. From a geological point of view, it is constituted essentially of the two types of eruptives noted, with the metamorphosed sandstone alongside.

All the samples of sandstone have, when freshly broken, a distinct petroleum odor and many of them show small cavities from which exude a heavy dark oil. Different samples of the compact black eruptive contained asphalt in the crevices; while the chocolate-colored rock showed not only asphalt, but a heavy oil that came out with effervescence when heat was applied. In a piece of quartz, almost hyaline, a cavity full of asphalt was found.

Having made excavations in all the hollows and ravines from the top of the Serra to the foot we concluded that the sandstone does not form a continuous bed. Probably a bed was broken into large blocks which were carried to different levels in the molten mass during the eruptions of the porphyrites. After an examination of a part of these rocks, Dr. Gonzaga de Campos decided that the occurrence of petroleum is in the contact zone of both the sedimentary rocks and the eruptives where these are completely modified by endomorphism.

INDICATIONS OF PETROLEUM MOST IMPORTANT IN BRAZIL

These indications of petroleum in the Serra da Baliza are the most important known in Brazil. Until now, the greater part of the known occurrences consisted of impregnations in clayey and calcareous beds of the Iraty shales and limestones or of asphalt, or its varieties, at different points in the states of São Paulo, Paraná, and Santa Catharina. Some rocks above the Iraty horizon also have a distinct petroleum odor when freshly broken. One such occurrence was found in São Paulo, by Dr. Gonzaga de Campos. But none of these rocks show petroleum immediately, when freshly broken, as do those of the Serra da Baliza. Owing to the nature of the rocks of the region, it does not seem wise to make borings for petroleum, however, until more minute studies have been made.

All the petroleum indications are found in the Iraty beds or in beds above these. As far as known, there are no indications of oil in the underlying Permian rocks nor in the Devonian beds, except a slight impregnation at the top of the fossiliferous shales of Ponta Grossa.

It is true that the Bofete well log, registered in Doctor White's report, mentions an oil horizon below this level; but it is quite possible that the heavy oil given as coming from this horizon really came from the Iraty, the rocks of which show oil only after the lapse of some time after breaking or boring—sometimes this occurs only after the application of heat.

Near the Colonia do Rio Claro, the existence of albertite penetrating the Estrada Nova and Rio Rasto beds has been known some years. This occurs near an eruptive contact. Albertite seems to be generally a good indication of petroleum, as shown by Doctor White in the same report: "That this was the origin of grahamite, albertite, uintahite or gilsonite is certain, since recent drilling near the Ritchie mine in West Virginia has revealed a productive oil sand (salt sand) at 1500 ft. below the valley, and what is most significant is the fact that only a little oil is found in the underlying sand until the wells are located from 500 to 800 ft. distant from the fissure, thus showing that the rock has been drained in the immediate vicinity of the latter." In the same report, he says: "Record of a well drilled within 300 ft. of the Ritchie mine (fissure holding grahamite), on the Macfarlan run. In this well only a small quantity of oil was found. This sand was good but the well acted as though the sand had been drained. Wells drilled farther from the fissure, however, secured good producing sand as shown by the following records. . . "

Thus drilling for oil in a region where asphalt occurs, as at Colonia de Rio Claro, is promising but should be located some distance from such veins of albertite, in order to avoid boring through rocks from which the petroleum has been drained.

In São Paulo, also, other borings in the Bofete region would be interesting. Though Horace E. Williams, who knows the region well, is of the opinion[1] that the bituminous sandstones of the Bofete region represents a fossil pool, eroded and oxidized, and that the existing strata above the Iraty in that immediate region are, perhaps, unfavorable for such accumulations. Whether or not this is true can only be determined by considerable drilling. Corroborating this point of view, in part, we find in the *Diario Official* the following considerations by Doctor White, written while he was on the ground: "When I learned that the boring

---

[1] Oil Shales and Petroleum Prospects in Brazil. See page 75.

near Rio Bonito, in São Paulo, had found some genuine petroleum, I was not surprised; but, as the boring had been made near a fissure in the rocks, which permitted a great quantity of the petroleum to escapé toward the surface and saturate the sandstone with its residual products (asphalt etc.) no oil might reasonably be expected to be found in commercial quantities in this boring. The drilling should be made some distance from this break in the rocks where the flow of eruptives has not emptied the deposits of the underlying rocks."

Analysis of Oil-field Water Problems*

By A. W. Ambrose,† Bartlesville, Okla.

(St. Louis Meeting, September, 1920)

The underground losses of oil exceed by hundreds of thousands of barrels all the oil that has been lost in storage, transportation, or refining. The quantity lost is, of course, indeterminate; but when it is considered that the contents of an entire oil field have been excluded from recovery by invading waters, some idea of the amount wasted may be gained. Similarly, enormous quantities of gas have been lost underground. Conservation of the oil, therefore, should start before it is brought to the surface rather than after it is placed in storage tanks.

Water is one of the most important causes for underground losses and the operator should give as serious consideration to an underground flood of water as he would to a destructive surface flood. The best insurance, of course, is to have the wells drilled in such a manner that water has no access to the productive oil and gas horizons, and on abandonment the wells should be properly plugged.

The encroachment of edge water and the occurrence of water in ;the base of an oil sand present a very serious problem to an oil company sooner or later these waters are bound to cause considerable damage, if they do not entirely destroy the possibilities of further production. Too often, however, a field has been considered to be in a hopeless condition, whereas wells in as bad a condition in other areas have been repaired and the life of the field appreciably lengthened. The corrections are very often suggested by technical study. Very successful results have been accomplished by detailed underground work in the California oil fields, in Cushing, Oklahoma, and in other areas.

The purpose of this paper is to outline briefly[1] general methods of analyzing oil-field water problems in a producing or in a producing and developing oil field, with a view to suggesting repair work on offending wells. Reference is continually made to producing oil wells; the same general method of procedure, however, should be adopted in a gas field.

---

* Published by permission of the Director of the U. S. Bureau of Mines.

† Superintendent, Petroleum Experiment Station, U. S. Bureau of Mines.

[1] A bulletin, "Underground Conditions in Oil Fields," prepared by the writer, will be published shortly by the U. S. Bureau of Mines, which goes into much more detail than this outline.

OBJECTIONS TO WATER IN PRODUCING WELLS

The prime necessity of mastering the underground-water problem is to prevent water from entrapping the major portion of the oil still underground. But if this point were entirely lost sight of, there are sufficient reasons for studying the water problems, provided the study is followed by corrective measures.

Water is objectionable because its presence in an appreciable amount means: (1) The ultimate loss of thousands of barrels of oil which may

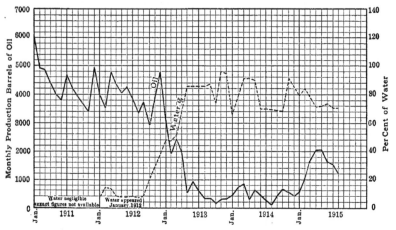

FIG. 1.—EFFECT OF WATER ON OIL PRODUCTION OF A WELL.

be trapped underground; (2) the loss of casing-head gas; (3) the increased lifting costs, as wells producing water cost more to pump and the life of the tubing, pump, and sucker rods is shorter, also the additional cost for replacement of corroded pipe lines and fittings; (4) the possibility of water flooding the sands and driving the oil to neighboring property; (5) the forming of emulsion, which necessitates expensive dehydrating plants to separate the oil and water.

Fig. 1 shows the effect of water on oil production in a well. Water appeared in this well in January, 1912. The oil production held up during 1912, but from December, 1912, to May, 1913, it declined from 4800 to 500 bbl. per month. Water constantly increased during 1912, and seriously interfered with the oil production during the first part of 1913.

DIFFERENT WATERS IN A WELL

In an area drilled with a hole full of mud or fluid, the operator should consider the contents of a sand as an unknown quantity, unless the sand has been tested in a neighboring well by bailing or pumping. In an area where the hole is filled with water while drilling, the hydro-static head of fluid in the hole is usually greater than that of the water or oil in the sand, hence oil or water will not come from the sand into the drilling well. Fig. 2 is a hypothetical sketch showing several possible waters in a producing oil field.

Those waters A and A_1 occurring in the sand above the producing oil horizon are generally known as top, or upper, waters. Top water may have access to the hole by: The shut-off being too high; the water leaking around the shoe of the water string; poor coupling connections, due to

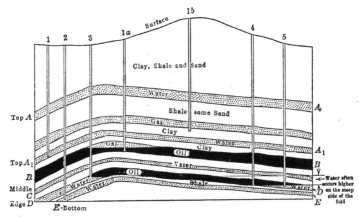

FIG. 2.—HYPOTHETICAL SKETCH SHOWING DIFFERENT WATER SANDS

cross-threading or the pipe not being screwed tight; collapsed casing; a split in the casing; pipe worn through by drilling-line wear; or corrosion of the casing due to strong corrosive waters in the sands.

Bottom waters E are those occurring in sand below the producing oil horizons. To avoid bottom water, it is necessary to learn the exact distance between the top of the water sand and the base of the oil zone, so that the operator can avoid drilling too deep.

Where there are several producing oil or gas horizons, the water C occurring between the producing sands is generally referred to as inter-mediate water.[2]

Edge water D occurs in the down-slope portion of an oil or gas stratum. Edge water may be middle water in one well $\mathcal{3}$ and bottom water in a well in another part of the field. It usually encroaches as production is

---

[2] If there are only two producing sands, the term middle water is often applied to the water occurring in a sand between them.

drawn from the wells up slope. The sand is termed an oil producer up slope, but wells drilled into the same sand down slope will produce water.

Water may occur in the base of an oil sand, although before drawing such conclusions it is advisable to consider carefully whether or not there is a small formational break of an impervious bed between the oil and water. Water, also, may occur in a lenticular body of sand and should be treated as top, bottom, or intermediate water, according to its location with respect to the productive sands.

DATA FOR ANALYZING OIL-FIELD WATER PROBLEMS

The following outline is suggested for the preparation and use of data in the study of water problems and corrective work in a producing field where drilling is still taking place and where little or no data have been compiled.

1. Prepare forms for recording important information.

2. Assemble all drilling and redrilling records, daily well reports, production records of oil and water, tubing depths, fluid levels, and other data, for the purpose of compiling a complete log for each individual well.

3. Obtain the elevation and location of all wells and prepare field maps.

4. Present underground conditions graphically by means of cross-sections, underground structure contour maps, convergence maps, peg models, stereograms, and miscellaneous graphic plots.

5. Study data on drilling and behavior of neighboring wells.

6. Collect and compile individual well records showing monthly production of oil and water.

7. Review the histories of abandoned wells.

8. Carry on certain field work, such as collecting samples of formation, water, and oil from drilling wells.

9. Conduct field tests for determining the contents of different sands, also test out "wet" wells for top, bottom, intermediate, and edge water and water in the base of an oil sand.

10. Study the chemical properties of the waters in different sands.

11. Investigate the possibilities of using dyes or other detectors to determine the source of the water in wells.

12. Study the indications of a field going to water.

13. Consider the source of water in individual wells or groups of wells.

14. Correct or repair wells making water.

Keeping of Records

Records form the basis for the successful operation of any property and may be considered the yardstick by which the past and present conditions and future possibilities may be measured. Where a company has no complete system of records, immediate attention should be given the preparation of forms upon which to record important data and to the collection of information for these forms. The forms used should be those necessary to keep the data brought out in the following pages.

Well Logs

A well log should contain, in addition to the formations, location, and elevation of the well, etc., a history giving complete data of the tests and all work done on the well that will in any way serve to show the contents of any of the sands or the condition of the casing, etc. This history should be arranged in a chronological form in which each piece of work is set out by itself.

Field Maps and Cross-sections

Field maps should be prepared, showing the elevation and location of wells, on such a scale that the wells can be measured or scaled off as accurately as the results of the survey.

Pins, with colored glass heads or with numbers on the head, may be stuck into certain well locations to serve as legends to designate the status of different wells. A certain color or number indicates the condition of the well, as drilling, redrilling, abandoned, etc.

In order that cross-sections may be of the greatest use, they should be clear and simple and should emphasize the important features and omit the unimportant. The occurrence of oil, gas, and water should be emphasized and the casing depths should be noted. All unnecessary figures, as depths of formations,[3] should be omitted, as these tend to obscure the more important data.

Considerable care should be exercised in the adoption of symbols to be used in cross-sections, as symbols avoid much lettering and furnish the basis for easy correlation of the different logs. The symbols selected should be in contrast with each other, easy to recognize and easy to plot. Once a satisfactory set is adopted, the same symbols should be used throughout the work.

The lines of cross-section selected should be such as will depict the underground structure. Every well on the property should be included in some cross-section, as one well may furnish a key to the situation. In case an isolated well is located off the line of a cross-section, considerable care should be exercised to see that it is projected to the line of section in a proper manner.

---

[3] It is often advisable to record the depths of the more important formations, as top and bottom of oil sands, markers, etc., but the practice of arbitrarily recording the depths of formations should be discouraged.

A satisfactory scale for plotting cross-sections has proved to be 100 ft. to the inch; this allows sufficient detail to show a 2- or 3-ft. change in the formation, which is about as close as is ordinarily detected by the drill. It is only in exceptional cases that there is justification for a different horizontal and vertical scale, because where the two scales are different the actual conditions are not properly shown.

Correlation is based on an identification of one or more identical strata in the logs of many wells. The term "marker" or "key bed" has been applied to formations that are constant in thickness and occurrence over large areas and can be recognized in most of the wells. An ideal marker is a formation that carries from well to well, is of uniform thickness, and can be readily recognized by its color, hardness, or toughness. From the cross-sections, the engineers should try to trace the marker from one well to another. In regions where there is no marked variation in the thickness of the formation, the producing sands are usually a certain distance below the marker. This interval will serve as a guide in new wells to indicate at what depth the water string should be landed and the producing horizons will be encountered. Similarly, bottom water and other features should be noted in relation to this marker. Faults and unconformities throw the beds out of their logical place and cause mistaken predictions. Formations may thicken or thin, causing irregularities in the occurrence of different beds.

Where few wells have been drilled, the surface structure may offer helpful suggestions in correlating the underground beds, as frequently the general surface dip will indicate the attitude of the beds underground. Unconformities and other irregularities, however, may cause underground and surface formations to dip at different angles, but until this has been established very often the engineer has no other guide at the start.

Underground Structure Contour Map

The underground structure contour map is useful for showing the attitude of beds in regions of low folds and faults of small throw. The structure is shown by means of contour lines, which are used to connect points of equal elevation on the surface, bottom, or other definite horizon of a key bed or marker. A structure contour map is made up, therefore, of a series of contour lines used to show the configuration of such a bed. A contour interval is the vertical distance between the different points of elevation as represented by contour lines. The accuracy of a contour map depends largely on the number and distributions of elevations over a given area, also on the accuracy of the well logs, their location, and elevation. The universal recognition of a marker or key bed in an area is also a factor. The main function of the structure contour map is to show broad structural relationships over a large area in a way that usually is not given by the most careful study of geological cross-sections; it is also used in selecting well locations, in the prediction of casing depths, well depths, etc.

Convergence Maps and Peg Models

Where the surface and underground beds are not parallel, as in some fields, a convergence map is necessary to take account of the convergence or divergence of different beds.

Peg models are used extensively in the California oil fields, also in the Gulf Coast, for showing structural conditions underground. These models are used for correlation, for the determination of proper water shut-off points, location of water, gas and oil sands, and for bringing out any marked irregularities of well depths, water shut-offs, etc.

Stereograms and Miscellaneous Graphic Plots

Stereograms are used to show graphically, and in three dimensions, the broad general relationships underground. They have been used very little to determine casing depths, water shut-offs, etc.

Miscellaneous graphic plots can be used to emphasize certain features. For example, if considerable material has been left in the hole, its location in relation to the producing sands may be shown by an individual graphic plot, which brings out in consecutive order the work done on the well at different times. The history may be shown on the same sheet; then, by a combination of the graphic drawing and the written history, the engineer may more easily realize the tests made and the work done on the well.

After the marker has been definitely established in a group of wells in a district, a set of graphic logs may be plotted to a common stratigraphic datum. The marker of different wells is plotted on a horizontal line and then the correlation should be along the horizontal, consequently any irregularities of well depths, water shut-offs, tops of plugs, etc. are readily noted.

Study of Neighboring Wells

Neighboring wells should be carefully studied, for they may furnish information that will help to solve the problem. Cross-sections, particularly of adjoining line wells, should be made, and the casing depths and histories of these wells carefully studied in order to obtain the same information as is gathered on the company wells. There should be a complete exchange of well data between neighboring companies, particularly line wells. It has been proved many times that the exchange of information is beneficial to both sides.

Monthly Individual Well Production Records

The production records of wells should be compiled in convenient form and should show the production of oil and water for each month during the life of the well. If there are no records, information should be collected from the foremen and pumpers. Information so recorded, however, should have a note telling the source of the information.

Abandoned Wells

In preparing data on different wells, special attention should be given to the histories of abandoned wells, because these may be allowing water to enter the producing sands. Where records of such wells are not available, it is often necessary to collect information from drillers, pumpers, etc. Every abandoned well should be properly plugged.

Collection of Samples of Formation, Water, and Oil

Samples of formation from different horizons should be collected from the drilling wells. These samples should be examined, marked, and saved for future reference. Glass bottles may be used as containers; they should be labeled to show the well number, depth, name of formation, and date collected. Samples of water representative of that in a sand should be collected, even though there may be no need for a chemical analysis at the time. Samples of showings of oil from any unexpected horizon should be collected for possible analysis or need later.

Field Tests

A series of field tests of wells making water and of drilling wells should be carried on simultaneously with the study of the data. The water strings of oil wells making water may be tested, bottom of wells may be plugged where bottom water is suspected, and the sands of a drilling well should be tested. The same sand need not be tested in several wells as one good test on a horizon in a certain area will often suffice. The necessity for knowing the results of former tests emphasizes the value of good records.

The location of oil sands and water sands can be determined most satisfactorily in a drilling well because it is possible to have only one sand exposed. The number of sands that can be tested in a drilling well are limited only by the practicability and expense of the operation. Once the sand is cased off, it is usually difficult to make a test of it. After a water string has been landed, a very careful test should be made by drilling a pocket below the casing shoe, bailing out water and allowing the hole to stand at least 6 hr., and preferably 12 hr., to see if any water enters.

In testing the water string of a producing well, first test to see if there is a leak in the pipe. If the casing does not leak, a bridge may be set a few feet below the casing and a test made to see if the water is coming around the casing shoe.

If a well is suspected of making bottom water, the bottom of the well can be plugged in successive stages, with cement, until some definite information is gained regarding the source of the water. Packers and lead plugs have also been used. In plugging up the bottom of the well to test for bottom water, it may be necessary to shoot the hole if there is any old side-tracked casing which may serve as a conductor for the water to work up into the well.

A bridge may be used to test out where desired. If a sand midway between the water shut-off point and the bottom of the well is suspected of making water, a bridge can be set in the sand suspected of carrying water, cement dumped in to fill the hole several feet above the sand and, after the cement has hardened, a bailing or pumping test made to determine whether or not the bridge has shut off the water. Very often a bridge saves a great deal of needless plugging. Often intermediate water can be located by deepening and testing successively lower sands in a drilling well.

Testing for water in the base of an oil sand is similar to plugging and testing for bottom waters, although much more care must be exercised. It is important to plug the wells with cement in successive stages in order to avoid shutting off the oil production. If, after plugging, the amount of water is retarded only temporarily, it is evident there is water in the base of an oil sand.

The field tests for edge water or water in a lenticular sand are necessarily guided by the suspected location of the water; that is, whether the water occurs in the top, bottom, or intermediate sands.

Water Analyses

The chemical analyses of oil-field waters can be used in solving oil-field water problems. They are particularly useful in distinguishing waters in different sands, and hence in determining the source of water in a "wet" well. Perhaps the most practical use of chemical analyses has been in the oil fields of the Gulf Coast and of California, where their use has saved costly work that otherwise would have been necessary to determine the source of the water in some of the wells.

The waters of each field are chemically different from those of another, and the engineer will find that the distinguishing features of different waters will probably vary in each field. He will undoubtedly find, however, some particular feature, as high chloride content, total solids, or primary salinity, that will serve to identify one water from another. It may be found that the properties, such as primary salinity, primary alkalinity, etc., are not the determining characteristics of a water in a field. In one field, the writer found that the top waters, in general, were high in sulfates. The sulfate content decreased as the sand approached the oil zone; and as the chlorides were negligible in the top waters, there was a decreasing primary salinity percentage with depth. The very bottom waters, however, had a high chloride content; hence the bottom and the top waters would have high salinity, because primary salinity is determined by adding the sulfate (SO_4) and chloride (Cl) percentages together and multiplying the resulting figure by 2. The operator might be confused if he relied only on primary salinity to determine the characteristics of the water.

In this work, the writer found it necessary to consider the chloride content in wells before giving too much value to primary salinity percentage. The most satisfactory results were obtained by using the percentage of reacting value for comparison rather than the figures of salinity and alkalinity. Again, it may be found that other factors will readily distinguish the waters. For example, the bottom water of the Augusta field, Kansas, shows total solids averaging about 36,000 parts per million, while the upper waters average nearly four or five times as much.

Collection of Samples of Water for Analysis.—A sample of water for analysis is of no value unless it is representative of the water found in the sand. The samples to be analyzed should not be mixed with drilling water and a sample is of little value where several water sands are exposed in the hole. When starting the work, the engineer should collect samples of unmixed waters from each sand, if possible, so that he may know the properties of the waters in definite water sands.

Where a producing well has made water for some time a true sample may be obtained from the flow tank or sump, as other water has been flushed out. If the well has just started to make water, and other water has been in the tanks, it is best to take a sample from the lead line.

Application of Water Analyses.—After a sample is collected and a chemical analysis made, the engineer should interpret the analysis according to Doctor Palmer's method. When there are several analyses, a tabulation should be made of the properties of the waters in known sands and of the distances of these sands from the marker. Then, when a well starts to make water, its source can be determined by comparison of the chemical properties of the water with those in the tabulation to see if it is the same water as any of those recorded in the tabulation.

To show the possibilities of using water analyses, the writer will cite one or two examples, taken from a report of the waters in the East Side Field, Coalinga, Calif., by the writer in September, 1916, to Mr. B. H.

van der Linden, field manager of the Shell Company of California. It was based upon forty samples of water taken from sands of different wells, the samples being collected from as many different sands as were accessible. By studying the analyses in connection with the graphic sections and well histories, it was possible to locate very definitely most of the water sands associated with production. The results of this work show how water sands may be definitely located; a prime necessity to avoid drilling difficulties and future water troubles. Table 1 demonstrates how it was possible to distinguish between the different waters by reference to the sulfate (SO_4) and carbonate (CO_3) columns.

TABLE 1.—*Characteristics of Water Sands, Arranged in Stratigraphic Sequence*

Well	Source of Water, in Feet below Marker	Na, Per Cent.	Ca, Per Cent.	Mg, Per Cent.	SO_4 Per Cent.	Cl, Per Cent.	CO_3, Per Cent.	S, Per Cent.
Record 5................	Above	47	1	2	39	2	9	0
Shell 31/34, sample No. 1..	355–365	48	2	0	28	10	12	0
Shell 31/34, sample No. 2..	416–418	46	2	2	8	6	24	11
Shell 31/34, sample No. 3..	420–438	49	1	0	1	5	29	14
Shell 10/2................	705–724	48	1	1	2	17	31	0

The water from Record 5, which is above the tar sand and producing sands, has 39 per cent. sulfates and 9 per cent. carbonates. The sulfates decrease in the successive lower water sands to Shell 31/34 No. 3 while the carbonates increase. This particular sand in Shell 31/34 No. 3 lies just above the producing sands but below the tar sands of that well. The sand of Shell 10/2 lies below the producing oil zones, the top of which is 267 ft. (81 m.) lower, stratigraphically, than the bottom of the water sand in Shell 31/34 No. 3. This bottom water is low in sulfates and high in carbonates, as would be expected in a bottom water in this field, but there is a chloride content of 17 per cent.

Another example of the practical application of water analyses to producing wells is shown in Fig. 3. This well was drilled to 2677 ft. (816 m.). All sands were perforated and the well put to pumping. The well produced for seven days, yielding an average of 83 bbl. of oil and 104 bbl. of water per day. It was expected to produce oil and no water, as the other wells in this area were producing from the same horizons and making no water. The water would have been much more with a larger pump, but the 2½-in. (6.4.-cm.) pump would not handle over 200 bbl. of fluid per day. If the water had not been shut off it would have .worked back into the oil sands and probably would have done great damage.

A chemical analysis of the water showed it to be a decided bottom water as the sulfate content was 0.1 per cent., the chloride content 24.4 per cent., and the carbonate content 25.5 per cent. The bottom waters in this area were high in carbonates and chlorides and low in sulfates. Accordingly, the very bottom sand was plugged off by ripping the casing and filling the hole with cement up to the base of the next sand—2620 ft. (798 m.). The cement was allowed to set eight days and the well again put to producing. The well then made 80 bbl. of oil per day and 1 bbl. of water.

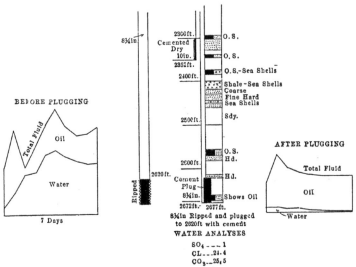

FIG. 3.—PRODUCTION OF OIL WELL BEFORE AND AFTER PLUGGING OFF BOTTOM WATER; SOURCE OF WATER WAS DETERMINED BY CHEMICAL ANALYSIS.

Use of Detectors for Tracing Movement of Underground Waters

Some idea of the rate of flow of water from one well to another may be gained by the use of dyes or other flow detectors. Water may come into a well from various sources and then get into an oil sand from which other wells are producing, thereby causing considerable damage. In an effort to trace this water from one well to another, several means have been used and others suggested. Certain flow detectors have been used with a fair degree of success in some oil fields. If the detector placed in one well appears in another well, it shows the direction and rate of travel of the water; but where the detector does not appear nothing is established. The best organic dyes are not infallible, primarily because their introduction into the oil sand through any well is not certain, rather than because the dye may be destroyed underground. It is of the greatest importance that any dye or detector be properly introduced into the water and mechanical means for insuring this can probably be developed.

There are two general uses of dyes or other flow indicators in determining the movement of oil-field waters: To determine whether or not water is migrating from one well to another; and to determine whether or not the water is entering the well through a leak in the casing or around the shoe of the water string.

In studying water migration from well to well, the dye is placed near the bottom of the well that seems to be flooding the other well or wells, generally in solution form, by means of a proper container in order to prevent dilution of the dye by its coming in contact with the long column of fluid in the hole. Often production is suspended at this well so that the dye will not be pumped out. Neighboring wells should be pumped vigorously and close watch made of the water for any evidence of the dye.

Dyes and detectors have also been used in an endeavor to find out whether or not the water string is leaking. In this case the dye, or other material, is placed on the outside of the water string and observations made of the fluid bailed or pumped from the wells to see whether or not the dye has worked its way into the well. Its appearance in the water in the well shows the existence of a leak in the water string; but its non-appearance does not prove the effectiveness of the shut-off, for if the formation has caved in against the outside of the pipe a few hundred feet below the surface, the dye may be held there.

Methods that have been used and suggested for determining the movement of underground waters are: Dyes and other materials recognized by their color; chlorides, nitrates, or other salts recognized by chemical analyses; lithium salts, which can be detected by the spectroscope; Slichter electrical method.

Value of Dyes.—It is customary for the operator to refer to fluorescein, eosine, and other organic dyes as "aniline dyes" although some of them are derived from substances other than aniline. Fluorescein, eosine, methylene blue, magenta or fuchsine, and Congo red have been suggested as dyes which could be used.

Fluorescein is perhaps the best organic dye that can be used, primarily because it is noticeable when present in very minute quantities and because it is not adsorbed by clays. Fluorescein will penetrate an acid solution further than eosine and will give a color reaction that eosine may fail to do. It will also stand sulfureted hydrogen and sulfurous acid. It can be detected with the naked eye when present in the proportion of one part in forty million; and by the aid of the fluoroscope when present in the proportion of from one part in five hundred million to one part in two billion. Congo red is too sparingly soluble. Methylene blue and magenta are basic colors and all basic colors are adsorbed by clays and are, therefore, unreliable. Fluorescein and eosine are not adsorbed by clays.

Fluorescein and other organic dyes have been used with success in certain cases and show that water travels from one well to another, but so far as the writer knows, the dye has failed, in the majority of cases, to appear in adjoining wells or in a well into which it is placed when it was inserted behind the pipe.

Certain inorganic substances, such as potassium dichromate and Venetian red, have been suggested as flow detectors. The use of potassium dichromate in oil-field waters is questionable, however, because many oil-field waters have a yellowish tint; it is decolorized by reducing agents, such as hydrogen sulfide; and it would require an exceedingly large amount of the compound to color such a large volume of water. The use of Venetian red also is limited because it would filter out quickly when passing through a sand; furthermore, it is not detected when present in as small quantities as is fluorescein.

Chlorides, nitrates, and lithium salts, also, have been suggested as flow detectors but, for various reasons, their use is limited.

Slichter Electrical Method

Slichter has described[4] an electrical method of measuring the velocity and direction of flow of underground water in shallow wells (about 50 ft. in depth). The method has been suggested as of possible use in detecting the movement of underground waters in oil fields, but it shows no promise of practical application in tracing the movement of underground oil-field waters in deep wells.

INDICATIONS OF A FIELD GOING TO WATER

The flooding of the oil sands of an area by top, bottom, or intermediate water can often be prevented by the correction of a few offending wells when the trouble starts. The operator should, therefore, investigate promptly any marked increase in the water content of a well.

The indications of a field going to water vary with each locality, but the most common and positive evidence is for the oil wells to start producing water. When a group of wells located high up on the structure, for instance on the top of a dome, show water while wells down slope do not, some well is at fault. In such a case the cause may be due to improper water shut-off points, leaky water strings, wells drilled into bottom water,

---

[4] Charles S. Slichter: Description of Underflow Meter Used in Measuring the Velocity and Direction of Underground Water. U. S. Geol. Survey *Water Supply Paper No.* 110 (1905) 17–31; or Field Measurements of the Rate of Movement of Underground Waters. U. S. Geol. Survey *Water Supply Paper No.* 140 (1905).

or wells improperly plugged when abandoned.   Top water, bottom water, and water in a lenticular sand may show in wells scattered irregularly throughout a field; these three waters usually lend themselves to repair work on the wells.

Water in the base of an oil sand and edge water present a much more serious problem, for as the oil and gas are withdrawn they will be replaced by water.   Water in the base of an oil sand often occurs in abundant quantities; as a hole is carefully plugged up with cement, and by stages, the water production is only temporarily retarded.   When the wells farthest down slope, located along a line parallel in general to the under-

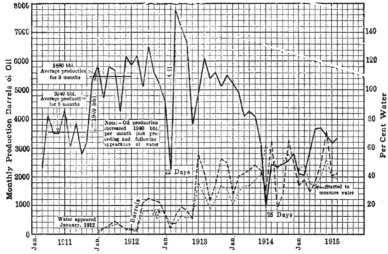

FIG. 4.—INCREASE IN OIL PRODUCTION FROM OIL WELL PRIOR TO ENCROACHMENT OF EDGE WATER.

ground contours, show an increased water content, there is suspicion of of the encroachment of edge water.

A sudden increase in oil production has been noticed in wells just before edge water appears.   This is shown in Fig. 4.   It will be noticed that the average production per month was 1940 bbl. more following the appearance of water in appreciable quantities in January, 1912.

SOURCE OF WATER IN INDIVIDUAL WELLS OR GROUPS OF WELLS

Determination of the source of water in a field is dependent on accurate and complete records.   Each well presents its own problem, but there are certain fundamentals that may be outlined here.   Problems should be attacked from two sides—study of old data and field tests. As the problem is studied from the records and graphic data prepared, the suggestions of the source of water in any well should be checked and tested by mechanical and field tests on the wells.

Production records indicate the wells that are making large amounts of water and complete records show when and where the water first appeared. Fluid-level records may possibly indicate what well is causing trouble and, often, water analyses will show immediately the source of the water. The history of an abandoned well may indicate that the well was not properly plugged, hence it may permit water to flood adjoining wells. It is advisable to consider whether a chemical dye has been used to trace the water, and, if so, what were the results? Careful consideration should be given to the field tests made on the different water strings.

With the correlated cross-sections before him, the engineer can make a detailed study of each well and prepare a tabulation showing: (1) The distance between the marker and the bottom of the hole; (2) the distance between the marker and the bottom of the water string; (3) the distance between the marker and the top of a plug; (4) the water production before and after the plug was put in; (5) the water production before and after any deepening job; (6) the initial and present production in oil and water; (7) date at which the well started to make a serious amount of water; (8) remarks as to what any field tests showed; and (9) the source of the water according to the analysis, etc. All of this information may be tabulated under each well on the cross-section, as well as on a sheet of paper, where there are many wells to investigate.

In studying histories, it may be noticed that water appeared at a time after the pipe was pulled from an adjoining well which was improperly plugged upon abandonment.

The question whether or not a well is making top water should be considered from two phases: First, whether the water string leaks and, second, whether the casing shoe has been landed too high. The history will also indicate whether a well made water when it was drilled in or whether the water started later. The history will also show what bailing tests were made on the wells at the time the water string was landed. If the original tests were satisfactory, the chances are that water has not broken in later.

The casing may be tested by a casing tester or by setting a plug in the casing shoe; again, a plug may be placed several feet below the casing shoe of the water string; then a bailing test would test not only the casing but the effectiveness of the water shut-off job as well.

In looking for top water, the engineer should first select a well at which the water string is landed highest stratigraphically, but still makes no water. After the proper landing point for the water shut-off strings has been determined, this distance should be expressed with reference to the marker, so that by the use of sections and tabulations it can be readily told whether or not the shut-off point is too high in other wells.

The possibilities of bottom water should be considered. To determine if the well has been drilled too deep, as indicated by the tabulation, showing the safe point to which wells may be drilled without encountering bottom water, the well that has been drilled deepest stratigraphically but still makes no water should be selected. This then determines a depth to which a well can be drilled with safety. The engineer must bear in mind that often comparison can not be made of wells located a great distance apart because, where there is an edge-water condition, the sand down slope may have water while up slope it contains oil. Plugging jobs also give information concerning bottom water. If the well has been plugged, the engineer should review the tests made after any plug was put in to see whether or not there is good evidence that the plug was tight. If bottom water is suspected, and it has not been plugged off, a test may be made with a plug, preferably cement. Bottom water may be indicated by deepening jobs shown in the history; if a well made no water until deepened, a marked increase in water afterwards would indicate that this well had encountered bottom water. Bottom waters usually have distinct chemical properties.

When all information and tests indicate that it is not bottom water, the possibility of water coming from a middle horizon should be considered. This is a difficult water to test. When middle water is present, it is necessary to make certain first that the water is not coming from top or bottom. If the water string is landed low enough and the ori inal bailing tests indicate a tight job, and, furthermore, if the well is not drilled deep enough for bottom water, evidently the water is coming from an intermediate source. Evidence is also gained by considering histories of adjacent wells, to note whether these wells have a similar water and, if so, if the bottom of any of the holes has been plugged. If a plug that should have held back any bottom water was once placed in the bottom of the hole, but the well still made water, there is indication that the water is coming from higher up the hole. In looking for the water of the middle horizon, it may be that adjoining wells were deepened in successive stages and the histories of these wells may indicate the depth below the marker at which middle water is encountered. The middle-water sand often has definite properties distinct from those of the top and bottom waters, which differences are brought out by water analyses.

Edge water may be suspected when a group of wells down slope show increased oil production. In addition, a group of wells located roughly parallel to a structure contour may show a sudden increase in water. It may be noticed that wells will produce oil from a certain sand in one locality while down slope this same sand contains water; some place between these wells there is an edge-water line and in time the water will encroach on the oil wells.

There remains, of course, the possibility of the wells making water from the base of an oil sand; a water analysis may indicate a new water which lies in the base of the sand. These wells usually turn from oil to water very suddenly. Where the water occurs in the base of a sand in a flowing well with large production, there is little evidence of gas and the well often flows very evenly. The production of water coming from the base of an oil sand is only temporarily retarded by plugging.

A lens of water will be detected by a different kind of water, as shown by water analyses; a study of well histories will show that only a small number of wells in a certain locality have this water. Inasmuch as a lens of water may occur in any part of the geologic column, it is often referred to as top, bottom, or intermediate water, depending on its location in reference to the oil sands.

Correction of Wells Making Water

In the various studies and field tests, the engineer should have ideas of the sources of water in the different wells and the location of the different oil sands. Recommendation should be made for correction of wells that may be letting water into any producing oil well.

Top Water

In case of a leak in the casing, one remedy is to place a packer between the tubing and the water string. Where there is a full oil string, it may be necessary to cut the casing and leave only a liner in the hole so that the packer may fill the annular space between the tubing and the water string.

The water string may leak because it was not screwed together. By screwing pipe together, it has been possible to shut off a leak in the casing.

If the casing leaks because of a collapsed water string, it may be swaged out; but very often it is difficult to repair such a leak. A packer may be used or some cement forced through the hole in the casing behind the pipe, but often another water string must be landed deeper to shut off the water, especially in a drilling well.

A leak in the pipe may be caused by a split joint in the casing or by corrosive waters eating through the casing or, possibly, line wear. If the split is large, a bridge capped with a cement plug may be placed in the pipe and cement forced through the leak. If the hole is small the casing may be ripped and cement forced in behind the pipe.

Where water is leaking around the shoe of the casing it may be possible to place a bridge several feet below the shoe of the water string and
then fill the top of the bridge with brick, stone, or cement.   After the
cement is set, cement may be forced behind the pipe under pressure,
although usually this is not an efficient and satisfactory means.   If circulation is possible, cement may be pumped in behind the pipe, as de-
cribed by Tough.[5]   Where circulation cannot be obtained, cement may
be forced through the tubing behind the pipe under pressure.   Often-
times when the water is leaking around the pipe it may be necessary to
cement a smaller-sized string of casing a few feet deeper, provided the
oil sand is not too close; if the oil sand is too close, redrilling is often
necessary, after shooting the bottom of the casing, and then recementing
the string at the same depth.   On some occasions, it has been possible
to drive the pipe several feet to make a lower formation shut-off; also
a liner has been landed with cement around the outside, so as to shut
off a water sand directly below the casing shoe.

Bottom Water

Cement is recommended for plugging the bottom of the well where
it has been established that the water is coming from bottom.   Mud-laden
fluid has been used, but is not to be recommended generally; likewise, a
lead seal or packer has shut off bottom water, but the writer prefers
cement.

Intermediate Water

When intermediate water is present, the operator must exercise great
care in protecting the upper oil sands while producing from the lower
sands or vice versa.   In producing oil from the lower sands only, the
upper oil sands should be protected either by the use of mud-laden fluid
behind the pipe or by pumping a liberal amount of cement behind the
water string.   The producer should make certain that a sufficient quantity is pumped in so that the top of the cement is actually above the
upper oil zone; this will prevent the middle water from entering the
upper oil horizon.
Where a well is making a large amount of water from an intermediate
sand and is producing from both the upper and the lower oil zones, the
well must be plugged up from the bottom to above the middle water and
production taken from the upper sand, or else redrilled and a string of
casing landed below the intermediate water.   In plugging, it is best to
use a large amount of cement to protect the lower sand; and where a
new string is landed below the intermediate water, sufficient cement or

---

[5] F. B. Tough: Methods of Shutting Off Water in Oil and Gas Wells. U. S. Bureau
of Mines *Bull.* 163.

mud-laden fluid should be placed behind the pipe to assure proper protection of the upper oil zone.

Goodrich suggests the use of a liner with packers on the outside of the pipe at the top and bottom to shut off an intermediate water. The liner would be set opposite the water sand, as the packers would be expected to confine the water to the sand. Where the water has any appreciable head, it is very doubtful if this method, in the majority of cases, will prove satisfactory.

Edge Water

The following are suggested methods for restraining encroaching edge water: (1) Use of compressed air to hold back the water by forcing air into those wells nearest the edge-water line, thus holding back the water while allowing increased production in the wells up slope. (2) Drilling ahead of the approaching water and plugging the well as soon as the water becomes troublesome.

For the purpose of obtaining a maximum production, a careful study should be made of drilling costs and production in order to arrive at an economic cost balance that will determine the maximum number of wells that can be drilled in order that the production may yield the largest profit possible. In short, the encroaching edge water will entrap much of the oil underground, so the operator should plan to get the greatest profit per barrel of production. In the case of edge water, this study should be made before water becomes the master.

Edge water may occur in the top, middle, or bottom oil sands. If edge water occurs in the top sand and the water has advanced to the well, it is, of course, a matter of treating the upper sand as a top-water sand and then making a shut off below it.

Where there are several producing sands and the edge water occurs in an intermediate sand, it may be handled by plugging, with cement, from the bottom to a point above the water sand, after which the operator can produce from the top sand in that well. In doing this, great care should be taken that no water is allowed permanent access to the lower oil zones. Another way is to redrill the well and land a water string below the edge-water sand and the well made to produce from the lower zone. It is important in such a case that the top oil sands be properly protected.

If edge water appears in the bottom sand of a well, it should be plugged off by cement and production taken from the sands above.

Water in the Base of an Oil Sand or in a Lenticular Sand

The operator should be certain that the water and oil are not separated by a small break before deciding that the water occurs in the base of an oil sand. It is very difficult to place a thin cement plug of an exact thickness by ordinary dump bailer methods, but cement should be used and the well plugged in stages and tested. In each test the operator should see if the cement is hard and should plug only a few feet at a time

The McDonald method of shutting off water in an oil well has been very successful in the Illinois field and has been described in a bulletin of the Illinois Geological Survey.[6] A description is also given in an article by Tough.[7] When the water and oil together occur in the same sand, the application of the McDonald method or any other can at best only delay its approach, for eventually water will cause much trouble and expense.

A lens of water may occur any place in the productive zone and should be handled as a top, bottom, or middle water, depending on its location.

DISCUSSION

R. A. CONKLING,* St. Louis, Mo.—We have found it more helpful to observe the amount of water by the sands above the oil than to analyze the water. A well in Texas came in at 1000 bbl. but in about a week began to show water. After the fourth day about 30 bbl. of water were produced. The field department thought it was bottom water and wanted to plug immediately. It has been producing for the last three months and has been making about the same amount of water that the water sands were thought able to produce.

In another case we went through a couple of water sands and ran into an oil sand about 200 ft. below.' We have been trying to get the field department to repair that well, because we know the water will go down when we get to the shallow oil sand. This shows that the geologist in the field should keep close records while drilling, for such records will help solve problems that will come up later.

E. DeGOLYER, New York, N. Y.—One point in connection with the question of bottom water that has not been considered much in American practice is keeping down the water by checking the flow of a well. In Mexico, especially in the southern part of the Tampico-Tuxpam region, oil occurs in very porous limestone and probably moves with an ease and freedom that is not equalled in any known field of the United States.

---

[6] F. H. Kay: Petroleum in Illinois in 1914–1915. Illinois State Geol. Survey *Bull.* 33 (1916) 87–88.

[7] F. B. Tough: Methods of Shutting Off Water in Oil and Gas Wells. U. S. Bureau of Mines *Bull.* 163, 82–85.

* Head Geologist, Roxana Petroleum Corpn.

Under the oil is the bottom water, which is practically the only water. Conditions are more or less artesian. If an oil deposit is trapped over water having an artesian head in an anticline, you have a condition similar to that existing in Mexico. In a certain field where the deposit of oil was possibly only a few feet thick, the entire field was practically ruined by trying to make 10,000-bbl. wells out of what were probably 100-bbl. wells. One well produced 6000 bbl. of clean oil in the first few hours, but water then broke through and in three or four minutes the product turned through the various shades from jet black to a dirty lemon yellow as the percentage of water increased and the well was ruined.

In one of the larger fields, when a well making 18,000 bbl. of clean oil began to show water, its production was reduced to 16,000 bbl., which checked the flow of water for a few months. Whenever water appeared, the production was checked and the oil cleared. The well is now producing 900 to 1000 bbl. of clean pipe-line oil. Over 2,500,000 bbl. of clean oil have been obtained since water first appeared by thus nursing the well along. We have made a set of curves showing temperature, water and sediment, flow-line pressures, etc., that demonstrates clearly the conditions governing occurrence of oil in Mexico.

Until the Potrero del Llano well began to show water there were only slight variations in the temperature of the oil. When water appeared, the temperature of the oil increased 18° to 20° F. within twenty-four hours.

Mr. Reilley.—Isn't it just as necessary to curb a well making a large volume of gas with the oil as it would be to curb a well with less gas making the same volume of water?

E. DeGolyer.—My whole consideration of this subject bears on the question of raising the critical cone in the water table and of lowering the top of that cone. I think that if there is a lot of gas with the oil, the cone is likely to be much sharper than with the dead, heavy oil. The worst condition resulting from water coming is when the crest of the cone reaches the bottom of the casing in a well and thus cuts off any remaining oil.

R. Van A. Mills,* Washington, D. C.—The Bureau of Mines has made a large number of experiments with oils of different viscosities under different rates of recovery that tend to substantiate many of Mr. DeGolyer's remarks. It is necessary to study the differences in the behaviors of oils of different viscosities as influenced by the rates of recovery under various conditions. In doing this work the Bureau of Mines is studying the relative times required to form water cones under different conditions of flow, together with the time required for the cones to

---

* Petroleum Technologist, U. S. Bureau of Mines.

flatten out under retarded conditions of flow, as well as during periods of rest. As a rule, water cones are accentuated by increased rates of flow, and decreased or eliminated by reductions in the rates of flow. My experiments indicate that under certain conditions water cones form more readily with dead oils than with oils heavily charged with gas. In attacking these problems it is dangerous to generalize because of the many factors and different sets of limiting conditions involved in the different fields.

R. A. CONKLING.—Our department has an exploration geologist, who has charge of the drilling of all wells; the total depth is always sent out from the St. Louis office. A geologist in the field sends in samples and works up as much data as possible. We have a lease 1 mi. long and $\frac{1}{4}$ mi. wide with four operators operating around, offsetting it. On the north, the wells began making water almost as soon as they were drilled in. Wells come in from 1000 to 3000 bbl. Until a week ago, we did not have 0.5 per cent. water, by analysis, in all of the oil on that lease; the adjoining leases have from 5 per cent. to 100 per cent. water; three wells are all water.

We simply stop the wells above water level when the edge water has just begun to come; the field department will plug back, because there is plenty of sand and the water will soon rise to that level.

At one time 30,000 bbl. of crude oil were turned out by one operator. He did not have any place to keep it so he had to turn it loose to take care of the other oil. That is what we save the company.

In another case the field department struck water and wanted to know whether to plug back. It was above our water level, and we had two other wells going down to deep sand nearby. Knowing that that was not the true water level, we told them to bail for a week. After bailing four days, the water was gone, so we deepened the wells to the proper horizon.

R. VAN A. MILLS.—In considering the determination of the source of oil-well water by chemical analysis, one must bear in mind the fact that the water produced from an oil-bearing horizon in a new field may be different from the water produced from that same bed a year or two later; not because water has leaked into that bed through wells, but because the water in that bed has undergone induced changes. Water associated with oil and gas in the pays undergoes induced concentration through the removal of water vapor in expanding gases, the concentration being accompanied by changes in the relative proportions of the dissolved constituents. The fact is emphasized that differences in the analyses of waters collected from the same well at different times do not necessarily indicate the infiltration of top or bottom water, especially if edge water accompanied the oil and gas in the pay when the well was brought in.

Again, we must consider the relations that the viscosities of oils, the pressures and proportions of gases accompanying the oils, and the textures and bedding of sands bear to the differential movements between oils and water. For instance, the differential movements between Appalachian crude oils of low viscosity and water are comparatively slight, whereas with oils of higher viscosities the differential movements are so pronounced as to lead operators to think that wells or entire fields have gone entirely to water, when in reality the wells are affected only by water cones, a large part of the oil still remaining to be recovered. Experiments show that an Appalachian oil of low viscosity migrates readily under the propulsion of hydraulic currents, whereas under the same conditions a California crude of high viscosity fails to migrate at all. Obviously the effect of water on oil recovery depends largely on the viscosities of the oils—the more viscous the oils, the more detrimental is the effect of water.

Oil is propelled to wells by the expansive force of gas. Under certain conditions the oil is thus propelled to the wells ahead of water, but as the gas is exhausted, this relationship may change so that the water advances to the wells ahead of the oil. This is illustrated by gushers in which we have slight, if any, shows of water until a large proportion of the gas is exhausted.

The sizes of pores through which the fluids pass also have a decided influence on the relations of water to the recovery of oil. Under various conditions, the differential movements between oil and water are accentuated as the sizes of the pores are diminished. Where the sizes of pores are sufficiently diminished by induced cementation, the recovery of oil may be greatly retarded or entirely prevented   It is imperative that we consider these fundamental principles in attacking oil-field water problems.

Oil-field Brines

By Chester W. Washburne, New York, N. Y.

(St. Louis Meeting, September, 1920)

Recently, Messrs. Mills and Wells[1] published a thorough chemical study of the waters associated with oil in parts of the Pennsylvania, Ohio, and West Virginia region. Many of their conclusions are of general application and the writer wishes to discuss some of these.

Messrs. Mills and Wells show that the composition of the deep brines of the Appalachian fields is such as would be produced by the evaporation of sea water and the precipitation of sodium chloride, combined with reactions with hydrocarbons and other substances. The brines are altered, also, by considerable mixing with meteoric water. They give good reasons for believing that the concentration of the brines was produced by evaporation in the rock pores induced by migrating gas, much of which probably escaped to the surface of the ground.

This hypothesis was considered by the writer in a former paper,[2] in which main stress was laid on a second hypothesis, that the excess of chlorine in the deep brines may have been due to the entrance of solutions rich in magnesium and calcium chloride which ascended from a deep, possibly intratelluric, source. Messrs. Mills and Wells present good arguments for the first hypothesis. Underground evaporation by ascending gases probably will be accepted as the best available explanation of the concentration and composition of deep well waters.

They have not considered the possibility that salt waters in deep sands may be concentrated by the diffusion of water vapor through gas into shale. The writer has given reasons[3] for believing that capillary forces concentrate gas and oil in the larger openings of rock, such as the pores of sandstones embedded in shale. The gas underground must be nearly saturated with water vapor at all times, because it always is in contact with the interstitial water of enclosing shales and of the sandstone. Experimental studies of soil moisture have shown that approxi-

---

[1] R. Van A. Mills and Roger C. Wells: The Evaporation and Concentration of Waters Associated with Petroleum and Natural Gas. U. S. Geol. Survey *Bull.* 693 (1919).

[2] C. W. Washburne: Chlorides in Oil-field Waters. *Trans.* (1914) **48**, 689, 690.

[3] C. W. Washburne: The Capillary Concentration of Oil. *Trans.* (1914) **50**, 829.

mately saturated soil air deposits its moisture on concave water surfaces of sharp curvature, such as the capillary surfaces in the pores of clay, while it is absorbing or evaporating water from concave surfaces of larger curvature, as in sands, where the water films are relatively broad.[4] In this way there is probably a slow migration of water vapor from sandstone into shale. The process is essentially a diffusion of water vapor through gas. It is effective only to the extent that the shale contains gas to exchange for the water it receives from the sand. Nevertheless, this process of vapor diffusion operating through geological periods would be a potent factor in evaporating the water in sands. It would operate in either stagnant or moving gas, and is regarded as supplemental to the process of evaporation by convection in moving currents of gas postulated by Messrs. Mills and Wells. This process of diffusion of water vapor and its condensation in shale would increase the concentration of the brines.

Origin of Salt Cores

Messrs. Mills and Wells carry their theory to its ultimate limit in trying to explain the origin of the salt domes of the Gulf Coast and other regions. They show that the volume of gas required to deposit salt under a pressure of 100 atmospheres and "under reasonable conditions" is from 145 to 1550 times the volume of salt deposited. The smaller figure is for temperatures of 100° C. and the larger figure is for 40° C. From the depth of the upper part of the salt cores, the temperature of deposition probably did not exceed 40° C. unless deep-seated hot waters were involved; hence, the volume of gas required would be about 1500 times the volume of salt.

Do they realize that the volume of most of the salt cores probably is over one cubic mile? Some of the domes, such as Humble and South Dayton, have proved volumes of five cubic miles or more, and possibly of many times this amount if they extend as far downward as commonly thought. Could the required 1500 or 7500 cu. mi. of gas be furnished by the thick sediments underlying the region tributary to any salt core? If this volume of gas, measured under 100 atmospheres pressure, escaped in one geological period at the site of any salt dome, it could be only through vertical channels which would necessarily be so free and open that there could have been no accumulation of oil and gas at these places. Moreover, the salt cores are at least a few thousand feet in height and cut so many water-bearing sands that the writer doubts if any process could concentrate the solutions to saturation. The water in these sands

---

[4] Lord Kelvin. Referred to by Lyman J. Briggs: The Mechanics of Soil Moisture. U. S. Dept. of Agric., Div. of Soils, *Bull.* 10 (1897) 12, with reference to Maxwell: Theory of Heat, 287. Important later references not at hand.

generally is only moderately saline, and commonly is under artesian head. At least some of the sands outcrop in higher country farther inland. Some of them furnish potable artesian water in the same general region, but not in the oil fields, where potable water occurs only in the shallower sands, although a few of the flows of deep waters are only moderately saline.

The region where the sands outcrop has not been submerged since they were deposited by fresh-water streams. The region of most of the productive salt domes probably was temporarily covered by the sea in Neocene time. Possibly the coastal parts of many of the oil sands are marine; marine shells occur in the lower water-bearing sands of the Goose Creek field. Some of the sands appear to be local and not to extend far inland, being possibly beach sands that spread laterally along the Tertiary sea shores. Other lenticular sands may occur. However, there are several wide-spread sands and it is probable that their outcrop always was higher than the region of productive salt cores. Hence, the water in the outcropping sands always tended to flow toward the sea, maintaining a certain degree of freshness in the sands. If this inference is correct, it would be impossible for salt cores to grow upward across the sands by any process of precipitation, because there must have been sufficient artesian circulation in the sands to keep the waters dilute. This artesian circulation would be set up as soon as fissures or other vertical channels were opened for the ascent of the hypothetical salt solutions.[5]

The mixing of the meteoric waters from the sands with the hypothetical rising salt solutions would keep the latter from reaching a state of saturation. All theories of the chemical origin of the salt cores postulate a period of free upward circulation at the loci of the domes. Such freedom to move upward would release the partly meteoric waters in the Tertiary sands and would let them circulate more rapidly down the dip, increasing their freshness. These waters would enter any vertical fissures and would prevent the deposition of salt therein; in fact, they probably would be fresh enough to dissolve any salt previously deposited. Hence it seems impossible that the salt cores could have been precipitated from waters that rose along fissures cutting all of these water-bearing sands. A better explanation is that presented by van der Gracht,[6] that the salt cores are essentially intrusive masses that were

---

[5] Vigorous natural artesian circulation of this type is taking place at the salt dome at West Point, Tex. This dome is surrounded by a ring-shaped valley full of fresh-water springs. The water comes from sands in the Wilcox formation. There are also some salty springs. See E. De Golyer, *Jnl. Geol.* (1919) **28**, 653.

[6] S. W. Assoc. Petrol. Geol. *Bull.* 1 (1917) 85. See also G. S. Rogers: The Origin of the Salt Domes of the Gulf Coast of Texas and Louisiana. *Econ. Geol.* (1918) **13**, 447; DeGolyer, *Trans.* (1919) **61**, 456; Rogers, *Econ. Geol.* (1919) **14**, 178.

squeezed up in semiplastic condition from hypothetical salt beds in the Permean or other underlying strata.

There seem to be only two dubious ways by which the fresh-water sands could be sealed sufficiently to prevent them from flooding the fissures. The first would be by clogging their pores with salts next to fissures in which large quantities of warm gases were rising from below. This method might effectively seal off the sands that contained nearly saturated solutions, but many sands are involved, and it seems probable that some of these that outcrop inland must have furnished strong flows of comparatively fresh water, which would prevent their becoming clogged by salt. The second way is as follows: Gas rising from great depths tends to maintain some of its original pressure by expanding. The gas rising in a fissure might be under higher pressure than the water in any artesian sand cut by the fissure. It would seem, therefore, that the gas would enter the water sands and drive back the water. The gas would enter the sands to a certain extent, especially through the larger pores where capillarity exerts less resistance to flow. The gas could not hold the water back in the finer pores of the same sand. Thus, circulations would be set up whereby water in the sand would be exchanged for gas in the fissure, until the pressure in the sand nearly equaled that in the fissure after which large quantities of water could enter the fissure from the sand. This method would be no more effective than an attempt to use gas to seal off a water sand in a deep well.

In other words, it seems impossible to postulate the precipitation of salt from solutions ascending in fissures many thousand feet across the broken edges of the Tertiary strata of the Gulf Coast. There are, and always have been, too many sands in these formations that contain fresh or moderately saline artesian waters, which would enter the fissures and would prevent precipitation of salt.

ORIGIN OF GYPSUM IN SALT DOMES

Thick bodies of gypsum are present near the tops of some of the salt cores of the Gulf Coast. These generally overlie the salt, having thus the same position that is commonly occupied by gypsum in the salt mines of Germany and other places. This suggests that the gypsum may have ascended *en masse* on top of the intrusive body of salt, but it is not a conclusive argument against other theories.

Messrs. Mills and Wells have observed the deposition of calcium sulfate in the bottom of wells, where sulfate waters had leaked down the casing from upper levels and had mixed with brines rich in calcium chloride. This suggests that the tops of the salt cores of the Gulf Coast may have been the places where descending sulfate waters mingled with brines of the oil-field type rich in calcium chloride, which would cause the precipitation of gypsum at that horizon. The latter explanation is necessary only if one adopts the theory that the salt cores were deposited from solutions. It is a possible source of the gypsum under either theory.

E. DeGolyer,[7] referring to the chemical work of Frank K. Cameron,[8] shows that the gypsum may have been deposited against the salt mass where the sulfate-bearing waters dissolved sodium chloride until they became highly concentrated therewith. Cameron showed that in a highly concentrated solution of sodium chloride gypsum is much less soluble than in a moderately concentrated solution. This doubtless would be an efficient method of precipitating the gypsum which had been dissolved from surrounding strata by moderately concentrated solutions of sodium chloride. However, a study of the following table by Cameron, copied from DeGolyer,[9] shows that the addition of sodium chloride, however great the amount, could not precipitate the gypsum dissolved

Solubility[10] *of Calcium Sulfate in Aqueous Solutions of Sodium Chloride at 23°*

NaCl, Grams per Liter	$CaSO_4$, Grams per Liter	Gypsum, Grams per Liter	NaCl, Grams per Liter	$CaSO_4$, Grams per Liter	Gypsum, Grams per Liter
0.99	2.37	2.99	129.50	7.50	9.42
4.95	3.02	3.82	197.20	7.25	9.17
10.40	3.54	4.48	229.70	7.03	8.88
30.19	4.97	6.31	306.40	5.68	7.19
49.17	5.94	7.51	315.55[a]	5.37[a]	6.97[a]
75.58	6.74	8.53			

[a] The solution in this case was in contact with both gypsum and sodium chloride in the solid phase.

from surrounding strata by solutions containing less than about 40 gm. per liter of sodium chloride. In other words, the original solution of the gypsum must have been effected by solutions of sodium chloride stronger than 40 gm. per liter. This restriction may limit the possible source of gypsum to a small zone immediately surrounding the salt mass, because there is reason to believe that most of the water in sands remote from the salt cores contain less than 40 gm. per liter of common salt. The table also indicates that anhydrite would be the mineral generally deposited, rather than gypsum, but the former readily converts into the latter under certain underground conditions. Both minerals occur in the cap rocks.

[7] Origin of the Cap Rock of the Gulf Coast Salt Domes. *Econ. Geol.* (1918) **13,** 618–619.

[8] Various papers, U. S. Dept. of Agric., Div. of Soils, *Bull.* 18, 33 and 49; also *Jnl. Phys. Chem.* (1901) **5.**

[9] *Loc. cit.*

[10] Frank K. Cameron: Solubility of Gypsum in Aqueous Solutions of Sodium Chloride. U. S. Dept. of Agric., Div. of Soils, *Bull.* 18 (1901) 25–45 (Table IX).

ORIGIN OF GYPSUM IN THE RED BEDS

The structure of the thick masses of gypsum found in the Red Beds of the Western part of the United States and other regions is strongly suggestive of a secondary origin. Some gypsum beds spread with fairly uniform thickness over broad areas, as the main gypsum bed of the Permean of southern Oklahoma. Beds of this type appear to have been deposited in semi-enclosed basins, possibly near the margins of the sea. In other cases, remote from known seas of the same age as the gypsum, as in the Big Horn Basin of Wyoming and south of the Owl Creek Mountains, the gypsum beds in the Triassic Red Beds are lenticular and irregular. There is much contortion of layers and considerable impurity.

Frequently the surrounding beds have forms which suggest that they have been shoved apart by the more or less concretionary growth of the mass of gypsum. The bedding is imperfect and the individual layers of the gypsum are imperfectly developed, and are exceedingly variable in thickness. It is quite possible that these gypsum beds have grown by accretion through the deposition of calcium sulfate precipitated by the mingling of underground brines rich in calcium chloride with sulfate waters of meteoric origin. Some of these lenses of gypsum give the impression of having been deposited in desert lakes and of having undergone a secondary growth after burial, in the same way that concretions of the pure mineral type grow in sedimentary rocks by shoving the matrix aside.

ORIGIN OF LIMESTONE CAPS

The uppermost level of a salt core commonly is a bed of porous limestone, which varies in thickness from 20 to over 100 ft. (6 to 30 m.) No fossils have been observed in this limestone. Fragments of it suggest that the limestone is of secondary origin and that it was deposited from solution. Its deposition might be explained either on the theory of the mixing of oil-field brines with waters containing carbonates or from the release of carbon dioxide carried in solution by ascending brines. It is possible also that the limestone caps may have lain above the gypsum of salt in their original positions in the Permean or other deeply buried formation. However, secondary limestones are rare in the Permean of western Texas.

There are difficulties in both of these explanations. If the limestone caps were deposited from solution, it is hard to see why they do not extend generally down the sides of the salt cores and why the deposition of lime did not spread laterally between the sand grains, converting the sandstones into solid bodies of calcareous sandstone or sandy limestone, instead of leaving them so completely friable and uncemented. In some cases, DeGolyer says, the limestone caps extend at least a little way down the sides of the domes and have what appears to be a "thimble shape."

The theory that the limestone caps rose on the top of intrusive masses seems hard to accept, because there is practically no evidence of the breaking up of the cap by faulting or brecciation, which would seem to be a necessary accompaniment of its ascent at the top of the intrusive salt plug. The cap rock, however, may be broken and fissured more than is commonly thought, since in studying Coastal well logs one can recognize only large displacements. The locally high production of oil or sulfur from the cap rock may be an indication of brecciation or extensive fissuring.

Limestone caps do not characteristically occur in great thickness above the salt and gypsum beds that have been explored in Germany and other foreign fields. Thin caps of this kind occur in some places above thick beds of rock salt, but I do not know of any salt mine in the world where there is a cap of secondary limestone that approaches the thickness of these caps on the salt cores of the Gulf Coast. The origin of the limestone caps of the salt cores remains an open question.

Lately, DeGolyer[11] has suggested that the calcium carbonate of the cap rock may have been precipitated by the action of sodium chloride, basing his suggestion on Cameron's observation that very concentrated solutions of sodium chloride can hold less calcium bicarbonate than weaker solutions. He quotes the following figures:

SOLUBILITY OF CALCIUM BICARBONATE IN AQUEOUS SOLUTIONS OF SODIUM CHLORIDE

SODIUM CHLORIDE, GRAMS PER LITER	CALCIUM BICARBONATE, GRAMS PER LITER
0.0	0.06
39.62	0.101
267.60	0.04

The experiments were made in equilibrium with atmospheric air. Their direct application to the problem is weakened by our ignorance of the amount of carbon dioxide in the underground gases that accompanied the solutions which deposited the limestone caps. The table seems to allow for the precipitation of 0.061 gm. of calcium bicarbonate per liter of salt solution when the sodium chloride in solution increases from 39.62 to 267.60 gm. per liter. A nearly equal amount, or 0.04 gm. of bicarbonate remains in solution. The general nature of the problem is similar to that of the precipitation of gypsum by salt, mentioned above. It requires the previous solution of calcium bicarbonate by rather strong salt solutions. Although the amount of limestone that would be precipitated by this process is small, it could build up thick

---

[11] *Econ. Geol.* (1918) **13**, 619. Reference to Cameron and Seidell, U. S. Dept. of Agric., Div. of Soils, *Bull.* 18, 58–64; also *Jnl. Phys. Chem.* (1901) **5**, 643.

limestone caps in geological time, if the salt solutions penetrated far enough through the sediments to gather nearly twice the required amount of bicarbonate.

In other words, all present theories of the secondary chemical origin of the substances in salt cores require extensive circulation. Extensive circulation, by releasing artesian waters, must promote solution rather than precipitation of salt. The gypsum and limestone may be precipitated *in situ*. The salt must be intrusive *en masse*.

SOLUTION AND DEPOSITION OF LIME IN SANDS

Messrs. Mills and Wells[12] describe processes by which calcium carbonate is deposited in the sands. They show that this may result either from the mingling of waters containing calcium chloride with waters containing alkaline carbonates, or through the liberation of carbon dioxide from the waters. The deposition of lime carbonates probably explains the occurrence of the numerous non-productive or slightly productive areas that occur in all highly calcareous oil sands. It is probably also the explanation of the presence of cap rocks on the tops of pay sands. Nearly all drillers believe that there is a hard streak, or cap rock, along the top of nearly all productive oil sands. This belief is so universal that it must be based on fact, but the origin of the cap rock remains to be explained. The writer is unable to advance any reason why lime cement should be deposited between the sand grains along the tops of the sands or in the immediately adjacent base of the overlying shale.

Calcareous sands, commonly, are firm hard rocks at the outcrop, but most oil sands in Oklahoma, California, and the Appalachian fields appear to be much softer and more easily drilled than the water-bearing parts of the sand or the cap rock. The writer has regarded this as at least an indication that the carbonate cement had been dissolved out by circulating water previous to or during the time of concentration of the oil in the sand. The effect may be due also to the deposition of carbonate in parts of the sand near the outcrop and in parts of the sand remote from production.

Further evidence of the solution of calcareous matter in oil sands is furnished by the general absence of fossil shells from these sands. It is often observed that fossil shells are absent in the outcrops of oil-saturated sands, although casts of fossils may be common. Hoefer has recorded this same fact as characteristic of all oil regions. Fossil shells appear to be found in the outcrops of oil-saturated sands only where the sand consists mainly of calcareous material. In these cases there was probably more calcite present than the solution could remove. The only explanation of this solution of calcite that the writer has observed in

---

[12] U. S. Geol. Survey *Bull.* 693 (1919) 50, 100.

previous literature is Hoefer's hypothesis that the carbonate is dissolved by the waters associated with the oil because of the liberation of carbon dioxide formed by slight oxidation of the oil, possibly aided by solution in the organic acids that exist in traces in many oils.

This explanation does not appear to be wholly adequate. The reactions given by Mills and Wells[13] for the hydrolysis of magnesium carbonate and magnesium chloride, resulting in the formation of free carbon dioxide and hydrochloric acid, would explain the solution of calcium carbonate very readily, as shown by the following equations:

$$MgCO_3 + H_2O\ \ = Mg(OH)_2 + CO_2$$
$$CaCO_3 + H_2CO_3 = CaH_2(CO_3)_2$$

Also,

$$MgCl_2 + 2H_2O = Mg(OH)_2 + 2HCl$$
$$CaCO_3 + 2HCl = CaCl_2 + CO_2 + H_2O.$$

Both of these reactions probably would result in the deposition of magnesium carbonate in the place of the dissolved calcium carbonate, but Messrs. Mills and Wells cite an example in which crusts of calcium carbonate were dissolved from the water jackets of gas engines by pumping brines rich in magnesium chloride through the jackets. The reactions take place readily in warm solutions and they probably operate very slowly in cold solutions.

There is a translation or diffusion of dissolved calcium carbonate through the water in an oil sand, from points of solution to points of deposition. As long as the water is in contact with calcium carbonate, it must be nearly saturated with that substance. It is probably capable of dissolving small particles of calcite and of depositing the dissolved carbonate on larger masses, as in the process of laboratory "digestion" to increase the size of precipitated crystals. This appears to be the main cause of the growth of concretions in shale and sandstone. That the process may be extensive is proved by the abundance of concretions in many formations, including shales that appear almost impervious. The magnitude of the process is demonstrated also by the occurrence of great concretionary masses, scores of feet across, found in some of our western formations. These must have drawn their supply of carbonates from considerable distances. The transfer of carbonate through sands may be due either to the movement of the water or to the diffusion of dissolved carbonates and bicarbonates through the water, or to both processes combined.

The solution of calcium carbonate would be promoted by the presence of magnesium chloride, as shown by Mills and Wells, and magnesium would be exchanged for calcium in solution. The formation of dolomite in this way has long been regarded as a cause of the porosity of the oil

[13] U. S. Geol. Survey *Bull.* 603 (1919) 72.

pay of the Trenton limestone of Ohio and of the Corniferous limestone of the Irvine field, Kentucky. Bownocker has demonstrated that the percentage of magnesium in the Trenton limestone increases as one approaches an oil field, and that it reaches a maximum in the productive area. The writer has observed that the limestone adjacent to oil-filled crevices in the outcrop of the Tamosopo limestone of the Sierra Madre Oriental, Mexico, is dolomitic, although the rest of the limestone, beginning a few inches away from the crevices, was practically pure calcium carbonate. This indicates that the original waters in oil fields were rich in magnesium and that the magnesium had been lost from solution by the replacement of calcium in solid carbonates.

This process does not explain all of the features of solution. Some fragments of the Corniferous limestone blown out of wells in the Irvine field, Kentucky, contain solution cavities an inch or more across. The same is true of the cap rocks of the Gulf Coast oil fields. The presence of these large cavities in solid limestone demonstrates that underground waters have dissolved much calcium carbonate besides that replaced as dolomite. It would be of great interest to students of oil geology if Messrs. Mills and Wells would devote attention to this subject of solution of calcareous cement.

In certain oil fields the process of solution has been very extensive. Some of the oil sands of California which are hard calcareous sands at their outcrops are only loose, friable, unconsolidated sands in the pay parts underground. The same is true of the Woodbine sand of Louisiana and of many other sands. Frequently these sands are so loose that they will flow into the wells with the oil and clog the holes. As a general rule, nearly all oil sands are so loose that sand grains work into the valves of the pumps and wear them out within a few months; this is more rarely the case with pumps that lift water. It appears to the writer that there is a general process of solution of carbonate cements in oil sands.

At the same time there is probably deposition of calcium carbonate between the sand grains along the top of the sand and in the parts of the sand that lie outside of the productive areas. Very commonly, the parts of a sand that lie structurally far below the oil-producing levels are so tightly cemented that they appear dry to drillers. There seems to be no reason to doubt that these barren parts of the sand are filled with water, rather than with oil or gas, and that the water is under pressure equivalent to that in the productive areas. The only reason why the water does not enter the wells in noticeable quantities must be because the pores are so fine and so clogged with mineral matter that it will not move with sufficient velocity into the wells to be noticeable. This condition is true of a broad area of the Bartlesville sand northwest of the Cushing oil field, Oklahoma. It is true also of most of the synclinal areas surrounding Ranger, and other fields of Eastland County, Texas, but it is not true of the more porous "lime pays" of Stephens County. In both the Cushing and Ranger regions, the water in the parts of the sand immediately surrounding the productive areas will flow rapidly into the well, but when one gets a few miles away from the productive areas the sand is so tight that the drillers call it "dry." The natural inference is that the solution of lime cement in the oil-producing areas of a sand is accompanied by the deposition of interstitial calcite in the non-productive regions.

Some persons doubt that fine pores can prevent the appreciable flow of water into a well under the pressures which exist at depths of 2000 or 3000 ft. (609 or 914 m.). That this is a fact they will probably appreciate when they consider that clay shales commonly have a porosity of 5 to 10 per cent., or about half that of the sands of the Appalachian fields. No water is observed flowing into the wells from these shales, and it is doubtful if any. does enter except from fissures. The main reason why fine pores prevent the flow of water probably is the fact that there are numerous bubbles of gas scattered through the pores. Each little bubble of gas is bordered by a capillary film of water, which clings to the walls of the pores and requires great pressure to move it. If one takes a fine capillary tube through which water will flow slowly and causes a few bubbles of air to enter, like a string of beads, the flow of water will stop, even though a much higher pressure is applied. This is the main reason why shales are capable of sealing deposits of gas and oil. If the pores of the shale were not filled with water, capillary action would quickly draw all of the oil out of the sand into the shale. Through capillary action and through the principle of the diffusion of water vapor into shale described above, all clay shales must be full of water. Migration of oil and gas across them can be only through fissures.

INDUCED SEGREGATION OF OIL ABOVE WATER

M. J. Munn and Roswell H. Johnson independently have shown that there will be no gravitative rearrangement or stratification of oil above water in a sand that contains water above oil, or in a sand containing a mixture of water and oil, unless the sand is shaken or unless movements of some kind are set up in the sand by external action. Recent experiments by McCoy strengthen this conclusion. Messrs. Mills and Wells[14] have shown that an induced segregation of this kind takes place when a mixture of oil and water flows through an oil sand into a well. As thought by Johnson, it is quite probable that the underground circulation of waters may promote this segregation of oil and gas above the water in productive sands.

---

[14] U. S. Geol. Survey *Bull.* 693 (1919) 94, 95.

This segregation promotes the anticlinal accumulation of oil. The process may be accelerated by leakage upward across the shale on the tops of anticlines. In many fields there is little indication of this leakage along the crests of anticlines, and if it occurs the oil probably is too widely scattered in minute joints to be noticed in drilling. In sharply folded regions, such as the Rocky Mountains, oil seeps are common on the axes of anticlines. In these regions, as in the Salt Creek and Grass Creek fields of Wyoming, one finds many shows of oil in drilling through the shale, and frequently there is sufficient oil in the shale crevices to make commercial wells. In such fields there is no accumulation of gas along the tops of the anticlines. Many volumes of gas must be formed for each volume of original oil; hence, all of the gas and some of the oil has leaked out of the productive sands. Some of the oil accumulated in higher sands, as in the Shannon sand of Wyoming; some of it is scattered through small fissures in the shale; some of it reached the surface of the ground. The paraffine wax and tar found along crevices at the ground surface of the Salt Creek field probably is the residue of oil that has leaked up from the Wall Creek sand.

Oil-field anticlines appear to be the result, mainly, of direct uplift from the folding of stronger formations that lie at greater depths. Evidently there must be much fracturing along their crests. The series of numerous small faults that cut across the axes of Rocky Mountain anticlines and die out on their limbs is an indication of the type of fissures through which most of the upward migration or leakage has occurred. Most of the accumulation occurs in the first sands above the source of oil. The accumulation in higher sands is in smaller quantity and the oil generally has become heavier because of the oxidation it has suffered en route. This upward leakage along the tops of the folds may have been more extensive than commonly thought. It would cause circulation, thereby promoting the gravitative segregation of oil above water in the sands, in a manner similar to that which Messrs. Mills and Wells have observed in producing oil wells.

Summary

The concentration of the brines in deep wells probably results in part from evaporation in gas that ascends through fissures or that comes in contact with the deep water in any way.

Water vapor, also, is transferred from sand to shale by diffusion, when the sand is partly filled with gas. The water vapor is evaporated from the broadly concave water surfaces in sands, and precipitated on the relatively sharply concave surfaces in shale. This process also concentrates the water solutions in the sands.

The salt cores of the Gulf Coast could not have been formed by the precipitation of salt from solution, because they cut across many sands that outcrop inland at higher elevations. Any upward movement of water at the site of a salt core would set up a vigorous artesian circulation of fresh water through these sands, which would destroy the concentration of the ascending solutions and would prevent the precipitation of salt. The salt cores seem to be intrusive plugs of salt.

The gypsum on top of the salt plugs may be uplifted parts of deeper gypsum beds, or may be secondary precipitates. The gypsum deposits in the Red Beds of our Western States appear to be primary precipitates in lakes, but some of them show structural indications of subsequent growth by the secondary precipitation of gypsum.

The origin of the limestone caps of the salt cores remains to be explained.

Carbonate cement has been dissolved from many pay sands, leaving them softer and more friable than neighboring dry areas of the same sand. The solution is probably due to the formation of bicarbonates from oxidation of the oil by sulfates, etc. Organic acids may slightly assist the solution of the cement. The entrance of brines containing magnesium chloride would cause solution of calcium carbonate by hydrolysis, as shown by Mills and Wells, but part of the calcite would be replaced by dolomite.

Carbonate cement appears to be deposited along the tops of oil sands, forming the so-called hard caps or cap rock of drillers. No reason for this action comes to mind.

DISCUSSION

R. Van A. Mills,[*] Washington, D. C. (written discussion[†]).—Mr. Washburne's paper is essentially a discussion of certain parts of a Geological Survey Bulletin.[15] The problems under discussion are so difficult to solve, and are of such scientific interest and economic importance as to demand our continued efforts toward their study. It is to be regretted, however, that the author has not presented more data resulting from his own investigations. Real progress in petroleum geology at the present stage of its development demands investigative rather than speculative study.

Messrs. Mills' and Wells' conception of the origin of the brines asso-

---

[*] Petroleum Technologist, U. S. Bureau of Mines.

[†] Published by permission of the Director of the Bureau of Mines.

[15] R. V. A. Mills and R. C. Wells: Evaporation and Concentration of Waters Associated with Petroleum and Natural Gas. U. S. Geol. Survey *Bull.* 693 (1919).

ciated with petroleum and natural gas in the Appalachian fields is summarized as follows:[16]

Marine water of sedimentation and ground water from other sources have been included and deeply buried in the sediments, where, in association with gas and oil, they have migrated and undergone concentration, accompanied by changes in the nature and relative proportions of the dissolved constituents. Concentration is due in part to the leaching of the sediments by the migrating waters, but mainly to the evaporation of water into gases that are moving and expanding through natural channels. Reactions between the dissolved constituents of different types of waters and between the dissolved constituents of the waters and the organic and inorganic constituents of the sediments, have been important factors in the formation of the brines, and so also have mass action and reactions due to deep-seated thermal conditions.

The fact is emphasized that deep-seated evaporation is only one of many factors entering into the formation of the brines. The factors governing the formation of the Appalachian brines cannot be the same as those giving rise to the primary alkaline waters of certain California fields or the sulfate-bearing brines of Wyoming, Kansas, and Oklahoma. Generalizations upon the formation of oil-field brines should follow rather than precede intensive studies in different fields.

The diffusion of water vapor through natural gas as an attribute to the deep-seated evaporation and concentration of the brines has also been considered.[17] Mr. Washburne's hypothesis upon the condensation of this diffused water vapor in shale and the influence of such a process upon the concentration of the brines remaining in the sands is too speculative to be accepted without substantiative field and laboratory data.

The writers of the Government Bulletin pointed out that deep-seated concentration and precipitation caused by the evaporation of brines in ascending gases, together with precipitation by the geochemical processes outlined in that paper, have probably played important roles in the formation of the salt masses and associated cap rocks. These conceptions hold true, no matter what theories may be regarded as best explaining the origin of the domes. Mr. Washburne's selection of the maximum volume of gas (at a pressure of 100 atmospheres) required to cause the deposition of a unit volume of salt by evaporation is somewhat misleading. The fact is emphasized that as gas expands from a pressure of 100 atmospheres to a pressure of 1 atmosphere (at constant temperature), the volume of the gas and hence its capacity to carry moisture is increased a hundredfold. A gradual, and possibly slight, lowering of temperature during the upward passage of the gas would make comparatively little difference in the evaporation effects of the ascending gas.

---

[16] *Bull.* 693, 6.   [17] *Bull.* 693, 80.

As pointed out[18] 1 cu. m. of gas expanding from 100 atmospheres to 1 atmosphere at a constant temperature of 40° C. would be able to evaporate 3800 gm. of water from a saturated solution of sodium chloride, thus causing the precipitation of 1400 gm. or 658 c.c. of salt. In this case the original volume of the compressed gas required to cause the precipitation of 1 cu. m. of salt at or near the earth's surface would be 1500 cu. m., but at deep-seated temperatures and pressures, the original volume of the compressed gases required to cause such a deposition of salt by evaporation would probably be less than 145 cu. m. Should the gas expand from an initial pressure of 200 atmospheres at a temperature of 100° C., only 72.5 cu. m. of compressed gas would be required to cause the deposition of 1 cu. m. of salt. At higher temperatures the initial volume of the compressed gas required to accomplish this evaporation would be less than 72.5 cubic meters.

To avoid the misunderstanding that may arise from Mr. Washburne's paper, the entire paragraph from which he takes his data on the volume of gas required to cause the deposition of a unit volume of salt is given:[19]

On the hypothesis of a cooling brine, then, we calculate that 1 cu. m. of saturated brine would deposit 11 kg. of salt on cooling from 60° to 20° C., whereas the same amount of salt could be deposited from such brine, through evaporation, by the escape of 790 cu. m. of gas at 40° C., 307 cu. m. at 60° C., or 74 cu. m. at 100° C. If the gas expands a hundredfold at the temperatures mentioned, the volumes of compressed gas required would be only about a hundredth of those mentioned. In short, the volumes of compressed gas would have to be from 24 to 260 times greater than a given volume of brine to leave salt as the final product under reasonably favorable conditions. The volumes of gas required are 145 and 1550 times the volume of salt formed at 100° C., and 40° C., respectively. Looked at in another way, 1 cu. m. of brine could deposit 11 kg. of salt by cooling or 330 kg. by evaporation.

Mr. Washburne's assumption that the salt masses were formed at temperatures not exceeding 40° C. is unjustified. Also, his statement that the escape of the gas causing evaporation "could be only through vertical channels which would necessarily be so free and open that there could have been no accumulation of oil and gas at these places," does not accord with present conditions in the salt-dome region where there are today many natural exudations of gas though the fields are productive. In Washington and Morgan Counties, Ohio, gas is escaping from sands only 30 to 100 ft. beneath the surface where little, if any, oil escapes except where the oil sands are actually exposed at the surface. Oil is being recovered from wells drilled to these oil sands only a few hundred feet from their outcrops and at similar distances from the places where the gases are escaping.

---

[18] Bull. 693, 84.   [19] Bull. 693, 92–93.

A concrete example of the amount of salt that may be deposited through the evaporative effects of expanding gas is given in the Government Bulletin. In a gas well that was "shut in" but in which there was underground leakage of both salt water and gas, more than two tons of salt were deposited during four months. The well cavity was filled with salt to a height of several hundred feet from the bottom.

Mr. Washburne's generalizations upon the effects of fresh water in the strata surrounding the salt domes do not seem to be warranted because there is no certainty as to how long these conditions have prevailed. They may not have existed when the domes were formed; also it is a matter of speculation as to whether the water conditions outlined are at all general. The subject requires intensive field and analytical study in conjunction with deep drilling.

In outlining the various theories advanced to explain the origin of the salt domes, the authors of the Government Bulletin say:

Several European geologists have recently revived the old and long-neglected view that salt-dome structure is due to the flow of salt made plastic by pressure.[20] Lachman[21] calls attention to the variety of deformations found in the German salt deposits and shows that the structural features range from those that are entirely conformable to the strata in which the salt deposits are found to those of domes which show practically no relation to the adjoining strata, having apparently been formed by the flowage of salt. Arrhenius[22] has discussed some of the physical and chemical problems involved in the formation of the German salt deposits and applies the principles of isostasy to explain the salt column in Drake's Saline, Louisiana. Before this explanation can be accepted, experiments upon the plastic flow of salt, with special reference to the effect of temperature and pressure, the action of water, and the possibility of flow by fracturing and granulation, followed by recementation and recrystallization are needed. Inasmuch as Arrhenius assumes that solutions have acted to some extent as a lubricant for the movement of the salt, and álso that many of the unusual structural forms found in the German potash salts are due to rearrangements brought about by water given off from hydrated minerals at depth, we feel that, even if the preceding views are accepted, the evaporation of solutions by gases is worthy of consideration.

Conditions of comparative weakness that might permit the plastic flow of salt under great pressure would also permit the movement and probably the escape of solutions and gases, especially where the movements of salt were accompanied by faulting and fracturing of the overlying strata. Probably no one of the theories we have cited

---

[20] O. Grupe: Zechsteinformation und ihr Salzlager im Untergrunde des hannoverschen Eichsfelds: *Zeit. prak. Geol.* (1909) **17**, 185. E. Harbort: Geologie der nordhannoverschen Salzhorste: *Deutsch. geol. Gesell. Monatsber.* (1910), 326; Richard Lachmann: Salinare Spalteneruption gegen Eksemtheorie: Idem, 697; H. Stille: Aufsteigen des Salzgebirges: *Zeit. prak. Geol.* (1911) **19**, 91.

[21] Richard Lachmann: Der Salzauftrieb, Halle, 1911. Separate from *Kali.* (1910) **4**, Nos. 8, 9, 22, 23 and 24. Studien ueber den Bau von Salzmassen: Idem. (1913) **6**, pp. 342–353, 366–375, 397–401, 418–431.

[22] Arrhenius, Svante, Zur Physik der Salzlagerstätten: *Meddelanden* k. v. Nobelinstitut (1912) **2**, No. 20.

will suffice to explain all the unusual phenomena of salt domes, but it is evident that in conjunction with any of the processes mentioned the evaporation of water into moving and expanding gas must be regarded as important.

In the light of present information, the origin of the cap rocks, gypsum and limestone, may well be regarded as geochemical, according to the principles cited by Mills and Wells[23] and by DeGolyer.[24] A geochemical theory for the origin of the cap rocks may also furnish an explanation for the failure of fresh waters to dissolve the salt masses. Salt deposits in oil and gas wells are frequently covered and protected against solution by calcium carbonate and calcium sulfate precipitated by infiltrating fresh waters. The introduction of primary alkaline and sulfate-bearing waters into wells to dissolve salt frequently fails to accomplish this purpose. Sometimes the reactions are such as to contribute more salt to that already in the wells, as in the reaction:

$$2\,NaHCO_3 + CaCl_2 = CaCO_3 + 2\,NaCl + CO_2 + H_2O$$

If the solutions are not saturated with sodium chloride, the calcium carbonate around deposited salt protects it against solution. These principles may well apply to the cap-rock phase of the salt-dome problem.

The writer cannot agree with Mr. Washburne's generalization that the tightly cemented sands along the tops and bottoms of productive sands are characteristically calcareous. The examination of several hundred samples of pay sands and their associated rocks of Pennsylvanian, Mississippian, and Devonian ages in Ohio, Pennsylvania, and West Virginia has shown the breaks and caps of pay sands to be characteristically quartzitic and lacking in lime. In samples of sandstone and shale formations from deep wells in those fields, most of the calcium carbonate found appears to have been deposited through induced cementation subsequent to the drilling and operation of wells. The analysis of oil- and gas-bearing sands and their associated rocks and the discussion of deep-seated waters as agents of cementation and of induced cementation[25] should make these points clear so far as the fields examined for that report are concerned. The original lime content of the deep-seated sands and shales has evidently gone into solution as chloride and bicarbonate.

While engaged in field work, during the preparation of *Bull.* 693, the writer observed that the sandstones exposed to salt waters pumped from old oil wells in Butler County, Pennsylvania, were being disintegrated and rendered more porous and friable by leaching and etching. Lumps of sandstone, the size of a man's fist, from the paths of the salt waters trickling from the tanks around these wells, could be crushed when slightly squeezed in the hand. The waters not only dissolved the cementing material but etched the quartz grains. The fact that oil-field waters are agents for the solution as well as the deposition of

[23] *Bull.* 693.   [24] *Econ. Geol.* (1918) **13**.   [25] *Bull.* 693, 16–18, 76, 44–50, 98.

mineral matter is thoroughly established but, except for the work of Chase Palmer, the geochemistry of some of the processes involved has yet to be made clear.

In regard to the unconsolidated sands of the California fields, has not Mr. Washburne put the cart before the horse? That sands at their out-crops have been cemented through surface agencies in no way signifies that the beds were ever similarly cemented at depth.

Recent experiments[26] made by the writer tend to disprove Mr. Washburne's opinion that oil does not segregate "gravitationally" above water under hydrostatic conditions in sands. Such segregation occurs very readily with certain oils and brines, even in sands of extremely fine texture. It is recognized, however, that where the segregation of oil above water is incomplete, currents induced by the drilling and operation of wells may cause the oil to migrate and to segregate more completely above the water; it is this that the writer has termed *induced segregation*.

The principles of induced segregation are worthy of consideration in the practical recovery of oil and gas as well as in the study of oil and gas accumulation. It seems probable that favorably situated parts of pay sands are enriched by induced migration and segregation. Again, the escape of gases, oils, and waters through natural passages such as fissures has evidently caused the migration and accumulation of the re-maining hydrocarbons into favorable entrapments.[27] Some of the ac-cumulations associated with faults have evidently originated in this way. Apparently Mr. Washburne accepts these views as he repeats them in his paper.

R. VAN A. MILLS (oral discussion).—The investigation outlined in U. S. Geol. Survey *Bull.* 693 was based largely on studies of the changes in Appalachian oil-field waters incident to the drilling and operation of wells. One of the principal changes is concentration due to the evapo-ration of the brines in expanding gas, the brines becoming sufficiently con-centrated to cause the precipitation of a part of their dissolved mineral matter. Realizing that the mineral deposits in wells could not be formed without changes in the proportions of the dissolved constituents in the waters that contributed the deposits, the next step was to determine the character of these changes. It was found that the changes occurred through chemical reaction as well as through concentration. The loss of sodium chloride is not the only change; various dissolved constituents are lost from solution, the changes are complex, many factors being involved.

I regard the hypothesis that diffused water vapor is condensed in the shales as rather speculative. More data are needed to establish such a theory.

As to the salt-dome problem; it was far from the intention of the

---

[26] *Econ. Geol.* (1920) **15**, 398–421.   [27] *Bull.* 693, 94–95.

authors of the Government bulletin to attribute the formation of the salt domes entirely to evaporation. We must attack this problem upon a basis of the multiple hypothesis, without restricting ourselves to any one theory. I believe that the theory of the deposition of salt due to the evaporation of brines by expanding gases is one of the theories worthy of consideration.

At present there is a considerable natural escape of gas in the salt-dome region where oil is being produced; consequently, Mr. Washburne's statement that the oil would have escaped with the gas causing the evaporation is not upheld by present conditions. In many fields where we now have production, we also have the natural escape of gas. In Morgan and Washington Counties, Ohio, within a few hundred feet of gas exudations, we have good oil production in the sands from which the gas is escaping. We also have good oil production within similar distances of the outcrops of the oil-bearing sands. In one case, where an operator has installed a barrel which catches oil (3 qt. in 5 hr.) from an outcrop of the Cow Run sand, he is also producing oil from wells tapping the same sand 400 or 500 ft. away from that outcrop.

The origin of the cap rocks overlying the salt masses appears, to me, to be distinctly geochemical. In various parts of the Appalachian field, waters from shallow beds, leaking into oil wells and coming into contact with deep-seated brines, cause the deposition of mineral matter not unlike that of the cap rocks of the salt domes. Thus in the mineral crusts formed in oil wells, we frequently have calcium carbonate and calcium sulfate associated with salt. Salt is occasionally coated with calcium carbonate and calcium sulfate. Under these conditions the dissolving of the salt might produce pores similar to those in the cap rocks of the domes. The failure to remove salt, by introducing fresh water into "salted-up" wells, is frequently due to the reactions between the dissolved constituents of the deep-seated brines and those of the fresh water introduced into the wells. The water introduced into the wells may not only cause the precipitation of carbonates and sulfates, but may also cause the formation of more sodium chloride according to the reactions quoted in *Bull.* 693.

I wish to emphasize the advisability of avoiding generalizations in attacking these problems. For instance, it is erroneous to assume that all of the relatively impermeable caps overlying oil pays are calcareous. Several hundred samples of oil-bearing sands and their associated rocks in the Appalachian fields have been found to be characteristically siliceous; carbonates are for the most part absent, even in the caps and breaks, except where the sands were very shallow or where they had undergone induced cementation subsequent to the drilling and operation of wells.

E. DeGolyer, New York, N. Y.—As I understand it, Mr. Washburne objects to Mr. Mills' theory, and Mr. Mills agrees with Mr. Washburne, yet he answers rather extensively Mr. Washburne's arguments. I would like to know to what extent Mr. Mills proposes his theory to account for the salt masses.

R. VAN A. MILLS.—The evaporation theory is advanced simply to supply one of the factors entering into the formation of the salt domes. The last paragraph of the discussion of the salt-dome problem[28] reads as follows: "Conditions of comparative weakness that might permit the plastic flow of salt under great pressure would also permit the movement and probably the escape of solutions and gases, especially where the movements of salt were accompanied by faulting and fracturing of the overlying strata. Probably no one of the theories we have cited will suffice to explain all the unusual phenomena of salt domes, but it is evident that in conjunction with any of the processes mentioned the evaporation of water into moving and expanding gas must be regarded as important." In the introduction we say that evaporation has played a large part in the formation of the domes, but we do not say that this one theory fully explains their formation.

E. DEGOLYER.—Mr. Washburne has objected to all theories of the precipitation of salt from solution by pointing out that the salt plugs cut through various sand strata; and by inference, if the salt was deposited from solution, the solution should have saturated these porous strata, deposition of salt would have occurred and the sands also should have been filled with salt. I think the point is well made, and the objection seems to me to be valid against theories of deposition.

R. VAN A. MILLS.—I do not remember that these points were raised in the paper under discussion. Several years ago Mr. Washburne suggested the hypothesis of concentration of oil-field brines by evaporation into ascending gases. One of his principal reasons for rejecting that hypotheses was the supposition that the interstices in the porous strata would be plugged by the deposition of salt incident to the concentration. Now that has not proved to be the case. If the salt solutions associated with oil and gas become concentrated sufficiently to deposit tons of salt in individual gas wells before the water-bearing strata are sealed, I think we have Mr. DeGolyer's objections answered, in some degree, by facts. The "salting up" of strata yielding unsaturated brines is a final stage in the plugging process. I understand that Mr. DeGolyer also supports Mr. Washburne's objection to the evaporation theory based upon the diluting and leaching effects of the so-called "artesian" waters.

E. DEGOLYER.—That was not what I intended to state; I was not talking about artesian waters but about the deposition of the salt masses

[28] *Bull.* 693, 94.

from any form of solution. Mr. Washburne states that we know that the salt masses pass through various porous strata which are now in contact with the salt and he contends that if the salt was deposited from solution, such solution would have saturated all of the porous strata which it penetrated and would have deposited salt in them. If you could get salt masses deposited from solution in a vertical core there would doubtless have been motion of the solution through the channel which is now occupied by the salt and which penetrates various porous sand lenses, thus giving the solution access to them. I am not talking about the solution moving through the sand, under ordinary conditions, but about its moving into the porous strata from such a channel.

R. VAN A. MILLS.—I cannot quite grasp Mr. DeGolyer's point of view. It must be remembered that theories on the deposition of salt from solution embrace only a certain group of the factors that probably entered into the formation of the domes. To assign undue weight to these contributory factors and then to reject them altogether because they fall short of their assigned values, is erroneous. It is my impression that fissures associated with the salt domes have constituted channels for the movements of solutions and gases toward the regions of least pressure which would be upward. New passages for these movements doubtless have been created as the salt masses forced their way upward through the overlying strata.

In regard to the failure of the shallow fresh waters to prevent the formation of the salt masses or to dissolve away these masses after they were formed, it should be remembered that when, in porous rocks, certain natural waters having different properties of reaction come into contact with one another, chemical reaction and precipitation frequently cause dense cementation along the zones of contact between these solutions. Thus the fresh waters coming into contact with salt brines or with the salt masses may cause barriers to form, through precipitation and cementation, that prevent further dilution or leaching.

W. E. PRATT,* Houston, Tex.—Mr. Washburne mentions a "shell" or hard upper crust on the top of many oil-bearing sands, for which he has no explanation. I understood Mr. Mills to say that he had not observed that condition.

R. VAN A. MILLS.—In the Appalachian field, the shells and caps are usually siliceous rather than calcareous.

W. E. PRATT.—I am under the impression that it is a general condition. Very often the "shell" is simply more firmly cemented with calcium carbonate than the lower part of the same sands. I wanted to ask whether other people's observations bore out my impression.

---

* Chief Geologist, Humble Oil & Refin. Co.

R. A. Conkling,[*] St. Louis, Mo.—The shell is usual. There may be a little poor sand almost on top of the oil sand. In other cases, the pay may be almost at the top of the sand. We usually have cementation, but the top part is cemented irregularly.

R. Van A. Mills.—We have waters and rocks of different types in different fields. I have accepted Mr. Washburne's paper as being essentially a discussion of *Bull.* 693, which was based on field and laboratory work in the Appalachian field, and I maintain my statement regarding the oil and gas-bearing sands of that region. In southeastern Ohio, we find the shallow pay sands carrying carbonates which may have been formed partly through the escape of gases and the infiltration of shallow ground waters which would be of the same order as induced cementation, that is, cementation subsequent to the drilling and operation of wells. The causes and effects of induced cementation by carbonates have been outlined in *Bull.* 693.

For the most part, the caps and breaks in the deep-seated sands of the Appalachian fields are densely cemented sands and shales that are characteristically siliceous. Carbonates are usually lacking. I think most of the carbonates you will find in these deep sands are due to induced cementation.

C. W. Washburne (author's reply to discussion).—It is surprising to find Mr. Mills disparaging speculation, as compared with investigation, because the part of the bulletin[29] under review is essentially a development of the idea of evaporation of oil-field water by underground gas, which idea first appeared as a working hypothesis in a purely speculative study.[30] Mills and Wells prove the value of speculation by testing this idea in the field and laboratory. The new facts they present will lead many geologists to seek explanations, or to speculate. Speculation has nothing to do with our daily work; yet it is the mind and soul of geology. Like all science, it feeds on facts.

Caution against over-zealous application of the idea of underground evaporation of water is found in the persistence of gasoline in crude oils; it is hard to evaporate water without evaporating the volatile parts of contiguous oil. Practically all crude oils retain volatile components. Light gasoline is absent in a few heavy crudes, such as the Topila and Panuco oils of Mexico and the Comodoro Rivadavia oil of the Argentine. The lightest commercial crude that has no gasoline probably is that of the Pine Island field, Louisiana, which has a specific gravity of about 28° Baumé. The light oil of the new Cat Creek field, Montana, which has a specific gravity of 50° Baumé, is said to contain no volatile gasoline, such as enters natural gas, but to have over 60 per cent. of heavy gasoline

---

[*] Head Geologist, Roxana Petroleum Co.

[29] R. Van A. Mills and R. C. Wells: U. S. Geol. Survey *Bull.* 693 (1919).

[30] C. W. Washburne: Chlorides in Oil-field Waters. *Trans.* (1914) **48**, 687.

of high initial boiling point but low ignition point, adapted to blending with casing-head products. There is little gas with the oil, which flows from the wells with a strange smoothness, like the oil in the southern part of the Peabody Pool, Kansas, and much like artesian water.

Gas probably accompanies the formation of all petroleums. Where it is absent, we may infer that the escaping gas carried away some of the volatile constituents; cases of this kind are rare. Moreover the underground waters of these fields are not very concentrated; the waters of the gasoline-rich Appalachian province are much more concentrated. The absence of gasoline in the rare exceptions mentioned may be explained by assuming that no gasoline occurred in the original oil, or else that it was of types that combined into heavier hydrocarbons.

Evaporation of oil doubtless is common at shallow depth, where the gasoline content generally is low. However, there is too much gasoline in deep oils to warrant the assumption that very much gas has passed through them. The deep oil sands contain the more concentrated brines. Hence the evaporation of water by gas passing through it in the oil sands cannot be a very important cause of its concentration. Most of the concentration probably took place in sands and other storage zones far below the present oil sands. The principles of capillarity and adsorption furnish good reasons for believing that the pores of clay shales generally are wet and incapable of penetration by gas and oil, except under unusual force, such as that of deformation. This view is confirmed by the geological preservation of gas and oil in sands protected by shale. The high temperature of great depth lowers the surface tension of liquids and weakens all effects of capillarity. So far as it goes, observation indicates generally greater dryness in the deeper sands and greater concentration of their brines. The exceptions seem to consist of continuous sands, such as the Saint Peter, that have artesian flow, which is diluted with comparatively recent surface water. Any complete explanation should take account of the fact that Lane and others have observed similar relations in the deep mines of the Lake Superior and other districts, where water decreases with great depth and becomes more concentrated. The apparent explanation is that the concentration was induced by evaporation at greater depth and that the brines largely ascended to their present position.

The water in shale pores seals them against penetration by gas and oil, but does not prevent the passage of water. Water, rather than gas or oil, escapes through shale pores. Vertical fissures, if present, would furnish the least resistance to the escape of water, oil, or gas; they would also permit the ascent of gas and oil, which is otherwise impossible, except under unusual force. Hence, the settling and compacting of strata by loading and deformation expels water, rather than gas. The water passes upward through the shale pores, at least until it meets a continuous sand through which it can move laterally toward the outcrop.

Any rise of temperature, as from deep burial, expands interstitial gas and forces more water upwards. The generation of natural gas and oil must displace water, driving it upwards.

The following factor is more hypothetical. Any leakage from abyssal crevices and any "sweating" through connecting pores would let the liquids of the earth's interior press upwards against the gas and brine in basal sediments. Juvenile water is of recognizable purity only in regions of marked diastrophism or of vulcanism, but it seems improbable that the rest of the earth is so tight that juvenile gas and water can not filter very slowly into the basal strata at many places. At the temperatures of great depths viscosity probably is more important than capillarity in resisting migration through pores. At depths of a few miles, such abyssal gases as helium-rich nitrogen and carbon dioxides are to be expected to escape in greater volume than water.

The ascending juvenile liquids would mix with the interstitial liquids of the sediments and would be altered chemically in the new environment. In the course of geological time, they would force all earth liquids some distance upwards, except where the latter are effectively sealed. Thus in some degree they add to the ascent of rock brines, which are driven upwards by the expansion of original gas from heating, by the generation of new gas and oil, and by the settling and compacting of strata from loading and deformation. A part of each brine is regarded as connate with unknown deeper strata, rather than with the sand in which it now occurs.

Mr. Mills says: "Mr. Washburne's assumption that the salt masses were formed at temperatures not exceeding 40°, is unjustified." This figure was used because it is the lowest for which Mills and Wells give the constants needed. It is too high, at least for the probable temperature of the hypothetical precipitation at the top of shallow salt cores; admittedly the temperature at the bottom of the cores was much higher. Many of the salt masses reach the present ground surface, except for a thin cover of recent clay. Most cores of this type are marked by a semi-circular lake or depression within an enclosing low ridge, suggesting that their tops have suffered solution. Many cores are marked by a low mound, due to recent settling of the porous sediments around the compact core, or else to uplift of the latter. The ridges encircling the lakes appear to be marginal remnants of collapsed mounds, undermined by solution.

Recent deposition has obliterated the surface manifestation of many domes. The region of the salt domes was characterized by an excess of deposition over erosion throughout most of Tertiary time, when they were formed. There is no very great break or hiatus in the record; every Tertiary epoch is represented. The formations are so uniform in thickness, distance considered, that we must conclude there was no great ero-

sion of the salt-dome region in Tertiary time. Deposition prevailed.
Post-Tertiary erosion is more important, but both stratigraphy and
physiography favor the idea that not over 100 or 200 ft. of cover had been
eroded from the salt domes near the coast. Some of the cores probably
reached the ground surface, as they do today; chemical deposition of
salt at the tops of such cores would have to take place nearly at surface
temperatures. The mean annual temperature of the region is about
25° C. so that those who favor the hypothesis of salt precipitation must
admit that much of the shallow precipitation took place at temperatures
below 40° C.

Mr. Mills thinks it too highly hypothetical to assume that when the
salt cores were formed the water of the artesian sands was essentially
of the same composition as today. What assumption could be less
hypothetical? The physiographic and structural conditions of the Texas
coast have undergone little change since that time. There has been no
diastrophism, other than small epirogenic movements. The salt cores
were formed in middle and late Tertiary time; the seacoast then lay
farther inland, but it did not cover the outcrops of the Cretaceous sands,
nor even those of the Wilcox formation. These outcrops being on land
and lying toward the source of the sediments, must have been higher
than the sea, which then covered some of the coastal belt. There is no
hypothesis involved in saying that the Wilcox sands under the salt domes
outcropped at a higher elevation inland at the time the domes were
formed. The present artesian condition of these sands is shown by many
fresh-water springs and wells. The same is true of the widespread Trinity
sand. The same artesian conditions must have existed in these sands in
mid-Tertiary time, because the character of the sediments prove that they
were derived largely from the northwest and were deposited on surfaces
that sloped in the same general direction as the present plains. There is
little hypothesis involved in the statement that, if these artesian sands
were punctured by fissures at the loci of salt domes in mid-Tertiary or
later time, they would let loose a flood of fresh water that would prevent
precipitation of salt. The first water released from a sand might be very
salty, but a continuation of the precipitation of salt long enough to produce salt masses of several cubic miles surely would draw much fresh
water from the higher inland parts of the sand. This statement seems
less speculative than that of Mills and Wells, that expanding gases
may have concentrated ascending brines, helping to precipitate them as
salt cores. The adverse conditions seem too strong for any theory of
precipitation.

Mr. Mills' attempt to nullify the dissolving effect of artesian water by
suggesting that it may have consisted of primary alkaline and sulfate
water in which common salt is not very soluble, does not help his case.
The climate of mid-Tertiary time was less arid than at present and the

ground water probably was fresher, otherwise I can see no reason for believing that its chemical nature was different. The same formations then surrounded the outcrops of the sands. The artesian water of the Trinity sand is not characterized by high primary alkalinity nor by high content of magnesium and calcium chlorides. I can find no analyses of the water of the Wilcox sands, but I have found it good to drink in many wells. Either water would dissolve common salt.

Mr. Mills is quite right in saying that my idea of the vapor transfer of water from sandstone to shale cannot be adopted as proved fact. Nothing ever is proved; things are established as true only for the time that they satisfy knowledge. I hope only that this working hypothesis of vapor transfer will be tested, as Messrs. Mills and Wells have tested that of underground evaporation. Analysis of the forces exerted on concave water films of very sharp curvature, as those of capillary water in shale, indicates that they promote condensation of water vapor to a greater degree than the relatively broad concave surfaces of water in sandstone. With reversed control, on account of reversed curvature, the principle is essentially the same as that of "digestion" and the precipitation of crystals in a laboratory beaker, or the growth of lime concretions with solution of disseminated calcite in rocks or the growth of raindrops in clouds. The idea is sound in theory. I believe the Bureau of Soils used it as a partial explanation of the transfer of soil water from and to clay. Any experimental test of the idea will be appreciated.

Secondary Intrusive Origin of Gulf Coastal Plain Salt Domes

By W. G. Matteson, E. M., E. Met., Fort Worth, Tex.

(New York Meeting, February, 1921)

THE origin of the salt domes of the Gulf coastal plain has been investigated by many of the most able geologists, but the problem cannot be said to have been satisfactorily solved. Since 1860, numerous theories have been presented, only to be discarded, at least in part, as more complete information revealed their fundamental weakness.

Real progress toward solution dates from 1902, when Hill[1] advanced the theory of secondary deposition of the domal matérials from saturated solutions of hot saline waters ascending from great depths under hydrostatic head along structural lines of weakness. Shortly thereafter, Harris,[2] using Hill's hypothesis as a basis, explained the doming and pronounced uplift associated with these salt cores as the result of forces exerted by growing salt crystals. This was a marked advance over the ideas of Coste[3] and Hager,[4] since no evidence of igneous intrusives, as they assumed to explain the uplifts, had been found associated with the domes. Harris[5] developed his theory until it offered such apparently plausible explanations of so many details of dome phenomena that his hypothesis and conclusions received widespread acceptance, despite some serious objections, and today his theory, somewhat modified, is considered by many able investigators to be the best explanation of the origin of these domes.

The immense production of oil per acre, the recognition of the high lubricating quality of the oil, the development and recognition of the efficiency and advantages of oil-burning vessels, with the subsequent exceptional demand for fuel oil, and the resultant advance in the price of coastal crude, have been responsible for a prospecting and develop-

---

[1] Robert T. Hill: Beaumont Oil Field with Notes on Other Oil Fields of the Texas Region. *Jnl.* Franklin Inst. (1902) **154**, 143.

[2] Gilbert D. Harris: Rock Salt in Louisiana. Louisiana Geol. Survey *Bull.* 7 (1907) 76.

[3] E. Coste: Volcanic Origin of Natural Gas and Petroleum. *Jnl.* Canadian Min. Inst. (1903) **6**, 73.

[4] Lee Hager: Mounds of the Southern Oil Fields. *Eng. & Min. Jnl.* (July 28, 1904) **78**, 137, 180.

[5] Gilbert D. Harris: Geological Occurrence of Rock Salt in Louisiana and East Texas. *Econ. Geol.* (1909) **4**, 12.

ment campaign throughout the coast country, during the last 5 years, that has seldom been equaled when the present depth of drilling is taken into consideration.   This drilling has produced much information relative to the peculiar characteristics of these salt domes, with the result that several new theories of origin have been promulgated.

One of the most ingenious of these new hypotheses is that of Norton,[6] who thinks that these salt masses and their associated materials, limestone and gypsum, have been deposited near the surface by highly saturated, thermal, spring waters, such deposition taking place contemporaneously with the sedimentation of the region.   He presents new ideas in contending that the limestone cap rock, associated with many salt domes, is due to deposition of calcareous sinter by these thermal springs; he also suggests that the gypsum may result from the alteration of this calcareous sinter through chemical reaction with acid sulfate waters and hydrogen sulfide.   He fails, however, to offer an adequate explanation of the factors responsible for the structural deformations connected with the salt cores so that his theories have not received the recognition they deserve.

Kennedy,[7] in 1917, advocated practically the same theories but he advanced a step when he contended that the domal uplift was due to increase in volume resulting from the conversion of limestone into gypsum. Mills and Wells,[8] shortly thereafter, supported Harris' theory, removing one of the chief objections to it by presenting evidence to show the effect of expanding gas on the deposition of sodium chloride from concentrated solution. Lucas[9] later maintained that the uplift was due to laccolithic intrusion at great depth.

Early in 1917, van der Gracht[10] called attention to the fact that salt domes in northwestern Europe, of somewhat similar character to those of the Gulf coastal plain, had been subjected to diamond drilling, mining, and such extensive development that their origin had been determined beyond much question. Their formation was ascribed to the intrusion en masse of solid rock salt into the overlying strata, the salt originating in deeply -buried, primary, bedded deposits, 10,000 (3040 m.), 15,000 up to 22,000 ft. below the present surface. His

[6] Edward G. Norton: Origin of the Louisiana and East Texas Salines. *Trans.* (1915) 51, 502.

[7] William Kennedy: Coastal Salt Domes, Southwestern Assn. Pet. Geol. *Bull.* 1 (1917) 34.

[8] R. Van A. Mills and R. C. Wells: Evaporation at Depth by Natural Gases. Abstract, Wash. Acad. Sci. *Jnl.* (1917) 7, 309.

[9] A. F. Lucas: Possible Existence of Deep-seated Oil Deposits on the Gulf Coast. *Trans.* (1919) 61, 501.

[10] W. A. I. M. von Waterschoot van der Gracht: Salt Domes of Northwestern Europe. Southwestern Assn. Pet. Geol. *Bull.* 1 (1917) 85.

suggestion of considering a similar origin for the American domes was not received with much enthusiasm until tentatively accepted by De-Golyer,[11] in 1918, after rejecting a volcanic origin. E. T. Dumble,[12] whose investigations in the Gulf coastal plain region have extended over 30 years and have made him an authority on this area, became converted at the same time as DeGolyer but it remained for G. Sherburne Rogers,[13] of the United States Geological Survey, to propound and apply in detail, through analogy and otherwise, the European theory to the American domes. Since then, nearly all opponents, and some advocates, of the theories of Hill, Harris, Norton, and Kennedy have accepted the primary intrusive origin so that opinion now seems to be about equally divided between this and the theory of secondary origin from ascending saline waters.

Rarely has any theory gained so many active supporters in so short a time, especially where there had been previously such a wide divergence of opinion. The primary intrusive theory eliminates some of the old difficulties of long contention connected with the previously accepted American hypotheses and to some, this has evidently been sufficient for its acceptance. Washburne[14] apparently has recognized some of the difficulties involved in this theory but his supporting argument fails to strengthen the case.

The purpose of this paper is to show that the European intrusive origin of salt domes, as applied to American occurrences by Rogers, does not comply with facts and does not satisfy fundamental conditions as observed in the field and is, therefore, not directly applicable; also, to propose a theory that apparently complies with all field observations and eliminates many of the objections to the present theories.

Intrusive Origin as Proposed by Rogers

Rogers[15] contends that the salt plugs of the Gulf coastal plain are off shoots of deeply buried bedded deposits of salt that have been subjected to great pressure or thrust, and have been partly squeezed upwards in a semiplastic condition along lines of weakness. He admits that he has no adequate explanation for the formation and intimate association of the cap-rock materials and the salt but suggests that the cap rock might have been formed subsequent to the salt or that an overlying anhydrite bed or

---

[11] E. L. DeGolyer: Theory of Volcanic Origin of Salt Domes. *Trans.* (1919) **61,** 456.

[12] E. T. Dumble: Discussion on paper noted in Footnote 11.

[13] G. Sherburne Rogers: Intrusive Origin of the Gulf Coast Salt Domes. *Econ. Geol.* (1918) 13, 447.

[14] Chester W. Washburne: Oil-field Brines. See page 269.

[15] Rogers: *Op. cit.*

block was brought up with the salt during intrusion. Since the presence of nearly a hundred salt domes has been recorded in the Gulf coastal plain province and practically all show varying thicknesses of cap rock, the theory which postulates that nature should perform with such harmony and coöperation as to provide an overlying bed of anhydrite at just the specific point of intrusion for every salt-dome occurrence is constructed on rather a precarious foundation of probability.

Rogers prefaces his discussion by admitting that any acceptable theory must consider and explain plausibly:

1. The source of the salt and the manner in which it attained its present position.

2. The sharp local upthrust of the sediments surrounding the salt core.

3. The source and relations of the gypsum, anhydrite, limestone, dolomite, and sulfur usually found above the salt.

4. The alignment of the domes and their relationship to the main structural features of the region.

5. The origin and mode of accumulation of the oil associated with most of the domes.

In attempting to prove his theories, four points are cited by Rogers[16] as favoring his hypotheses; namely:

1. The sharp local doming of the sediments above the salt—doming of a type that several writers have stated is known to have been produced elsewhere only by (igneous) intrusion.

2. The flow structure, crystal orientation, and cleavage of the salt itself, indicative of pronounced movement in a vertical direction.

3. The plasticity of salt, which is considerable at ordinary temperatures and increases rapidly with heat.

4. The clear evidence that similar domes in other countries have actually been formed under the conditions postulated.

In addition, Rogers admits that the proof of his conclusions and the acceptance of the primary intrusive theory depends on the ability to show: A reasonable possibility that bedded salt deposits exist at depth beneath the Gulf Coast and the possibility that forces competent to produce the results observed have been operative.

Possibility of Bedded Salt Deposits at Depth Within Gulf Coastal Plain Province

The reasonable possibility of the existence of bedded deposits of rock salt at depth underlying the present Gulf coastal plain province is the foundation on which the present accepted intrusive origin of these salt domes has been erected. Eliminate this possibility and the superstructure of the theory crumbles.

---

[16] *Op. cit.*, 468.

Rogers[17] assumes the existence of Permian salt deposits in Permian strata underlying the Gulf coastal plain province. He concedes that positive evidence to indicate even the existence of Permian rocks has not been forthcoming although over a thousand deep tests have been drilled, some of which, in the Cretaceous belt bordering the coastal plain, penetrated the full measure of Cretaceous rocks but failed to find underlying beds of Permian age. The Texas Panhandle region, an entirely different province with different conditions of sedimentation and 200 mi. (321 km.) removed, is made up of Permian strata, which contain beds of gypsum and salt. Rogers[18] cites N. H. Darton, who has studied the deposits of the Panhandle, to the effect that he regards the existence of another and similar salt-bearing basin to the southeast as conjectural but entirely possible. In addition thereto, Rogers adds:

There is but little positive evidence either for or against the supposition that deep-seated bedded salt deposits exist in the coastal region. Wells penetrating 4000 ft. of Tertiary sediments have found no salt and it is evident that if any exists it is in the Mesozoic or Paleozoic rocks. No bedded salt deposits of any consequence are known to occur in the Cretaceous or Triassic beds and there seems little real basis for assuming their presence. In view of the lack of positive evidence, it is perhaps permissible to beg the question and argue that the best evidence of buried salt beds is the domes themselves.

This seems to be arguing in a circle and is proof of the insecure and uncertain foundation on which the theory is built. E. T. Dumble,[19] in a paper written three years previous to that of Rogers, discusses and makes comparative notes on the occurrence of petroleum in eastern Mexico and the Gulf coastal plain, as follows:

To the southward in Central Mexico, very complete sections are found of both Trias and Jura, but if the waters of those periods ever reached the Texas coast, no evidence remains to prove it.

Referring to the oil-bearing Woodbine sands in northern Louisiana, he says:

Between this great Louisiana field on the north and the greater Mexican field on the south, there is an interval of more than 600 mi. in which these formations are not found within the Coastal area, unless some portion of the basal Eagle Ford shale may represent a time equivalent, and even if that be the case. no oil deposits are found. We have no evidence whatever of any Permian deposits southeast of the Lampasas geanticline nor of the continuation of the Woodbine as an oil horizon as far southward as the coast.

Dumble's statements have an important bearing because if conditions in the Texas-Louisiana region prevented the deposition of the Triassic and Jurassic observed farther south, the most reasonable deduction points to

[17] *Op. cit.*, 476–477.   [18] *Op. cit.*, 476–477.
[19] E. T. Dumble: Occurrences of Petroleum in Eastern Mexico as Contrasted with those in Texas and Louisiana. *Trans.* (1915) **52**, 250.

such conditions maintaining during Permian times, thereby eliminating the possibility of the presence of Permian strata.   This conclusion seems inevitable when the negative character of hundreds of well records are considered.

In a much later paper, Dumble[20] agrees with Rogers as to the intrusive origin of the salt domes from bedded deposits of rock salt at depth, but realizing the extremely weak nature of the foundation on which Rogers builds his theory by assuming the presence of Permian beds, Dumble cites what he believes is more logical evidence as to the possibility of buried salt strata.   From general stratigraphic considerations, he believes there were three periods in Mesozoic and Tertiary times that were favorable for the development of salt deposits:

The association of the gypsum, salt, and anhydrite suggest their derivation from sea water by evaporation.   The Trinity in Arkansas carries considerable beds of gypsum, a condition which was duplicated in west Texas, where, in the Malone Mountains, we have hundreds of feet of gypsum of Lower Cretaceous age.   There is, therefore, no reason why salt and gypsum deposits of this age may not be expected in the area of northeast Texas occupied by the interior domes.

A second period favorable for such deposits is found in the interval between the Comanchean and the Upper Cretaceous.   While we have no such positive evidence of the accumulation of such deposits of sea salts at this period, the fact that for hundreds of miles the contact between the Buda Limestone, which marked the close of Comanchean deposition, and the Eagle Ford shows no sign of erosion proves that during the long period that elapsed between them, the top of the Comanchean must have remained at or near sea level and in such relation to it that no terrigenous sediments could be laid down on it.   In the more littoral zone of northeast Texas, the Buda is represented by clays and the conditions would be even more favorable for the formation of salt basins and the accumulation of gypsum and salt prior to the beginning of Upper Cretaceous sedimentation.   There is every reason to believe, therefore, that the gypsum and salt found in connection with the interior domes may have been deposited during the Lower Cretaceous or in the Mid-Cretaceous interval.

That the withdrawal of the sea at the close of the Eocene was accompanied by the deposition of beds of massive gypsum is clearly shown at the southern end of the belt of the Gulf Coast Eocene on the Conchas River in Mexico.   Here the Frio clays, which are the uppermost Eocene beds and probably of Jackson age, form a large portion of the Pomerane Mountains.   They carry in their upper portion heavy beds of gypsum, alabaster, and selenite, interbedded with clays.   While similar conditions are not known to have positively occurred in eastern Texas, it is probable that they did, and that salt and gypsum, which occur in connection with the coastal domes, was deposited at the time of this emergence and prior to the deposition of the Corrigan sands.

Since the publication of the papers by Rogers and Dumble, the development in these areas has furnished sufficient reasons for believing that salt deposits do not exist and were not formed in the epochs outlined by Dumble:

1. Numerous wells, scattered over the coastal plain province and drilled

[20] E. T. Dumble: Discussion on Theory of Volcanic Origin of Salt Domes.   *Trans.* (1919) **61,** 470.

as deep as 5000 ft. (1520 m.), have failed to record the presence of salt beds although the formations in which Dumble predicts their presence have been penetrated.

2. Deep wells, miles from the Trinity contact, have encountered neither gypsum in quantity nor rock salt in eastern Texas.

3. Deep wells, penetrating the Lower Cretaceous on the Sabine uplift, have given no indications of gypsum and rock salt.

4. The Cretaceous deposits of the Malone Mountains are in a belt where Permian rocks, carrying salt and gypsum, occur in vast quantity; no such association is known in eastern Texas. Moreover, the Malone Mountains are in a different physiographic and stratigraphic province and such deductions as are suggested by Dumble are dangerous and not justified. Intimate studies of the sedimentation processes affecting the Gulf coast province show such processes to be complex and varying in character to such an extent that even the same formation, within short distances, may be hardly recognizable from its lithologic character. A good illustration is the Fleming clays, which are palustrine at their outcrop and non-bituminous and marine a few miles south of their outcrop and bituminous. Therefore, even if certain conditions existed in Arkansas or western Texas, this fact is no safe basis for assuming similar conditions to exist in adjacent territory owing to the factors influencing sedimentation.

5. Recent deep tests have proved that the conditions obtaining at the end of the Eocene in Mexico probably did not continue northward into Texas. A well recently drilled by the Kleberg County Oil & Gas Co., 7 mi. (11 km.) south of Kingsville, Tex., to a depth of nearly 4000 ft. (1200 m.) penetrated the Yegua clays probably at around 1850 ft. and probably stopped in the Marine beds or the Wilcox; it showed insignificant quantities of gypsum and no beds of rock salt. Two tests of 3300 and 3500 ft., drilled by the Gulf Production Co. at White Point, near Corpus Christi, were abandoned close to the bottom of the Yegua, half way through the total thickness of the Eocene formations; they showed insignificant amounts of gypsum and no rock salt. The evidence is clear that the gypsum is thinning rapidly to the north from the Mexican border and the salt beds presumed by Dumble do not exist. The author recently collected fossils from the well of the Texas Oil, Gas & Mineral Products Co., in Grimes County, which Kennedy and Dumble identified as probably of Cook Mountain age. The author examined the samples of cuttings taken from this well but found absolutely no evidence of rock salt. Another well in the same county and several hundred feet deeper records the same results. If the quantity of salt necessary for the formation of so many salt domes existed in the form of bedded deposits, some of the numerous, deep, wildcat tests drilled during the past 5 years throughout the Gulf coast province would record the fact. It should also

be borne in mind that gypsum and salt are not always associated together, even if the combined occurrence is common. The presence of gypsum, therefore, does not necessarily signify the presence of salt.

In presenting this controvertive evidence, over a thousand well logs have been examined by the writer. Beginning with the 5000-ft. deep test, starting at the Fort Worth limestone horizon of the Lower Cretaceous formation at Polytechnic near Fort Worth, Tex., and including an area as far as New Iberia, La., to the southeast, and Brownsville, Tex., to the southwest, not one deep test, except those on defined salt domes, has encountered deposits of salt so as to warrant the reasonable conclusion that extensive bedded deposits of such substance existed at depth. Hence the foundation of the intrusive theory of salt domes, as promulgated by Rogers, must be rejected.

Intensity of Forces Producing Intrusion

Rogers,[21] in developing the theory of intrusive origin, states that it is necessary to show that forces competent to produce the results observed have been operative, and suggests three possible causes that might produce the pressure and force demanded, namely, igneous intrusion at great depth, the weight of the overlying sediments, and lateral or compressive thrust. He dismisses the first two causes as either improbable or not sufficiently competent to produce the results observed but concentrates on the third cause as the most plausible, basing his belief on analogy with European conditions. Van der Gracht[22] describes the salt-dome area of northwestern Europe as a geosynclinal basin, intensely folded and faulted. So intense has been the folding that some of the folds have been overthrust. In discussing the Roumanian salt domes, van der Gracht states "that orogenetic pressure was the cause of these upthrusts is evident from the whole structure of the region. We find, however, that fairly often the continuing lateral compression has squeezed out the stem of the salt core, perhaps even to the extent of separating the saline mass at the surface from its roots in the Miocene."

Evidently the lateral compression forces brought into play in the European areas have been enormous and much more intense and complex than anything observed in the Texas-Louisiana area, which is also monoclinal in structure as compared to the geosynclinal condition in Europe. Drilling adjacent to the American domes shows conclusively that these domes are not situated along the axes of highly compressed and arched folds, as van der Gracht describes for Germany, Holland, and Roumania. The disturbances in the Gulf coastal plain province apparently partake largely of the nature of block faulting. Lateral compression on a small scale has undoubtedly been a complement. While these disturbances

---

[21] Op. cit., 481. [22] van der Gracht: Op. cit.

are probably sufficient to cause intrusion of salt masses to some extent, the evidence is strongly against such stresses being sufficient to cause movement of a salt plug from a depth of 10,000 to 20,000 ft. through thousands of feet of overlying strata, as claimed by Rogers. On this point, Norton[23] says, "If salt in regular bedding exists in Louisiana and Texas, it has never been penetrated by the drill. If we assume that it exists at depths greater than have been reached, and has been elevated to the surface by an anticlinal development, the assumption is not supported by evidence that such mountain-building forces have been at work."

Analogies between European and American Domes

So much has been said about the European domes and such constant reference has been made to the origin, character, and the similarity of these domes, in many ways, to the American occurrences, that it might be well to present briefly the evidence on which the origin of the European domes has been formulated.

1. The undoubted presence of a great thickness of upper Permian red marls and dolomites, lying at a depth of a few thousand up to 22,000 ft., and containing a nearly continuous deposit of rock salt averaging about 1000 ft. in thickness in the southern area of its deposition but increasing to more than 2000 ft. farther north, and possibly considerably more in deeper basins, where the original mother bed has not been reached.

2. The presence in and throughout the salt cores of the domes of Europe of anhydrite and intercalations of valuable potassium salts such as is found in the original beds.

3. Areas and blocks of Permian, Triassic, Jurassic, and Cretaceous rocks exposed at or lying close to the surface, having been pushed up through thousands of feet of overlying Tertiary and Quaternary deposits.

4. The domes occur in and along folds or faults in geosynclinal basins where compression and folding have been very intense.

5. Lines of weakness have been developed in two main directions, northwest to southeast and northeast to southwest; wherever these intersect high uplifts occur.

To quote van der Gracht:[24]

As a rule, red Permian marls and often blocks of massive Triassic sandstones or limestones have been pushed upwards with the salt, and now appear near, or sometimes even at the surface, sometimes standing out in relief as curious red rocky hills amidst the Quaternary plain. The most striking of these is the red, rocky island of Helgoland in the North Sea, off the mouth of the Elbe River. . . . The main point is, however, that invariably the rocks associated with the saline core prove that the plug has its roots in the Permian rock salt, however great the depth of this latter may be, even up to 20,000 feet.

---

[23] Norton: *Op. cit.*, 503.   [24] *Op. cit.*, 88.

In comparing the evidence presented by the Texas-Louisiana domes, it is to be noted that:

1. There is no direct, positive, or probable evidence of the presence of deeply buried bedded deposits of rock salt nor has deep and extensive drilling revealed a reasonable possibility of the existence of the same.

2. Potassium salts, such as are commonly associated with bedded deposits of rock salt, are practically missing and only small quantities of anhydrite, not at all comparable with what should accompany bedded deposits, are observed. This anhydrite is practically always found as part of the cap rock and not embedded in the salt.

3. Areas and blocks of older, underlying formations have not been upthrust so as to be exposed at or to lie near the surface. The only foreign material observed in connection with the American domes was a small mass of red sandstone at Avery's Island, the presence of which can be explained in various ways.

4. The American domes are on a gently dipping monocline and, in general, orogenic disturbance has not been intense.

5. Lines of weakness have been developed in the American domes in two main directions, northwest to southeast and northeast to southwest; wherever these intersect, doming is apt to occur.

Thus of the five fundamentals affecting and determining the origin of the European domes, only one is in any way similar and is duplicated in the Texas-Louisiana area. It is true that the American domes have cores of salt as in Europe, that the salt is overlain by gypsum and limestone as in Europe, and that the deformation partakes of a quaquaversal nature; in other words, the form and the material of the domes bear close and striking similarity in general features, like two veins of copper, but as the veins of copper may have widely divergent origin, so does the origin of the European domes, according to evidence presented, vary considerably from what facts observed in the field must establish for the American occurrences.

Conclusions

The intrusive origin of the Gulf coast salt domes, as promulgated by Rogers, is not tenable, in that sufficient, well-established, definite data cannot be adduced to substantiate the fundamental requirements constituting the basis of the theory, and the theory does not conform to nor satisfy the numerous facts and details observed in association with the coastal domes.

SPECIAL CHARACTERISTICS ASSOCIATED WITH SALT DOMES OF THE GULF COASTAL PLAIN PROVINCE

Any acceptable theory relative to the origin of the Texas-Louisiana domes must conform to and explain, even to a reasonable degree of detail the unusual structural, stratigraphic, and mineralogical peculiarities of these domes. A tabulated, descriptive review of their characteristics shows the following features:

1. A core of domal materials which includes rock salt, gypsum, anhydrite, limestone, dolomite, and sulfur.

2. Pure rock salt forms the greater portion of the core. This salt is generally overlain and in direct contact with a thick, massive core or bed of gypsum. Small quantities of anhydrite are found occasionally in this gypsum. Sometimes the gypsum is capped by a thin to thick deposit of limestone, and limestone is generally found intermixed, included within, and scattered throughout the gypsum. The sulfur occurs as crystals or crystalline masses in cavities in the limestone and gypsum. The dolomite is due to the alteration of limestone masses.

3. The salt is relatively pure; it carries no potassium compounds, which are characteristic of bedded deposits from sea water; neither are broken boulders or strata of anhydrite and limestone found within the main salt mass.

4. A microscopic examination of crystals of the upper portion of the salt sometimes shows included gypsum crystals.

5. These cores of domal materials are roughly cylindrical or elliptical in outline, their vertical dimensions often exceeding their diameters or longer horizontal axes.

6. The cap-rock material of gypsum and limestone does not overlap the salt mass to any extent.

7. The domal core is gently rounded to flat on top with dipping sides of 60 to 90 degrees.

8. Stringers, sheets, or pencils of salt and gypsum are sometimes found projecting from the main core; sometimes these minor deposits appear to be completely disconnected or severed from the main core.

9. Domal materials appear to be localized and are not found in isolated quantity at any distance from the domal core with the possible exception of limestone. Limestone so found is part of a stratigraphic unit and is not of the same lithologic character as the domal material.

10. Crystals of salt sometimes show elongation in a vertical direction, the salt often shows pronounced cleavage and plication, and the gypsum is cavernous and often fractured, shattered, and broken.

11. Strata immediately adjacent to the domal core are abruptly domed or upturned and highly inclined, and show considerable deformation. The strata dip from 20° to 50° but appear practically undisturbed and flat lying within relatively short distances from the core.

12. The domes apparently have a definite alignment in northeast to southwest and northwest to southeast directions.

13. Often the top of the domal material lies at or close to the surface; at other times at considerable depth. In some instances, the salt has

never been penetrated and in a few cases, not even the main gypsum mass has been encountered.

14. The presence of certain recognized horizons at or near the surface, which normally should be found at considerable depth, indicates uplift in the vicinity of the core of 1000 to 3000 feet.

15. Several of the domes show faulting, and radial faulting from the core is believed to be more general than has been indicated.

16. Some domes are featured on the surface by slight to abrupt more or less circular mounds or elongated ridges; others, by central depressions surrounded by hills; while others are absolutely lacking in topographic characteristics.

17. In domes reproductive of oil, the oil is found on the east, southeast, south, or southwest side in most instances.

19. No foreign material in quantity, such as blocks and boulders of deeply buried strata, is found in the salt.

EVIDENCE OF SECONDARY DEPOSITION

After studying these domes in the field, reviewing the features associated with them, and noting especially from examination of numerous logs and some surface excavations, the intimate contact relationship of the domal materials, there remains practically no doubt that the domal materials are secondary in character; that they have been formed under similar conditions of time and deposition, and that any acceptable theory of origin must explain this type and intimacy of relationship and the processes whereby it has been developed. In other words, no theory is acceptable that explains the origin and position of the salt and not that of the cap rock.

Attention has been called to the fact that practically no potassium or allied salts, common to original bedded deposits or deposits resulting from the evaporation of sea water, are present and alternations of beds of salt, gypsum, anhydrite, and limestone are unknown. Neither is the salt core to any extent contaminated or intercalated with silts, muds, or any foreign deposits or substances but is so relatively pure that deposition from solution appears to be conclusive. The author knows of no single instance where the cap-rock material is completely and wholly separated from the main salt core by intervening sedimentation or strata. There are instances, as at Barber's Hill, where the drill has encountered boulders of gypsum in sand when drilling through the cap rock and also salt, intermixed with sand, but more extensive drilling has shown the cap-rock material to lie in direct contact with the salt, pointing to deposition and formation in one period.

THEORY OF SECONDARY INTRUSIVE ORIGIN

Although the sponsors of the intrusive origin of the Gulf coast salt domes did not specifically qualify the same, the author has taken that liberty here in order to make the distinction between this and the theory about to be proposed absolutely clear. The designation of the older theory as primary intrusive is in conformance with the statement of those proposing the theory that the salt so intruded was an offshoot of an original bedded or primary salt deposit existing at great depth. After presenting a concise statement of the secondary intrusive origin of the Gulf coast salt domes, it is proposed to discuss the facts and arguments supporting and proving the same.

The secondary intrusive origin of salt domes states that hot, saline, saturated to supersaturated solutions or brines, accompanied by vast quantities of gas, ascending along lines of structural weakness, deposited, by the action of various and several agencies hereinafter discussed, the domal materials relatively near the surface; that the initial period of movement and uplift, caused by the force of growing crystals and the increase in volume in the conversion of limestone to gypsum, occurred contemporaneously with the formation of the domal materials and sedimentation, causing gradual uplift locally as the surrounding area was sinking; that erosion over a considerable time interval ensued, removing part or all of the sediments capping the domal materials and, in some instances, portions of the domal material itself, to be followed by another period of sedimentation, deposition, and uplift, with several minor phases, during which time sufficient lateral thrust and compression was operative to cause gradual upward movement or intrusion of the domal materials en masse into the overlying strata, producing thereby, together with the first period of uplift, the deformation, doming, and general conditions as observed at the present day.

Origin and Formation of Domal Materials

Kennedy,[25] Washburne,[26] DeGolyer,[27] Deussen,[28] Norton,[29] and others have conceded the secondary nature and origin of the gypsum and limestone cap rock overlying the salt of the coastal domes. The intimate contact relationship, and other evidence, gained from several years of personal examination of these domes and from the study of hundreds of well logs, indicates that all the domal materials, including

---

[25] Op. cit., 56.

[26] Op. cit., 4–8.

[27] E. L. DeGolyer: Origin of the Cap Rock of the Gulf Coast Salt Domes. *Econ. Geol.* (1918) **13**, 616.

[28] Alexander Deussen: The Humble Oil Field. Southwestern Assn. Pet. Geol. *Bull.* 1 (1917) 74.

[29] Op. cit., 508.

the salt, were deposited under similar conditions, by similar agencies, and closely following one another. It appears most likely that the original domal materials consisted only of limestone, probably in the form of travertin, and salt, the gypsum being the result of the conversion of the limestone through the action of sulfuric acid and hydrogen sulfide-bearing waters and gases. There may be some question as to which was deposited first, the limestone or the salt. Norton[30] gives strong evidence to support his contention that the limestone was precipitated and then the salt, but Kennedy takes the opposite view. Norton states the reasons for his conclusions as follows:

Hot ascending solutions, containing calcium and magnesium carbonates, sodium chloride, carbon dioxide, with varying amounts of hydrogen sulfide, mingled with the artesian saline waters of the Cretaceous beds. These waters were forced to the surface by the hydrostatic head of the region, through channels that were opened by faulting, etc.

Great deposits of travertin or calcareous sinter, similar to the deposit at Winnfield, La., were formed around the thermal springs that issued from these openings, the sinter continuing to build as long as the hydrostatic head was sufficient to maintain the flow . . . . Contemporaneously with the building of these sinters, sands and clays were deposited around their bases. At times, owing to the suddenly increased activity of these springs resulting from downward movement and relative increase of hydrostatic head, the sinter accumulation encroached upon the marsh; at other times the accumulation of sediment encroached upon the sinter.

As the sinter continued to build, coincident with the subsidence and sedimentation of the region, the same excess of carbon dioxide in the ascending waters that prevented a deposition of carbonates in the channel below, attacked and redissolved the bottom layers. By the periodic rapid deposition of the sinter above and its slow, constant dissolution below by the carbonated, saline waters, open spaces were developed that were carried upward, in which the salt was deposited from ascending solutions that were supersaturated with saline contents by the release of pressure, as well as by evaporative losses these waters must have sustained at the surface, as the rapid sinter accumulation checked the flow from the springs.

Kennedy[31] states:

We may reasonably suppose that the deposits carry a great many times more the quantity of saline matter than calcic matter and this, as well as the more ready solubility of the salt, would give the salt a greater preponderance in the percolating solutions and under these conditions it is probable a large proportion of salt had reached the depression or basin in which it was deposited before the less soluble lime carbonate began to move. Evidently the two ingredients reached the basin together in unequal proportions and then, due to this inequality, the lime remained longer in solution than the salt. Very little lime occurs intermingled with the salt but considerable salt remains in the lime or its gypsum condition. This no doubt accounts for the presence of gypsum above the salt.

Several factors were concerned in the precipitation or deposition of the salt. In the order of their probable importance, they may be stated as follows:

---

[30] *Op. cit.*, 507, 508.   [31] *Op. cit.*, 58.

1. Deposition and precipitation due to the evaporative effects of expanding gases on concentrated brines and supersaturated saline solutions.

2. Precipitation caused by the presence of a common ion.

3. Precipitation due to chemical reaction between brines of varying concentration and slightly different composition.

4. Deposition and precipitation due to lowering of temperature.

5. Deposition and precipitation due to lowering of pressure.

6. Deposition and precipitation due to other evaporative effects near or at the surface.

Opponents of the secondary origin of the domal materials based their arguments on the vast quantity of salt known to underlie these domes, and the enormous and seemingly improbable quantity of brine required to yield such masses through the processes of precipitation. Rogers[32] sums up some of these objections as follows:

> The effect of decrease in temperature and pressure on the solubility of salt is small. One hundred parts of water can carry 45 parts of sodium chloride in solution at 180° C., 39 parts at 100°, and 36 parts at 15°. If it be assumed that the solution became saturated and started to rise from a depth of 7200 ft., where its temperature would be 100° C., it would lose only 3 parts, or about 8 per cent., of its total load, and for every ton precipitated, over 11 tons must have escaped.

> The escape of the bulk of the saturated solution is not explained. No smaller bedded deposits of salt from evaporation of these solutions are observed in the vicinity of the domes. As some of the domes have grown within recent years and are probably still growing today, we should expect to find great volumes of saturated salt solutions issuing from them, yet only minor seeps of relatively dilute character are known.

The volume of ordinary sea water required to produce the salt would be extremely great. Sixty domes, each containing only the quantity of salt already blocked out at Humble (66 billion tons) require 4000 billion tons, which represents the complete evaporation of about 25,000 cu. mi. of sea water. Assume that one-third of the rocks consist of material coarse enough to allow an appreciable movement of water and that this material has a porosity of 30 per cent., then the whole section has an average porosity of 10 per cent. The 25,000 cu. mi. of sea water would saturate 250,000 cu. mi. of rock, or a block 2 mi. deep, extending from Matagorda, Tex., to Assumption, La., and stretching from the coast northward to central Arkansas and Oklahoma, where the Paleozoic rocks outcrop. On the other hand, if we accept Harris's theory of deposition from concentrated brine, enough saturated brine to fill the pores in an equal block of strata would be required.

Any legitimate objection raised by the above reasoning has been eliminated by the recent investigations and researches of Mills[33] and

---

[32] *Op. cit.*, 451.

[33] R. Van A. Mills and R. C. Wells: Evaporation and Concentration of Waters Associated with Petroleum and Natural Gas. U. S. Geol. Survey *Bull.* 693.

Wells. They recite instances where gas wells in the Appalachian field under 810 lb. rock pressure have gone "dead" in 24 hr., and, on pulling tubing and cleaning out the well, 4000 lb. of salt were removed after which production was again obtained. The deposition of barium sulfate and calcium carbonate in wells is also recorded and specific attention is called to the banded nature of these deposits. The bands look much like what Rogers[34] terms stratification planes in the salt core at Avery's Island, and suggest the probability that such so-called bedding is due to similar causes and the contorted nature of these bands is the result of slight, local, differential movements. Mills and Wells[35] note the greatly decreased solubility of salt in the presence of a common ion showing that, whereas a solution free from calcium chloride and having a specific gravity of 1.202 at 25° C. can dissolve and contain 26.43 per cent., by weight, of sodium chloride, when calcium chloride is added up to 24.58 per cent. by weight, only 5.63 per cent. by weight of sodium chloride remains in solution. These investigators likewise show that a cubic meter of gas, confined at a pressure of 100 atmospheres in contact with a saturated salt solution at 40° C., in expanding to atmospheric pressure, would evaporate 3800 gm. of water, which would cause the precipitation of 1400 gm. of salt. Continuing, they make the following conclusions:[36]

Although it is true that the solubility of salt decreases with falling temperature, the change is small. The amount of salt that will precipitate from 1 cu. m. of saturated brine on cooling from 60° to 20° C. is about 11 kg. The deposition of 11 kg. of salt would leave 883 kg. of water saturated with 317 kg. of salt as a brine taking no part in the process, so that the amount of brine necessary to form a dome by deposition due to cooling would be very large. Looked at in another way, 1 cu. m. of brine could deposit 11 kg. of salt by cooling or 330 kg. by evaporation (through expansion of gases).

It is not necessary that the gases escape to the surface in order to cause evaporation, for in deep-seated strata, under certain conditions, especially where the beds have undergone fissuring, gas may flow from one bed where the pressure is higher to another bed where a lower pressure prevails. The evaporative effects of the migrating gas would, under these conditions, be none the less important. The deposition of constituents other than chlorides, such as carbonates or sulfates, might be caused by evaporation, so as to produce the unusual relations sometimes observed in salt domes. It also seems probable that where the salt masses are associated with deposits of calcium sulfate and calcium carbonate, geochemical processes yielding sodium chloride, together with other compounds, have been brought about through the mixing of solutions that have different properties of reaction or through reactions between constituents of certain solutions and those of the containing rocks.

Adequacy of Gas Supply to Cause Great Salt Deposition

The gases necessary to produce the enormous evaporative effects noted may be derived from: (1) Dry marsh gases, peat and lignitic gases, derived from the decay of swamp matter and other vegetation; (2) hydro-

---

[34] Op. cit., 469. [35] Op. cit., 73. [36] Op. cit., 92, 90.

gen-sulfide gases resulting from the oxidation of iron pyrite; (3) gases developed in the metamorphism of carbonaceous shales frequently found associated with oil and gas deposits.

There is sufficient evidence of gas in the Gulf coast province to indicate that the quantity demanded for the great evaporative effects necessary has been, and probably still is, present. Thousands of seeps, emitting marsh and hydrogen sulfide gases, are recorded over the thousands of square miles of the Gulf province; and these seepages have been in operation for a long period. The 10,000 ft. (3040 m.) of Tertiary and Quaternary deposits is featured by vast quantities of iron pyrite, peat, and lignite in well-defined beds or disseminated throughout the formations. In addition, the Yegua formation alone has developed enormous quantities of gas from the Rio Grande River to the Louisiana border and beyond.

Thus, the researches of Mills and Wells have eliminated the last barrier to the acceptance of deposition from solution as the origin of the salt and associated materials. The deposition of the greater percentage of salt from a saturated solution due to the evaporative effects of expanding gases has been proved both as a possibility and a probability and the presence of sufficient quantities of gas for the purposes in view has been indicated. Such gases, combined with the five other factors enumerated, acting on a continuous supply of strong, saturated or supersaturated brine over a great period of time, could unquestionably produce the enormous salt masses observed in the domes of the Gulf coastal plain province.

Adequate Source of Supply of Salt and Limestone

The Quaternary and Tertiary sediments of the Gulf coastal plain province have an estimated thickness of 10,000 ft. (3040 m.) with 8000 to 10,000 ft. of underlying Cretaceous formations. Thirty to forty per cent. of the Cretaceous deposits consist of limestones and the remainder is composed of marls, shales, and sands. Several investigators have called attention to the disseminated saline character of the Cretaceous deposits. In this connection, Hill[37] made the following notation:

The fact that the water increases in temperature and salinity is conclusively proved by a line of wells, 100 mi. in length, between Comanche and Marlin, in the lower portion of the Cretaceous series. The same stratum which furnishes water at both places outcrops at Comanche and supplies good potable water at almost every atmospheric temperature. At Marlin, 100 mi. eastward, this water comes from a depth of 3200 ft., has a temperature of nearly 140° F., and is excessively saline and sulfurous.

A careful, lithologic examination of the Tertiary strata, composed almost entirely of unconsolidated sands and clays, reveals numerous

---

[37] Hill: *Op. cit.*

horizons, from the oldest to the youngest, containing numerous concretions and nodules of limestone, testifying to the calcareous and alkaline character of the sediments.   Even greater is the evidence of vast quantities of saline material, disseminated throughout the formations.   Investigators have recorded the presence of hundreds of salines and salt licks from the Cretaceous-Tertiary contact southward.   These licks, destroying all vegetation over them except the salt grass, are featured by the white crusts formed at the surface in a dry period following a wet spell, when capillary processes and evaporation bring the saline material to the top of the ground.   Generally, bare spots in an otherwise grass-covered prairie testify to their presence.   No better evidence of the widespread, disseminated, saline character of these Tertiary sediments is desired than the presence of these salt licks.   Kennedy[38] early called attention to the saline content of the Yegua formation and, in a more recent publication, made the following statements:[39]

The Fleming shales carry large quantities of saline and other mineralized wateis and probably such waters have something to do with the formation of the mound. These shales also carry lime plentifully in a carbonate form as well as gypsum. . . Carbonate of lime goes into solution when associated with carbon dioxide in alkaline solutions.   An examination of the analyses of the soils, subsoils, clay, and waters of the rivers and deep wells shows the presence of alkalies in considerable quantity. The lime of the domal materials was obtained from the leaching of the various beds from the Upper Cretaceous to the Miocene and probably Pliocene. . . We know that most of our Miocene deposits carry large percentages of salt, carbonate of lime, and organic remains. . . Moreover, there is an abundance of saline matter throughout the Gulf coast Tertiaries to account for the salt found in these mounds, enormous as it is.

Direct field observations and theoretical considerations present conclusive evidence of the existence of conditions favorable to the deposition of lime and salt.   DeGolyer,[40] in calling attention to the experiments of Frank K. Cameron, notes that calcium bicarbonate has a solubility of 0.06 gm. per li. in solutions with no sodium chloride, a solubility of 0.101 gm. per li. in a solution containing 39.62 gm. of sodium chloride per liter, and a solubility of only 0.04 gm. per li. in a solution containing 267.6 gm. of sodium chloride per liter.   Thus calcium carbonate is less soluble in concentrated brines than in pure water.   There is no doubt that these conditions obtained during the deposition of the mound materials. The high alkaline and carbon dioxide content of the hot, ascending brines, due to the leaching of thousands of cubic miles of soils, rich in alkaline earths, would hold the lime in solution until near the surface, when the

[38] William Kennedy: A Section from Terrell, Kaufman County, to Sabine Pass on the Gulf of Mexico. *Third Annual Rept.* Geol. Survey of Texas (1891) 43.

[39] William Kennedy: Coastal Salt Domes. Southwestern Assn. Pet. Geol. *Bull.* 1 (1917) 54, 57.

[40] E. L. DeGólyer: Origin of the Cap Rock of the Gulf Coast Domes. *Econ. Geol.* (1918) **13**, 616.

escape of the carbon dioxide and the supersaturated saline condition of the brine would cause rapid deposition of the lime, probably in part as travertin. In discussing the rapidity of such sinter accumulation, Norton[41] quotes Geikie as follows:

> The travertin of Tuscany is deposited at the Baths of San Vignone at the rate of 6 in. a year, at San Filippo 1 ft. in 4 mo. At the latter locality, it has piled up to a depth of at least 250 ft., forming a hill 1¼ mi. long and ⅓ mi. broad. Another illustration of the rapidity with which the travertin may be deposited is furnished by the Eocene sinter of Sezanne, Marne. This deposit contains hollow casts of flowers which fell on the growing sinter and were crusted over with it before they had time to wither.

Norton[42] refers to a notation by Veatch of an outcrop of gray, granular, sandy limestone, containing very imperfect plant impressions at Drakes's Saline and to Vaughn's statement that near Atlanta, in Winn Parish, there outcrops a hard, blue limestone which is traversed by minute fissures and in which Veatch found impressions of dicotyledonous leaves, Norton regarding all such limestone with leaf impressions as non-marine and not having been formed in the ordinary way. The great deposits of limestone and sinter at Winnfield, La., have already been cited and Kennedy,[43] in 1902, called attention to the deposition of calcium carbonate from salt springs at High Island. It should be noted that no such quantities of limestone observed capping some of the coastal plain salt domes are found anywhere associated with the European salt domes—strong testimony of different conditions of origin and formation.

Frank K. Cameron,[44] during his soil researches for the Government, conducted a series of experiments which proved that beyond a certain point an increase in the sodium-chloride content of salt brines caused a decrease in the solubility of gypsum. On this basis, DeGolyer suggests that the cap rock was precipitated from solution by circulating calcium-sulfate-bearing waters coming in direct contact with the salt plug after the formation of the latter, the salt content of the calcium-sulfate waters being thereby increased to the point demonstrated in Cameron's experiments as necessary to cause deposition. The vital defect in this process is the formation of a protective covering of gypsum cap rock over the top of the salt in a short time, after which the circulating calcium-sulfate waters would have difficulty in coming in direct contact with the salt core and effecting the necessary concentration to cause the precipitation of sufficient gypsum to account for the masses 300 to 1000 ft. (91 to 304 m.) thick now overlying the salt in the various domes.

[41] Norton: *Op. cit.*, 510.

[42] *Op. cit.*, 507.

[43] C. W. Hayes and William Kennedy: U. S. Geol. Survey *Bull.* 212.

[44] Frank K. Cameron: Solubility of Gypsum in Aqueous Solutions of Sodium Chloride. U. S. Dept. of Agriculture, Division of Soils No. 18, 25–45.

Rogers[45] also questions this theory due to the fact that examination of the analyses of waters in the Gulf coast province shows them to be surprisingly low in calcium sulfate.

In supporting the primary intrusive origin of the salt domes, Washburne[46] recently gave several reasons why the formation of the salt cores through deposition from supersaturated brines is impossible: That the escape of gas at the locus of any dome in sufficient quantity to account for the deposition of salt through the evaporative action of expanding gas would keep the vertical channels so free and open that there would be no accumulation of oil and gas at these places; also, that since the salt cores are at least a few thousand feet in vertical dimension, they cut so many water-bearing sands that the possibility of ascending, saturated solutions maintaining sufficient concentration to effect precipitation is doubtful due to the mingling of fresh artesian waters or dilute brines from these water-bearing sands with the saline-bearing brines. He even states that the cutting of fresh-water sands by these vertical channels would permit such meteoric waters to enter the channels and not only prevent deposition of the salt but would probably dissolve any salt already deposited.

Washburne's contentions can be easily eliminated from the problem. The oil and gas did not accumulate on and around these salt domes in all probability until some time after the core was formed, so that free and open channels would in no way affect the problem of petroleum accumulation. Neither would these vertical channels cut fresh-water-bearing sands capable of releasing artesian flows that would dilute the ascending solutions of concentrated brines beyond the point where precipitation of salt and other materials would be possible. Meteoric waters are seldom found in the Gulf coastal plain province below 1500 ft. (457 m.) and all water-bearing sands to this depth immediately adjacent to these domes are lenticular in character, and, therefore, strong artesian flows are not observed. Even if such artesian flows occurred, their volume would be so small, in comparison with the volume of saturated brine ascending from a depth of 5000 to 20,000 ft. under several thousand feet of hydrostatic head, that such fresh waters could have little quantitative effect. Mills and Wells[47] have shown that these dilute, meteoric waters often contain calcium chloride, which decreases the solubility and causes the deposition of sodium chloride. These investigators also quote instances where fresh water, introduced into gas wells plugged by salt, failed to dissolve materially this substance.

The lenticular nature of the water-bearing sands is acknowledged by Washburne who states, however, that there are some sands of widespread

[45] G. Sherburne Rogers: Intrusive Origin of the Gulf Coast Salt Domes—Discussion. *Econ. Geol.* (1919) **14**, 179.

[46] Chester W. Washburne: Oil-field Brines. See page 269.

[47] *Op. cit.*, 73.

continuity which outcrop much farther to the north and which would answer his purpose. Hill has shown that fresh waters entering such sands at their outcrop become highly saline in character in a relatively short distance from the outcrop. Washburne suggests two additional reasons that might prevent meteoric waters from entering the open channels along which the supersaturated brines might be ascending: First, the pores of the sands next to the fissures might be clogged with salts; Mills has shown that this actually happens. Second, the gas pressure may keep the water from entering the fissures. Harris[48] adds a third reason: "The compacting and slickensiding of the deposits about the lower main part of the core would tend to debar the close approach of fresh waters, and yet leave a suture line for the ascension of brines." Likewise the importance of explaining logically the presence of the limestone in any theory accounting for these domal materials is conceded by Washburne, who abandons such explanation as hopeless, however.

Alteration of Limestone to Gypsum

The limestone of the salt domes is found in irregular masses throughout the gypsum, and occasionally as a cap rock of limited thickness on top of the gypsum. When a cap rock, it varies in thickness from a few to a hundred feet, and occasionally more, and is decidedly sandy in character, which may account in part for its resistance to alteration and its capacity to hold oil. Spindletop and Humble have a typical limestone cap rock. Considerable thicknesses of massive gypsum, however, almost always overlie the salt, throughout which limestone is always found, bearing such relationship to it as would be expected as a result of alteration and replacement. Cores from Sulphur Mine, La., show the cap rock to be composed of gypsum and limestone with cavities filled by crystalline sulfur. At Damon Mound, according to Kennedy,[49]

The line of separation between the limestone and gypsum is an extremely irregular one. In places, large blocks of gypsum extend many feet into the limestone and in others the lime descends into the gypsum. Although the massive beds are designated as gypsum, they are by no means wholly sulfate of lime. Tests made of cores brought out of a number of wells show them to be a mixture of sulfate and carbonate with the carbonate usually in contact with the sulfate.

Examination of the cores at Bryan Heights shows the gypsum to contain numerous, irregular masses and nodules of limestone and occasional barite. The limestone is soft and porous and both the limestone and gypsum contain cavities filled with crystalline sulfur. Enormous quantities of hot waters containing free sulfuric acid and hydrogen sulfide are observed at this dome. Hot water, carrying free sulfuric acid, and great volumes of hydrogen sulfide are known at Sulphur Mine, La.

---

[48] Harris: *Op. cit.*   [49] *Op. cit.* (Assn. Pet. Geol.) 50.

There is little question that the sulfur is the result of secondary action and has been deposited from solution.

Constantly accumulating data present convincing evidence that the gypsum of these salt domes has resulted largely through the alteration of limestone due to the action of waters carrying free sulfuric acid and hydrogen sulfide. The writer has examined many cuttings from various formations of the Gulf coastal plain province and has been impressed with the quantity of disseminated pyrite present. Sour springs and waters carrying free sulfuric acid are observed in connection with these domes today. Hill[50] has mentioned the sulfurous character of the Cretaceous waters at Marlin. Hydrogen sulfide, in great quantity, is also of general record throughout the coastal plain region.

There is absolutely no question of the superabundance of sulfides which may, through oxidation, supply vast quantities of acid-bearing waters, suitable in character to convert the limestone into gypsum. One is impressed with the correctness of this conclusion when recalling that investigators of international repute have shown that massive gypsum, so extensive and in such great bodies as observed in connection with these coastal domes, has been formed generally through the alteration of limestones. Grabau[51] agrees with Dana that the gypsum masses of the Salina formation of New York result from the alteration of limestones by acid-sulfate waters, which abound in the formation and which have resulted from the oxidation of iron pyrites in the rock. He adds, "the occurrence of gypsum in the dolomites overlying the salts at Goderich, Canada, is probably explained in a similar manner and the gypsum deposits of Nova Scotia have been attributed to the action of sulfuric acids on marine limestones." Kennedy[52] states that "at Damon Mound, the gypsum has the appearance of being the altered end of the massive limestone beds found at the southern end of the mound. Two limestone beds having a thickness of 70 and 650 ft., respectively, are reported from the southern end, lying between 260 and 1180 ft.; $\frac{3}{4}$ mi. north, the gypsum beds are 378 and 409 ft. thick with 30 ft. of sulfur and sand between."

Of course there is the possibility that the gypsum has been deposited from the same brines carrying the salt. Attention has already been called to DeGolyer's statements that the solubility of gypsum in aqueous solutions decreases with a constantly increasing concentration of sodium chloride. Harris also points out that the solubility of gypsum in aqueous solutions is rapidly decreased when the salt content is reduced below 14 per cent. The writer has recently obtained data at High Island which suggest that the precipitation of salt from saturated to supersaturated

---

[50] Op. cit.

[51] A. W. Grabau: "Principles of Salt Deposition" McGraw-Hill.

[52] Coastal Salt Domes. Southwestern Assn. Pet. Geol. Bull. 1, 50.

brines is so complete that such dilution, favorable to the subsequent precipitation of gypsum, is actually obtained.

However, the cavernous condition of the gypsum, its spongy character in many instances, and the presence of cavities lined with sulfur crystals, point to the action of waters, and the presence of waters of the necessary character has been recorded. These facts in connection with the disseminated condition of the limestone in and throughout the gypsum, both as small grains and irregular masses of varying size and indefinite outline, and the irregular projection of gypsum into limestone and limestone into gypsum where a definite limestone cap rock is observed, point strongly to alteration and replacement processes; in fact, no other theory could logically explain such conditions and association.

Initial Period of Movement and Uplift

Abundant evidence exists that the salt domes have been subjected to more than one period of movement. Deussen[53] states that the evidence is plain that the growth of the salt core is not one growth, but the result of several movements at different times, as shown in the Anderson County domes. The coastal domes also have two distinct movements.

The initial movement probably occurred contemporaneously with the formation of the domal core and continued for some time thereafter. Three factors were involved in this uplift; namely, the increase in volume resulting from the alteration of limestone to gypsum, the forces of growing salt crystals, and the general subsidence of the region.

An increase in volume of 32 to 50 per cent. occurs when limestone is converted into gypsum. Since the gypsum shows a varying thickness of a few to several hundred feet, and since part of the gypsum has probably been removed by erosion in some instances, it can be estimated conservatively that an uplift or doming of 200 to 300 ft. would result from this factor alone. That the forces of growing crystals would also be a factor in doming is evident but it is quite probable that the extent of uplift due to this cause alone has been greatly exaggerated.

There is slight doubt that the domal materials were originally deposited relatively near the surface. The thickening of the same formations away from the domes, which can hardly be accounted for except that the first two factors of uplift were probably sufficiently active to cause uplift, which prevented deposition of sediments to some extent on top of the dome while subsidence of the immediately adjacent areas permitted continuous sedimentation and deposition and at least apparently increased the extent of the initial uplift, is direct evidence of near surface precipitation. Conditions observed at many of these domes indicate an uplift

[53] *Op. cit.*, 84.

of 1500 to 2500 ft. (457 to 762 m.) yet nothing in the character of the orogenic movements affecting the coastal plain province points to forces capable of producing such displacement being generally active.   It therefore seems probable that part of this uplift is the result of nicely balanced conditions of sedimentation wherein initial forces of uplift were sufficient to keep pace or overcome the general subsidence of the region; this resulted in formations near the domes being considerably elevated compared with the same formations away from the domes.

Recently the writer made an examination at High Island.   On the south side, an area of one or two acres has been designated as the "trembling marshes," because the crust, when traversed, would shake or tremble over a considerable distance in a manner similar to the shaking of quicksand.   Eighteen years ago this area was a slight depression in which animals would bog; today, it is a mound, about 3 ft. higher than the surrounding territory, composed of crystals of salt.   Salt is being deposited there by springs and the run-off, after deposition, is much less saline than ordinary ocean water.   The writer regards this evidence as most important.   Kennedy[54] records the deposition of carbonate of lime from springs on the west side of the Island together with the salt.   Is not this an illustration of the processes which resulted in the formation of the domal materials and caused in part the initial uplift?

Subsequent Periods of Movement and Uplift

The formation of the domal materials and the initial period of uplift in the Texas-Louisiana coastal plain province was followed by general subsidence of the entire region and a long period of sedimentation. During this time, minor movements of uplift might have occurred but on such a small scale as to be hardly noticeable.

Subsequently a series of movements and uplifts occurred as the direct result of isostatic readjustments and the action of orogenic forces of some magnitude.   These forces acted along old fault planes and other lines of weakness and undoubtedly were accompanied by considerable lateral thrust.   As a result, the salt cores were thrust upwards and intruded into the overlying strata.   The entire period of movement was gradual and probably extended over a considerable time interval.   The direct result was the fissuring of the cap rock, the possible production on a small scale of plications within the main salt mass, the possible recrystallization of the salt, and the production of cross and radial faulting around the dome.   It was these subsequent periods of uplift which produced the abrupt upturning of the strata as well as the shearing and piercing of the same now observed around these salt cores.   The combined uplift undoubtedly produced marked surface doming.   During

---

[54] Hayes and Kennedy: U. S. Geol. Survey *Bull.* 212.

this period, erosion was again active and continued some time thereafter. Subsequently, subsidence occurred and recurred, and sedimentation was resumed on such a scale as to eliminate completely the remaining topographic features of these domes. These subsequent periods of uplift varied for different series of domes but occurred at various intervals from the end of the Cretaceous era to the present time. In the case of the commercially important oil-bearing domes immediately adjacent to the present Gulf coast, the major portion of the subsequent uplift occurred near the end of the Miocene and into Pliocene time as we find the Miocene formations pierced and deformed with the domal materials often intruded into the overlying Pliocene, which has been steeply arched.

During the subsidence that followed, Beaumont clays of Pleistocene age were deposited over this area, completely obscuring the topographic features of these domes. Another period of gradual uplift then began and has continued up to the present time; it has been general in character and is still in its early stages. Its important significance has been the development of slight mounds, hills, or ridges at the loci of domes in a number of instances where the forces have been locally more intense and especially where the domal materials are close to the surface. At points where the domal materials are deep lying, local mounds or ridges due to uplift are generally absent, suggesting that the forces of uplift have so far been absorbed by the unconsolidated overlying material without deformation. Erosion may account for the absence of mounds and ridges in some instances. This topographic expression has been one of the most reliable characteristics in determining the locus of a salt dome and domes are now classified in the field according to topography, as:

Domes of the first class with definite, characteristic, topographic expression in the form of low, roughly circular to elliptic mounds or ridges, rising 10 to 80 ft. (3 to 24 m.) abruptly above the adjacent prairie, as illustrated by Damon Mound, Hull, and High Island.

Domes of the second class with slight topographic expression in the form of low, roughly circular to elliptic mounds or ridges, rising 2 to 10 ft. (0.6 to 3 m.) somewhat abruptly above the adjacent coastal prairie, as illustrated by Spindletop, North Dayton, etc.

Domes of the third class with no domal or ridge topography but occasionally with sunken depressions, featuring the locus of the salt core, as illustrated by Goose Creek, Edgerly, or South Dayton.

OIL ACCUMULATION

The oil associated with those salt domes where commercial production has been obtained is of Tertiary origin. This conclusion is based on the grade of oil, the character of the accumulation, and the fact that the

Tertiary formations are petroliferous.   The oil of the Texas-Louisiana coastal plain has an asphalt base, whereas the great percentage of oil obtained from the Cretaceous formations in this province is largely of a paraffine base, while oil from the Pennsylvania horizons is practically always of paraffine character.

On 90 per cent. of the producing domes, the oil is obtained from the east, southeast, south, or southwest sides.   As this is the side corresponding to the normal dip of the formations and is generally the side of gentlest dip, this fact is strong evidence of the Tertiary origin of the oil. If the oil came from the underlying Cretaceous and Pennsylvanian formations, migrating upward through the same channels with the saline brines, the oil should be found as often on the northern and northwestern sides of the domes as on the normal sides.

Data gathered point to the Fleming clays of Upper Miocene age as the original source of the oil.   These clays are bituminous in character in their seaward, marine phase underlying the present area adjacent to the Gulf shore line.   Oil in considerable quantity has been found only in regions underlain by marine Fleming clays.   Tests west of the Colorado River in Texas, where the equivalent of the Fleming clays is a fresh-water deposit, have failed to find oil in commercial quantity.

In discussing the Humble oil field, Deussen[55] describes the oil there as originating in the Yegua clays and migrating upwards into the overlying Oligocene sands.   The fossil evidence that he has presented, though, appears open to question, and the failure of the Yegua to yield commercial oil from hundreds of tests that have penetrated this formation adds to the difficulty of accepting Deussen's conclusions without more data.   However, the Yegua, Cook Mountain, and Mt. Selman formations of Eocene age are partly or entirely marine in character, and have exhibited bituminous phases.   The Yegua has yielded dry gas in great quantity and numerous oil showings up to 1 or 2 bbl. wells.   Although general conditions would indicate that these formations have oil-producing possibilities, the universal failure of numerous wells penetrating these formations where structural conditions are known to have been favorable is discouraging.

Conclusions

The intrusive origin of the Gulf coastal plain salt domes, which postulates that the salt is of primary origin and that its configuration, character, and position are due primarily to intrusion *en masse* from bedded deposits below, is untenable, as the fundamental principles constituting the basis for this theory are not substantiated by data gathered from exhaustive detailed field examinations.   Since the theory

---

[55] *Op. cit.*, 73.

was first proposed, numerous deep tests have penetrated the formations in which conditions were most favorable for the development of beds of salt but no such beds have been encountered over a wide area.

The analogy between the European and American domes, which has been used to establish similar modes of origin, is one of form only and is more apparent than real. Details, fundamental in character, indicate that these domes must have had a widely different origin. Bedded salt deposits of great thickness and extent are known to underlie great areas in northwestern Europe. The salt cores there have brought up great blocks of Permian, Cretaceous, Triassic, and Jurassic strata that normally occur at great depths; the salt masses contain numerous inclusions of silt, clays, sandstones of the typical bedded variety, alternating beds of anhydrite, limestone, and gypsum, while analysis of the salt shows the presence of potassium and other salts typical of original bedded deposits. The gypsum and limestone cap rock of the American domes is much thicker. The orogenic forces in the European area have been most intense and, with the geosynclinal structure of the region, have undoubtedly been of sufficient magnitude to produce the results observed. No bedded deposits of salt have been found; no blocks of deep-lying strata have been upthrust into younger strata, no inclusions of bedded formations, silts, clays, etc., nor of anhydrite, limestone, potassium and allied salts are generally observed in the American domes; the greater portion of the salt shows a purity that can be logically explained only by precipitation from solution, definite in chemical composition and subjected to uniform conditions of chemical reaction. The weight of the overlying strata and such orogenic forces as were operative in the Texas-Louisiana region are insufficient to bring the salt cores to their present position from depths of 10,000 to 20,000 ft. through intrusion en masse.

The secondary intrusive origin differs from the previous theory in that intrusion of the domal materials en masse is only one of the several factors accounting for the present position of the salt cores and structural deformation produced, the domal materials being first regarded as secondary products, deposited relatively near the surface directly from solutions of secondary origin and character. This theory is based on fundamental details observed in the field; it is in conformance with various processes still observed in operation in and around these domes; it satisfies every detail in occurrence, character, and shape of the dome and domal materials, and their relative positions, and offers a satisfactory explanation of the formation of the cap rock, which the European theory fails to do.

This secondary intrusive origin theory is a combination of several theories. The writer has taken the acceptable portions of such theories and, from data gained in several years of detailed field investigation, has supplied the connecting evidence necessary to mold the facts and principles into a concrete, coherent expression or explanation of the mode of origin, in such a way as to eliminate apparently the objections and shortcomings of previous statements on this subject, to account for chemical, stratigraphic and structural conditions, and to satisfy all observations of fact recorded in the fifty odd years of investigation of these domes. The Gulf coastal plain province is one of the most difficult areas to analyze geologically. Each dome exhibits certain peculiarities so that extensive data become essential before general deductions can be made and theories propounded and data of this nature can be accumulated only by careful, diligent, and exhaustive investigation and study on the part of numerous geologists.

If the secondary intrusive theory of origin should be generally accepted, it should be remembered that such a theory has been made possible only by the investigations of Lucas, Kennedy, Hill, Dumble, Veatch, Rogers, and Lee Hager; the researches of Harris, Norton, Mills, Wells, and Washburne; the intelligent and constructive criticism of Rogers, Woodruff, and others; and the able observations of Deussen and DeGolyer. To these geologists especially, and to many others, credit is due for whatever intelligent understanding we possess of these peculiar structural entities, and of this the writer has ever been mindful and appreciative, not only in the present discussion but in the years of personal field observation conducted with a similar object in view.

DISCUSSION

Eugène Coste,* Calgary, Alberta.—The author's argument is weak in that he attributes the enormous masses of secondary salts, liquid hydrocarbons, and natural gas and sulfuret of hydrogen found under the domes to the leaching of sediments. It is absolutely impossible to admit that these large quantities of salt, limestone, silica, sulfur, sulfuret of hydrogen, and gaseous and liquid hydrocarbons have been leached out of the sediments in which these vertical chimneys of secondary products, known as salt domes, are found. It is just as untenable to believe that as it is to believe that the salt masses under the domes are derived from big beds of salt in the lower rocks, which it has been proved do not exist. The circulation of meteoric waters in the sediments takes place only in a few porous beds, mostly sandstones, and the great quantities of salt, sulfur, and other products mentioned cannot possibly come from such a source.

---

* President and Managing Director, The Canadian Western Natural Gas, Light, Heat and Power Co., Ltd.

We will, therefore, have to return to the solfataric volcanic view regarding the formation of these salt domes that I advocated before this Institute[56] about 17 years ago. We know that all through that district and extending south to Mexico and beyond there have been orogenic movements at different geologic periods and some in recent time. These have resulted, in Texas and Louisiana, in great deep faults, hundreds of miles long, from which at separate points along their course gaseous and liquid solfataric or juvenile emanations have come up from the interior. The faults are miles deep and have brought up these gaseous and hot waters from great depth; in fact, from the volcanic magma below. This is clearly the origin of all of the secondary products under the salt domes; and in some of the domes in Texas and Louisiana, as well as in Mexico, many of the salt waters and oils are still at high temperature.

We cannot imagine that the great masses of sulfur in Calcasieu parish, Louisiana, for instance, can come from the leaching of sediments, or that the natural gas can come from beds of carbonaceous shale that have not been distilled and exist in the sediments as undistilled shales. Such sediments will not give these products at all, especially in anything like the enormous quantities, recognized by the author, it is necessary to admit must have come up these chimneys, or domes, to help produce such huge deposits of salts, etc. by the concentration of brines by evaporation. These enormous quantities of gases and heavily charged brine cannot be obtained except from the interior magma along deep faults and these are remarkably well indicated all through that country by the linear directions of all these deposits or salt domes.

E. W. Shaw,* Washington, D. C.—We have great need for more detailed information concerning individual salt domes. If those who have studied them for years, have examined hundreds of well records, and studied the outcropping rocks would tell us, for example, just what domes are known to have salt, if all of these domes have cap rocks, if the cap rock is ever around the sides or on the slopes of the dome, as well as on top, and whether or not it is certain that all that is called cap rock is of secondary origin, the height and shape of each dome as shown by structure and convergence maps, how much indurated rock is involved in each uplift, the composition of the cap rock from dome to dome; and other details, we might find the basis for one or more inferences of value. For example, if the cap rock is due to diffusion and represents a chemical reaction arising out of the introduction of salt or some other substance into the strata, one might expect the process to affect everything around the deposit of salt—sides as well as top.

---

[56] Volcanic Origin of Oil. *Trans.* (1905) **35**, 288.
* Geologist, U. S. Geological Survey.

The speaker has stated that there is evidence of sufficient gas in the region to meet the quantity demand for the evaporation of enough salt water to yield the enormous quantities of salt in the domes. I have not had time to go over the figures lately, but two or three years ago I tried to estimate about how much gas would be required and found that the amount required for one dome was considerably more than all of the natural gas that the United States has produced to date, as a matter of fact several hundred thousand times as much. That is over and above the problem of whether the salt water and necessary migration of the salt water to the points of deposit has been, or is likely to have been, of such a nature as to have yielded the quantities of salt.

On page 311, it is stated: "Thus, the researches of Mills and Wells have eliminated the last barrier to the acceptance of deposition from solution as the origin of salt and associated materials." It seems to me that this statement is quite erroneous for two reasons. One is that when gas evaporates salt water and deposits salt, the salt closes the pores and the process is self-inhibiting, whether it occurs in the bottom of a well, or in fissures or other cavities at great depth or near the surface. The other is that enormous quantities of gas and salt water are required at the point where the salt is being deposited.

JOSEPH E. POGUE, New York, N. Y.—A few years ago I made a hasty examination of an old salt mine in Colombia. This is a great mass of salt near the surface, and at the time I was impressed with the resemblance of this deposit to the Gulf salt domes in a general way. This salt deposit was quite evidently an intrusion in the surrounding strata, the evidence of that conclusion being the nature of the contact, the inclusion of shale and other sediments along the borders of the salt mass, and the general contorted character of the salt mass. My impression at the time was that the salt was forced into the surrounding rock in much the same condition as a plastic mass of tooth paste extrudes out of the container; in other words, the material came in from a lower level under pressure.

F. G. CLAPP, New York, N. Y.—In discussing the origin of the salt domes, we have made little effort to take into account all known factors concerning salt domes, to say nothing of the unknown factors. For instance, in the case of Louisiana domes, we know that regardless of whether or not the alignment is important and whether or not the phenomena are deep-seated or superficial the domes are arranged in lines. Going north into central Arkansas, we find an intersection of two lines of volcanic phenomena, or intrusion of igneous rock, two of which rise several hundred feet above the earth's surface. The third igneous intrusion is about the same distance from the second that the second is from the first, but not conspicuous, although fragments of volcanic rock of similar character are found in the bottoms of gullies traversing the region. At a greater distance from the two known intrusives is the Arcadelphia salt dome or salt marsh. In considering the origin of the salt domes, it seems necessary to consider whether this feature only 100 mi. or so north of Louisiana may have some bearing on the Louisiana question.

The parallelism of salt-dome distribution with the distribution of volcanic phenomena has been contradicted in Mexico, though not so convincingly but that some truth may exist in this parallelism; and it is worth while to consider whether there may not be some connection between the two classes of phenomena.

It is a question whether all European geologists have accepted the theory of salt having been pushed in from underlying and pre-existing salt beds. I think the theory originated in the German fields and possibly with the German geologists; but in the Transylvania fields many conditions appear discordant with that hypothesis. There seems to be no more evidence than in Louisiana that the strata were underlain by salt beds.

In the same field there is evidence that some domes may still be rising and I have heard that even in the Louisiana fields evidence exists that some salt domes may still be rising. In Transylvania, one evidence is the existence in places of salt at the surface—salt not covered by superficial deposits but by vegetation, old trees, etc., which show signs of movement during the time of growth which at most cannot be more than fifty years. Evidence that the folding was not limited to ancient periods seems to exist in the pinching out of strata as they approach the dome in some of the Red Beds in southwest Oklahoma and in the Tertiary deposits of Transylvania. We must consider the dome problem from a worldwide viewpoint, rather than on the basis of evidence furnished by one field like Louisiana.

E. DeGolyer. New York, N. Y.—The indicated intention of the author is twofold: to disprove the theory of intrusive origin of the gulf coast salt domes and to propose a more acceptable theory. To my mind he accomplishes neither. The bulk of his argument against the theory of intrusive origin concerns itself with critical discussion of the fact that we know of no bedded salt deposits in the rocks of the coastal plain which might have served as a source for the intrusion of the salt core or stock of the dome. This lack of knowledge of the existence of salt deposits of sufficient magnitude has always constituted a weakness of the tectonic theory, as well recognized and stated by its adherents as by its opponents.

The simple fact is that we **do not** know whether there are bedded deposits of salt or not.   Any required by the theory of intrusive origin would obviously lie below the salt of our known domes.   The deepest wells in the Gulf coast that have penetrated the domes have not gone below the salt of the stock.   As an example, the Producers Oil Co. well No. 17, Block 29, of the Wheeler & Pickens Fee at Humble, penetrated the salt stock of the Humble Dome at a depth of 2342 ft. and continued in it to 5410 ft., where the well was abandoned without having gone through the salt.   Manifestly, according to the theory of intrusive origin, any bedded salt deposit necessary as a source of the salt of this stock would lie below this depth of 5410 ft. and clearly, below our zone of knowledge.

It may be objected that the stratigraphic horizon equivalent to a 5410 ft. depth at Humble, lies at much shallower depths farther northward, that it outcrops in fact and consequently lies well within our zone of knowledge.   This must be admitted, but other salt domes show the salt below still lower horizons stratigraphically.   In the Palestine, Texas, dome, certainly Eagleford and probably Woodbine, the base of the Gulf series of the Upper Cretaceous, has been recognized.   The source of the salt then must have been below this horizon and, consequently, throughout most of the coastal plain, entirely below our zone of knowledge.   The information yielded by Mr. Matterson's examination of over 1000 well logs is not of value in this investigation because it is not from the horizons in which we are interested.

It seems quite probable, if the intrusive theory is correct, that the salt domes, as we at present know them, had their origin in folded and faulted pre-Cretaceous rocks upon which our Cretaceous and more recent rocks lie unconformably as a masking mantle and that the salt stocks themselves were forced upward into this more recent mantle.   This is pure speculation, but so is every other theory of salt-dome origin that has come to my attention.

The tectonic theory is not more deficient in its failure to account for a source of the salt than is any of the various theories of deposition from solution.   Simple explanations of deposition from natural brines, particularly connate waters, whether with the assistance of gas evaporation, as suggested by Mills and Wells, is not enough.   There are many places in the world where faults, even cross faults, are common and where brine and natural gas are also as common as on the Gulf Coast, yet where salt domes do not exist.

Many of Mr. Matteson's other arguments are quite faulty.   His attempt to discredit the apparent analogy between American and European domes will not hold.   The resemblances between the two types is much more marked than is their differences and in the Isthmus of Tehuantepec region of Mexico, the differences are still further sunk since good examples of both types are present in the same area.

It is not true in the American domes that "areas and blocks of older, underlying formations have not been upthrust so as to be exposed at or to lie near the surface" as is the case with certain European domes. The Cretaceous is thrust up through the Tertiary in many of the north Louisiana and Texas domes. The Palestine dome shows such an upthrust of the order of 3500–4000 ft. beyond any question. At Tonolapa, on the Cececapa dome, in the Isthmus of Tehuantepec, Jurassic limestones have been thrust upward several thousand feet, coming to the surface through Miocene or Miocene-Pliocene marls.

Nor is Mr. Matteson more fortunate in his conclusion that the salt of American salt domes cannot be free from bedded deposits because "potassium salts, such as are commonly associated with bedded deposits of rock salt, are practically missing." Phelan[57] gives analyses of salt, brines, and bitterns from various bedded deposits and domes in the United States. The chemical composition of the various salts is remarkably uniform and there is no variation between salt-dome salt and bedded salt deposits greater than the variation between various samples from bedded deposits. The same is true with regard to natural brines except that brines from Humble and Sour Lake, both salt domes, show slightly higher potassium contents than other brines—the direct opposite of Mr. Matteson's argument.

The theory that Mr. Matteson proposes as a substitute is one that includes bits from all previously proposed theories. This is its strength as well as weakness, since many forces have probably combined to produce a salt dome. We are interested, however, principally in the main force forming the salt dome. We might consider a dome as a structural feature alone. Of course, the salt is common to all of them, but there are domes that have no cap rocks, sulfur, or oil. What theory will explain the tremendous upthrust by which blocks of sediments ½ or 1 mi. in diameter are thrust upward 4000 ft. to perhaps much greater distances, through rocks of younger age? I do not believe that Mr. Matteson's theory will meet this test nor do I believe that any other theory of deposition from solution is sufficient.

Of course, much can be said for every theory. I do not believe, for example, that the volcanic theory is acceptable, but it explains fairly well some facts difficult to explain by any other theory. It explains the hot waters associated with the domes and it is more satisfactory in explaining the extremely high sulfur oils of coastal Texas and Louisiana, the Tampico region, and the Isthmus of Tehuantepec region, two salt dome regions and one region of volcanic activity, than is any other theory.

W. G. MATTESON.—As to the impossibility of admitting that such great quantities of salt could be leached from the various sediments under-

---

[57] U. S. Geol. Survey *Bull.* 669.

lying the coastal plain area, the researches of Mr. Kennedy have proved conclusively that just in the Miocene deposits alone there is enough disseminated salt content to account for the salt in these coastal domes. Besides, there are the underlying formations with disseminated salt in the Cretaceous deposits, which we know may be considerable.

Mr. Shaw questioned the advisability of having more detailed information on the cap rock. The major operating companies in the Gulf coastal plain area recognize the importance of geological application, especially since the bringing in of oil around some of the older formerly abandoned domes. Some companies have accumulated a great mass of valuable detailed evidence. It is difficult for a consulting geologist to get this information in every instance but this would be a good opportunity for the United States Geological Survey to bring up to date its data on the Gulf coastal plain area. It is quite possible that several companies would turn over a considerable portion of this information to the Survey. The author has had access to some of this information.

Mr. Shaw raises the question of gas being present in sufficient quantity to cause deposition through evaporation. The Yegua formation has yielded enormous quantities of gas, and in many cases where we drilled into that formation all we obtained was gas. The fact that gas has been found from the Mexican border across the entire states of Texas and Louisiana shows that the amount of gas in this one formation alone is great; in addition there is the gas in the Cretaceous and underlying Pennsylvanian formations. There is an adequate supply of gas to account for great evaporative effects on salt brines with the subsequent deposition of vast salt masses.

In the preparation of this paper I was able to consult with Mr. Kennedy, who has recently investigated the logs of wells of the Freeport Sulfur Co., at Bryan Heights and some of the material presented herein was the result of that conference.

Mr. DeGolyer speaks of the tremendous upthrust of strata at the loci of domes. First, the changing of the limestone into gypsum results in an increase in volume of from 32 to 50 per cent. As we find a gypsum cap anywhere from 200 to 700 ft. in extent, and probably considerable of the gypsum cap has been eroded, this alteration alone indicates an initial uplift of a few hundred feet.

In presenting this paper I admit that a considerable part of the uplift has been due to intrusion en masse, but that intrusion has occurred after the salt has been deposited relatively near the surface along with the limestone and other materials. The upthrust of those domes, amounting to close to 4000 ft. in the interior domes, to which Mr. DeGolyer calls attention, seems to be greater than the average of 1500 to 3000 ft. which we find along the Gulf coastal plain proper, but might not that be due to the fact that the interior domes are nearer the great Balcones fault?

H. W. HIXON, New York, N. Y.—There was great pressure of gas below the cap. Where gas could not escape water certainly could not, which shows that the seal was complete. Therefore, I do not believe it would be possible for much water to have escaped from these domes without leaving evidence of its having been present.

My theory of the origin of these domes involves a fundamental conception of the physical condition of matter in the interior of the earth. The temperature at a moderate depth, say 150 mi., will be a critical temperature for all the matter in the interior of the earth; after it passes the critical temperature it is in a gaseous condition, and when matter is above its critical temperature, it is capable of high compression. Gravitational compression unrestrained is capable of producing, in a gaseous core, matter that is denser than the solids that will form out of them. Admitting, for the sake of argument, that the gaseous core can be denser than the solids that will form out of it, we have a solid crust floating on the gaseous core, like ice floats on water. Likewise, when matter changes from the condition of gas denser than solids, to that of a solid, it expands. If the gaseous core is expanded by loss of temperature, the cold crust above will be fractured, and great fault planes, which are called orogenic in nature, should be formed and more or less parallel.

These fault planes may be of two series, which intersect at approximately right angles. These salt domes occur at the intersections of those faults, which furnished a passage for volatile material from the interior. Whatever shale beds covered those fault planes did not rupture clear through to the surface, but stretched under pressure of the load and completed the seal. The volatile material in these domes rather confirms that. Salt is volatile at a moderate temperature, so is sulfur, and these gases come up in a dry condition. I do not believe there is any water to speak of in connection with them, except that necessary to alter the limestone cap into gypsum. But these fault planes, created by the expanding force, have been the determining cause of the domes, and the gradual increase of the size of the plug of the salt domes is due to the cumulative effect of volatile matter coming up through the cracks.

R. VAN A. MILLS,* Washington, D. C. (written discussion†).—Mr. Matteson not only contributes new and valuable data, resulting from his investigations, but he attacks the problem on the basis of a multiple hypothesis. By recognizing the grain as well as the chaff in several hypotheses and by applying such parts of these hypotheses as accord with the facts thus far established, he has adopted the most promising method of attacking the salt dome problem.[58]

---

* Petroleum Technologist, Bureau of Mines.
† By permission of Director of Bureau of Mines.
[58] R. Van A. Mills: Discussion on Oil-field Brines. See page 281.

That we are in no position to restrict ourselves to any one of the theories previously advanced is recognized from several facts. The presence of deep-seated salt beds in the Gulf coastal region from which intrusive masses of salt might originate has never been established; the evidence presented by Mr. Matteson, so far as it goes, is rather against the presence of such beds. Again, there has been no systematic geochemical study of the domal materials or associated waters, gases, and oils, to determine their relationships together with the geochemical processes that have probably been operative in the dome building. Strange to say, the only published results of a systematic investigation of this kind come from the Appalachian region[59] where there are no salt domes.

We agree that there has been intrusion by salt but we know neither the origin of the salt nor the causes for the intrusion. Our theories upon these phenomena constitute little more than working hypotheses through which to attack the problem. Accepting this view, real progress must come through systematic and laborious investigation rather than by the easy and alluring road of speculation.

Advocates of the theory of primary intrusion of the salt masses have endeavored to substantiate that idea by inadequate data and also by eliminating all theories upon the deposition of salt from solution. They have also attributed undue importance to the so-called flowage lines in the salt masses; first, because the lines may be caused by secondary intrusion, and second, because the lines may be of depositional origin. Such lines commonly appear in specimens of the mineral salts deposited through the agency of water in oil and gas wells. The lines are especially common in deposits of sodium chloride. Laboratory experiments indicate that lines of this kind in masses of salt deposited from solution may be largely, or wholly, of depositional origin. Irregular bands or lines in the salt and gypsum deposits of southwestern Virginia are attributed to secondary depositional phenomena by Stose.[60] Conditions in that locality apparently preclude any probability that the masses of salt and gypsum attained their present positions and irregularly banded characteristics through intrusion.

To what extent the growth of crystals may have contributed toward the intrusion of the salt and uplifting of superencumbent strata in the Gulf coastal region is problematic. It is recognized that enormous forces are exerted in the growth of concretions and that certain failures of concrete are caused by the forces of crystallization, but data upon the

---

[59] R. Van A. Mills and Roger C. Wells: Evaporation and Concentration of Waters Associated with Petroleum and Natural Gas. U. S. Geol. Survey *Bull.* 693 (1919).

[60] George W. Stose: Geology of the Salt and Gypsum Deposits of Southwestern Virginia. Virginia Geol. Survey *Bull.* 7 (1913), 70–71.

forces exerted through the crystallization of salt are meager. Rogers[61] cited data indicating that such forces would be inadequate to cause the intrusion of the salt with the consequent uplifting of superencumbent beds. He did not however, show that such forces were inoperative. They have undoubtedly played their part. In laboratory experiments upon the cementation of sands and the exclusion of water from oil wells by plugging the interstices of the sands through chemical precipitation, numerous instances of the displacement of sand by crystalline growths have been observed. The crystaline precipitates formed masses that displaced the loose sands. In these experiments, repeated failures of the glass fronts of the apparatus were caused by the expansive effects of the crystallization of chlorides, sulfates, carbonates, and silicates.

Recognizing that in the light of the meager information now available, the effects attributed to the forces of growing crystals by Harris[62] constitute a weakness in his theory, but also recognizing that the intrusion of salt from one cause or another has played a major role in the dome building, it is logical for Mr. Matteson to postulate intrusion through the agency of dynamic forces acting from without the salt cores themselves. The hypothesis of secondary intrusion, together with that of geochemical origin of the cap rocks, constitute valuable working hypotheses for future investigations.

Where so much has to be taken for granted and so much more has yet to be learned through deep drilling, accompanied by systematic investigation, it is not to be assumed that Mr. Matteson has presented the ultimate explanation for the origin of the Gulf coastal domes. We must, however, recognize that his method of attacking the problem is an important step toward the ultimate solution.

W. G. MATTESON (author's reply to discussion).—The director of the U. S. Geological Survey recently assigned M. I. Goldman to the Gulf Coast province for the purpose of making a thorough study of the presence, variation in composition, physical character, and so forth of the cap rock of the coastal plain salt domes.

Mr. E. W. Shaw says that "we have great need for more detailed investigation concerning individual salt domes;" yet much of the specified information he desires has been obtained. Unfortunately, however, such data have not been gathered within one cover, for this would require a monograph that only special organizations like the Survey have the facilities to produce. As Messrs. Shaw and Mills state, an extensive geochemical investigation of the coastal plain domes is needed, yet such

---

[61] G. Sherburne Rogers: Intrusive Origin of Gulf Coast Salt Domes. *Econ. Geol.* (1918) **13**, 447–485.

[62] Gilbert D. Harris: Rock Salt in Louisiana. Louisiana Geol. Survey *Bull.* 7 (1907).

an investigation is not without its difficulties. To be of greatest value, it should be as extensive as outlined by Mr. Shaw, but this would require several years of field work, considerable expense, and probably constant change in the personnel of the investigators. Indeed, it is doubtful if some of the information can be obtained. An examination, last year, of a proved dome required several months of persistent effort. The dome had never produced, yet twenty wells had demonstrated its character and some of these wells had excellent oil showings. In preparing the map, it was necessary to find the drillers and then have them locate, in the dense underbrush, the abandoned wells. Six months were required to secure, from varied scattered sources, the logs of these twenty wells. Some of the large companies that had drilled on the dome had no logs, so that information could be obtained only by finding some outside person who was interested in the tests at the time. If only the present producing domes were thoroughly investigated geochemically, much new and valuable information would certainly be procured. In this connection, it is to be earnestly hoped that the U. S. Geological Survey will increase the scope of its present activity in the Gulf Coast region by having Messrs. Mills and Wells continue their researches, which have been so fruitful in the Appalachian province.

It is unfortunate that Mr. Shaw did not accompany his statements respecting the quantity of gas necessary to cause vast salt deposition through evaporation with adequate data and an outline of the method by which he arrived at such conclusions, as his assertion is debatable. Records show that the Yagua formation, of Eocene age, has yielded many billions cubic feet of gas in the Gulf Coast region and has an estimated possible potential production even more vast. Yet this is only one formation of several that are gas bearing. Mills and Wells[63] call specific attention to the fact that relatively small quantities of gas were capable of causing deposition of several tons of salt in 24 hr. in certain wells in the Appalachian province. The statement that "when gas evaporates salt water and deposits salt, the salt closes the pores and the process is self-inhibiting," is merely speculation on Mr. Shaw's part. The only published data relating to this phase are those of Mills and Wells[64], and apparently they find nothing to warrant the conclusions of Mr. Shaw.

The author was much surprised that Mr. DeGolyer should select the Palestine salt dome in Anderson County, Texas, as an example of upthrust similar to that observed in European occurrences and, on this basis, question the reliability of the writer's conclusions. The writer made a thorough investigation, a few years ago, of the Palestine dome and

---

[63] R. Van A. Mills and R. C. Wells: Evaporation and Concentration of Waters Associated with Petroleum and Natural Gas. U. S. Geol. Survey *Bull.* 693.

[64] W. G. Matteson: *Op. cit.*, 3.

the Keechi dome, several miles to the northeast. The Austin Chalk and overlying formations of Upper Cretaceous age are here found at the surface entirely surrounded by the Wilcox formation, of Lower Eocene age. There is no question that faulting has occurred, yet the evidence shows the Wilcox to be tilted at angles of 30° to 40° and sloping from the center of the dome on all sides. Likewise, the Wilcox is the next younger overlying formation of the geologic series in this specific area, the Midway being absent. In few instances is the evidence of uplift (and not upthrust) accompanied by faulting and followed by erosion more positive than here, this erosion revealing the presence of the immediately underlying Cretaceous formations. Few experienced Gulf Coast investigators will agree with Mr. DeGolyer that these occurrences in Texas and North Louisiana are real upthrusts of older formations into younger and especially of the nature and extent to justify their classification with the European occurrences, where formations have been completely sheared from their parent beds and upthrust many thousands of feet into much younger horizons, which horizons in no way show the deformations characteristic of the older rocks. While the presence of the Austin Chalk at the surface at the Palestine and Keechi domes seemed to indicate a local uplift of 3000 to 4000 ft., with the information available at that time, recent data point to the Sabine uplift being much more extensive than commonly supposed while the influence of the Balcones fault must also be considered. It is quite probable, therefore, that the Cretaceous formations here are not so deep lying as supposed and that the amount of uplift will necessarily have to be modified.

The negative character of the results obtained by Phelan in his analyses of bedded salt and salt-dome deposits and brines does not justify any positive conclusions tending to disprove the writer's statements. While potassium salts may not have been characteristic of the bedded deposits analyzed by Phelan, it is generally agreed that extensive bedded deposits are often so characterized by such salts along with anhydrite. Van der Gracht[65] and Kennedy[66] have admitted the difficulty of assigning an exactly similar mode of origin to both the European and the American salt domes, due to the general absence of potassium salts in the latter instance, and the opinion of these authorities cannot be lightly disregarded. The potassium content of Humble and Sour Lake, mentioned by DeGolyer, is very slight and might be due to several factors. Such occurrences are of little value in the present discussion.

The writer finds no basis to justify Mr. DeGolyer's conclusion that the 1000 logs examined in the present instance are of no value since they

---

[65] W. A. I. M. von Waterschoot van der Gracht: Salt Domes of Northwestern Europe. Southwestern Assn. Pet. Geol., *Bull.* 1 (1917).

[66] Personal interview.

represent wells that have not penetrated below Tertiary horizons. These logs, selected with extraordinary care, include a series of wells penetrating varying formations from the lower Pennsylvanian to the Recent, yet in not a single instance has the presence of bedded salt deposits been detected in the Pennsylvanian, Cretaceous, or Tertiary. DeGolyer says "the deepest wells in the Gulf Coast that have penetrated the domes have not gone below the salt of the stock" but Kennedy[67] records a well that passed completely through the salt into the underlying formations. DeGolyer asks, "what theory will explain the tremendous upthrust by which blocks of sediments, $\frac{1}{2}$ to 1 mi. in diameter, are thrust upwards 4000 ft. to perhaps greater distances?" This question of uplift and not upthrust is logically explained on pages 317 to 319 and the explanation herein offered accords strictly with all facts. It is quite evident, however, that the primary intrusive origin that Mr. DeGolyer is attempting to substantiate will in no way adequately answer his question, for there is absolutely no evidence of the action of orogenic forces of sufficient intensity in the Gulf Coast region to push salt masses and accompanying sediments upwards through thousands of feet of overlying materials. Moreover, why should we not find Cretaceous and older formations overlying the salt in the domes along the Gulf Coast proper, or at least fragmentary evidence of the same, if there is any merit in Mr. DeGolyer's contention that "if the primary intrusive theory is correct, the salt domes had their origin in folded and faulted pre-Cretaceous rocks upon which our Cretaceous and more recent rocks lie unconformably as a masking mantle and that the salt stocks themselves were forced upwards into this more recent mantle?"

In concluding the discussion, it might not be remiss to utter a word of caution against the tendency to speculate without sufficient basis of fact. The Gulf Coast salt domes offer a most alluring field in this respect and the temptation has been too great for some of our most able authorities to resist. After all, a theory is only a suitable working hypothesis conforming to reason and observed conditions. As Mr. Mills says about the problems of these domes, "real progress must come through systematic and laborious investigation rather than by the easy and alluring road of speculation."

---

[67] William Kennedy: Coastal Salt Domes. Southwestern Assn. Pet. Geol., *Bull.* 1, and personal interview.

Application of Law of Equal Expectations to Oil Production in California*

By Carl H. Beal† and E. D. Nolan,‡ Washington, D. C.

(Chicago Meeting, September, 1919)

In February, 1918, the conclusion was published by Lewis and Beal "that wells of equal output on the average will produce equal amounts of oil in the future, regardless of the ages of the wells." This conclusion was based upon the study of data collected principally in Oklahoma and was not known at that time to be true for other oil fields. An abundance of statistical proof was later collected by the senior author of the present paper, which showed that the conclusion was undoubtedly well founded and that it applied to other fields as well. Accordingly, it was later restated[2] as the "law of equal expectations" as follows: "If two wells under similar conditions produce equal amounts during any given year, the amounts they will produce thereafter, on the average, will be approximately equal, regardless of their relative ages."

Although only scanty data from the California oil fields were available at the time this publication was prepared, sufficient information was analyzed upon which to base the insert on Fig. 80, which showed beyond a doubt that the law held at least for a part of the Midway oil field in California. Recently the authors have collected more complete data in California, and it is the purpose of this paper to explain the method used in demonstrating the truth of the law and, in addition, to give several methods by which curves constructed in accordance with this law can be used in a practical way with ease and accuracy.

The Family Curve

The law of equal expectations means that each individual of a group of wells producing under similar conditions will decline along approximately the same type of curve, the rapidity of decline varying with the output

---

* Published by permission of the Director, U. S. Bureau of Mines.
† Petroleum Technologist, U. S. Bureau of Mines.
‡ Consulting Petroleum Engineer, U. S. Bureau of Mines.
[1] Some New Methods of Estimating the Future Production of Oil Wells. *Trans.* (1918) **59**, 492.
[2] Carl H. Beal: The Decline and Ultimate Production of Oil Wells with Notes on the Valuation of Oil Properties. U. S. Bureau of Mines *Bull.* 177 (1919) 36.

of the well. For instance, if the first year's production of a well is very large, its decline will be much more rapid than that of a well having a smaller output. Furthermore, the second well will produce oil at the same rate as the first well after the latter has declined to the same output as the second. Inasmuch as the wells in a group, under similar conditions, produce oil along a certain curve, if this curve can be made up from decline records of wells of different size, we are able to forecast with accuracy the decline of normal wells of different size in that area. Such curves have been built up for different fields in California. They have been called "family" curves for lack of a better name and because the decline of wells of different output will follow the same curve.

The use of the family curve is not claimed to be original in the present paper, as its possibilities were given by Lewis and Beal,[3] and one method of preparing such a curve and its advantages were later briefly given by Beal.[4] The particular method of building up the family curve, however, is unique, and the various methods of using the curve for estimating the life and future production of wells are new.

CONSTRUCTION OF FAMILY CURVE

In preparing family curves for other oil fields, it has usually been necessary to use the production records of tracts, for in most fields the output of all wells on a tract is gaged in the same tank. The use of such records has some advantages and some disadvantages. If the records of individual wells are used, there will be smaller chance of the entrance of such complex factors as the undue maintenance of production by the bringing in of new wells on a tract. In the oil fields of California the production records of individual wells are usually available, and the following curves are based entirely on such records.

Fig. 1 shows a family curve based on the production records of wells in a California oil field. Briefly, the preparation of such a curve consists of, first, choosing the records of all normal wells—such as those unaffected by redrilling, cleaning out, deepening, water encroachment, etc.; second, plotting the yearly decline of the largest well A and joining the points showing the production per year by straight lines; and, finally, taking successively smaller wells and plotting the decline of each well, the initial or first year's point being located on the production curve of the largest well at the proper point, and subsequent points at spaces to the right representing years. For instance, in Fig. 1, the points marked A represent the decline of the largest well, those marked B represent the

---

[3] *Trans.* (1918) **59**, 512, Fig. 9.
[4] U. S. Bureau of Mines *Bull.* 177 (1919) 198.

decline of the second largest well, and points C, D, E, F, G, and H the declines of the smaller wells. The initial point, or the first year's production, of well B is located on the curve of well A at a distance of 90,000 bbl. (the first year's production) above the horizontal axis. This procedure is repeated for the smaller wells.

After the declines of two or three wells have been plotted the average line can be drawn by determining the numerical average of points within adjoining vertical segments of the cross-section paper and drawing the curve through the average points. From this time on, the decline of the smaller walls may be begun on the average curve. For instance, in Fig. 1, after the records of wells A and B were plotted, the heavy average line was drawn to point X and the first year's production of wells C, D, and E

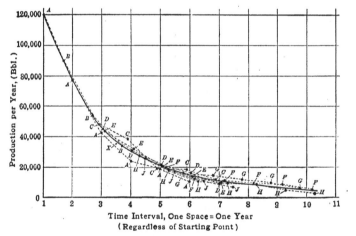

FIG. 1.—SHOWING METHOD USED IN CONSTRUCTING A FAMILY CURVE FROM PRODUCTION RECORDS OF INDIVIDUAL FIELDS.

were plotted on the curve. The process of gradually extending the family curve and plotting on it the initial year's production of smaller wells is continued until all the data are plotted. Rarely will a case be found where the plotting of more than three or four wells is necessary to determine the beginning of the average curve. The greatest difficulty is usually experienced in determining the proper rate of curvature of the decline curve when it begins to flatten out. This part of the curve usually represents the exhaustion of the high gas pressure, which is closely associated with the rate of expulsion of oil from the well. After part of the gas pressure is released, the curve representing the decline of practically any well, unless changed by some mechanical accident, trends only slightly downward at a rate decidedly less than its previous rate of decline.

It should be noted that the entire length of the average, or family curve, as shown by the heavy line in Fig. 1, is the result of the plotting of past production. Sufficient data are usually available so that the curve can be carried even to the point representing minimum economic production, so that the necessity of projecting the curve to represent production in the future is obviated. The family curve in this case is based absolutely on actual performance. The objection to some curves is the necessity of projecting them, the projection in many cases varying with the person who makes it. This is not true, however, with the family curve, especially when the records of several wells representing different outputs are available.

In a new field, such as the Montebello field, it might be found advantageous to construct a curve with a monthly time interval as the horizontal scale. Then with wells but a few months old, a family curve may be prepared with but little difficulty.

Another method[5] of preparing a family curve is to divide the production records into classes representing different productivity. The yearly output of all wells in the highest class (those that made 110,000 to 120,000 bbl. the first year) is averaged; then the yearly output of the next highest class (those that made 100,000 to 110,000 bbl. the first year) is averaged, and the average points plotted and so on until all the averages are obtained.

Use of Family Curve

Because wells of different size decline along the same type curve, the work of making estimates of future production is greatly simplified. The life of the average well can also be quickly determined and the limits of decline may be shown graphically. Furthermore, the future yearly production of a well of any output may be read directly from the average family curve.

Future Production Curves

In Fig. 2, curve A, above the family curve B, was determined by adding the future production of wells of different output as shown by the family curve and then plotting these future production estimates vertically above the point on the family curve representing the first year's production of a well. For instance, assume a point on the family curve, representing 21,000 bbl. to be the first year's production of a well of that output. The second year's production will be one year to the right, the third year's production two years to the right, etc., to the point of minimum economic production. These estimated annual productions, with the exception of the first year, as shown on the family

---

[5] Carl H. Beal: U. S. Bureau of Mines *Bull.* 177 (1919) 198, and Fig. 80.

curve, are added together and plotted vertically above the point on the family curve representing the first year's production, and a curve *A* drawn through the points. In the present instance, the yearly production, as represented by the family curve, is as follows:

YEAR	PRODUCTION, BARRELS	YEAR	PRODUCTION, BARRELS
1	21,000	6	5,500
2	15,000	7	4,000
3	11,500	8	2,000
4	9,000	9	1,000
5	7,000	Total............	76,000

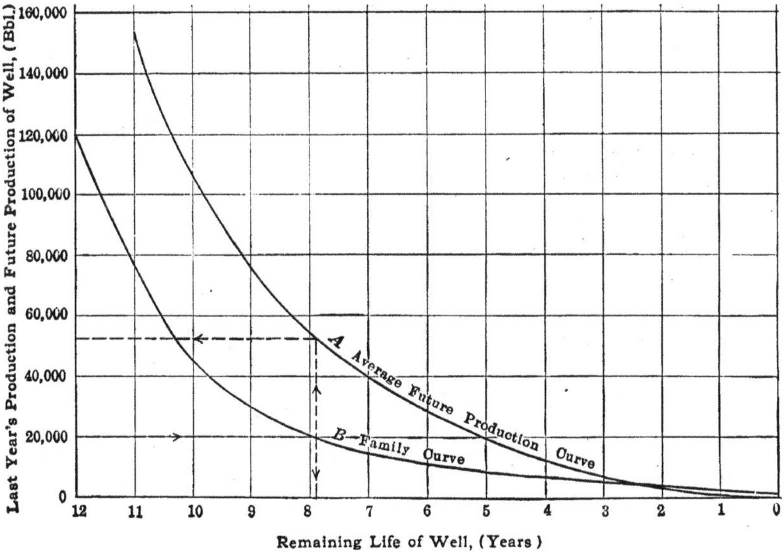

FIG. 2.—AVERAGE FUTURE PRODUCTION AND "FAMILY" CURVES OF A CALIFORNIA OIL FIELD.

The average ultimate production of the well will be 76,000 bbl., and the future 55,000 bbl. (76,000 − 21,000). The future production, 55,000 bbl., is plotted vertically above the point on the family curve representing 21,000 bbl. Because of the law of equal expectations, the production of 21,000 bbl. could represent the most recent years' production if desired; that is, suppose the well during 1918, its third year, produced 21,000 bbl. Then if the well is an average well, it will produce 55,000 bbl. from 1919 to 1926, inclusive, and it produced 30,000 and 45,000 bbl. during 1917 and 1916, respectively.

By the use of curve *A*, the future production of a well at any period of its life may be determined by selecting the production for the last year. Find this amount on the left margin and trace the line to the right to a point where it intersects the family curve, follow the vertical line through this point upward to its intersection with the future production curve, thence to the left margin of the figure. The reading is the future production of the well selected. For example, take a well that made 20,000 bbl. during the first year, follow the horizontal line to the right to its intersection with the family curve, thence upward to its intersection with the future production curve and thence to the left margin where 53,000 bbl. is indicated as the average future production of a well of that output. The estimate would have been correct if the production of 20,000 bbl. represented the most recent year's production instead of the first year's production.

Determining Average Life of Wells of Different Size

In the lower margin of Fig. 2 will be found figures that decrease to the right. These figures represent the remaining average life of wells and are determined by counting the years of remaining life for wells of different output, as shown by the family curve. For instance, the remaining life of the well that made 20,000 bbl. during the first year, by reading downward on the vertical line passing through the point on the family curve representing 20,000 bbl., is found to be 8 years. From this curve, it is evident that the lives of oil wells vary directly as the volume of production, for the larger the production, the longer the remaining life.

Ultimate Production Curves

If desired, the future production may be added to the last year's production, which will give the ultimate production direct. These statistics may be plotted for wells of different size and curves thus constructed. Both this and the average future production curves may be plotted if desired, although the ultimate production may readily be obtained by first determining the future production and adding to it the past year's production.

Another Method of Showing Future Production Curve

The curve representing future production may be expressed as an average appraisal curve if desired. The appraisal curve was named by Lewis and Beal,[6] and consists of showing the relation between the first year's production of a well and its ultimate production. The suggestion was made that additional curves showing the actual future could be plotted by subtracting from the ultimate production the past year's

---

[6] *Trans.* (1918) **59**, 492.

production. The future production curve, as arrived at by the family curve, may be expressed in the same way; that is, the past year's production may be used as the abscissa and the future production may be shown as the ordinate. In this way, the curve begins a distance to the right of the lower left-hand corner, which represents the minimum economic production to which the average well in a district can be pumped, and rises gradually to the right, having the same form as appraisal curves. There is no particular advantage in this form of curve over curve A, Fig. 2, which likewise represents future production directly.

Erratic Wells

In any field, certain wells will be found having a decline wholly different from that of the family curve of that group; these usually are con-

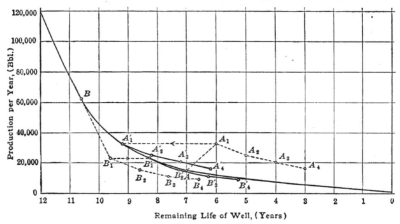

Fig. 3.—Production of erratic wells plotted on "family" curve to show that such wells usually decline along some part of the curve after erratic period ends; records indicated by A and B are wells in California field.

sidered abnormal wells. The causes of these wells may be divided roughly into three classes—geological, accidental, and lack of histories of wells. Geological causes may be either a very thick series of oil sands with varying gas pressures or the comparatively sudden invasion of edge water. In certain small districts, such as the area in Sec. 27, T. 19 S., R. 15 E. in the Coalinga field, or that near Fellows in the North Midway field, the thickness of oil-bearing series is from 500 to 700 ft. (152.4 to 213.3 m.). Wells drilled through this thick series of alternating oil sands and shales often show an increasing production for 2 or 3 years after inception. After having reached their maximum, however, their decline follows the family curve. The probable explanation of this increase is as follows: Such wells penetrate a number of rich oil sands, but under varying gas pressures. When first brought in only those sands with the higher gas pressures are able to produce but time permits a lessening and readjusting of the pressures and all sands are able to contribute to the well's production. Curve A, Fig. 3, shows the production of a well of this type and its relation to the family curve. Wells producing from a sand suddenly invaded by water may show an increase in production just prior to the appearance of the water, but almost invariably show a rapid decline and a sudden end.

Accidental causes of erratic wells might also be called mechanical causes. The "oil string" may collapse, shutting off its production, or a redrilling job may be a failure, causing the abrupt ending of the well's life. In the loose unconsolidated sands of the California fields, shale may cave in, shutting off the perforations. These accidents usually cause a sharp break in the decline of the well and a consequent dropping away from the family curve. After this initial break, its decline through the remainder of its life usually follows some other part of the family curve. The decline indicated by B, Fig. 3, is of such a well.

Another class of erratic wells that often cause trouble are those that have been deepened. When a well is deepened into lower sands or is redrilled, with a consequent opening of new sands, or possibly shutting off other oil sands, it must be treated as a new well, and accordingly a new part of the family curve selected as its decline curve.

Wells varying from the family curve sufficiently to be termed erratic wells are rare, certainly less than 10 per cent. of the total wells in the California fields. The divergence in most erratic wells takes place during the first 2 or 3 years of the life of the well. From that time on, the output of the well follows some part of the family curve.

ESTIMATING FUTURE PRODUCTION OF WELLS ABOVE AND BELOW THE AVERAGE

Most wells in a field will follow the family curve with fair exactness. Some will trend slightly above it, follow it for a year or so, and finally fall below. One is usually safe, however, in making estimates of the future production, if he assumes the well to be an average well; he is unwise, however, if he makes no effort to determine the amount a well is above or below the average, for if it deviates far from the average the estimate may and should be modified accordingly. Fortunately, as most estimates of future production are made by using the last year's production, the curve tends to correct itself by automatically shifting the point on the family curve at which the estimate is made to the right or left, according to whether the curve is below or above the average. This may be more clearly shown by taking an example. Suppose well A, Fig. 3, has produced 2 years, as shown by A and A_1; the estimate of future production is made by applying the last year's production (indicated by A_1) to the family curve, thus shifting the point A_1 to the left to where a horizontal line through it intersects the family curve. As subsequent production from this particular well has proved (see points A'_2, A'_3 and A'_4), the estimate of future production would have been slightly above the average curve.

Another example will serve to show the method by which the estimates of future production of wells below the average will tend to correct themselves. Suppose well B, Fig. 3, has produced 2 years (B and B_1), an estimate of its future production will be made from point B'_1 on the family curve. Subsequent production would indicate that the well produced along a curve (B'_2, B'_3 and B'_4) almost coincident with the family curve. If estimates are made yearly, they become closer and closer even though the well may produce along a curve considerably above or below the average.

FAMILY CURVE APPLIED TO TRACT OR PROPERTIES PRODUCTION

Where the individual well records are lacking or where the average well production is quite small, it may be either necessary or convenient to construct a family curve for a group of tracts rather than for a group of individual wells. Such curves when constructed from a number of properties and applied to properties that are sufficiently drilled are quite accurate.

VALUE OF FAMILY CURVE

The greatest advantage of the family curve is the fact that it is based entirely on history; it usually has no projections and it is not difficult to prepare. Furthermore, its advantage over the appraisal curve is that it can be prepared with less data. In fact, the statistics representing the decline of a dozen wells might suffice for the preparation of a curve, the decline of which represents the decline of wells of different size in an area where conditions affecting production are practically equivalent. The accuracy of the curve, however, is increased in direct proportion to the number of records used in its preparation.

Another advantage over appraisal curves is that the future production of a well from its first year can be estimated more readily when the decline of the well is above or below the average. Owing to the fact that the last year's production is used and that erratic wells after their abrupt change follow a portion of the family curve, the curve reduces error to a small amount, and tends to correct errors due to its own limitations. The simplicity and completeness of the curve are the principal arguments in its favor. One may read direct the future production of a well, its probable life in years and its probable production in any year in the future.

Essential Factors in Valuation of Oil Properties*

By Carl H. Beal,† M. A., San Francisco, Calif.

(Chicago Meeting, September, 1919)

The most important factors that should be given consideration in the valuation of oil lands are: (1) the amount of oil the property will produce; (2) the amount of money this oil will bring (based upon the future prices of oil); (3) development and production costs; (4) the rate of interest on the investment; (5) the retirement or amortization of invested capital; and (6) the salvage or "scrap" value of the equipment when the property is exhausted. These factors are of varying importance and some of them may not enter all valuation problems, but most of them should be given consideration in any valuation even though only a rough estimate of the value of the property is desired.

The value of a property may be changed over night by the completion of important test wells, by the sudden water flooding, or by a change in the price of oil. The best a petroleum engineer can give is the value of the property under the conditions existing at the time the appraisal is made with a fair forecast of future action of the wells and of the price of oil.

Our experience in the scientific valuation of oil lands is not broad and there is very little published information on the subject; it, therefore, becomes necessary in studying such problems to form comparisons with the factors involved in the valuation of mines—the closest parallel. One of the reasons for the lack of substantial progress in oil-land valuation methods has been the necessity of making an estimate of the future production of the oil property to be valued. Oil men and accountants have not generally conceded that such estimates could be made with any degree of accuracy. It has been shown, however, in several recent publications that with certain data available reasonably close estimates can be made. The accuracy of an appraisal depends chiefly on the accuracy of the estimates of future production and of the future price of oil. The accuracy of the former is sometimes necessarily based on geological inferences. Geology is not an exact science and geological data in connection with oil production cannot always be mathematically evaluated.

---

* Published by permission of Director, U. S. Bureau of Mines.
† Petroleum Technologist, U. S. Bureau of Mines.

FUTURE OIL OR EXPECTATION

In considering the factor of future oil, two related questions must be answered: How much oil will the property produce? At what rate will the oil be produced? If we can determine the future annual production of an oil property, we may easily determine the total future production by addition, so we will consider only the question of rate of future oil production.

A satisfactory answer to this question is the keynote to the whole valuation; for, although our work, has, by no means, been completed after the question has been disposed of, the work of determining the value of the property is greatly simplified, for on the yearly output of oil depends the yearly gross income. From the gross income the annual net return is computed, each year's return being considered in the light of a profit available at a future date. The present value of these deferred profits is then determined by discounting them at a rate of interest compatible with the risk involved.

No uniform yearly revenue can usually be expected from an oil property, for the annual output, and thus the annual income, depends on the rate of production. Only under exceptional conditions can a steady oil production be maintained for long unless the property is old and production well settled. The future annual oil output hinges on the rapidity with which new wells are drilled and on the rate of production of the individual wells which, with very few exceptions, always declines.

Rate at Which Oil Will Be Obtained.—The rate of production of the wells will affect not only the rate of output of the old wells, but will regulate that of the wells to be drilled. Furthermore, the decline in the initial output must be considered; the longer the development of the proved acreage is deferred, the less will be its ultimate production, for, under usual conditions, the wells on the drilled acreage cause a decrease in gas pressure over the undrilled acreage, which results in decreased initial production of the wells eventually drilled there. The rate at which oil wells will produce is the resultant of many complex factors, which will not be discussed here. For more information on this subject, the reader is referred to a bulletin by the author.[1]

The most trustworthy method of determining the rate of production of the wells of a group is to prepare a production curve that will give the average yearly output of wells of different initial yearly output. It is necessary to determine this for wells of different initial production, because wells of different output decline in production at different rates—other factors being equal.

---

[1] Carl H. Beal: Decline and Ultimate Production of Oil Wells with Notes on the Valuation of Oil Properties. U. S. Bureau of Mines *Bull.* 177 (1919).

Drilling Program.—The rate of the production of the property depends not only on the rate at which the individual wells will produce oil but also on the rapidity with which new wells are added to the producing list; this depends on the drilling program. The valuation should not be attempted until a drilling program is decided upon. But before a drilling program can be determined, it is necessary to know the amount of land that certainly will support commercially productive wells; trustworthy estimates of future oil production can be made only for the drilled acreage and for the undrilled proved acreage. Only such land furnishes a concrete basis of value, for the annual production of oil can be estimated; other land has a speculative value that varies with the uncertainty of obtaining oil in commercial quantities. These tracts, if included, should be valued separately and on a different basis.

Although there is no case exactly parallel in metal mining, the metal-mining engineer refuses to commit himself on the value of a prospective mine. The petroleum engineer may determine the magnitude of the risk and compute mathematically the probability of obtaining oil on a tract of land; but the author is inclined to agree, in a measure, with Rickard[2] that "the doctrine of probabilities has been stultified too often to allow of its being stated as a scientific thesis."

In valuing the proved oil land, the engineer should compute the value of the output of the property based on a drilling program that will bring the maximum return in profits to the investor. It is true that a variation in the drilling program sometimes will greatly reduce the profits eventually gained from a property, but there can be only one maximum value and this is the one to be determined.

CLASSIFICATION OF LAND TO BE VALUED

Before the future annual production can be estimated, it is necessary to classify the land to be valued, to determine the amount of acreage that will support new wells. For this purpose the land is first divided into drilled and undrilled. These two classes of acreage must be valued separately.

Estimating the future production of the old wells usually is not difficult, if production curves are available. Our greatest difficulty lies in making estimates of the probable future production of the proved undrilled acreage. Here we must be guided by underground geologic conditions and by what the new wells probably will produce by comparing the conditions under which they are to produce with the conditions under which the nearby old wells are producing. The undrilled oil

---

[2] T. A. Rickard: Valuation of Metal Mines. International Engineering Congress, 1915.

land may usually be divided into the following four general classes: Proved acreage, probable acreage, prospective acreage, and commercially non-productive acreage. Some engineers use much more detailed classifications. These, the writer believes incompatible with the uncertainty of underground conditions. The following definitions are advanced tentatively:

Proved acreage should include that in which drilling involves practically no risk. The following definition is proposed, which has been modified from that given by R. P. McLaughlin.[3] "Proved oil land is that which has been shown, by finished wells supplemented by geologic data, to be such that other wells drilled thereon are practically certain to be commercial producers."

Probable oil land includes those areas generally adjacent to producing oil and gas wells where the existence of oil is not proved, but where geologic evidence indicates a good chance of obtaining oil in commercial quantities.

Prospective oil land includes those areas usually not adjacent to producing oil and gas wells, where the existence of oil is not proved, but where geologic data justifies drilling a test well. Land in this class is distinguished from the probable oil land by the greater uncertainty of obtaining oil owing, usually, to its location some distance from producing oil and gas wells.

Commercially non-productive oil land is that on which commercially productive wells cannot be drilled at present. The existence of oil under the areas of this class may be proved, probable, or prospective.

Exceptions undoubtedly will be found in every class. For instance, under some conditions, a person may feel warranted to place land in the probable class when it is favorably located geologically, even though it is several miles from producing wells, for the reason that the occurrence of oil and gas with relation to certain geologic structures in that region may be so certain as to make the chance of not obtaining some oil very small. Furthermore, the classification of land may change rapidly, owing to the drilling of new wells, damage by water, or change in price. For example, an area that may be rated as commercially non-productive may become commercially productive and proved with an increase in the price of oil.

FUTURE PRICE OF OIL

The accuracy of any valuation depends on the price that is to be received for the oil, for on it depends the net profit per barrel of oil marketed. A small variation in the price of oil may mean the difference

---

[3] R. P. McLaughlin: Petroleum Industry of California. California State Mining Bureau *Bull.* 69 (1914) 13.

between gain or loss. In fact, since the working out of new and more trustworthy methods for more accurately estimating future oil production, the estimation of the future price has become one of the most uncertain elements to be contended with in oil land valuation.

The engineer, to make sound predictions as to the probable price of oil, even during the immediate future, must possess a broad knowledge of the petroleum situation as regards supply and demand. Either prices will be allowed to adjust themselves in accordance with the law of supply and demand, or they will be manipulated by monopolies or controlled by the Federal Government. If manipulation or government control exists, or if there is a strong probability of their coming into existence, the engineer should be guided accordingly. Otherwise, the question of price must be answered solely by the domestic and foreign oil situation. The past range of prices has often been great, but the future probably will never see such low prices of oil. The market is now more stable because the demand for the commodities made of petroleum is greater and new oil fields are much more scarce and more costly to develop.

The reason for the great demand for oil is primarily because of the great demand for one of its products—gasoline. The great demand for gasoline is created by the phenomenal development of the internal-combustion engine. This development is, probably, by no means, completed. The adoption of oil as fuel by the great navies of the world and the development and adoption of the Diesel engine have greatly increased the demand for the heavier products of petroleum. Very likely the future demand for oil and its products will not decrease.

The upward limit of prices is set by the cost of importing oil and the cost of developing a supply of oil from oil shales, of which there are immense deposits in this country. By considering the status of the industry at the present time and these two limiting factors, the engineer should be able to make reasonably sound estimates of the price of oil for the next few years. Some engineers find it advisable to use the present prices as a basis of estimating the value of the property or to determine the value of the property at several different prices of oil, and thus allow the investor to select the one that, in his judgment, will best meet future conditions.

COST OF PRODUCTION AND DEVELOPMENT

In determining the future net receipts from each barrel of oil, the cost of producing the oil must be subtracted from the gross income or selling price. For the purpose of estimating future production costs, including drilling charges, tankage, and, in fact, every charge that contributes to the final total cost of production, the appraiser should refer to trustworthy statistics and should be able to interpret these statistics in terms of probable future conditions. This, again, requires not only a broad knowledge of the oil industry but also detailed knowledge of costs in the locality where the property is situated.

INTEREST ON INVESTMENT

The proper rate of interest to be received from an investment must be such that capital will be attracted to the enterprise. If the risk attached to the investment is great, the rate of interest on the money invested must be high or investors cannot be found. The returns from oil investments are always speculative to some degree, so the interest demanded is usually high. If there is no risk, the investor can afford to invest his money at the same rate as if he put it in the savings bank at 4 per cent.

The basis of value in oil.lands is net income. The net income for each future year of the productive life of the oil property must be estimated and these future values compared with their real values at the present moment by reducing them to present value at a given rate of interest. This is discount and is the reverse of compound interest, the factor used in the reduction of future values to present values being called the discount factor, which is a very important element in oil-land valuation. By the reverse of discount, or compound interest, the future value of a present income may be determined.

Present value of a future income may be defined as that sum which, when placed at interest at a stated per cent., will equal the income at the date when it is to be realized. Thus, the longer the deferment of an income the less it is worth at the present time, for which reason one can afford to pay more for income to be obtained from the oil from a well drilled now than for the same well drilled a year hence, providing the price of oil remains constant and equal amounts of oil are produced. Furthermore, the longer drilling is postponed the less the net proceeds from the wells are worth to a prospective purchaser at the present time. Other things being equal a property should be drilled as quickly as possible, if the maximum income is to be derived from it. This may not be best from the standpoint of the public, and, if generally practiced by oil producers, would eventually work to their advantage.

The interest required on the investment must be high because risk is attached to the venture. Some engineers consider that the discount used in reducing future income to present value, however, should not be compounded for the reason that to compound a certain present sum to determine its future value means the first year to determine the interest on the principal and thereafter to compute yearly the interest on the principal and accumulated interest earnings. The rate used is a high rate because the capital, or principal, is being risked. This rate should not be applied to the accumulated interest earnings, however, because these are not risked capital. They are earnings and should be considered as such.

The computed maximum value of the property may be considerably less than what actually could be paid for the property for as the returns in the investment are realized they may be reinvested in gilt-edged securities at an accumulative rate of interest.

Amortization of Investment

In investing in an exhaustible resource, the investor expects not only the return of a certain interest on the investment, but also the return of the principal by the time the resource is exhausted. This is called amortization, or retirement of capital, and may be effected by a sinking fund into which annual contributions are made. The sinking fund may be placed at interest, so that the sum of the annual contributions may not be required to equal the total original capital. Although sinking funds may not be established, some attempt must be made to return capital uniformly and justly, where it is possible to estimate the amount of oil recoverable and the hazard of the investment is not too great to make such calculations useless.

A method often practiced by oil companies to determine the rate of retiring the capital invested in both physical property and in the resource is called the "settled production method," and consists of applying a unit value per barrel of settled daily production. The value of the property at any time is the daily production mutiplied by the unit value. The difference in the value determined at any two periods is the depreciation or appreciation according to whether the value has gone down or up.

A modification of this method for the purpose of determining the depletion deduction in connection with the computing of taxable income, is called the "reduction in flow method." The method has been authorized by the Treasury Department, but obviously is unfair when it is remembered that the basis of the method depends on a reduction in the output of an oil property from the existing wells only. No depletion is allowed and, therefore, no capital is retired unless production is decreasing. If production decreases 5 per cent. during the taxable year, 5 per cent. of the capital invested is retired. During the next year, if the decrease is 10 per cent., that percentage of the unretired capital is "written off." As a general rule, the output of an oil property increases for a few months, at least, while drilling of new wells is in progress, and in some fields, production may increase for several years. Still, by use of this method no capital can be retired until the production of the tract begins to decrease. Production of oil means depletion of its recoverable content and every barrel of oil taken from a property exhausts it just that much, and brings it just that much nearer the end of its life. To retire no capital while production is largest and then when production begins to decline, to retire large amounts against a decreasing income not only is inequitable to the oil operator but places the whole enterprise in jeopardy by deferring the amortization to a period when the field is rapidly approaching exhaustion and too late to cover the return of capital.

The producer has made a definite investment in each barrel of recoverable oil. If he can estimate the amount of recoverable oil, he can easily determine the cost per barrel. For every barrel of oil produced, he should retire an amount of capital equal to the original investment in that barrel of oil. This is called the "unit cost method," by which a fixed charge per barrel of oil produced, based on quantity, is assessed. It is sound in principle, not difficult of application, and has been adopted by the Treasury Department in the determination of the depletion deduction in connection with the administration of the income and excess-profits tax laws. This undoubtedly is the fairest and most equitable method of amortizing an investment in a mineral property. The method is suggested in several publications on mine accounting,[4] so has the added weight of precedence.

The basis of this method is to determine the total capital invested in the oil and then divide the estimated recoverable oil into the capital invested; the result is the unit cost. For instance, if the sum of $1,000,000 is invested in the oil under a property, estimated to produce ultimately 10,000,000 barrels of oil, the unit cost per barrel is 10 c. The producer has paid this sum for each barrel of oil under the property. If he sells each barrel of oil for $1.50, his net income for each barrel will be determined by deducting all charges from $1.50. Suppose all charges, excepting unit cost, amount to 40 c. per barrel, his net income is, therefore, $1.

Estimates of future production may be revised each year and a new "unit cost" obtained by dividing the unretired capital by the remaining recoverable oil. The amount of capital to retire during that year on account of depletion will be the unit cost multiplied by the production.

Many oil companies have adopted this system because by its use they are enabled not only to determine the depletion deduction equitably and justly, but also because they are enabled to retire the capital investment at the same rate at which the oil is produced. The only unknown factor in the determination of unit cost is the amount of recoverable oil, and

---

[4] F. Hobart in "The Economics of Mining," by T. A. Rickard and others, 223, 1905.

this can be estimated with a reasonable degree of certainty by the use of methods outlined by Lewis[5] and the author.[6]

Depreciation refers to the wear and tear on physical property and capital invested in it must be retired in addition to the capital invested in the exhaustible resource. The methods of retiring such capital will not be discussed in this paper. The amount retired should be equal to the capital invested minus the salvage or "scrap" value of the equipment.

SALVAGE VALUE OF EQUIPMENT

When the oil is exhausted, a certain amount of physical property will be on hand. The investment in this physical property should have been completely amortized, with no investment remaining except that which can be realized from the disposal of the equipment. This sum is called the salvage, or "scrap," value of the equipment. Ordinarily this "junk" value is not great at the exhaustion of the oil. Furthermore, the proceeds derived from the sale of the "junk" must be discounted to the time of the valuation at a certain rate of interest. Usually the property will have a life of more than 20 years, and the present value of the junk, even at a comparatively low rate of interest, is rather small when compared with other sources of income that must be present before the investment is a good one. Occasionally, the expenses in connection with the abandonment of the property, such as properly plugging the wells, will cost as much or more than the present value of the junk, so that this item in oil-land valuation is ordinarily not important.

---

[5] J. O. Lewis and Carl H. Beal: Some New Methods for Estimating the Future Production of Oil Wells. *Trans.* (1918) **59**, 492.

[6] U. S. Bureau of Mines *Bull.* 177.

Appraisal of Oil Properties

By Earl Oliver,* Ponca City, Okla.

(New York Meeting, February, 1920)

THE term oil property, in this discussion, includes any type of ease-
ment or grant under which petroleum might be produced; it ranges from
the mere right to drill on undeveloped wildcat acreage up to a fully
developed oil property. The values of an oil property, as thus defined,
vary widely according to the use for which it is intended, whether it is
from the viewpoint of the speculator, the fraudulent stock promoter,
the refiner and pipe-line owner, or the oil producer.

The market value of a property is usually a combination of some
two or more of these influences, and occasionally a combination of all
of them, but we prefer to treat each viewpoint as distinct from the others,
allowing the reader to make his own combination in such proportions
as his inclination and property seem to require. This paper will treat
the subject from the viewpoint of the oil producer. However, the other
influences have such bearing on the cost of property to the producer
and on the price he might obtain by its sale as to warrant brief discussion
of them.

Speculation, particularly lease speculation, is a parasitic growth on
the oil industry, healthy enough, but of no economic value. The class
of property usually dealt in for this purpose is the undeveloped lease.
Its speculative value may have some relation to its probable productive
value, but most frequently this value is the product of temporary
excitement due to local development. The lease speculator, seeking out
the trend of development or securing early information regarding a proposed test or a new discovery, immediately secures leases whose market
value will be increased when the existence of such development becomes
more widely known. His profits are purely unearned increment.
Not proposing to spend money in exploration, he can afford to compete
in purchase with the operator, who, in addition to bonus paid, must
spend large sums in testing out his own acreage and that of the nearby
speculator as well. It has a tendency to compel unduly high bonuses.
The so-called "checker-board" system of leasing wildcat acreage by the
larger companies is on the same principle of attempting to secure the

---

* Member Executive Staff, Marland Refining Co.

benefit of money expended by others in testing, but most of them remove the obnoxious holdup feature by contributing toward such testing by others in proportion to the benefit they themselves secure.

Fraudulent stock promotion, as contrasted with lease speculation, is an unhealthy disreputable growth and justifies mention only that attention may be called to the burden it places upon the oil industry. The value of an oil property for this purpose has no real relation to its productive value, the property being selected for its adaptability to a fantastic tale of fabulous earnings to draw money from people with small savings. Consequently, the promoter pays for leases suitable to his purpose prices entirely out of proportion to their probable economic value. Such prices, however, immediately influence owners of surrounding unleased areas when dealing with the legitimate operator.

The lease speculator and fraudulent-stock promoter, while entirely dissimilar in their standards of ethics and respectability, have, therefore, this in common—they tend to place on prospective oil-producing areas, before these get into the hands of the oil operator, values that have little relation to their economic value. The impossibility of determining the exact economic value, the knowledge that other men are paying like prices, and the optimistic spirit of the operator generally lead him to meet these fictitious values and pay prices for undeveloped leases, especially in the vicinity of a new discovery, entirely out of proportion to their chances of being profitably productive.

Prior to 1888, the American oil industry was operated in two distinct divisions: first, that of the producer, who brought the oil to the surface; and, second, that of the refiner, who purchased the crude oil at the well and carried it through all remaining phases, including, in some instances, its sale to the consumer. However, about the date mentioned, would-be competitors of the then dominant refining interest, recognizing the necessity of owning and controlling producing properties as an insurance against disturbance of their crude supply, began to acquire oil-producing properties. The dominant refining interest quickly adopted the same practice, until now the greater percentage of production is owned and controlled as a necessary adjunct to the refining and transporting branches of the industry. New pools, when opened, have diverse ownership but eventually gravitate, to a great extent, into the character of control mentioned. Thus, while to the oil producer, as such, an oil property has value only to the extent of the margin of profit between the cost of the oil as produced and his receipts for it as sold in its crude state, to the refiner and transporter it has a double value—one of which is identical with that of the producer, while the second is that it stabilizes and makes secure his more important business of transporting and refining, provided his producing properties are so situated as to make his production available for his facilities. The value of the property for the latter purpose is frequently of much greater importance than for the former.

VALUE OF OIL PROPERTY TO THE PRODUCER

The value of an oil property to a producer as such (stripped of its speculative feature) is simple in principle and is nothing more nor less than the present worth of the aggregate margin of profit between the expenditures for producing, saving and disposing of the oil in its crude state and the price received for the same. To ascertain such value is a simple matter in accounting—of balancing expenditures against receipts and introducing interest charges. It differs little from the average bookkeeper's problems except that it is reversed. The bookkeeper balances expenditures against receipts on business that is past; the oil-property appraiser balances expenditures against receipts on business of the future. To his bookkeeping ability he must, therefore, add the ability to see into the future—and underground as well.

His principal difficulties are to determine the number of barrels of oil the particular property will produce and the price per barrel he will receive. He has several other factors to consider, such as the cost of developing, and of maintenance and operating, but these are easily disposed of if he is able to accurately forecast the first two named.

Until within very recent times there has been no one well-established, widely used, generally accepted method of determining the probable future production of an oil property. Scientists have evolved general theories as to the laws controlling the accumulation of petroleum that have much merit; in a large way, they are helpful. It must, however, be confessed that in their early application, or rather misapplication, to individual small tracts for valuation purposes the results obtained by them have been disappointing. These theories introduce too many assumptions to make the appraisals safe as a basis upon which to invest large sums of money. Consequently, an experienced oil producer, who had the details of a few properties stored in his mind for the purpose of comparison, was more successful in judging values of oil properties.

But the average oil producer, whose real business is operating the properties he controls, has no inclination nor time to collect data that would give him a wider vision than his own experience furnishes him, and is handicapped to that extent. These two men, the scientific theorist and the experienced oil producer, represent the two types of appraisers that have been known to the oil industry for many years. The difference between them might be better understood by stating that the early petroleum engineer, in purchasing a race horse, would have investigated carefully the history of the dam and sire for several generations back, ascertained what food and treatment the dam had prior to the foaling, would have given a hasty glance at the horse itself and concluded, that in view of the record of its progenitors and the early influence brought to bear upon it, it should be able to do a mile in 2 min. flat, and, accordingly, purchase it on that basis. Whereas, the producer, purchasing the same horse, would send it around the track, time it, and buy it on the speed shown, in ignorance of inherited and prenatal influences.

There is now being evolved a new type of petroleum engineer. He knows the theories of the scientist, but he demands that they shall be measured up with actual results. He knows what percentage of structures in a given locality have been found productive or barren upon being tested. He knows, from compiled data, where oil accumulates. He calculates oil content per acre by counting barrels actually produced from similar areas already exhausted. He measures the numerous scientific theories with actual results. His business is the collection of complete data from innumerable properties so that he may know the habits of petroleum, instead of assuming for it certain habits in keeping with his ideas of what they should be. This new type of petroleum engineer is expected to develop methods of determining the probable productiveness of certain areas with much greater accuracy than the methods now available will permit. However, until such methods and data are available, it is necessary to use the older general principles.

Method of Appraising Oil Property

The property is first divided into developed and undeveloped portions. The developed portion is then subdivided into "settled" and "flush" productions, provided the old and new wells are not so intermingled as to make this impracticable. For both classes it is desirable to have production figures extending back month by month to the completion of the wells. Should these figures not be available, they should at least run back one or two years, if the wells are old enough. There will, of course, in so far as it is available, be a complete history of each well and full data regarding it, including, among other things, date of completion, initial and present productions, thickness and depth of sands, casing record, gas and water production, vacuum application, etc. One purpose is to ascertain whether the composite production is made up of comparatively uniform wells, and whether the production figures as shown month by month represent the regular and natural rate of decline, or whether some unusual condition might have changed the past production from the natural rate. Should there be such condition, its influence is given such consideration as it appears to warrant, and it is eliminated, if possible, from the figures.

On the "settled" production figures, a curve is then constructed covering the entire past life of the property in so far as such figures are available. While properties of the same age differ greatly in their rates of decline, each property, throughout its entire history, is characterized by the same rate of decline; *i.e.*, if during the first part of its history it shows a rapid rate as compared with other properties of like production per well and age, it will have a rapid rate up to the point of exhaustion. After the flush production is off, provided the wells are not unusually large, the rate of decline is so uniform as to make possible, for all practical purposes, the use of some definite percentage each year from the previous year. Thus, some properties will decline at the average rate of 40 per cent. each year from the previous year, while other properties will decline only at the average rate of 15 per cent. Consequently, when on settled production the figures are available, by constructing a curve running back several years, it is comparatively easy and safe to project the curve to the point of exhaustion. Of course, it must be seen that some extraneous influence does not cause the decline to deviate from its natural rate.

For the appraisal of individual settled properties, where something more reliable than mere generalities is desired, a general production curve should not be used; instead the curve of the property itself should be projected. The writer has before him curves of several Bartlesville sand properties 6 to 8 years old in substantially the same district, and which would ordinarily be considered of the same type and subject to the same decline curve. Yet they range from 15 per cent. annual decline on some properties to 40 per cent. on others, which means that the first named, although having no more present daily production than the latter, will produce three times as much as the latter before exhaustion.

Every producing property is a type unto itself and where reliable appraisal is desired no property can be thrown into any general class at so much per barrel. Each property has characteristics that place it above or below the average and as stated on settled production, located apart from new wells, it is desirable that its own production curve be projected.

With flush production, however, this cannot be done since there will not have elapsed sufficient time to indicate a rate of decline. Reliance upon general decline curves cannot be avoided, but care should be taken to select a curve compiled from properties as similar in type as it is possible to secure. As a check, data compiled from similar properties as to yield per acre-foot is helpful for "flush productions," although of too general nature to be of assistance in appraising "settled" production. Briefly, therefore, for "settled" production the curve of the property examined should be projected while for "flush" production the use of a general decline curve compiled from similar properties is permissible.

The undeveloped portion of a property should be viewed from an entirely different angle. This will range from so-called rank "wildcat" acreage up to that which is sufficiently surrounded by production as to be substantially proved. It is rare, however, that any undrilled acreage will be so thoroughly proved as to justify a method of appraisal that will include assumption of a certain number of locations with an assumed initial production per well and with application thereto of a decline curve. Theoretically such method appears satisfactory but it does not work out well in practice. However conservative the appraiser attempts to be such method of appraising generally leads to overvaluation. It is poor practice. Such a method starts with the assumption that the undrilled acreage will be productive and then attempts to call to mind all factors of uncertainty for which discount should be made but some of these will be missed and the property be, thereby, given a higher rating as to certainty of production than, as a rule, it justifies.

The basic assumption on undrilled acreage should be the reverse of the above; *i.e.*, that it is barren, and then such factors should be assembled as will tend to take it out of that class. Perhaps this distinction has not been made clear and it has reference only to the state of mind of the appraiser toward the property, but this tendency toward optimism regarding the probable productiveness of acreage has caused much more money to be spent in the attempt to produce oil than the industry has paid back in the aggregate to the producer. By the very nature of the industry there must be great individual gains and losses but the industry as a whole should pay its own way; and the fact that it does not in the aggregate do so should be more widely recognized and values adjusted accordingly.

It is in this field of undeveloped acreage that the work of the petroleum engineer is in most need of extension. Comprehensive data showing the percentage of seemingly favorable structure that has proved profitably productive, the conformity of subsurface to surface structure in given districts, the percentage of production seemingly off structure, the persistence of sands, the yield per acre-foot, and thorough study of the accumulation of oil as it is actually found to exist rather than a preconceived theory of how it should accumulate, together with a more widely spread knowledge of the results will be helpful to the industry and place undeveloped acreage on a proper footing. Such factors together with many others must be taken into consideration by him who would appraise undeveloped acreage with any degree of safety. This investigation must, however, be made with the cold calculating analysis of the engineer— who looks only at facts, rather than by the scientist whose province is farther afield.

A factor in the appraisal of oil properties of almost equal importance to the amount of oil secured is the price to be received for the oil. Regardless of the views sometimes asserted, the market price of crude oil and the usual fluctuations are influenced by the law of supply and demand. In attempting to forecast the price that might reasonably be expected for the oil output of a property it is, therefore, necessary to assemble and consider the factors that will influence supply and demand. This is a large field; there is no intention to say that an actual price can be forecast, but that the market trend can be reasonably foreseen over at least the next thereafter succeeding 2 or 3 years. However, this phase of the subject must be left for future discussion.

Having determined upon extent and rate of production that might reasonably be expected from a property and the probable trend of the market price, there then enter the cost of development and maintenance and the consideration that should be given the possibility of increasing the amount of oil to be produced by the application of different methods. This, again, is an uncertain field and if the property is fully drilled such possibility frequently no more than offsets the dangers unseen. The equipment that is needed for the permanent operation of the property should be given no credit unless the property is relatively near the point of abandonment.

FACTORS INFLUENCING VALUATION OF OIL PROPERTY

A few of the questions that must be considered by one who would eliminate as much as possible the factors of uncertainty in the purchase of properties, together with the character of information that would be helpful are here given, although the outline is by no means complete.

I. Plans of Purchaser
    1. General scope and business of company
    2. Purpose for which property is needed
    3. Amount of oil needed
    4. Amount of money available to secure it

II. Probable Oil Market Trend
    1. Production: (a) World's past and present production, in detail by countries and districts, and by whom controlled, in each instance giving wells completed and producing. (b) World's probable future production by countries and districts, and by whom controlled.
    2. Consumption: (a) World's past and present consumption, showing distribution by countries and districts and distribution as to use. (b) The world's probable future needs as to countries and as to uses.
    3. Prices: (a) Of crude oil. (b) Of refined products. (c) Margin of profits. (d) Prices to which crude petroleum might go before other products become competitor.
    4. Graphic charts of prices, production, and consumption
    . Possible substitutes and probable influence of same
    5. General consideration of factors that might influence price, and survey of the entire field of market trend

III. Selection of Regions for Exploitation
    1. Geographical location
    2. Geology: (a) General petroleum possibilities. (b) Laws controlling local accumulations. (c) Degree of conformity of production to structure. (d) Persistency, thickness, and characteristics of petroleum-producing strata. (e) Comprehensive data showing actual petroleum extracted per acre-foot from all types of producing strata. (f) Rate of decline of various types of producing properties, with described conditions. (g) Water conditions.
    3. Development: (a) Past history. (b) Maps marked up to date.
    4. Oil: (a) Quality. (b) Amount produced by periods. (c) Market price of same by periods. (d) To whom sold.
    5. Relative cost to operate: (a) Depth and cost of wells. (b) Method of operation. (c) Proximity to supplies and markets.
    6. Relation to transportation systems
    7. Relation to company's plans and its existing properties
    8. Ownership compilations: (a) Production figures and comprehensive ownership data on producing properties. (b) Comprehensive ownership data on non-producing lands.
    9. Records of sales of both producing and non-producing properties to indicate prevailing prices

IV. Selection of Properties. This section deals with especially selected properties that might be worthy of consideration as contrasted with regions dealt with in section III and therefore goes much more into detail in each case.
    1. Geographical and legal description
    2. Geology: (a) Surface. (b) Underground. (c) Persistence, thickness and characteristics of possible producing strata. (d) Water conditions. (e) Application to this property of laws of accumulation peculiar to region. (f) Data showing production per acre-foot from similar already exhausted properties. (g) Data showing probable rate of decline deduced from past history of this property, also from similar already exhausted properties.
    3. Development: (a) History. (b) Well records. (c) Maps marked up to date, including adjoining properties.
    4. Oil: (a) Quality. (b) Amount by periods produced from beginning of development. (c) Market price. (d) Possible markets. (e) Probable amount to be produced. (f) Probable rate of decline.
    5. Relative cost to operate: (a) Depth and cost of wells. (b) Production expense. (c) Method of operation. (d) Equipment. (e) Proximity to supplies, labor, and markets.
    6. Relation to transportation systems
    7. Relation to company's plans and its existing properties
    8. Records of sales of similar properties to indicate prevailing prices

V. Terms of Lease
    1. Rate of royalty   3. Development requirements
    2. Term to run   4. Unusual conditions

VI. Taxation questions and general governmental conditions and safeguards

VII. Conditions of title

DISCUSSION

CARL H. BEAL, San Francisco (written discussion).—I heartily endorse the statement that the petroleum engineer should look at oil properties, if possible, from the viewpoint of the practical operator. Too much appraising has been based on theory and not enough on facts. If the facts were not available, this process of appraising might be condoned; but the files of practically every oil company contain considerable data of importance that should be studied in connection with oil production.

Mr. Oliver has brought out one important fact; that is, the influence of the speculator and the fraudulent stock promoter on the selling price of leases. These men greatly increase the amounts that must be paid for proved and wildcat land. Incidentally, this fact makes difficult the adoption, by the Treasury Department, of sales values of actual transactions as a method of limiting the values placed on developed and partly developed oil land. Some time ago this method was suggested as one that would be used as a possible limitation by the Department in checking valuations made for the purpose of determining the depletion deduction. High market values nearly always prevail in a new field where the excitement is running high. Ranger a year ago is a good example. The actual amounts of oil that could be obtained from some of the partly developed leases would lack much of measuring up to the average sales values of surrounding properties. Market values much higher than intrinsic values indicate a speculative period, whereas intrinsic values in excess of market values indicate a period of stagnation, when oil-property transactions are rare. The market values may fluctuate rapidly because of new discoveries, but the actual value reposing in the oil does not change, except as the factors influencing such values change; for instance, an increased demand, lower or higher drilling costs, etc. An average of exchange values over a long period of years would equal the actual value of the properties, for the prices paid during periods of excitement and periods of depression will result in an average that fairly represents actual value.

Mr. Oliver separates the developed from the undeveloped parts of the property and then the "settled" from the "flush" production. It often has been very difficult for me to separate the flush and settled productions, for there seems to be no particular line of separation between them. In some places the wells may be called settled after 6 months, in others after 1 year, whereas other wells may be called settled from the very beginning. The age of a well when its production is settled varies with the initial production and the conditions under which the oil is produced. A well drilled in the Lima-Indiana field coming in at 10 or 15 bbl. a day very likely will have but a short period of flush production;

whereas a 5000-bbl. well in California may be irregular in its production for several months, or even 2 or 3 years. Flush production is a relative term—of use in describing production of individual properties, but difficult to define.

The future production of wells still in their youth may be estimated from data showing the relation of the production of a well the first day or month to its output during the first year. Most production curves used in estimating the future production of oil properties are based on the rate of decline of wells, which produce different amounts the first year. If the relationship between the production of the first day or the first month and the production of the first year can be determined, the estimate of the future production of new wells may be much more accurately made than by the method suggested by Mr. Oliver. I found by studying many records in Oklahoma that the average well in the fields east of Cushing would produce daily, in the first year, about 25 per cent. of its initial daily production. In other words, if the initial production of a well were 100 bbl., its first year's production would be about 9000 bbl.; although this ratio changes for wells of different initial output, it may be used in roughly estimating the amount a new well would make the first year. In California, I find this ratio to be much different; for instance, in some fields a 100-bbl. well will make between 50 and 75 bbl. daily the first year.

The statement that where individual properties are being appraised a curve of the property itself should be projected, for estimating the future production of that property, cannot be too strongly emphasized. Some of us are prone to apply average curves, but the curves of the property itself are very much more trustworthy and simpler to use. The details of preparing such a curve are simple. The annual production should be obtained, if possible, for the life of the property, and the average number of wells producing divided into this annual production. The resulting amounts are the annual production per well of the property. In almost any property where drilling has been carried on at a normal rate the peak of production of the property will occur only a few years after the first well has been drilled; often it will occur in the first or second year. From that time on, regardless of the rate at which the property is drilled, the annual production per well decreases. This curve, projected into the future, will show the estimated annual production per well. The only remaining step in the problem is to estimate the number of wells that will be producing each year in the future. After these annual amounts have been determined, they are multiplied by the estimated annual production per well; the product is the estimated future annual production.

Possibly one reason that this method has not been used more is because some persons believe that, as all wells in the field have not been drilled, the daily production per well will be upheld by the yield of new wells. This objection is not valid, especially if the production per well in the district has begun to decline on account of interference. After a field or property has attained a certain age, the decline in the daily production per well remains practically unchanged, regardless of the number of new wells drilled. It is necessary, however, that the wells shall be drilled close enough to be affected by drainage. In a field where the productive sand is lenticular, or made up of several disconnected lenses, or if the wells are widely spaced, this method cannot always be used.

Referring to Mr. Oliver's statement that yield per acre-foot should be obtained, I have not been able to obtain sufficient data to prove that such statistics are of any value. Production per acre, I believe, is very much preferred, for the reason that it is practically impossible for anyone to determine the portion of a sand that produces the oil. The thickness is not measured accurately in the first place. Some parts of the sand are more porous than others, and some parts produce water. It is possible that the pressures in the various members of an oil sand or zone may be different. The first production of a well may come from two-thirds of the sand, until the pressure is reduced to that of other members of the sand, and the next portion of the production may be expelled from all parts of the sand. Statistics on production per acre from a sand like the Bradford sand in northwestern Pennsylvania probably would be of some value, but most of the oil sands with which I am acquainted are so irregular that statistics of production per acre-foot cannot be compiled that are of any particular use.

The reference to the necessity of estimating future price of oil in oil-land appraisals brings up one of the most difficult factors in valuations. Although, as Mr. Oliver says, the market trend can be reasonably foreseen over at least the next 2 or 3 years, to forecast future price for several years, as is necessary in oil-land appraisals, is not in accordance with the exactitude that engineers desire in their profession. The fact remains, however, that some estimate must be made, and the engineer is the man who must make it. In making the estimate, it is necessary for him to consider the economic side of the oil industry; a small variation in the price of oil may mean the difference between gain or loss. In fact, since the working out of new and more trustworthy methods for more accurately estimating future oil production, the estimation of the future price has become one of the most uncertain elements to be contended with.

R. H. JOHNSON, Pittsburgh, Pa.—Stress on division of the flush and settled periods is an unfortunate habit. It is far better for us to think of the thing as a curve than to get this other impression. Mr. Oliver takes a regular percentage decline for several years after he says the well is settled; in the flush period we are not offered any particular guidance. That is unfortunate. It is in the flush period

that a great many purchases are made; and the man who is trying to
find out the values of properties, the man who wishes the services of the
appraiser, is very much concerned with that period. If we think
of this whole thing as a curve and study the problem as a whole, we are
on a more healthful basis than this artificial distinction.

Mr. Oliver has exaggerated the period of time that one is safe in using
the same percentage of decline year to year. It is true that we approach
such a curve in the old age of the well, but Mr. Oliver starts too early
to assume that we can take a constant percentage of decline.

Mr. Oliver should have given some attention to the compound
discount factor in working out present worths. He says that the prices,
in the long run of exchange values, will average the productive value, that
is, the periods of inflation will cancel the periods of depreciation, so that
in the long run productive values must be the same as exchange values.
This is not correct. Except in periods of excitement, we generally make
a considerable allowance for risk. I never like to recommend the purchase of a property at exactly what I think it is worth. I always feel that
we should recommend a liberal allowance for risk to a purchaser. To
be sure, there are inflated values, but they are recognizable. In fact, it
is not uncommon to find old operators who say that they buy on the
basis of paying out in four, or some other number of years. To be sure,
these men have not made a regular allowance for compound discount.
It is probably partly involved in this expression of theirs, although
they do not realize it, but it is also partly the fact that they wish to
allow an ample amount for risk. That should be the custom of all
buyers, with the possible exception of the one who has a large refinery
that must be kept going and who must protect himself from loss at other
points.

Variation in Decline Curves of Various Oil Pools

By Roswell H. Johnson, M. S., Pittsburgh, Pa.

(New York Meeting, February, 1920)

The Manual of the Oil and Gas Industry, under the Revenue Act of 1918, published by the Treasury Department for the guidance of oil companies in preparing their estimates of future recoverable oil for the purpose of calculating depletion, gives the first large public collection of comparative decline curves for the whole country. It is a matter of both scientific and practical interest to so arrange these data that the pools can be readily compared. There are certain difficulties in such a comparison, however: (1) The economic limit varies from 50 to 2000 bbl. a year in different pools taken. (2) Because of this variation in economic limit, the range of data shown makes comparison possible only for wells of intermediate size.

In order to be as inclusive as possible, I have taken as an expression of the rate of decline the amount of oil produced in the period during which a well drops from 3000 to 2000 bbl. a year. No period of smaller production could be used because of the high economic limit in California; and no period of larger production and yet include the small well areas of the Appalachian. As it is, the Lima-Indiana wells are excluded. In a few instances, curves were extrapolated to obtain the reading, where the curve seemed regular enough to warrant it.

In general, the amount of oil, in barrels, produced while a well declined from 3000 bbl. to 2000 bbl. a year is shown in Table 1.

TABLE 1.—*Production of Wells During Decline 'from* 3000 *to* 2000 *Barrels a Year*

Field	Minimum Barrels	Median Barrels	Mean Barrels	Maximum Barrels
California...................	2,500	5,800	6,260	10,200
Gulf Coast (Saratoga only).....	2,160		3,930	5,700
South Mid-Continent..........	1,700	2,700	2,666	3,600
Gulf Cretaceous..............	1,000	1,500	2,175	5,600
Mid-Continent	600	1,350	1,785	3,400
Rocky Mountains.............	700	1,310	3,170	7,500
Appalachian.................	450	1,225	1,489	2,750
Illinois.....................	410	1,288	1,285	2,200

Amount of oil produced while well declined from 3000 to 2000 bbl. a year.

Appalachian Field

	BARRELS
Big Injun sand, Roane Co., W. Va.	2550
Gordon sand, Greene Co., Pa.	2300
Berea sand, Lincoln Co., W. Va.	1900
Gordon sand, Wetzel Co., W. Va.	1850
Shinnston pool, Harrison Co., W. Va.	1400
Clinton sand, Wayne and Hocking Co., Ohio	1300
Wayne Co., Ky.	1150
Gore pool, Perry and Hocking Co., Ohio	1050
St. Mary's pool, Washington Co., Ohio	850
Dorseyville, Allegheny Co., Pa.	500
Ragland, Ky.	500
Irvine, Ky.	450

Illinois Field

	BARRELS
Dennison pool, Lawrence Co., Ill.	2200
Siggins pool, Cumberland Co., Ill.	2075
Robinson pool, Crawford Co., Ill.	1550*
Carlyle & Sandoval pools, Clinton and Marion Co., Ill.	1500*
Upper Lawrence Co., Ill.	1500
Birds-Flatrock pool, Crawford Co., Ill.	1425*
Pike Co., Ind.	1150
Kirkwood pool, Lawrence Co., Ill.	1150
Johnson pool, Clark Co., Ill.	1140*
Sullivan pool, Sullivan Co., Ind.	700
Plymouth pool, McDonough Co., Ill.	630
Westfield pool, Clark Co., Ill.	410*

* Extrapolated.

Oklahoma-Kansas Field

	BARRELS
Bird Creek-Skiatook district, Okla.	3400
Glenn pool, Okla.	3200
Okesa district, Okla.	2850
Avant-Ramona district, Okla.	2850
Cleveland district, Okla.	2800
Bartlesville-Dewey district, Okla.	2000
Okmulgee district, Okla.	1800
Blackwell district, Okla.	1500
Muskogee-Boynton district, Okla.	1350
Eldorado district, Kans.	1300
Augusta district, Kans.	1200
Cushing district, Okla.	1110
Nowata district, Okla.	1100
Garber pool, Okla.	1100
Adair district, Okla.	1000
Neodesha district, Kans.	600

South Mid-Continent Field

BARRELS

Burkburnett pool	3,600
Electra district	2,700
Healdton pool	1,700

Gulf Cretaceous Field

Corsicana pool	5,600
Mooringsport pool	2,100
Marion Co., Tex	1,600
DeSota parish, La	1,400
Vivian pool, La	1,350
Red River district, La	1,000

Wyoming Field

Salt Creek pool	7,500
Grass Creek pool	1,310
Elk Basin pool	700

California Field

BARRELS

Kern River pool	10,200
McKittrick district	10,000
Olinda district	10,000
Old Santa Maria pool	9,500
Fullerton-La Harba pool	9,000
Twenty Five Hill pool	9,300
Maricopa Flat pool	6,700
West Side Coalinga pool	6,000
Salt Lake pool	5,800
East Side Coalinga pool	5,500
Belridge pool	4,500
Lost Hills pool	4,000
Buena Vista Hills pool	4,000
Whittier district	3,900
West Coyote pool	3,500
Fellows-Midway district	3,000
Shields Canyon district, Ventura Co.	2,500

The Gulf Coast data are given with two different bases of reference, so that all the districts are not mutually comparable. Those under each head may, however, be compared. Amount of oil produced while well declined from 3000 to 2000 bbl. a year: Saratoga Rio Bravo normal spacing, 5700 bbl.; Saratoga town lot spacing, 2160 bbl. Amount of oil produced while well declined from 500 to 300 bbl. a month:

BARRELS   BARRELS

Batson	4780	Humble	1680
Evangeline	4230	Vinton	1530
Sour Lake	3075	Goose Creek	1230

For the sake of finding how the Lima-Indiana pool ranks with other small-well pools, Table 2, based on the oil produced while the wells are declining from 500 to 100 bbl. a year, was prepared.

TABLE 2.—*Production of Wells During Decline from 500 to 100 Barrels a Year*

Field	Minimum Barrels	Median Barrels	Mean Barrels	Maximum Barrels
Appalachian	450	1700	1655	3150
Lima-Indiana	965	1278	1315	1990
Illinois	400	1255	1171	2100

Amount of oil produced while well declined from 500 to 100 bbl. a year in the various pools is as follows:

Appalachian Pool

	BARRELS
Fifth sand, Pa.	3150
Keener sand, Jasper Ridge pool, Monroe Co., Ohio	2850
Speechley sand, Pa.	2800
Bradford sand, Pa.	2550
Floyd County, Ky.	2050
Ragland, Ky.	2000
Gordon sand, Allegheny Co., Pa.	2000
Berea sand, Lincoln Co., W. Va.	1800
Hundred foot sand	1750
Gordon sand, Greene Co., Pa.	1700
Gordon sand, Wetzel Co., W. Va.	1700
Berea sand, Jefferson Co., etc., Ohio	1600
Big Injun sand	1400
Keener sand, St. Mary's pool, Washington Co., Ohio	1100
Wayne Co., Ky.	850
Shinnston, W. Va.	600
Clinton sand, Perry and Hocking Co., Ohio	600
Irvine, Ky.	600
Dorseyville, Pa.	450
Clinton sand, Hocking and Wayne Co., Ohio	450

Trenton Pool

	BARRELS
Hancock Co., Ohio	1990
Wood Co., Ohio	1410
Seneca Co., Ohio	1380
Adam Co., Ind.	1300
Ottawa and Lima Co., Ohio	1255
Sandusky Co., Ohio	1230
Grant Co., Ind.	990
Mercer Co., Ohio	965

Illinois Pool

	BARRELS
Siggins pool, Ill.	2100
Robinson pool	1850
Johnson pool, Ill.	1780
Gibson Co., Ind.	1425
Westfield pool, Ill.	1270
Birds-Flatrock pool	1240
Pike Co., Ind.	835
Carlyle and Sandoval, Ill.	410
Plymouth pool, Ill.	400
Sullivan Co., Ind.	400

Caution must be used in avoiding conclusions based on differences not clearly in excess of probable error, yet the following conclusions seem to be warranted from the differences which are large and consistent.

1 The widespread impression of the much greater persistence of the California field is fully borne out in general, yet a considerable variation is shown.

2. The great importance of the thickness of pay is well borne out by the great contrast, in Wyoming, between Salt Creek on the one hand and Grass Creek and Elk Basin on the other.

3. The Gulf Cretaceous shows a great contrast between the persistent Corsicana and much less persistent northern Louisiana fields.

4. The south Mid-Continent field shows an excellent persistence at Electra, Burkburnett, and Healdton. The more recent close drilling in Burkburnett and the inclusion of the Ranger, Desdemona and Caddo, Texas, pools will cause this field to show less favorably in the future.

5. The three fields of oldest geological age all show poor persistence. The hypothesis that this is characteristic of fields older than Devonian is proposed. This should be expected theoretically, as there should be an increased cementation in older beds, so that the low pressures in old wells become impotent to expel oil. The fields in question are the Lima-Indiana of Ordovician age and the Clinton and Hoing sands (Plymouth pool) of Silurian age.

6. The poor showing of the Appalachian fields, which are relatively younger (Dorseyville and Shinnston), is probably due to the method employed, which bases the future history of all wells on the performance of the small wells in the early history of the pool. These small wells, having in general a thinner pay, have a more rapid decline than the typical wells after they have reached the same size.

7. The most interesting result is that the Appalachian is apparently not more persistent than the Mid-Continent, as had been supposed, when wells of the same size are compared. The long life of the Appalachian wells is mainly the result of the lower economic limit. The Mid-Continent wells will show a longer life as the price in that field rises and so puts down the "economic limit" of the wells. This consideration is a favorable one in the appraisal of Mid-Continent properties.

8. There is a great variation from pool to pool within the field. It follows that the attempt to appraise a property by applying to it the barrel-day price of a property in some other pool or sand or an average from many pools or sands is unwarranted where the data permit an analytical appraisal.

9. Since the rate of decline is not constant, the barrel-day price unit changes during the life of a property, therefore barrel-day prices are not comparable except for properties of similar age and size as well as the same pool.

DISCUSSION

M. L. REQUA, New York, N. Y. (written discussion).—In paragraph 7 of his conclusions, the author states a fact, that is well known to many people, but which I think, is overlooked by the general public; that is, that the line of an oil field bears direct relation to the price of the product. In other words, with every advance in the price per barrel there will be more barrels made available in the form of oil that cannot be produced except at high prices. There is a dead line, of course, beyond which no production will take place. This dead line is well illustrated by the practice, in certain fields, of "flooding" with water and driving the oil to the surrounding wells. When the surrounding wells begin to pump water, the end has been reached, of course, regardless of price. Again, regardless of price, I think it entirely feasible to construct a decline curve to the point of exhaustion. Whether the property will operate in the latter years of this curve is dependent entirely on the price at which the product can be sold.

CHESTER W. WASHBURNE, New York, N. Y. (written discussion). Further explanation of Table 1 would be appreciated. It is not clear how a well that declines from 3000 to 2000 bbl. a year could produce less than 2000 bbl.; probably I do not understand just what the author's figures represent. The comparisons in these tables and the resulting conclusions are most useful in appraisal work. Most scientific investigators, doubtless, already have reached Professor Johnson's seventh conclusion: that Mid-Continent wells will show as long life as Appalachian wells of equal size. Moreover, there is every reason to expect a closer approach of price per barrel in the two fields, and an increase in the "economic limit" of Mid-Continent wells. That region is today the best part of the United States in which to buy oil lands on current market prices of production.

CARL H. BEAL,* San Francisco, Calif. (written discussion).—The value of oil properties depends principally on the amount of oil they will produce and the rate at which this oil is to be obtained. This is a self-evident fact and requires no proof. Consequently, such studies as those made by Mr. Johnson are of the greatest value and interest, for they furnish a definite comparison of the rates at which wells in different fields will give up their oil. Studies of this kind show the importance, furthermore, of studying the effect of various factors influencing and controlling the amount of oil that may be obtained from an oil sand, and the rate at which this oil will be given up.

Mr. Johnson's comparison of the rate at which oil is obtained in different fields would be more illuminating and valuable if augmented

---

* Petroleum geologist and engineer.

with a similar comparison of the amounts that the different fields will produce per acre. Other factors being equal, the amount of oil that may be produced by an acre of land depends on the initial production of the wells drilled to this sand, and the rate at which the oil is produced, for ultimate production, under such circumstances is a function of initial production. These statistics may be obtained without great difficulty from the same data used by Mr. Johnson in the preparation of his paper; in fact, such a comparison could be made much more easily than the one of the decline curves.

This study shows not only the necessity of more investigation along this line, but the necessity of getting at the fundamental influences that control production in different oil fields. Certainly the variation shown by Mr. Johnson to exist can be laid only to the factors influencing production, or the different conditions under which the oil is produced in the oil fields of the United States. If all fields existed under the same conditions originally and development and production were carried on in the same manner, the decline curves would be identical. As they are not identical, the conditions affecting the production of oil must be different, for development and production conditions are usually approximately the same in all fields.

As each decline curve possesses its particular shape because of the composite results of the factors influencing oil production, it is essential that we strive to learn the effect of these different factors on the decline of oil wells, so that, with a few of the more important factors known, we can predict with approximate accuracy the decline of the wells in a new field. For instance, let it be assumed that the composite effect of all the important factors influencing the rate of production in a certain field is known, the thickness of the oil sand is not variable, the wells are spaced a certain distance, and the depth is fairly uniform. By an analysis of these data one can determine in what way almost any important factor affects the ultimate production per acre, and the rate at which the oil is obtained. If the influence of each production factor in the field can be measured, the estimation of the possibilities of properties in other fields, if one or more of the important production factors are similar, will be greatly facilitated. The problem is to determine the individual influences of the different factors. We already know that certain production factors have certain specific influences upon the output of wells; for instance, large initial production will create a tendency toward a large ultimate production, whereas a small initial production will create a tendency toward a small ultimate production. Thick and thin sands react in the same manner, respectively.

If the individual effect of these conditions, or production factors, can be determined, there should be no great difficulty in estimating the general tendency that will be followed by producing wells in new fields, providing some of the more important factors are known. For instance, take the new Ranger field in north Texas. More than a year ago it was evident to anyone familiar with the influences of different production factors that the wells of Ranger would have a rapid decline and produce but small amounts per acre. In fact, the particular decline that these wells would follow could have been forecast with fair certainty, for the reason that the conditions in southeastern Ohio in certain localities were practically equivalent. In the latter district, the high rock pressure, probably due to the great depth, and the thin and rather porous sand favored rapid decline and small ultimate production. Unquestionably, if any new field were found to exist under approximately the same conditions, we could expect approximately the same production rate. As thick sands were not common in Ranger up to a year ago and as the depth was approximately the same as in the southeastern Ohio fields, the rate of decline could be expected to be approximately the same as that found in southeastern Ohio.

Mr. Johnson proposes the hypothesis, in his fifth conclusion, that it is characteristic of fields older than Devonian to show poor persistence. Any such hypothesis is unnecessary. It usually is a self-evident fact that these fields will not produce as much oil per unit as the fields of younger geological age. As a rule, the oil is less viscous, the sands thinner, less porous, and more compact.

The word "persistence" is misleading, for it indicates length of life and the length of life depends, to a certain extent, on the price of oil. It would be better to express the productiveness of fields in proportion per acre. Furthermore, the word persistence, indicating length of life, when applied to fields is very misleading, for the reason that the life of a field or of a tract of land is roughly proportional to the size of the field, or to the size of the tract. Life or persistence should not be expressed for a field, for the life of a field or of a tract depends on the rate at which that field or tract is drilled up and on the margin of profit derived from the oil. These terms may easily be expressed to signify the length of life of wells, for the production of individual wells of limited life make up the production of the tract or of the field. The great length of life in some eastern fields is due, first, to the price of oil and, next, to the size of the fields and the slowness of drilling. The life of individual wells possibly has been lengthened by the price of oil. Probably in very few cases will the life of a well be as long as that of the field or of the tract.

Mr. Johnson's conclusions on the fallacy of the barrel-day price of a property cannot be emphasized too strongly. This method of purchasing producing properties is a gage of doubtful value, and has no engineering basis. As a general rule, it will be found that, as the prospects of future production become better, the value of the property, as determined by the barrel-day price, will automatically be reduced, for the reason that the users of this method of valuing properties do not accurately gage the quantity of future production available. Even though the possibilities are accurately gaged, the value of the oil in the undrilled portion of the tract cannot be easily expressed in the barrel-day price without making an engineering appraisal. If it were possible to gage accurately the prospects of obtaining oil in the drilled and undrilled portions of the tract, and the barrel-day price were raised or lowered accordingly, such a method would be worth while. It is not to be denied that the method has some merit as a rough gage of oil-land values.

R. H. JOHNSON.—Mr. Washburne asks how it is possible that any of these figures should be less than a year's production. The reason is that a year might be represented by one section of the curve and another year might be represented by a section of the curve that overlaps the first, instead of leaving more or less of a gap outside. It means that less than a year suffices to bring about the stated drop in average.

When Mr. Beal asks for the acre-yield data, he is asking for the impossible, so far as the manual is concerned, because it does not give the acreage. It was desirable to have that information in the manual, but those who got out the manual had a large task to accomplish in a limited period of time, so that the acre-yield together with other desirable data could not be obtained.

I want to make a plea for that word "persistence;" we need a word for this attribute, which is extremely important. It is an attribute we have to talk about and handle, and it seems to me we should have a name for it. What would be helpful would be a better name, but until we get a better one, we need this one.

THE CHAIRMAN (RALPH ARNOLD, Los Angeles, Calif.).—The term "persistence of a well" or "persistence of a field" would overcome Mr. Beal's criticism.

Application of Taxation Regulations to Oil and Gas Properties

By Thomas Cox, New York, N. Y.

(St. Louis Meeting, September, 1920)

This paper makes no claim to any new idea; it simply reviews the Treasury Department Regulations pertaining to the practical application of depreciation and depletion and other allowances governing taxation of oil and gas properties. Other methods may be existent, but as they may not conform to the legal status they must be discarded.

In complying with the present laws governing the industry with regard to taxation and the allowable deductions therefrom, the following considerations are essential: Depletion, depreciation, amortization, other allowances, and items not deductible.

It is definitely understood that *depletion* is the loss or exhaustion sustained in the continuous operating of an oil and gas property, and that each unit of oil or gas taken out reduces the value of the property until its final exhaustion. Depletion applies only to the natural deposits of oil and gas due to their removal in the course of exploitation of any property.

Depreciation is defined to cover the waste of assets due to exhaustion, wear and tear, and obsolescence of the physical property, and is separate and distinct from depletion; its allowance is that amount which should be set aside for the taxable year in such sums as for the useful life of the property will suffice to repay its original cost—or its value as of Mar. 1, 1913, if acquired by the taxpayer before that date—less the salvage value at the end of such useful life.

Amortization is allowed for such facilities as were built or acquired on or after Apr. 6, 1917, for the production of articles contributing to the prosecution of the war and, in the case of vessels, those built and acquired after that date. The amount to be extinguished, in general, is the excess of the unextinguished or unrecovered cost of the property over its maximum value under stable post-war conditions.

Claims for amortization must be unmistakably differentiated in the returns from all other claims of depreciation. The taxpayer is also required to furnish full information with the claims for amortization to the full satisfaction of the Commissioner. Further reference is directed to the specific rules and regulations for making these claims.

Other allowances are: cost of development, all operating expenses, repairs, taxes, losses, personal services, bonuses to employees, damages, abandoned wells, same as individuals.

Items not deductible are: donations to employees, losses in illegal transactions, indeterminate oil losses, accrued deductions not charged in prior years, depletion for past years.

ACCOUNTS

In order to carry out the intention of the Government regulations, and to render the returns properly, it is essential that books of accounts be kept to conform to the schedules issued by the Treasury Department.

Every taxpayer claiming and making a deduction for depletion and depreciation of mineral property shall keep accurate ledger accounts in which shall be charged the fair market value as of Mar. 1, 1913, or within 30 days after the date of discovery, or the cost, as the case may be, of the property, and of the plant and equipment, together with such amounts expended for development of the property or additions to plant and equipment since that date as have not been deducted as expense in his returns.

These accounts shall be credited with the amount of the depreciation and depletion deductions claimed and allowed each year, or the amounts of the depreciation and depletion shall be credited to depletion and depreciation reserve accounts, to the end that, when the sum of the credits for depletion and depreciation equals the value or the cost of the property plus the amount added thereto for development or additional plant and equipment, less salvage value of the physical property, no further deduction for depletion and depreciation with respect to the property will be allowed.

Because of the fact that depreciation and depletion deductions are applied against different capital sums, which are usually returnable at different rates, it is essential that these accounts be kept separately; that is, the cost or value of the physical property subject to depreciation with deductions for depreciation enter into one account, while the cost or value of the property (exclusive of physical property) together with additions for such development costs as have not been charged to current operating expenses or deducted as depletion, enter into a separate account.

If dividends are paid out of a depletion or depreciation reserve the stockholders must be expressly notified that the dividend is a return of capital and not an ordinary dividend out of profits.

It is, therefore, necessary to reflect in the books of accounts and records such items as are required to be filled in on the Treasury Department questionnaire, in so far as it pertains to the taxpayer.

Maps that accompany the records and statements must be sufficient to show the property in relation to section, township, and range lines, and should have the name of the state, company, or individual, scale of map, date of survey, and points of compass. All wells should be located and company property designated to distinguish it from the property of adjacent owners. The character of the wells should be properly indicated by standard symbols explained in marginal note.

ESTABLISHING VALUE OF PROPERTY

Determination of Cost of Deposits

In any case in which a depletion or depreciation deduction is computed on the basis of the cost or price at which any mine, mineral deposit, mineral rights, or leasehold was acquired, the owner or lessee will be required, upon request of the Commissioner, to show that the cost or price at which the property was bought was fixed for the purpose of a bona-fide purchase and sale, by which the property passed to an owner, in fact as well as in form, different from the vendor.

No fictitious or inflated cost or price will be permitted to form the basis of any calculation of a depletion or depreciation deduction, and in determining whether or not the price or cost at which any purchase or sale was made represented the actual market value of the property sold, due weight will be given to the relationship or connection existing between the person selling the property and the buyer thereof.

Determination of Fair Market Value

A determination of the fair market value of an oil or gas property (or the taxpayer's interest therein) is required:

1. In connection with the computation of depletion allowances: (a) As of Mar. 1, 1913, in the case of properties acquired prior to that date; and (b) at the date of discovery, or within 30 days thereafter, in the case of oil and gas wells, discovered by the taxpayer on or after Mar. 1, 1913, and not acquired as the result of purchase of a proven tract or lease where the fair market value of the property is disproportionate to the cost.

2. In connection with computing the amount that may be included in paid-in surplus, as of date of conveyance, where the tangible property has been conveyed to a corporation by gift or at a value accurately established or definitely known as at date of conveyance clearly and substantially in excess of the cash or of the par value of the stock or shares paid therefor.

3. In connection with the computation of profit and loss from sale of capital assets in the case of properties acquired prior to Mar. 1, 1913.

Where the fair market value of the property at a specified date, in lieu of the cost thereof, is the basis for depletion and depreciation deductions, such value must be determined, subject to approval or revision by the Commissioner, by the owner of the property in the light of conditions and circumstances known at that date, regardless of later discoveries or developments in the property or in the methods of extraction.

No rule or method of determining the fair market value of mineral property is prescribed, but the Commissioner will lend due weight and consideration to any or all factors and evidence having a bearing on the market value, such as: (a) Cost, (b) actual sales and transfers of similar properties, (c) market value of stock or shares, (d) royalties and rentals, (e) value fixed by the owner for the purposes of the capital stock tax, (f) valuation for local or state taxation, (g) partnership accountings, (h) records of litigation in which the value of the property was in question, (i) the amount at which the property may have been inventoried in probate court, (j) disinterested appraisals by approved methods, (k) other factors.

The decline curve method is one of the most reliable for making appraisals of oil properties. By this, one can estimate and compute the recoverable oil contents of the property and thus arrive at a reasonable unit cost for making the proper annual depletion charge. This method has been tested and has proved efficient and acceptable. A constant record is thus provided for all future variations and additions to the property.

REVALUATION OF PROPERTY NOT PERMITTED

The cost of the property or its fair market value at a specified date, as the case may be, plus subsequent charges to capital sum not deductible as current expenses, will be the basis for determining the depletion and depreciation deductions for each year during the continuance of the ownership under which the fair market value or cost was fixed; and during such ownership there can be no revaluation for the purpose of this deduction. This rule will not forbid the redistribution of the capital sum over the number of units remaining in the property, where erroneous estimates have been revised with the approval of the Commissioner.

Valuation of Fee under Lease

The valuation of a fee ownership in oil or gas land under lease acquired prior to Mar. 1, 1913, will have to do with the equity in its oil and gas contents remaining to the owner of the fee title after deducting the value of the lessee's rights. But subsequent investments or discoveries by the lessee will not affect the lessor's valuation.

Proof of Discovery and Allowances

The following articles in Regulations 45 have been amended in Treasury Decision 2956, to read as follows:

Article 220, *Oil and Gas Wells.*—Section 214 (a) (10) and section 234 (a) (9) provide that taxpayers who discover oil and gas wells on or after Mar. 1, 1913, may, under the circumstances therein prescribed, determine the fair market value of such property at the date of discovery or within 30 days thereafter for the purpose of ascertaining allowable deductions for depletion. Before such valuation may be made the statute requires that two conditions precedent be satisfied:

(1) That the fair market value of such property (oil and gas wells) on the date of discovery or within 30 days thereafter became materially disproportionate to the cost, by virtue of the discovery, and

(2) that such oil and gas wells were not acquired as the result of purchase of a proven tract or lease.

Article 220 (a) *Discovery, Proven Tract or Lease, Property Disproportionate Value.*— (1) For the purpose of these sections of the Revenue Act of 1918, an oil or gas well may be said to be discovered when there is either a natural exposure of oil or gas, or a drilling that discloses the actual and physical presence of oil or gas in quantities sufficient to justify commercial exploitation. Quantities sufficient to justify commercial exploitation are deemed to exist when the quantity and quality of the oil or gas so recovered from the well are such as to afford a reasonable expectation of at least returning the capital invested in such well through the sale of the oil or gas, or both, to be derived therefrom.

(2) A proven tract or lease may be a part or the whole of a proven area. A proven area for the purpose of this statute shall be presumed to be that portion of the productive sand or zone or reservoir included in a square surface area of 160 acres having as its center the mouth of a well producing oil or gas in commercial quantities. In other words, a producing well shall be presumed to prove that portion of a given sand, zone or reservoir which is included in an area of 160 acres of land, regardless of private boundaries. The center of such square area shall be the mouth of the well, and its sides shall be parallel to the section lines established by the United States system of public-land surveys in the District in which it is located. Where a district is not covered by the United States land surveys, the sides of said area shall run north and south, east and west.

So much of a taxpayer's tract or lease as lies within an area proven either by himself or by another is a "a proven tract of lease," as contemplated by the statute, and the discovery of a well thereon will not entitle such taxpayer to revalue such well for the purpose of depletion allowances, unless the tract or lease had been acquired before it became proven. And even though a well is brought in on a tract or lease not included in a proven area as heretofore defined, it may not entitle the owner of the tract or lease in which such well is located to revaluation for depletion purposes, if such tract or lease lies within a compact area which is immediately surrounded by proven land and the geologic structural conditions on or under the land so enclosed may reasonably warrant the belief that the oil or gas of the proven areas extends thereunder. Under no circumstances is the entire area to be regarded as proven land.

(3) The property which may be valued after discovery is the well. For the purposes of these sections the well is the drill hole, the surface necessary for the drilling and operation of the well, the oil or gas content of the particular sand, zone or reservoir (limestone, breccia, crevice, etc.) in which the discovery was made by the drilling and from which the production is drawn, to the limit of the taxpayer's private bounding lines, but not beyond the limits of the proven area as heretofore provided.

(4) A taxpayer to be entitled to revalue his property after Mar. 1, 1913, for the purpose of depletion allowances must make a discovery after said date and such discovery must result in the fair market value of the property becoming disproportionate to the cost. The fair market value of the property will be deemed to have become disproportionate to the cost, when the output of such well of oil or gas affords a reasonable expectation of returning to the taxpayer an amount materially in excess of the cost of the land or lease if acquired since Mar. 1, 1913, or its fair market value on Mar. 1, 1913, if acquired prior thereto, plus the cost of exploration and development work to the time the well was brought in.

Article 221, *Proof of Discovery of Oil and Gas Wells.*—In order to meet the requirements of the preceding article to the satisfaction of the Commissioner, the taxpayer will be required, among other things, to submit the following with his return:

A map of convenient scale, showing the location of the tract and discovery well in question and of the nearest producing well, and the development for a radius of at least 3 mi. from the tract in question, both on the date of discovery and on the date when the fair market value was set.

A certified copy of the log of the discovery well, showing the location, the date drilling began, the date of completion and the beginning of production, the formations penetrated, the oil, gas and water sands penetrated, the casing records, including the record of perforations, and any other information tending to show the condition of the well and the location of the sand or zone from which the oil or gas is produced on date discovery was claimed.

A sworn record of production, clearly proving the commercial productivity of the discovery well.

A sworn copy of the records, showing the cost of the property.

A full explanation of the method of determining the value on the date of discovery or within 30 days thereafter, supported by satisfactory evidence of the fairness of this value.

INVESTED CAPITAL

The invested capital is defined in section 326 of the Revenue Act of 1918, as: Actual cash bona fide paid in for stock or shares; cash value of property, other than cash, bona fide paid in for stock or shares (as limited by the statute); and paid-in or earned surplus and undivided profits, not including surplus and undivided profits earned during the year. The surplus and undivided profits, if not correctly reflected in the taxpayer's accounts, may be adjusted in accordance with the regulations. These considerations are shown in the paragraphs relating to surplus and undivided profits.

Surplus and Undivided Profits, Allowance for Depletion and Depreciation.—Depletion, like depreciation, must be recognized in all cases in which it occurs. Depletion attaches to each unit of mineral or other property removed, and the denial of a deduction in computing net income under the Act of Aug. 5, 1909, or the limitation upon the amount of the deduction allowed under the Act of Oct. 3, 1913, does not relieve the corporation of its obligation to make proper provision for depletion of its property in computing its surplus and undivided profits.

Adjustments in respect of depreciation or depletion in prior years will be made or permitted only on the basis of affirmative evidence that at the beginning of the taxable year the amount of depreciation or depletion written off in prior years was insufficient or excessive, as the case may be. Where deductions for depreciation or depletion have either on the books of the corporation or in its returns of net income been included in the past in expense or other accounts, rather than specifically as depreciation or depletion, or where capital expenditures have been charged to expense in lieu of depreciation or depletion, a statement indicating the extent to which this practice has been carried should accompany the return.

Surplus and Undivided Profits Reserves for Depreciation or Depletion. If any reserves for depreciation or for depletion are included in the surplus account, the account should be analyzed so as to separate reserves and leave only real surplus. Reserves for depreciation or depletion cannot be included in the computation of invested capital, except to the following extent: Excessive depletion or depreciation included therein and which, if charged off, could be restored under article 340 may be included in the computation of invested capital; and where depreciation or depletion is computed on the value as of Mar. 1, 1913, or as of any subsequent date, the proportion of depreciation or depletion representing the realization of appreciation of value at Mar. 1, 1913, or such subsequent date may, if undistributed and used or employed in the business, be treated as surplus and included in the computation of invested capital.

For the purpose of computing invested capital, depreciation or depletion computed on the value as of Mar. 1, 1913, or as of any subsequent date, shall, if such value exceeded cost, be deemed a pro rata realization of cost and appreciation and be apportioned accordingly. Except as above provided, value appreciation (even though evidenced by an appraisal) that has not been actually realized and reported as income for the purpose of the income tax cannot be included in the computation of invested capital and, if already reflected in the surplus account, it must be deducted therefrom.

The term *capital sum* is here applied to the total amount returnable to the taxpayer through depletion, depreciation and obsolescence allowances. It is to be clearly distinguished from the term invested capital, which is the basis for the determination of war-profits credits and excess-profits credits of corporations. Invested capital is the actual cash, or its equivalent, paid in plus undistributed surplus profits, and no appreciation in the value of any asset may be included except as provided in article 844 (2).

Where amortization is allowed, such sum cannot be restored to the invested capital for the purpose of the war-profits and excess-profits tax, nor any portion of the amount covered by such allowance.

Capital Sum and Invested Capital

The capital sum has no necessary relation to the invested capital. It may represent the investment of funds belonging to the taxpayer, or the investment of borrowed funds, which have no relation to invested capital; under the provisions of the law and regulations, the capital sum may include amounts based on the right of valuation as of Mar. 1, 1913, or within 30 days after the discovery of oil or gas by the taxpayer.

Where such valuations are allowable, they have no application to invested capital, except in accordance with subdivision (2) in the preceding paragraph pertaining to surplus and undivided profits reserve for depreciation or depletion, and may not be used for any purpose other than as a basis on which to determine the gain or loss arising from the sale or surrender of property acquired prior to Mar. 1, 1913. With respect to any allowance for amortization, the basis is the cost of the property acquired after Apr. 5, 1917, and no amount may be added on account of revaluation.

Certain deductions from gross income are based on the capital sum; credits are based on invested capital. It is necessary that these terms be clearly understood by the taxpayer in order to avoid confusion in making returns. In general, the deductions from gross income allowed corporations are the same as allowed individuals, except that corporations may deduct dividends received from other corporations subject to the tax and may not deduct charitable contributions.

DETERMINATION OF QUANTITY OF OIL IN GROUND

In the case of either an owner or lessee, it will be required that an estimate, subject to the approval of the Commissioner, shall be made of the probable recoverable oil contained in the territory with respect to which the investment is made as of the time of purchase, or as of Mar. 1, 1913, if acquired prior to that date, or within 30 days after the date of discovery, as the case may be. The oil reserves must be estimated for undeveloped proven land as well as producing land. If information subsequently obtained clearly shows the estimate to have been materially erroneous, it may be revised with the approval of the Commissioner.

The estimate of probable recoverable oil in the ground is fundamentally necessary if a reasonable deduction for depletion is to be calculated and, while it may be impossible to determine exactly the future production of a well or tract, it has been found possible to predict future productions with a comparatively narrow limit of error. The result of analysis of a great volume of production records has led to the development of the methods suggested in the following paragraphs. It is good practice to reduce estimates to the per acre basis of the contents of the well; this affords a reasonable check on such estimates.

METHODS OF ESTIMATING RECOVERABLE RESERVES

The Treasury Department does not prescribe any particular method of estimating recoverable reserves, but the methods described herein are suggested as applicable to a wide variety of conditions. The underlying principle of the methods outlined is that the best indication of the future production of any well is to be found in the history of similar wells in the same or similar districts, and that, other things being equal, a well's production is more likely to approximate the production of a similar well in the tract or district than to deviate widely from the average. The method may be summarized as follows:

1. Plotting the record of production of individual wells, or, lacking such detailed information, the average production per well for each tract.

2. Deriving from these graphical records an average or composite production decline curve for the district.

3. Estimating from the last year's average production per well the probable future production, based on the average production decline curve, or a future production curve derived from the production decline curve.

4. Ascertaining probable total future production of producing wells by multiplying average future production per well by the number of wells producing at the end of the year.

5. Estimating the probable future production of undeveloped proven land on the basis of nearby production, making due allowance for the decline in pressure due to the extraction of oil from the pool.

It is to be emphasized that the value of estimates will depend almost entirely on the skill with which the method is carried out and the character of the production records on which they are based.   Where accurate detailed records are not kept, it may be difficult to determine a reasonable allowance for depletion.

The taxpayer may estimate his recoverable reserves by any method that can be shown to be well founded, but in all cases the data on which such estimate was based must be submitted, with a description of the method employed, and a résumé of the calculations.

COMPUTATION OF ALLOWANCE FOR DEPLETION OF OIL WELLS

When the cost or value as of Mar. 1, 1913, or within 30 days after the date of discovery of the property, shall have been determined and the number of mineral units in the property as of the date of acquisition or valuation shall have been estimated, the division of the former amount by the latter figure will give the unit value for the purposes of depletion, and the depletion allowance for the taxable year may be computed by multiplying such unit value by the number of units of mineral extracted during the year. If, however, proper additions are made to the capital account represented by the original cost or value of the property, or circumstances make advisable a revised estimate of the number of mineral units in the ground, a new unit value for purposes of depletion may be found by dividing the capital account at the end of the year, less deductions for depletion to the beginning of the taxable year which have or should have been taken, by the number of units in the ground at the beginning of the taxable year. This number, unless a revision of the original estimate has been made, will equal the number of units in the ground at the date of original acquisition or valuation, less the number extracted prior to the taxable year. If, however, recalculation is made, the number of units at the beginning of the year will be the sum of the gross production of the year and the estimated mineral reserves in the property at the end of the year.

Each barrel of oil or unit of gas extracted and marketed must, before a profit can be realized, pay not only its proportionate share of the operating expense and deductions for depreciation and obsolescence of physical property, but also must pay its proportionate share of capital sum returnable through depletion allowances. This proportionate share of capital sum returnable through depletion allowances that each unit of oil or gas must pay is unit cost.

Unit cost is obtained by dividing the capital sum returnable through depletion by the estimated recoverable reserve at the beginning of the taxable year. The depletion deduction is computed by multiplying the unit cost by the number of units produced during the taxable year.

It is to be noted that the estimated recoverable reserves and the number of units produced are used in estimating the depletion deduction for both lessor and lessee. Since, however, they are applied to different capital amounts returnable through depletion deductions, the unit costs for lessee and lessor are not identical, and the deductions bear the same ratio as the capital sum of lessor and lessee. Usually the lessee's investment is greater than the lessor's and his deductions are correspondingly greater. Stated in another way, if a certain proportionate part of the lessee's capital returnable through depletion deductions is deducted in a given year, the same proportion of the lessor's capital sum returnable through depletion will be deducted.

Computation of Depletion Allowance for Combined Holdings of Oil Properties

The recoverable oil belonging to the taxpayer shall be estimated separately on the smallest unit on which data are available, such as individual wells or tracts, and these, added together into a grand total, are to be applied to the total capital assets returnable through depletion. The capital sum shall include the cost or value, as the case may be, of all oil rights, freeholds, or leases, plus all incidental costs of development not charged as expense. The unit multiplied by the total number of units of oil produced by the taxpayer during the taxable year from all of the oil properties will determine the amount that may be allowably deducted from the gross income of that year. In the case of sale of particular tracts, full account must be taken of the depletion of such tracts in computing profit or loss thereon.

A convenient summary record may be kept of acreage and production with average decline curves of wells if leases are contiguous, or if property consists of many separate leases or districts, one curve should be made for each. Such a summary form would be a permanent record and greatly assist in making up the annual returns. Each lease having its own decline curve and production can be balanced out with the reserves at the end of each year. New additions brought in during the year must be added and carried out in accordance with the general plan.

COMPUTATIONS OF ALLOWANCE FOR DEPLETION OF GAS WELLS

The deductions allowed in computing income from natural-gas properties are in general similar to those allowed oil operators, but the method of computing the deductions and the various assets differ in certain particulars, the most notable of which are involved in the problems of estimating the probable reserves and computing the depletion. On account of the peculiar conditions surrounding the production of natural gas, it is necessary to compute the depletion allowance for gas properties by methods suitable to the particular cases. Usually the depletion should be computed on the basis of decline in closed or rock pressure, taking into account the effects of water encroachment and any other modifying factors. In many fields, more or less additional evidence on depletion is to be had from such considerations as: Details of production and performance records of well or properties; decline in open flow capacity; comparison with the life histories of similar wells or properties, particularly those now exhausted; and size of reservoir and pressure of gas.

Methods of Computing Gas Depletion

Gas depletion may be computed from the details of production or the performance record of the well or property, estimating, from its best records, the quantity the well may be expected to produce and also the rate of production. The decline in open flow capacity indicates the rate of exhaustion.

Depletion may also be computed by a comparison with the life history of similar wells or properties, particularly those exhausted or nearing exhaustion; also by comparing the size of the reservoir and the pressure of gas or by the pore space method. The factors that make this method difficult to apply are the difficulties of accurately ascertaining the thickness of pay, limits of pool, percentage of pore space, effect of encroaching oil or water, and the quantity of gas remaining when production is no longer possible.

Other indications of depletion are the decreasing supply by general observation, by minute pressure changes, and by line pressure observed at compressing stations. The appearance of water or oil in a gas well may be the significant symptom of the approaching termination of the life of the well. Clogging by paraffine, salt, or other deposits may demand the modification of depletion estimates.

Closed, or Rock, Pressure Method

This is the best method of estimating the depletion of gas properties as the rock pressure can be ascertained with a fair degree of accuracy, and the pressure decline established, based on Boyles' law. In gaging, care must be taken to insure that the gage is accurate; it should be tested both before and after being attached to the well. Care must also be taken to empty the well of oil and water and the well should be closed long enough to allow the pressure to build up to its maximum.

Several corrections and refinements are made in applying this method to the computation of depletion; it does not afford data on the amount of gas originally in the pool, but only the portion of the gas that has been removed. The atmospheric pressure must be taken into consideration when taking the difference of gage, adding the same to each condition in making the fraction remaining in the ground. Account should also be taken of pressures at which wells are abandoned in the district.

Unit Cost as Applied to Natural Gas

The unit-cost method can be used by regarding pounds of closed pressures as units, for the actual quantity of gas commonly varies with the decline in pressure. The relative quantities at the beginning and end of the tax year, and at the time of abandonment, may be used for tax purposes when better information is lacking.

Apportionment of Depletion Among Various Sands

Where more than one sand under a property is yielding gas, the problem arises as to how to weight or evaluate the decline in pressure in the different sands. The depletion sustained is not indicated by the average decline in pressure, but is more nearly proportionate to the decline in the good sand. If accurate figures on capacities of wells are obtainable, it will be possible to make a fairly accurate weighting of the pressure declines; or if facts indirectly indicating capacity of individual wells are obtainable, some light may be thrown on the question. But, as a general rule, it is necessary to average the decline of wells drawing from different sands as though they were drawing from the same sand.

Testing is recommended in summer or early fall. Abandoned wells may be regarded as fully depleted and their pressure counted as zero in computing depletion. It is suggested that the capital sum at the beginning of each year be treated as 100 per cent., for the average pressure at the beginning of the year and the average decline during the year will then furnish a readily usable basis for computing the depletion allowance.

The following formula has been recommended for use by the Treasury Department:

$$\frac{x}{y} \times z = \text{Depletion allowance}$$

in which x = capital sum to end of the year; y = total future pressure decline, or difference between sum of pressures at beginning of the tax year and the sum of pressures at time of expected abandonment; z = pressure decline during year as obtained by adding to sum of pressures at begininng of year the sum of pressures of any new wells completed during year and subtracting sum of pressures at end of year.

The formula may also be written as follows:

$$\frac{\text{Capital sum to the end of tax year}}{\text{Sum of pressures at beginning of year} - \text{sum of pressures at time of expected abandonment}} \times \frac{\text{Sum of pressures at beginning of tax year} + \text{sum of pressures of new wells} - \text{sum of pressures at end of tax year}}{} = \text{Depletion allowance.}$$

The regulations require gas-well pressure records to be kept, and where the field is too new to determine the quantity in reserve a tentative estimate will apply until production figures are available from which an accurate estimate may be made.

Computation of Depletion Allowance for Combined Holdings of Gas Properties

In the case of gas properties, the depletion allowance for each pool may be computed by using the combined capital sum returnable through depletion of all tracts of gas land owned by the taxpayer in the pool and the average decline in rock pressure of all the taxpayer's wells in each pool in the formula given in article 211. The total allowance for depletion of the gas properties of the taxpayer will be the sum of the amounts computed for each pool.

The depletion of gas supplies belonging to a taxpayer may be more accurately computed by making estimates for each tract, though it is quite possible that the expense of making separate estimates for individual tracts may be greater than the benefits arising from such a procedure.

DEPRECIATION

The Treasury Department has issued many suggestions pertaining to depreciation of physical property. Individual companies may apply different rates of depreciation on equipment of a similar nature if the rates are derived from reliable records kept by the respective companies. It is specifically stated in the regulations that each claim for depreciation must show facts upon which such claim is based. Special claims receive special consideration.

Depreciation deductions are to be charged to a reserve fund, and are in addition to any regular charge for repairs and operating maintenance. For the general equipment of a producing property, depreciation may be charged at the same rate as depletion because the general well equipment is serviceable only as long as the life of the wells. This method makes a simple and consistent form for such depreciation charges.

The summary of the suggestions in Table 1 is given as a convenient method of reference, and is taken from the Treasury Department Manual for the Oil and Gas Industry (see p. 1718).

OTHER ALLOWANCES

Development costs, except the cost of physical property, may be deducted as an expense in the year in which they are paid out or, at the option of the taxpayer, may be charged to capital returnable to the several allowable deductions. Election once made under this option is final and will control the returns for all subsequent years.

Cost of development comprises all payments made for and incident to the drilling of wells, such as cost of:

Physical property, geological and other surveys made subsequent to acquisition, roads, water supplies, hauling, wages, drilling, shooting, overhead charges (incident to drilling of wells), fuel and all other similar expenditures.

Both cost of property and cost of development, in so far as they have not been decreased by allowable deductions, are chargeable to capital sum and are returnable through the several allowable deductions. Structures and equipment may also be included in capital assets and are returnable through depreciation. In the case of revaluations as of Mar. 1, 1913, or within 30 days of a discovery by the taxpayer made subsequent to Feb. 28, 1913, the value thus established plus subsequent costs not otherwise deducted becomes the total of capital sum. This revaluation, however, does not affect the invested capital, as previously noted.

TABLE 1.—*Summary of Suggestions from Treasury Department Manual*

Class	No.	Reference Page		Useful Life Years	Annual Depreciation, Per Cent.
A	1	57	Drilling equipment	4	40–25–15–10
	2	57	Wells		
	3	57	Dehydrators		
			Electric	5	20
			Pipe and tanks	2	50
	4	58	Tanks		
			Steel 5000–55,000 bbl.	20	5
			2500–5000 bbl.	12	8⅓
			Galvanized-iron 500–2500 bbl	12	8⅓
			Less than 500 bbl.	8	12½
			Wood	5	20
A	4	58	Movable tanks		
			Galvanized-iron 500–2500 bbl	9	11⅑
			Less than 500 bbl.	6	16⅔
			Water tanks		
			500–2500 bbl.	8	12½
			Less than 500 bbl.	5	20
	5	58	Tools	3	33⅓
	6	58	Transportation equipment	3	33⅓
	7	58	Water plants	10	10
	8	58	Electric equipment	10	10
	9	59	Machine shops	7	14
	10	59	Buildings		
			Small wood	10	10
			Frame structure	15	6⅔
			Corrugated-iron siding	6	16⅔
			Concrete	25	4
			Brick	25	4
			Steel	25	4
B	1	59	Pipe lines		
			Mains over 6 in. diameter	20	4½
			Mains under 6 in. diameter	16	5⅝
			Gathering lines	10	9
			Less 10 per cent. salvage		
			Pump stations	10	10
C		60	Tank cars	20	5
	1	60	Refineries		
			Class 1. Located at point assuring a long supply of crude oil; or well-constructed plants.	20	5
			Class 2. Located at points assuring supply of crude oil for several years.	10	10
			Class 3. Skimming plants and small refineries of poor construction, or located at points where supply of crude oil is not assured for a long period of time.	6	16⅔
D	1	62	Sales or marketing equipment		
			Tankers	20	5
			Barges	5	20
			Filling stations		
			Class A. Ordinary wood or corrugated-steel construction.	5	20
			Class B. Brick and concrete or extraordinary construction.	10	10
			Distributing stations	10	10
			Tank wagons		
			Motor	4	25
			Horse	6	16⅔
			Steel barrels	7	14 2/7
			Track and switches	8	12½
E		63	Natural gas (utility companies)		
	1		Drilling equipment (see A–1)		
	2		Wells (see A–2)		
			Gas pipe lines		
			Mains	12	8½
			Gathering lines	10	10
			City lines	10	10
	4		Compressor stations	7	14 2/7
	5		Gathering stations	6	16⅔
	6		Field stations	4	25
	7		Meters and regulators	5	20
			Considered as a whole plant	10	20
F	1	64	Natural gas gasoline		
			Plant, compression with 20 per cent. salvage value	4	35–20–15–10
			Absorption plants, with 20 per cent. salvage	4	35–20–15–10

Operating Expenses

Expense includes all amounts paid out (exclusive of amounts paid for physical property and development charged to capital sum) incident to the development and operation of producing properties and the preparation of their product for market, such as costs of pumping, cleaning, reshooting (including cost of torpedoes), gaging, storing, treating, reducing, repairs and maintenance, transporting, refining, conserving, marketing, overhead expense, insurance, etc. The cost of repairs and replacements made necessary through deterioration of equipment may be charged off as expense, but if this is done the amount allowed as a depreciation deduction will be reduced. In all cases, items of expense must be charged off as such for the year incurred and can neither be deducted from the income of subsequent years as expense nor added to capital sum.

Repairs

The cost of incidental repairs that neither materially add to the value of the property nor appreciably prolong its life, but keep it in an ordinary efficient operating condition, may be deducted as expense, provided the plant or property account is not increased by the amount of such expenditures. Repairs in the nature of replacements, to the extent that they arrest deterioration and appreciably prolong the life of the property, should be charged against the depreciation reserve.

Amounts expended for additions and betterments or for furniture and fixtures that constitute an increase in capital assets or add to their value are not a proper deduction, but such expenditures when capitalized may be reduced through annual depreciation deductions.

Taxes

Federal taxes (except income, war-profits, and excess-profits taxes), state and local taxes (except taxes assessed against local benefits of a kind tending to increase the value of the property assessed), and taxes imposed by possessions of the United States or by foreign countries (except the amount of income, war-profits, and excess-profits taxes allowed as a credit against the tax) are deductible from gross income.

Postage is not a tax. Amounts paid to states under secured-debts laws in order to render securities tax exempt are deductible. Automobile license fees are ordinarily taxes.

Losses

Losses sustained during the taxable year and not compensated for by insurance or otherwise are fully deductible (except by non-resident aliens) if: incurred in the taxpayer's trade or business; incurred in any transaction entered into for profit; or arising from fires, storms, shipwreck, or other casualty, or from theft.

They must usually be evidenced by closed and completed transactions. In the case of the sale of assets, the loss will be the difference between the cost thereof, less depreciation sustained since acquisition, or the value as of Mar. 1, 1913, if acquired before that date, less depreciation since sustained, and the price at which they were disposed of.

When the loss is claimed through the destruction of property by fire, flood, or other casualty, the amount deductible will be the difference between the cost of the property, or its value as of Mar. 1, 1913, and the salvage value thereof, after deducting from the cost or value as of Mar. 1, 1913, the amount, if any, which has been or should have been set aside and deducted in the current year and previous years from gross income on account of depreciation, and which has not been paid out in making good the depreciation sustained. But the loss should be reduced by the amount of any insurance or other compensation received. Losses in illegal transactions are not deductible.

Losses of oil and gas are of two kinds: Those that are unforeseen or unavoidable, such as losses sustained through fire or accident; and those that are anticipated and recognized as unavoidable under operating conditions, such as evaporation of oil in storage, ordinary leakage, re-refinery losses, etc. Usually losses of the latter class are indeterminate as to amount and are absorbed, either implicitly or explicitly, in current operating expenses or in cost of the oil or gas. Indeterminate losses may be deducted from gross income.

Compensation for Personal Services

Among the ordinary and necessary expenses paid or incurred in carrying on any trade or business may be included a reasonable allowance for salaries or other compensation for personal services actually rendered. The test of deductibility in the case of compensation payments is whether they are reasonable and are in fact payments purely for services.

Bonuses to Employees

Gifts or bonuses to employees will constitute allowable deductions from gross income when such payments are made in good faith and as additional compensation for the services actually rendered by the employees, provided such payments, when added to the stipulated salaries, do not exceed a reasonable compensation for the services rendered.

Donations to employees and others, which do not have in them the element of compensation or are in excess of reasonable compensation for services, are considered gratuities and are not deductible from gross income.

Damages

Any amount paid pursuant to a judgment or otherwise on account of damages for personal injuries, patent infringements, or otherwise, is deductible from gross income when the claim is liquidated or put in judgment or actually paid, less any amount of such damages as may have been compensated for by insurance or otherwise.

If subsequent thereto, however, a taxpayer has for the first time ascertained the amount of a loss sustained during a prior taxable year, and not deducted from the gross income therefor, he may render an amended return for such preceding taxable year, including such amount of loss in the deductions from gross income, and may file a claim for refund for the excess tax paid by reason of the failure to deduct such loss in the original return. Provided, that no such credit or refund shall be allowed or made after five years from the date when the return was due, unless before the expiration of such five years a claim therefor is filed by the taxpayer.

Abandoned Wells

When wells collapse, become wet or otherwise unprofitable producers, and are abandoned, the cost of such abandonment is chargeable to current operations. Usually the value of the recovered material is credited to its investment cost and the difference, not already depleted, is deductible as being fully depleted.

In general, the deductions from gross income allowed corporations are the same as allowed individuals, except that corporations may deduct dividends received from other corporations subject to the tax and may not deduct charitable contributions.

ITEMS NOT DEDUCTIBLE

Donations to employees or others that are not compensation or are in excess of reasonable compensation for services are considered gifts and are not deducted from gross income.

Losses

Losses in illegal transactions are not deductible.

Losses of oil and gas are of two kinds: (a) unforeseen or unavoidable, as through fire or accident; (b) anticipated and recognized as unavoidable under operating conditions, as evaporation, leakage, refinery losses, etc Usually the latter are indeterminate as to amount and are absorbed either implicitly or explicitly in current operating expenses or in cost of oil or gas. Indeterminate losses may not be deducted from gross income.

Accrued Deductions not Charged in Prior Years

The expenses, liabilities, or deficit of one year cannot be used to reduce the income of a subsequent year. A person making returns on an accrued basis has the right to deduct all authorized allowances, whether paid in cash or set up as a liability; it follows that if he does not within any year pay or accrue certain of his expenses, interest, taxes or other charges, and makes no deduction therefor, he cannot deduct from the income of the next or any subsequent year any amounts then paid in liquidation of the previous year's liabilities. A loss from theft or embezzlement occurring in one year and discovered in another is deductible only for the year of its occurrence.

Depletion for Past Years

Where under the Act of Oct. 3, 1913, or of Sept. 8, 1916, a taxpayer has not been allowed to make a deduction for the full amount of his depletion, the amount of such deficiency cannot be carried forward and deducted in any later year. Depletion attaches to each unit of mineral or other property removed, and a taxpayer should make proper provision therefor in computing his net income. Under the Revenue Act of 1918, the amount recoverable through depletion will be the cost, or the value as of Mar. 1, 1913, or within 30 days of the date of discovery, as the case may be, less proper allowance for the mineral or other property removed prior to Jan. 1, 1918.

RÉSUMÉ

The foregoing generally embraces a résumé of the Regulations and methods of applying the valuation, and also the depletion, depreciation, amortization, and other deductions from gross income, of gas and oil properties, and are either copied or briefly condensed from the Treasury Department Regulations 45. In the general application of these, the taxpayer will, through his proper books of accounts, record all transactions of capital, assets, reserves for depletion, depreciation, or amortization and other deductions, also distributions of investments to the various facilities and cost of all buildings and equipment that will fully reflect the business conditions. In order to set up the proper depletion and depreciation deductions from gross income, it is necessary that an investment be set up on each lease or property.

RECORDS OF PRODUCTION AND ESTIMATED RECOVERABLE OIL

Individual small tracts can be more readily made up than for very large ones; in fact, for small properties computation by well areas make a desirable and complete record.

Production records should be kept by individual wells, if possible, or as few as are operated in a group. Complete records for each lease or subdivision are desirable, as copies of such data are requested with the questionnaire. If records of production of individual wells were kept, decline curves would be both accurate and easy to produce.

All gage tickets should be preserved for a check of oil run and balance with production and stocks. These, too, record the gravity of the oil. The posted prices are recorded in the settlements for such oil.

Logs of wells should be filed and preserved and a proper working map is necessary to show the location of each well and the position of the property in relation to all adjoining leases. Water records should be kept of each well and also time and method of each well's operations; also records of suspensions or abandonment. These data are very useful in connection with figuring depletion deductions.

If the property is large, the Geological Department defines the classes of land and directs the calculations of oil reserves.

The difference between invested capital and capital sum is clearly defined in the Regulations, as also the method for discovery revaluations. In practical operation, the chief items in making up the tax returns necessitate that the investments of each tract be properly set up; that the reserves be figured, methods submitted with records of all productions, checking or balancing the reserves both of past and end of current year so that unit costs can be readily and accurately obtained.

The deductions for depletion, depreciation, and others are also readily obtained from the proper method of charging, through the books of accounts, supported by the usual records of production, shipments, well data, acreage, royalties, and a general systematic business routine.

Generally the petroleum industry has adopted most of these methods, and is conforming to the new orders and conditions. The Regulations are drawn up with clarity to aid operators in making their returns, and are worthy in their intent.

This paper is submitted through a desire to arrange the laws and rulings as a concise reference and not with any intention of presenting anything new.

Acknowledgment is made to Mr. F. J. Hoenigmann for assistance and aid in compiling these pages.

DISCUSSION

RALPH ARNOLD, Los Angeles, Calif.—The subject of taxation must be considered from the standpoint of both the tax official and the taxpayer. The needs of taxing jurisdiction are paramount in communities dependent on mining or oil. When a man says, "Let us use the last production tax or income tax," he is looking at the question from his own standpoint. When there is no income or gross production, the needs of the community in which the mine or well is situated are practically the same, so that taxes must be paid or the government must fail. In such cases the ad valorem system is better fitted to conditions.

The question as to whether this discourages development has been asked. In Wisconsin, where the ad valorem system is used for valuing iron-ore properties, ore has been developed until about two billion tons are now in sight. In Arizona, also, this system has not interfered with the development of new deposits.

In Minnesota, where the dominant electoral element is agricultural, the taxes are based upon a fair market value of all properties, but the assessment is 33⅓ per cent. of the value for agricultural property, 40 per cent. for urban properties, and 50 per cent. for mining properties. This is a clear discrimination against mining. In Montana and Idaho, where the dominant influence is mining, the system of taxation puts its burden on the agricultural and other industries of the state. In California, the assessment in Orange County is based on the fictitious value established on the production of the previous year. It is assumed that the property will last ten years and produce at the same rate so that value is multiplied by ten to get full value and 40 per cent. of that is taken as the assessable value of the property. If there should be a big production one year and a small one in the next, as is often the case, the taxes the second year would be out of all proportion to one's ability to pay, and have no relation at all to the taxes on the surrounding real estate.

In its report, the Mine Taxation Subcommittee of the National Tax Association advocated the placing of all taxes, especially for local purposes, on the ad valorem basis; that is, treating mines, oil, and gas properties the same as other classes of real estate.

The reason that this question of taxation is of great importance to engineers and geologists is: If the ad valorem method is adopted, and it probably will be adopted in many places, oil and gas properties must be valued for purpose of taxation; that valuation will have to be done by an engineer, it is not work for the ordinary assessor.

In the work for the government, it was necessary to employ engineers to solve the tax problems. This question of valuation and taxation is not a subject for lawyers, but for engineers. Just at the present time the lawyers are handling most of the cases, which in many instances could be better done by engineers.

What is the fair market value of the property? The tendency now is for oil and mining companies to try to base the value on the engineer's report. It is a hypothetical value, based on the present worth of the estimated amount of mineral in the ground. The regulations call for consideration of a number of factors in reaching this fair valuation. All of these must be taken into consideration, and I do not believe we are going to arrive at a fair market value by making any one of those factors dominant for all localities, or for all types of property.

Valuation Factors of Casing-head Gas Industry

By Oliver U. Bradley,* Muskogee, Okla.

(St. Louis Meeting, September, 1920)

The utilization of casing-head gas in the manufacture of casing-head gasoline by both the absorption and the compression method is a most important factor in the conservation of our natural resources. Any industry connected with the oil business, in general, possesses particular attraction for a large number of people not conversant with its basic principles, for the reason that the large fortunes made in the production and utilization of petroleum and its products have been given undue prominence. The general impression of the public that enormous profits are to be realized in the casing-head gas industry with minimum expenditures of both capital and effort has, in a large measure, accounted for the phenomenal expansion of the industry in recent years and, likewise, has resulted in many mistakes and loss of investment funds. It is true that many installations have been very profitable, but such instances are always the result of careful planning, experienced judgment and conservative estimates.

The inception and subsequent activity in the manufacture of casing-head gasoline, enabling the business to assume an important position in the petroleum industry, are of comparatively recent origin, as its greatest growth, particularly in Oklahoma, occurred during the years 1917 and 1918. Much information must yet be secured and systematized concerning the methods of manufacture of gasoline from high-yield casing-head gas, and a large field is still open for the application of accumulated experience and good engineering practice in devising better methods of extracting gasoline from casing-head gas of the poorer grades.

The absorption process is coming into general use as a most efficient system of treating casing-head gas, and even so-called dry gas. In fact, there is a decided tendency toward the universal adoption of the absorption process as against compression methods. However, a general discussion of the relative merits of these two systems is not within the scope of this article.

A few of the facts that must be given consideration in arriving at a fair and impartial estimate of the actual investment value of the casing-head gas business are the quantity of gas available, the quality and

---

* United States Oil and Gas Inspector.

composition of the gas, accessibility of plant to railroads and water supply, efficiency of operation of oil leases connected to plants, plant efficiency, estimates of production and marketing costs, contract for purchase of gas, and market price of casing-head gas.

QUANTITY OF GAS AVAILABLE

The most important factors are the quantity of casing-head gas available and the conditions that will have a material bearing on its future supply, such as location of field, depth of oil wells, initial rock pressure, thickness and porosity of oil sands, relative position of oil and gas strata in the sand, grade of oil, life of oil wells, location and rapidity of water infiltration, vacuum carried, and regularity of its application. More mistakes have been made in the estimation of the available supply of gas than in any other feature of the business. It is at once appreciated that as close a determination as possible of the marketable quantity of casing head gas is of extreme importance. When volume tests are made, it should be remembered that orifice tests of built-up pressure of casing-head gas on individual wells do not necessarily indicate the performance of these wells under vacuum conditions. The application of the vacuum frequently increases the volume of both oil and casing-head gas temporarily, but the effects of the continuous operation of wells under a vacuum cannot be clearly defined, as it is an open question as to when and under what conditions a vacuum should be applied to oil wells in order to produce the maximum extraction of both oil and casing-head gas.

The exploitation of casing-head gas is quite different from ordinary mining operations, as available sources of supply are not susceptible to exact measurements, like ore in a mine, for example, which may be developed by shafts and drifts, blocked out by raises and winzes, sampled and assayed, and the mineral content closely estimated. Casing-head gas, technically speaking, is not in place, cannot be stored, and, therefore, must be treated and disposed of at once after being brought to the surface. Many casing-head gasoline plants have been designed and erected for the treatment of a certain estimated quantity of gas, which after two or three months have found that the supply of gas has decreased more than 50 per cent., necessitating the dismantling and removal of several units of the equipment, or having on hand surplus machinery, which imposes a considerable handicap on the profitable operation of the business. In the case of many plants in Oklahoma, if conservative engineering estimates had been made at the beginning of operations, a smaller plant would have been installed and additions made thereto, in case the supply of gas justified them. In this way, the equipment could have been enlarged to meet the requirements of the gas supply instead of reversing the process.

The location of oil leases, with reference to the general producing area of the pool, is important, as investigation has shown that when leases are located on the edge of the pool the casing-head gas frequently fails to maintain its usual volume, and its richness is much less than that from wells in the main or central portion of the field. Consideration should also be given to underground conditions in estimating the possibilities of the supply of casing-head gas.

QUALITY AND COMPOSITION OF GAS

A chemical analysis of the gas should be made in order to determine its actual physical characteristics, as a basis for applying a method that will obtain a maximum yield of casing-head gasoline. Furthermore, a practical field test should always be made, so as to secure dependable information regarding the results that may reasonably be expected in the operation of a plant. A demonstration of the desirability, as well as the necessity, of applying both chemical and practical tests to casing-head gas, is clearly shown in the accompanying data, giving percentage loss due to evaporation in conducting tests to determine its correct productivity.

No. of Test	Cubic Feet Used	Gasoline Un-weathered, Cubic Centimeters	Gasoline Weathered, Cubic Centimeters	Cubic Centimeters Lost	Percentage of Evaporation	Productivity per 1000 Cu. Ft.
1	200	2,985	1,960	1,025	34.0	2.59
2	200	3,830	1,960	1,870	48.8	2.58
3	200	640	635	5	0.9	0.83
4	200	2,475	2,070	405	16.3	2.73
5	200	2,405	2,040	365	15.1	2.69
6	167	1,345	1,165	180	13.4	1.84

These tests were all made from casing-head gas from the Bartlesville sand in the Cushing Field and illustrate the variability in the composition of such gas, the higher fractions sometimes predominating and sometimes, the lower.

Conditions that may produce a considerable variation in the results of tests may be summarized as follows: (1) The time of the year taken, as climatic conditions and temperature have a bearing on the results. (2) Conditions on the lease, such as wells on the pump or off, cleaning out wells, and other lease work. (3) Point of sampling the gas and conditions under which the sample is taken. (4) Improper design of machine, such as lack of cooling surface, inefficient compression, faulty manipulation, poor connections, and defects in mechanical equipment designed to make these tests. (5) Natural error creeping in when small quantities of gas are tested, together with incorrect meters. (6) Excessive evaporation in open-air field tests.

Because of the presence of one or more of these conditions, the results of field tests are frequently too high or too low and, in calculating the value of the gas, proper allowances should be made after a survey of all the facts. If careful attention is given to the chemical analysis of the gas and an effort is made to get a practical field test under as nearly as normal conditions as possible, the chances of error in figuring commercial yields are greatly reduced.

Accessibility of Plant to Railroads and Water Supply

Plants are sometimes located unfavorably with regard to supply of casing-head gas. It is frequently a debatable question as to whether the plant should be located close to railroad facilities, with the supply of gas several miles away, or close to the supply, with railroad facilities several miles distant. The general factors relative to loading losses, cost of upkeep of field lines, and general efficiency of plant operations should be considered in selecting the location of a plant. Furthermore, a dependable water supply is always important. Numerous plants have been located where the initial expense of installing a suitable and adequate water supply and its subsequent maintenance have been excessive, thus imposing a heavy charge on the future profits of the business.

Efficiency of Operation of Oil Leases Connected to Plants

Serious friction may often arise between the operator of an oil lease and the manufacturer of the casing-head gasoline. This contingency is of particular importance, though it is frequently given no attention, because the close relationship between the production of oil and casing-head gas is not fully appreciated. Considerable inroads on the profits of a casing-head gasoline plant may be made by undue irregularities in the operation of oil leases, such as disconnecting wells at inopportune times, cleaning out same, admission of air into lines through leaking stuffingboxes and defective lead lines. Many difficulties of this sort may be eliminated by the incorporation of certain provisions in casing-head contracts. Practically all of the larger companies operate their own casing-head gasoline plants, or this work is done by closely affiliated or subsidiary companies, which is far more satisfactory from the standpoint of efficiency, as there will be close coöperation between the oil-producing department and the casing-head gasoline division.

PLANT EFFICIENCY

There are many methods of cooling the gas and its treatment under varying pressures; also, many systems of blending are in use, all of which have a material bearing on results. A résumé of the numerous practices will not be given at this time. However, casing-head gasoline manufacturers should be willing to coöperate in comparing the various methods employed, to the extent of giving independent investigators as much information as possible, as the collection of reliable data on the efficiency of different methods of handling the various grades of gas would benefit the entire industry and need not necessarily make public the particular trade secret of any company. Under the most careful management, there will still remain considerable variations in plant operation, and sometimes these differences will result in changes of production ranging from 15 to 20 per cent. during any one month.

Some of the causes, not associated with the efficiency of plant operation, that will produce substantial changes in monthly productions of plants are as follows:

1. Climatic conditions. An examination of the records of monthly production of casing-head gasoline plants will show changes corresponding to the seasons of the year, the production in the spring and fall months usually being greater than that of the summer and winter months.

2. Frequently, one or two of the wells will produce a different quality of gas, when considered in connection with its gasoline productivity. If the pressure on one well should be greater than on others, it will naturally force proportionately more lean gas into the plant. This will often result in a great difference in the daily production; on some days, this high pressure will put a greater quantity of gas into the plant than on others. The mixture of lean gases with the regular gas coming into the plant will reduce the productivity of the entire volume of gas in a considerably larger ratio than would be revealed if a test were made of the individual productivities and an average taken. It has been found necessary, in many plants, to cut out these lean wells in order to secure a reasonable degree of uniformity in the average daily production.

3. It is necessary, in the operation of casing-head gasoline plants, to guard against excessive amounts of air in the lines. Daily tests should be made of the gas mixture entering the plant and the presence of excessive amounts of air should be investigated and faulty conditions remedied. Air not only has a direct bearing on the output of the plant but is a source of considerable danger from explosion, when it reaches a high percentage in the mixture.

The varying monthly results of plant operation may be shown by the following tabulation:

Month	Total Gas Consumed Cubic Feet	Total Condensate Produced Gallons	Gallons Per 1000 Cubic Feet
April.....................	8,727,000	30,034	3.44
May......................	9,106,000	29,382	3.23
June.....................	9,389,000	18,630	1.98
July.....................	9,877,000	20,741	2.10

ESTIMATES OF PRODUCTION AND MARKETING COSTS

Careful estimates should be made of the cost of labor and supplies, superintendence, insurance, taxes, yearly depletion of gas supply, depreciation of equipment, the unavoidable shipping losses, and the general hazards of the business, such as inability to find a ready market for the product, due to different specifications of purchasers as to gravity and blending material.

In reality, the marketing factor frequently becomes a question of vital concern. Most manufacturers of casing-head gasoline must now supply their own cars, specially designed at considerable expense, not only in order to comply with the Federal shipping regulations but to avoid excessive evaporation losses and leakage.

CONTRACTS FOR PURCHASE OF GAS

Contracts for the purchase of casing-head gas have gone through the various stages of development, or evolution, corresponding rather closely to the expansion of the industry. In a general way, such contracts may be divided into several distinct classes, viz.:

(a) The flat-rate contract in which there is a specified fixed rate per thousand cubic feet for the gas, extending over a period coinciding with the terms of the lease. These flat rates were made in the infancy of the industry and, compared with present conditions, are extremely low, as most of the instruments drawn for the purchase of casing-head gas in the early days show a price ranging from 3 to 5 cents per thousand cubic feet.

(b) Sliding-scale rate in which a certain price is specified for the gas, based on the Chicago tank-wagon price for casing-head gasoline, or f.o.b. loading rack price at plant, or a designated local market; that is, 3 cents per thousand cubic feet for the gas when the price of gasoline is 10 cents per thousand cubic feet, with $\frac{1}{2}$ cent increase in the price per thousand cubic feet for the gas for every 1 cent increase in the price of gasoline. These sliding-scale contracts range from 3 cents on 10-cent gasoline to 8 cents on 12-cent gasoline, with the percentage increase feature. It will be noted that no mention is made of the productivity of the casing-head gas.

(c) A fixed percentage of the gross proceeds derived from the sale of casing-head gasoline produced. Contracts of this character, varying from 25 to 50 per cent. of the gross proceeds are considered fair, as they show exactly what the plant produces and the settlement for the gas is made on such basis. Provisions are frequently incorporated in these contracts, charging up the proportionate cost of the blend and its transportation against the seller of the gas, particularly if the percentage of gross proceeds is above 40 per cent. Some difficulty is encountered, at times, in making settlements with royalty owners on the basis of plant production, but from the standpoint of the lessee, who usually owns a group of leases, the contracts are equitable.

(d) A test of the productivity of the gas and the Chicago tank-wagon price per gallon for gasoline. The price of the gas is determined by a schedule showing the yields of gasoline from the gas on a scale of ½-gal. units, arranged in a horizontal column, and the Chicago tank-wagon price of gasoline, in a vertical column. Several kinds of schedules are in use and are included in contracts, but the principle involved in each is the same; the schedule shown in Table 1 was suggested to the Department of the Interior by different casing-head gasoline producers and was approved by that Department.

In contracts providing for a test of the gas, it is rare that any method of procedure is prescribed for making the same. The ordinary equipment and requirements of a field test of the productivity of casing-head gas that will give reasonable satisfactory results may be specified as follows:

1. A fairly dependable testing machine usually consists of a small gasoline-engine unit, belted to a compressor, with coil racks, cooling tanks, accumulator tanks, gages, meters, pipe connections and necessary fittings, the entire equipment being portable. Coil racks should contain at least 18 ft. (5.49 m.) of ⅜-in. (9.53 mm.) galvanized-iron pipe in the form of a spiral, and all lines from the compressor to the coils and from the coils to the accumulator tank should have a natural drain so that all condensate will move to the accumulator tank by its own gravity The testing machine should be placed and jacked up with this end in view.

2. The compressor should make 250 r.p.m., in order to do the most efficient work.

3. The casing-head gas to be tested should be taken from the discharge of the vacuum pump or from the discharge of the low side of the plant compressor at a pressure of 4 oz. at the intake of the meter.

4. All leases connected to the vacuum pump should be shut off except the lease to be tested, when such is possible, and the main line should be given time to clear itself of all mixed gases; or, a vacuum-pump unit may be installed on the testing machine, thus enabling a sample of gas

TABLE 1.—*Schedule for Determining Price of Gas from Chicago Tank-wagon Quotation*

Gallons per 1000 Cu. Ft. — Cents per 1000 Cu. Ft. of Gas

Sale Price per Gallon or Less (Cents)	½	1	1½	2	2½	3	3½	4	4½	5	5½	6	6½	7	7½	8	8½	9	9½	10	10½	11	11½	12
6	¾	2	3	4	5	6	7	8	9	10	11	12	13	14	15	16	17	18	19	20	21	22	23	24
7	¾	2	3	5	6	7	8	9	11	12	13	14	15	16	18	19	20	21	22	23	25	26	27	28
8	¾	3	4	5	7	8	9	11	12	13	15	16	17	19	20	21	23	24	25	27	28	29	31	32
9	¾	3	5	6	7	9	11	12	14	15	16	18	20	21	23	24	25	27	29	30	32	33	34	36
10	1	3	4	6	8	10	11	13	15	17	18	20	22	23	25	27	28	30	32	33	35	37	38	40
11	1	3	4	7	9	11	13	15	17	18	20	22	24	26	28	29	31	33	35	37	39	40	42	44
12	1	3	5	8	10	12	14	16	18	20	22	24	26	28	30	32	34	36	38	40	42	44	46	48
13	1½	3	5	9	11	13	15	17	20	22	24	26	28	30	33	35	37	39	41	43	46	48	50	52
14	1½	4	5	9	12	14	16	19	21	24	26	28	30	33	35	37	40	42	44	47	49	51	54	56
15	1½	4	6	10	12	14	18	20	23	26	27	30	33	35	38	40	42	45	48	50	53	55	57	60
16	1½	4	6	11	13	16	18	21	24	26	29	32	34	37	40	43	45	48	51	53	56	59	61	64
17	2	5	7	11	14	17	19	23	26	29	31	34	37	40	43	45	48	51	54	57	60	62	65	68
18	2	5	7	12	15	18	21	24	27	30	33	36	39	42	45	48	51	54	57	60	63	66	69	72
19	2	5	8	13	16	19	22	25	28	32	35	38	41	44	47	50	53	57	60	63	67	70	73	76
20	2	6	8	13	17	20	23	27	30	33	37	40	43	47	50	53	57	60	63	67	70	73	77	80
21	2	6	9	14	18	21	25	28	32	35	38	42	46	49	53	56	60	63	67	70	74	77	81	84
22	2	7	9	15	18	23	26	29	33	37	40	44	48	51	55	59	63	66	70	73	77	81	84	88
23	3	7	10	15	20	24	27	32	35	38	42	46	50	54	58	61	65	69	73	77	80	84	88	92
24	3	7	11	16	22	25	28	33	37	40	44	48	52	56	60	64	68	72	76	80	84	88	92	96
25	3	8	12	17	22	27	30	35	39	42	46	50	54	58	63	67	71	75	79	83	87	91	95	99
26	3	8	12	17	22	27	30	35	39	43	48	52	56	60	65	69	73	76	82	87	91	95	100	100
27	3	9	13	18	23	28	32	36	41	45	50	54	59	63	68	72	77	81	86	90	95	99	104	104
28	3	9	13	19	23	28	33	37	42	47	51	56	61	65	70	75	79	84	89	93	98	103	107	108
29	3	10	14	19	24	29	34	39	44	48	53	58	63	68	73	77	82	87	92	97	102	106	111	112
30	3	10	15	20	25	30	35	40	45	50	55	60	65	70	75	80	85	90	95	100	105	110	115	120

to be taken from any point on the field lines under vacuum, but such a unit must be operated efficiently in order to get satisfactory results.

5. The usual vacuum should be pulled at the time of the test so that the quality of the gas tested will be similar to that ordinarily utilized in the plant in the manufacture of casing-head gasoline.

6. The temperature of the cooling coils should be between 50° and 60° F. (10° and 17° C.).

7. Scrubber tanks and lines at vacuum stations should be blown out so as to eliminate waste oil and all foreign matter before making the test.

8. Gasoline should be drawn from the accumulator tanks at atmospheric pressure. After measuring the contents, Baumé and temperature readings should be taken.

9. A cubic centimeter jar should be used when weathering; warm water is a satisfactory medium for slowly raising the temperature of gasoline to normal, or 60 degrees.

10. The pressure on the accumulator tank at the time the test is run should be the same as the pressure carried in the gasoline plant. After the sample of gasoline is weathered to 60° F., the Baumé reading should be noted.

11. In starting the test, build up the pressure of gas in the machine and the accumulator tank to 300 lb., and note any leakage in the line or connections of the machine. If no leaks appear, retain the pressure at 300 lb. and blow off the accumulator tank until all liquid is discharged, but do not let the pressure go below 250 lb., on the gage. Close the valve and start reading the meter for the test.

12. If a scrubber tank is located between the compressor and the accumulator, it should be drained of gasoline upon the completion of the test, as some gasoline will always condense in it; this gasoline should be added to the volume drawn from the accumulator, in order to get the full volume of casing-head gasoline coming from the gas that has been metered.

(e) Contracts in which the productivity of casing-head gas is determined by the results of plant production; that is, the total number of gallons of condensate produced during the month is divided by the total volume of casing-head gas utilized, which shows the average productivity in gallons per thousand cubic feet of gas. The schedule shown in Table 1, or one similar to it, may then be used in determining the price of casing-head gas per thousand cubic feet.

(f) An ascending flat scale of prices on a yearly basis. For example, 15 cents for the first year, 20 cents for the second, 25 cents for the third, and so forth, no reference being made to the productivity of the gas.

A casing-head gas contract constitutes a vital part of the investment in the business and, therefore, the terms should receive careful attention. The more important items, such as initial supply of gas, richness, and estimated percentage of yearly decline will certainly not be overlooked, but minor considerations, such as the regularity of the vacuum carried on wells, upkeep of field lines, and return of dry gas to the lease for operating purposes frequently are not given sufficient consideration. Instances are numerous where a provision for the return of a certain amount of dry gas for lease purposes has made it necessary for the gasoline manufacturer to purchase dry gas and supply it at considerable expense to the operating company, in order to fulfill the terms of the contract.

Market Price of Casing-head Gasoline

Market quotations for casing-head gasoline are a controlling factor in the profitable or unprofitable aspect of the casing-head gas business; it should be pointed out that casing-head gasoline is considered in a different class to straight-run gasoline. The various methods of handling this product—blending into different grades, requirements of shipping in order to make same acceptable to certain market demands, and the commercial connections enabling a company to get its output before the public—are matters of grave concern to producers of casing-head gasoline. In conclusion, therefore, the general conditions in the business make it necessary to take a long-range view, including an estimate of the probable effect of future demands and trade conditions, as related to possibilities of motor-fuel substitutes, from the standpoint of efficiency and cost of production.

Modified Oil-well Depletion Curves

By ARTHUR KNAPP, M. E., SHREVEPORT, LA.

(New York Meeting, February, 1921)

OIL-WELL depletion curves, to be of value, should show when a well or lease may no longer be operated at a profit. The difference, at any time, between the total expenditures and the total income of a lease or well may be called the lease status. Plotting this lease status against time will give a curve subject to more accurate and different interpretations than the barrels-time curves.

According to the hypothetical barrels-time curves shown in Fig. 1, well A, at the end of 16 mo., is producing twice as much oil as well B and will continue to produce for another year whereas well B will cease producing in about 6 months.

DATA FOR LEASE STATUS-TIME OR DOLLARS-TIME CURVE

	TIME MONTHS	LEASE STATUS
Lease purchased for $3000...............................	0	−$ 3,000
No expenses chargeable to lease for 3 mo................	2	− 3,000
Well A is started; cost of drilling for month is $16,000.....	3	− 19,000
Well is completed at additional cost of $10,000...........	4	− 29,000
Well A is brought in and flows 10,000 bbl. first month (see Fig. 1). Necessary to invest in tanks, boilers, pipe lines, etc.: difference between expenditures and receipts gives profit for month of $5,000............................	5	− 24,000
Investment is small, cost of operating well A is small, so profits are $15,000...................................	6	− 9,000
Well ceases to flow so there is additional investment for pumping equipment and additional operating expense; net earnings from 6,000 bbl. of oil produced is $4,000....	7	− 5,000
Lease operation now becomes normal and curve becomes smooth	8	− 500
	9	2,000
	10	4,500
	11	6,000
	12	7,700
	13	8,700
	14	9,000
	15	9,500
	16	10,000

A hypothetical lease status-time, or dollars-time, curve of this lease is shown in Fig. 2. At zero time, the lease is purchased for $3000, which is plotted below the zero dollar line. All subsequent entries are plotted below this line until total receipts exceed the total expenditures, when the curve crosses this line and shows a credit, or profit.

Fig. 1 shows that well A made 600 bbl. during the sixteenth month, or 20 bbl. per day, and it indicates that an average daily production of

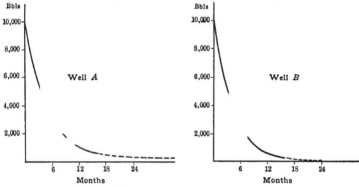

FIG. 1.—HYPOTHETICAL BARREL-TIME CURVES.

18 bbl. per day may be expected during the seventeenth month. But using this curve to determine the probable profit leaves out of account the gradual increase in the operating cost, which occurs as the well becomes older, due to the increased water to be handled, wear on ma-

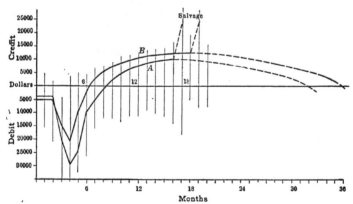

FIG. 2.—HYPOTHETICAL LEASE STATUS-TIME, OR DOLLARS-TIME, CURVE OF LEASE CONTAINING WELL A.

chinery, etc. So that while the well may be profitably operated for 100 bbl. per day, it cannot be profitably operated for 5 bbl. per day.

According to the hypothetical lease status-time curve, though well B was depleted more rapidly than well A, it was the more profitable well at the end of the sixteenth month. It cost less to drill and the difference between income and investment was greater after the well was completed. This curve had not reached the apex at the end of the sixteenth month, although the well was producing only one-half the quantity of oil produced by well A so that well B could have been profitably operated until the seventeenth or eighteenth month. The flatter curve of well B may be

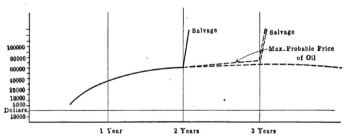

FIG. 3.—TYPICAL CURVE OF LONG-LIVED WELL FOR GULF COAST FIELD.

due to the fact that the operating expenses were uniformly lower than those of well A because they were shared by several properties, while well A was so far from other production as to necessitate its being operated by itself. A difference in the amount of salt water handled would influence the curve.

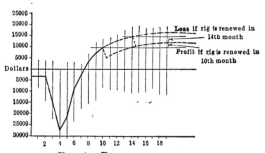

FIG. 4.—DOLLARS-TIME CURVE.

As stated, these curves are hypothetical. Few wells would show a profit for one month and a sufficient loss the next month to' warrant abandoning the well. In a great many cases, the apex of the curve is flat and a small profit will be shown for a period extending from several months to several years. In order to show the method to be followed, the simplest case has been taken.

OTHER USES OF LEASE STATUS-TIME CURVES

Depreciation of the derricks, pumping rigs, and machinery do, not materially affect the lease status-time curves of short-lived wells. If, however, the wells are long lived the salvage value may determine the point at which wells may be profitably pulled. Fig. 3 shows a typical curve of a long-lived well for the Gulf Coast field. This well shows a good profit up to the end of the second year. During the third year there will be a small profit but the curve shows that the salvage value at the end of the second year is greater than the probable salvage value at the end of the third year plus the probable profit for the year. It would, therefore, be more profitable to abandon the lease at the end of the second year than to operate during the third.

In attempting to analyze such curves for as long a period as a year account must be taken of the probability of a fluctuation in the price of

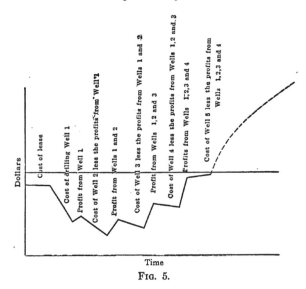

Fig. 5.

oil. Fig. 3 shows that even taking into account the maximum probable increase in the price of oil the maximum profit from this lease will be obtained by pulling the wells at the end of the second year. If considerable repairs are necessary this form of curve is valuable for deciding whether or not the investment is warranted.

The recovery value of wells A and B added to Fig. 1 show that while well B showed a greater profit than well A when both reached the apex of their curves, the final profit from each well was the same, as the salvage from well A was greater than that from well B.

In the case of accidents, such as fire, it is hard to determine whether or not the investment in new rigs and machinery will be profitable. With the dollars-time curve, as shown in Fig. 4, this question may be decided with some degree of accuracy. If the fire occurred in the tenth month and the renewal was estimated at $5000, transposing the probable lease status curve shows that the additional investment is warranted, for the apex of the curve finally rises above the point at which the fire occurred. If, however, the fire occurred in the fourteenth, the curve will not rise above the profit shown at the time of the accident, which means that it would not be profitable to renew the machinery.

LEASES WITH MORE THAN ONE WELL

While the majority of leases have more than one well and the several wells are not all drilled at one time, this does not affect the curve after the drilling program is complete and the production is settled. While drilling the second and subsequent wells the lease status curve may run

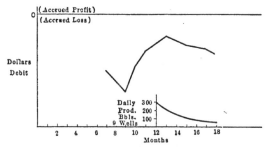

FIG. 6.—LEASE STATUS-TIME CURVE FROM A PROPERTY IN PINE ISLAND, LA., DISTRICT. RECORDS NOT AVAILABLE PREVIOUS TO FOURTH MONTH. NINE WELLS DRILLED ON 160-ACRE LEASE BETWEEN FOURTH AND NINTH MONTH. DECREASE IN PRODUCTION AND INCREASE IN AMOUNT OF SALT WATER WAS SO RAPID THAT FOUR MONTHS AFTER DRILLING PROGRAM WAS COMPLETE THE LEASE CEASED TO BE PROFITABLE WITHOUT HAVING PAID OUT. AN INCREASE IN PRICE OF OIL IN FIFTEENTH MONTH SERVED TO CHECK THE LOSS BUT DID NOT TURN IT INTO A PROFIT. CURVE SHOWS CONCLUSIVELY THAT REGARDLESS OF AMOUNT OF PRODUCTION OR AGE OF WELLS, THE LEASE SHOULD BE ABANDONED. INSERT SHOWS DEPLETION CURVE OF LEASE. PREVIOUS TO TWELFTH MONTH, NO GAGE OF DAILY PRODUCTION COULD BE TAKEN, AS WELLS PRODUCED INTO EARTHEN STORAGE.

nearly horizontally, the income from production of the wells drilled offsetting the cost of wells drilling. When all the wells on a lease are pumping into the same tank and all of the wells have been drilled within a reasonable length of time of one another, the depletion curve of the lease is a fair gage of the depletion of each well.

PROPERTY VALUATION

The value of a property is its salvage value plus its probable earnings up to the time it reached the point of maximum profit, interest, taxes, and insurance. If the lease status-time curves have been properly and accurately drawn, the value of any property may be taken directly from its curve. As interest, taxes, and insurance are constants, they do not affect the shape of the curve and need not be included for ordinary analysis.

VARIATIONS OF DATA

Variations in the systems of bookkeeping may produce slightly different curves but they usually allow for the same analysis as outlined. If interest and overhead are charged monthly to the expense of the lease they make a more accurate curve but do not affect the shape. If depreciation is charged off yearly against the lease, it should be prorated monthly at the end of the year and the lease status curve redrawn.

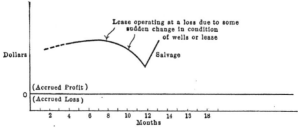

FIG. 7.—END OF A LEASE STATUS-TIME CURVE IS GIVEN ABOVE. IF CAREFUL ATTENTION HAD BEEN PAID TO STATUS OF LEASE, IT WOULD NOT HAVE BEEN OPERATED AT A LOSS FOR SO LONG A TIME.

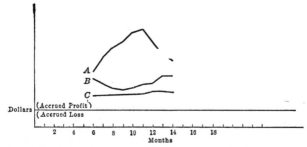

FIG. 8.—A. LEASE WAS BEING OPERATED AT GOOD PROFIT UNTIL SALT WATER BROKE IN AND RUINED ONE OF THE WELLS.
B. ADVANCE IN PRICE OF OIL CHANGED A SMALL LOSS INTO A SMALL PROFIT.
C. LEASE WAS OPERATED FOR NINE MONTHS AT TOTAL PROFIT OF $48. DEPRECIATION ON MACHINERY WOULD MORE THAN OFFSET THIS.

The easier way, provided the bookkeeping system will allow, is to charge the total investment against the lease and credit the salvage to the lease when the operation is stopped.

EXAMPLES FROM PRACTICE

The curves shown in Figs. 6, 7, and 8 are taken from data of actual operations. In one case, the full curve is given and in the rest just enough of the curve is shown to illustrate the point desired. The values of the dollar ordinate have not been given, the various curves having been reduced to the same size as various scales on the dollar ordinate would be confusing.

DISCUSSION

Roswell H. Johnson,* Pittsburgh, Pa.—The curve suggested differs from many other curves in being what might be called synthetic; it shows in one line the net result of several items to get the result in which the executive is particularly interested. Another feature is that it is simply calculated, so that it can be made cheaply. If the executive does not care about graphs, the same data in tabular form can be a guide. The graph should carry the numerical data on the sheet for the use of the executive.

It would be a mistake for the executive to feel that all the information he needs is in this one graph. He should also watch graphs showing his cost per well for maintenance, in order to check the relative efficiency on his leases and personnel; he should also keep a graph on decline but the ordinary decline curve is not helpful in old-age wells. The slope of the decline on an old-age well does not show up fine differences at all easily so that a graph showing the proportion of one year to the previous year (persistence factor) will show readily the changes in his rate of decline.

---

* Professor of Oil and Gas Production, University of Pittsburgh.

Barrel-day Values

By Glenn H. Alvey and Alden W. Foster, Pittsburgh, Pa.

(New York Meeting, February, 1921)

THE measure of value of an oil property is approximated by the length of time it takes to "pay out;" viz., the time required for it to return the original investment. This time varies in different fields. In the Appalachian and Mid-Continent fields, a good investment pays out in about four years; in California, it requires a slightly longer time; and·in the Gulf field about two years.

The two principal methods for establishing these values are based on acreage, as in California, and on production, as in the Appalachian field. The method of establishing values based on production[1] was worked out in the Appalachian fields and approximates the value remarkably well in some fields; but it is a rule of thumb and should be used intelligently. Briefly, the method is as follows: An arbitrary number of dollars (called the barrel-day price) multiplied by the number of barrels of settled daily production of the property .gives the value. In its crudest application, the barrel-day price is determined by the "$10 to $0.01" rule, which means that the barrel-day price is one thousand times the prevailing price of oil. For instance, with oil at $3.50 per bbl. the barrel-day price would be $3500 per bbl. This rule, however, is not strictly applied, for the barrel price is varied according to whether or not the wells "hold up." But when a barrel price for a district has been fixed it is quite general to raise or lower the price with the fluctuation in the price of oil; here the "$10 to $0.01" rule is used extensively. However, the prospective purchaser takes into account tangible equipment upon the property, the extent to which the drilling program has been carried out, whether or not the wells are "shot" or natural, depth of wells, spacing of wells, paraffin trouble, etc.

The basis of this method is settled production. As soon as production is considered settled, a flat barrel-day price is generally applied to all properties, no matter what their age. If it were possible to show that this flat rate was erroneous and that the value of the property depended on the point on the decline curve at which the wells happened to be (in other words, their age), and also on the operating costs, the future price

---

[1] Acknowledgment is due to Roswell H. Johnson, at whose suggestion this problem was undertaken.

of oil, and the discounted value of the dollar, the buyer who recognized these facts and bought accordingly (also observing depth and spacing of wells, etc.) would be in an advantageous position; and the seller (although he does not usually have sufficient data at hand to make a complete appraisal) would be able to know when to sell his property to the best advantage.

It is the purpose of this paper to show that there is a best time to buy or sell production; in other words, that the value of property varies with a number of complex factors, most of which may be used in making an analytic appraisal of a property. To demonstrate the possibility of doing this, two problems are presented:

First.—In a given pool, one well was brought in May 1, 1920, one was one year old on that date, another two years old, still another

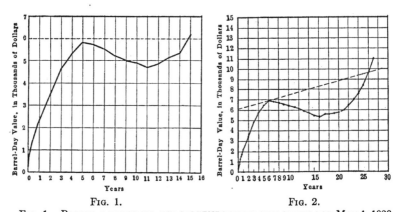

FIG. 1.

FIG. 2.

FIG. 1.—BARREL-DAY VALUES AND CURRENT BARREL-DAY PRICES FOR MAY 1, 1920, FOR AVERAGE WELLS OF DIFFERENT AGES IN CAMERON DISTRICT, PA.

FIG. 2.—BARREL-DAY VALUES AND CURRENT BARREL-DAY PRICES, ASSUMING THAT WELL CAME IN WITH OIL AT $6 PER BARREL, FOR DIFFERENT YEARS IN LIFE OF AVERAGE WELL IN CAMERON DISTRICT, PA.

was three years old, and so on. Which of these wells is it best to buy, the one that just came in, one four years old, or one seven years old?

Second.—To get the best return of the money invested, should a well that comes in on May 1, 1920, be purchased on that date, or two years from that date, or some years later?

For the first problem, the barrel-day value of an average well was determined on the assumption that it came in May 1, 1920; also, the barrel-day value as of May 1, 1920 if the well had come in a year prior to that date, two years prior to that date, and so forth, until a sufficient number of years prior to make it ready for abandonment on May 1, 1920, had been considered. These barrel-day values were plotted and curves drawn. Similarly, for the second problem, analytical appraisals were made for each year in the life of an average well, assuming that the well came in May 1, 1920. That these values are different from those of the preceding problem is due to the advancing price of oil. In the first case, each appraisal starts with the price on May 1, 1920; in the second case advanced prices of oil were used. In working out these appraisals, the following form was used:

TABLE 1

YEAR	NET YEARLY PRODUCTION	PRICE PER BARREL	GROSS INCOME FOR YEAR	WELL COST FOR YEAR	NET YEARLY INCOME	COMPOUND DISCOUNT FACTOR 7½ PER CENT.	DISCOUNTED INCOME
A	B	C	$B \times C = D$	E	$D - E = F$	G	$F \times G = H$
1							
2							
3							
4							
....							
....							
n					(I) Total		

Economic limit
J = salvage,
K = compound discount factor for n years,
$J \times K = L$ = discounted salvage,
$I + L$ = present worth,
M = daily production at beginning of year,
$L \div M$ = barrel-day value.

These values were worked out for the Cameron district, Pennsylvania, of the Appalachian field, and the Osage Nation of the Mid-Continent field. The production data used in the former case were secured from a company operating in that district; in the latter case, from Beal's Decline Curve.[2] An advancing price of oil was predicted; 2 per cent. for the Cameron district (which was chosen in order to give a constant advance up to $10 during the life of the well), and 10 per cent. for the Osage until the price reached $10 for the remaining years (which was taken so as to accord more closely with the price predictions that are considered to be correct for the Mid-Continent field). Costs were taken at $600 per year for the Cameron district and $1 per well day for the Osage. A 7½ per cent. compound discount factor was used. The salvage value used for the Cameron district was $2000, and for the Osage $1000 (see Table 2).

Fig. 1 shows the curve for wells at different ages in the Cameron district; the barrel-day values are plotted against the ages of the wells. According to this curve the highest barrel-day value is for a well five years old; this, therefore, would be the best well to buy because the present

---

[2] U. S. Bureau of Mines *Bull.* 177, 108.

worth of the future production of this well, compared to its daily present production, is the highest. The barrel-day values of older wells decrease until those ages are reached at which the well is near abandonment, when the barrel-day values rise sharply. This is because of the higher present worth of the salvage (the salvage value is discounted less and less with the advancing age of the well). It would be attractive to buy an old well on the barrel-day rate because one can sell the salvage and make a profit.

TABLE 2

DISTRICT	PRICE OF OIL AS OF MAY 1, 1920	WELL COSTS PER YEAR	COMPOUND DISCOUNT FACTOR, PER CENT.	SALVAGE	COMPOUND DISCOUNT FACTOR FOR SALVAGE. PER CENT.
Cameron, Pa......	$6.00 with 2 per cent. rise (yearly)	$600.00	7½	$2000.00	7½
Osage, Okla.......	$3.50 with 10 per cent. rise up to $10 and then a flat rate of $10.	$365.00	7½	$1000.00	7½

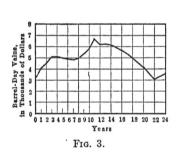

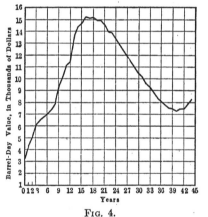

FIG. 3.

FIG. 4.

FIG. 3.—BARREL-DAY VALUES FOR AVERAGE WELLS OF DIFFERENT AGES IN EASTERN PART OF OSAGE INDIAN RESERVATION, OKLA.

FIG. 4.—BARREL-DAY VALUES FOR DIFFERENT YEARS IN LIFE OF AVERAGE WELL IN EASTERN PART OF OSAGE INDIAN RESERVATION, OKLA.

Fig. 2 shows the changes on barrel-day values for different periods in the life of a single well in the Cameron district. According to this curve, the best time to buy is six years from the date the well was brought in, for then the well has its highest barrel-day value.

Figs. 3 and 4 are similar to Figs. 1 and 2, respectively, but are for the Osage district. The most interesting fact is that the highest barrel-day values come much later than for Cameron. This is due to a flatter decline curve and a steeper predicted price curve. The significance of this is that no generalization can be drawn for the best year to buy that will apply to all pools. It must be worked out for each pool and the year determined will be influenced by the price of oil predicted by the appraiser.

CONCLUSIONS

The barrel-day value for settled production is not flat, as is generally supposed, but from completion, increases with the age of a well up to the maximum and then decreases; it increases again when the salvage value becomes attractive.

There is a method by which the age at which a well has the highest barrel-day value can be determined.

This age varies with the different pools and is due to five causes: the decline curve of the pool, future price of oil, well costs, compound discount factor, and salvage value.

DISCUSSION

ROSWELL H. JOHNSON,* Pittsburgh, Pa.—The belief that one may appraise on a flat barrel-day value is one of the most dangerous blunders in the oil business. It is properly merely a method of expressing prices. Some appraisers start with it as a basis and then work plus and minus from it. Such a procedure is crude and objectionable.

These curves of Alvey and Foster show characteristically three stages. In the first stage the decline rate is the dominant factor, and throughout it the unit value advances; the rapid increase in unit value is the outstanding characteristic. The one exception is where we have wells that are very short-lived, such as some Ranger and Gulf Coast wells, where it is quite possible that the second year would show a poorer value than the first, because the whole thing has been shortened and the year unit is too large a unit for such a curve.

The next stage is where the cost is the dominant factor; the rate of decline is less weighty and the cost is steadily becoming a more important factor.

The third stage is dominated by the foreshadowing of the salvage due and represents the final up turn at the end. A curve, such as is made here, can be constructed without salvage value, treating the salvage value as a separate unit; but it seems best to put in the salvage value consideration.

In Table 1 is the best working formula for oil appraisal that has yet appeared in literature. In column E, well costs are handled by the year, after yearly production has been multiplied by the price per barrel. That is the place to take out the cost, because if taken earlier it will be

---

* Professor of Oil and Gas Production, University of Pittsburgh.

taken out on the barrel basis, which is so variable as it depends on production. Costs then ought to be taken out on the basis of the well cost, not the barrel cost. This formula puts in discounted salvage, which feature has not always been recognized.

The formula makes obsolete the time-to-pay-out and the acre-yield methods, in their usual forms except for the most rapid work. Both suffer so severely by their non-recognition of the time element of compound discount, that they are fallacious. It is surprising to note, for instance, that in a long-lived curve, such as Salt Creek, the compound discount, if one takes it at 10 per cent, cuts down the value to only 45 per cent of the uncorrected acre-yield value. With such an enormous range as that, the danger of acre-yield methods is seen.

The time-to-pay-out methods are so crude, in that they ignore the shape of the curve after the well is paid out, that they cannot be considered. Acre-yield and time-to-pay-out methods can be used for quick appraisal, however, if one works a series of annual analytic values and uses these values to set up tables using various assumptions.

Another feature of this paper that demands attention is the method of predicting a price advance. It is here taken on a percentage basis to a plateau, and then flattened. This is better than the method of fixed advance in cents per barrel; because, first, the price of different grades of crude do not fluctuate with fixed differentials in cents, but by percentage of the highest grade although these percentages are not absolutely fixed. For instance, all of us have been looking at these recent cuts in prices. We noticed that Kansas-Oklahoma oil· was cut 50 per cent., and so predicted a drop in Pennsylvania of 50 per cent.; and we knew before the last cut that there was another cut due, and still another is due.

Furthermore, the theory of making price advance based on the fixed amount per barrel, would have to be dependent on the thought that as the price rises the demand is shortened. But in the case of oil and gas we have some peculiar conditions. There is a nearly constant expansion of the market for oil. We have oil going into new things—the tractor, the motor boat, the Deisel engine—so that these larger needs postpone saturation.

Isostatic Adjustments on a Minor Scale, in their Relation to Oil Domes*

By M. Albertson,† E. M., Shreveport, La.

(New York Meeting, February, 1921)

At Cobalt, Ontario, Canada, a lake was drained to facilitate mining, by the Mining Corpn. of Canada, during the spring and early summer of 1915. Previous to pumping out the water, great quantities of sands and slimes from concentrating plants had been discharged into the lake and during and after the lake's drainage, its basin was a receptacle for tailing products. As the writer was at work along the shore line as the lake was being drained, he had a good opportunity to observe the changes that took place as the water was withdrawn. Some adjustments between the incoming sands and the mud in the lake had taken place before pumping was commenced. One of the most interesting results was the appearance of a small dome in a path the writer traversed twice a day for several months; he remembers distinctly the difficulty of crossing this.

Cobalt Lake owed its existence to the gouging out of a rock basin by glaciation. The long axis of the basin closely follows the strike of a thrust fault of about 500 ft. (160 m.) vertical displacement. The lake was originally shaped somewhat as shown. The length was about 3000 ft. (914.4 m.), the width at the lower lobe was about 1000 ft., and the width at the narrows probably 400 to 500 ft. The original depth of water varied from 20 to 30 ft. (6.1 to 9.1 m.), near where the island later was formed, to 60 to 70 ft. (23.6 to 27.6 m.) in the widest part of the lower lobe. At the narrows, the depth was 30 to 40 ft. Above the bed rock was sand, with boulders near the bottom, and mud.

During the building of the railroad in 1903, considerable filling was done along the right of way. With the commencement of mining operations, about 1905, waste rock and mill tailings were dumped into the lake. One of the mining companies operated a hydraulic giant to remove the glacial debris from several hundred acres of rock surface; much of the sand and clay from this operation was deposited in the lake. About 1,500,000 tons of mill tailings, composed of sands and slimes, were discharged into the lake's waters previous to 1915. Most of this material

*Published by permission of R. O. Conkling, chief geologist, Roxana Petroleum Corpn.

† Geologist in charge Louisiana Division, Roxana Petroleum Corpn.

settled near the point of discharge but the fine slimes spread throughout the lake basin Four artificial deltas were formed by tailings from various concentrating mills, somewhat as shown along the upper end of the lake.

The mud of the lake bottom was a thin black oozy slime, much too thin to support a man's weight and too thick to swim in. Its specific gravity was much less than that of the sands and slimes coming in; probably it was not much greater than that of water. The incoming

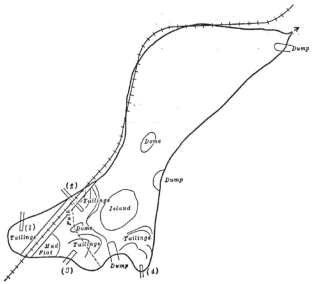

FIG. 1.—PLAN SHOWING LOCATION OF DOMES, ISLAND, TAILING PILES, ETC.

material pushed it aside in places and caused it to bow up as islands and near islands in other places. The main part of the slimes from the tailing discharge at 2, settling over the narrow part of the lake bottom, strengthened the mud layer. Much of the mud from the upper end of the lake was forced out toward the center of the upper lobe. After the pumps had lowered the water level a few feet, an island a few hundred feet in diameter appeared.

Small domes appeared near the shore at several points. After the upper end of the lake was entirely drained, a large mud dome appeared in about the middle of the narrows. It is perfectly clear that the weight of the sands and slimes became too great to be supported by the thin oozy mud and consequent buckling resulted in the formation of a dome.

The process of adjustments that brought the lake domes into existence is conceived to have gone on somewhat as follows: At the points where the tailings were discharged into the lake, the mud, which was less dense and in a jelly-like condition, was forced aside. Since the tailings entered the upper basin of the lake from several well distributed points along its shores the mud was forced toward the center of the lake. At the same time a thin layer of fine sands and slimes was deposited over the whole of the lake bottom, but chiefly near the shore and in a fan-like arrangement from the points of tailing discharge. The effect of this was to strengthen the mud layer near the shore and to weigh it down so that the mud layer was weakest in the center of the lake. When the weight of the sands became sufficient for the sand to displace the mud the displacement occurred where the mud layer was weakest.

It is the writer's conclusion that domal structures have originated in this manner in the Tertiary deposits of the Mississippi embayment region and that these structures of the Tertiary are, in many cases at least, non-existent in the more compacted Cretaceous formations under them. Thus a dome structure in surface formations does not necessarily mean a dome in the Cretaceous oil-bearing sediments.

If this process is active in one region it must be considered as a structural factor in all areas of sedimentation. It is suggested as a factor in the formation of certain domes observed in the Pennsylvanian area of Missouri.

During 1911, 1912, and 1913, the writer, then a geologist for the Missouri Bureau of Geology and Mines, became well acquainted with minor dome structures that characterize the Pennsylvanian strata of northern Missouri. Some of these domes are shown on a structural map of Kansas City.[1] Many others are known in the coal mines of the state. The origin of these domes has long been a puzzle. It is of course possible that they are entirely the result of regional folding stresses, but this view does not appear entirely logical.

---

[1] McCourt, Albertson and Bennett: Missouri Bur. Geol. and Mines, *Geology of Jackson County* (1917) **14** [2] Pl. xvi.

Anthony F. Lucas

ANTHONY F. LUCAS died suddenly at his home in Washington, D. C., on Sept. 2, 1921. Captain Lucas, as he was known to us, was born in Dalmatia, Austria, in 1855, of Montenegrin ancestry. He was graduated as an engineer at the Polytechnic of Gratz and served in the Austrian Navy as second lieutenant. In 1879, he obtained leave of absence and visited an uncle in the United States. After an extension of this leave of absence, in order to undertake an engineering engagement in the lumber district of Michigan, where he resided, he decided to become an American citizen. He was naturalized in May, 1885.

His name was Luchich, but as his uncle had adopted the name of Lucas, which was more easily pronounced by Americans, from his entrance to this country, he used this Anglo-Saxon form. Without knowing this fact, upon first meeting him a person was sometimes surprised to note the rather Germanic pronunciation of the Captain.

Although he subsequently revisited Austria with Mrs. Lucas, he made his permanent home at Washington, D. C. His son served with distinction in the A. E. F. during the World War.

His activities in this country as a mining engineer were at first in Colorado and later in salt mining at Petit Anse and Belle Isle, La. During his salt investigations, his attention was directed toward the possibility of oil in the Gulf Coast region and in January, 1901, his well, the "Lucas Gusher," on Spindle Top, Tex., started a new era in the oil business and his reputation as discoverer made him famous throughout the world.

Captain Lucas became a member of the A. I. M. E. in 1895. During 1914, 1915, 1918, and 1919, he was chairman of the Petroleum and Gas Committee of the Institute and was at all times prominent in Institute affairs.

As to the personality of Captain Lucas, the lasting impression is of a courteous hospitable gentleman, genial, affable, obliging, and helpful with his advice or assistance to any colleague. He was sincere, honest, firm against all obstacles, backing his judgment with his own hard work along any course which he had determined to be correct.

His value in the engineering world lies mainly in the petroleum industry. In the oil business all wild-catters are pioneers that deserve credit and gratitude upon their success. There are, however, names that particularly stand out in our history. Drake conquered such obstacles as ridicule, lack of finances, and started the oil business. In 1901, Captain Lucas had the conviction that Spindle Top, a dome rising about 12 ft. above the coastal prairie south of Beaumont, Tex., contained commercial oil. He was scoffed at by practical oil men of the East. Noted geologists were condemnatory on the ground that such an occurrence was unprecedented. It was a rank wildcat. Savage, Sharp, and others had tried to drill wells, one at least, with cable tools, and had given up, but Captain Lucas put down one well, which was ruined at about 600 ft., having had a showing of heavy oil. Obtaining financial support, with the J. M. Guffey Petroleum Co., he drilled another well to the depth of less than 1100 ft. In January, 1901, this well came in at a rate estimated as high as 125,000 bbl. per day and flowed wild for ten days before it was finally controlled.

ANTHONY F. LUCAS.

The discoverer estimated the flow at 75,000 to 100,000 bbl., but the above figure is an estimate of a civil engineer who gaged it by a full 6-in. stream of oil 200 ft. high and what tests could be made of runoff of the oil.   This discovery astounded the oil men of the world.

It should be recalled that for the drilling of this well a rotary rig was used; this method was then in its infancy (having been used only in Corsicana and in some water-well drilling) so that the wild-catter had but slight benefit of experience of others.   He was obliged to devise his own methods of combating drilling difficulties, and in doing so earned rights to patent, of which I do not believe he availed himself.

I quote here a question and answer published in the *Mining and Scientific Press* (Dec. 22, 1917):

T. A. RICKARD.—I hope, Captain, that you received a proper financial reward this time?

A. F. LUCAS.—I did, but my chief reward was to have created a precedent in geology whereby the Gulf Coast of the Coastal Plain has been and is now a beehive of production and industry.

We would all have asked the question and must regret that, while his later career was undoubtedly financially successful, at Damon Mound and some other localities where our deceased fellow member wild-catted, he was too far ahead of his time to make further successes. The answer was characteristically sincere and intrinsically true. Old timers remember Beaumont, a small lumber town with mud streets, becoming a regular beehive during 1901. We recall the forest of derricks with overlapping legs on the 300-acre Spindle Top; the lakes of oil lying unused on "the Hill" without proper transportation for removal; the hurly burly where land was sold and paid for at the rate of one million dollars per acre; the wild stock-selling schemes that filled the daily press. These are disagreeable though interesting sides of the feverish and foolish oil stampede. On the opposite and wonderful side, as a direct result of Captain Lucas' perseverance, was the rise of legitimate operators to success; namely, the J. M. Guffey Petroleum Co. (The Gulf Refining Co.) and The Texas Co. Further development was stimulated at Sour Lake, Batson, Saratoga, Humble; later, at Damon Mound, Goose Creek, Hull and other fields.

This all refers to the Gulf Coast, but why should Captain Lucas have confined his influence to that region? There are today, in the Mid-Continent and other districts, veritable powers in oil production who had their lessons on the derrick floors of Spindle Top rigs subsequent to the "Lucas Gusher."

Along the scientific side, there has been much discussion of Coastal Plain problems. By such free exchange of knowledge, advance has been made toward the truth, not only as applicable to Texas and Louisiana but to Oklahoma, California, and the East; to Mexico, South America, and other foreign fields.

This is what Captain Lucas was after in starting his oil venture; this is what we are after—the truth. We owe his memory gratitude for starting a new era in oil production twenty years ago, which has had tremendous effect in the professional and business lives of all of us down to the present time.

Although he considered himself "properly rewarded," we may believe that even though he was beyond want, he was entitled to much greater financial reward than he received. Certainly he deserves a prominent and permanent place in oil history because of his Spindle Top discovery.

H. B. GOODRICH.

Rock Classification from the Oil-driller's Standpoint

By Arthur Knapp, M. E., Shreveport, La.

(New York Meeting, February, 1920)

The ordinary well log is subjected to a great deal of criticism, much of which is well founded. Sometimes, though, the difficulty in interpreting the log is due to the fact that the geologist or engineer using the logs does not know the limitations of the drilling method used. The rotary drill, especially, has inherent limitations that make it difficult to secure definite information at all times. The identification of well-defined key beds is about all that can be expected from the rotary log. The formation in a drilled hole, as reported by the driller, has a direct relation to the speed with which the drill makes the hole or to the reaction of the various strata on the bit, called the "feel of the bit." When this is not thoroughly understood by the geologist or engineer endeavoring to interpret the log, the result is an erroneous correlation with other wells or a discarding of the log as worthless.

General Terms

Hard and Soft.—Hard and soft are relative terms. In the case of well logs, they are very misleading as they are used in connection with both resistance to abrasion and resistance to percussion. In technical rock classification, hardness is relative resistance to abrasion. The term brittleness is used in connection with resistance to blows. These terms are misleading to the geologist or engineer who is not familiar with both the cable-tool, or standard tool, method of drilling and the rotary method. In the case of the standard tools, the driller's report of the hardness of the formation is in terms of its resistance to blows. For instance, a cable-tool driller might be able to make from 30 to 50 ft. a tour in a brittle limestone, which he would call soft and at the same time he might call a relatively soft (from a purely mineralogical standpoint) gypsum hard, because it is somewhat elastic and is not readily broken by blows. The rotary driller would reverse the terms. The limestone is hard in that it resists the abrasive action of the bit, while the gypsum might be soft in that it is readily cut by the rotary bit. It is rare that wells drilled by the standard tools are correlated with those drilled by the rotary, but the technologist who has worked with well logs from one system might be misled when working with the other.

Sticky.—With the rotary drill, a formation is sticky which cuts in large pieces that adhere to the bit and drill pipe. A formation that is sticky with the rotary is usually sticky with the cable tools. On the other hand, formations are encountered in which the cable tools stick, either owing to the elasticity of the formation or to the fact that the drilled-up particles do not mix readily with the water in the hole and settle so quickly as to stick the bit. These formations might not appear sticky to the rotary driller.

Sandy.—This term may be used quite accurately by the cable-tool driller. He obtains samples of the formation through which he passes, of sufficient size to determine the relative amount of sand to clay or sand to shale in any formation. In the case of the rotary drill, this term is misleading.

The rotary well is drilled with the aid of a "mud" of varying density It is usually thought of as a mixture of clay and water with a small amount of suspended sand. As a matter of fact this mud often contains as high as 40 to 50 per cent. sand. This sand tends to destroy the colloidal properties of the mud and the action of the mud on the walls of the well is the same as a thin mud with less fine sand. The water would tend to exchange the suspended sand for mud from the walls of the well, thus thinning the well wall. It is impossible to settle out the very fine sand in any rotary mud. An easy and quick way to separate the two for examination is to fill the glass of a centrifugal separator half full of mud and add a saturated solution of common salt. The sand will be thrown to the bottom when the machine is turned for a short time. The mud alone can be turned indefinitely without any appreciable separation.

Any change in the density of the mud changes its capacity to carry sand. Even a small shower falling on the slush pit will change the density enough to cause some of the suspended sand to be precipitated. These properties of the mud lead to error in the observation of the formation. If a clay formation containing a moderate amount of sand is encountered while drilling in a mud low in sand content, the mud will absorb most of the sand, which will not settle out in the overflow ditch and its presence in the formation will not be noted, if not felt by the action of the bit in drilling. If, some time later, the mud is thinned by adding water this sand will appear in the overflow and may be attributed to a formation many feet below the one from which it actually originated.

The so-called "jigging" action of the rising column of mud on the sand or cuttings also leads to misinterpretation. I have often heard drillers remark that the deeper you drill, the finer the sand. This is not true, but it is true that the deeper you drill, the finer the sand or cuttings brought to the surface by the mud. The coarser particles have been pounded into the walls of the well or broken and the deeper the well, the more opportunity the drill pipe has had to do this.

A change in the speed of pumping the mud also causes a change in the amount and size of the cuttings that appear at the surface. Thus, in the case of the rotary, "sandy" may have little or no meaning when applied to a formation. The term sandy is often used in contradistinction to sticky. A formation that drills easily and is not sticky is often put down as sandy because sand tends to interfere with the stickiness. Sand does not always account for the lack of stickiness but the latter is often attributed to its presence.

Dark and Light.—This brings up the subject of color. The first question is the age of the specimen when the color is determined. A wet specimen, fresh from the hole, has an entirely different color from the same specimen dried. Specimens, when dried, bleach and deteriorate. Many of them air slack or oxidize and change composition altogether. The terms light and dark should be used only for the extremes. They are, in general, relative and therefore very indefinite and misleading. A sample of wet shale examined under an electric light might appear many shades darker than in day light. It is better to use a definite name than the words light and dark; such as slate-colored or chocolate-colored shale. On the other hand, color is not very important except in key beds, which are usually of extreme shades, either very light or very dark.

FORMATIONS

Clay, Gumbo, Tough Gumbo.—Clay is readily recognized by the "feel of the bit" while drilling with either cable tools or rotary. To some drillers all clay is gumbo while to others gumbo is only sticky clay. Some clays have the property of cutting in large pieces but do not adhere excessively to the bit and drill pipe and are designated as "tough."

Sand, Packed Sand, Water Sand, Quicksand, Heaving Sand, Oil Sand, Gas Sand.—Free, uncemented sand is easily recognized by the feel of the tools in both systems of drilling. In rotary territory, we often run across the term "packed sand." This is a sand that is slightly cemented with some soft easily broken cementing material, such as calcium carbonate. It cuts, when drilled with a rotary, with much the same feeling as when cutting crayon with a knife. The cementing material is dissolved by the mud or the sand grains are all broken apart before reaching the surface, so that the driller finds only sand in the overflow. A microscopic examination of sands from the overflow often shows cementing material to be present when not suspected by the action of the bit.

Water sand is a sand containing water. There is no specific sand associated with water; any porous formation may or may not contain water. In the case of both rotary and cable tools, sand that is fresh and bright and has a clean appearance when taken from the well impresses one as being a water sand and probably does come from a wet stratum. If it so happens that the sand has been thoroughly mixed with the mud in the hole so that each particle is colored by a film of mud, it does not appear fresh and clean and does not give the impression of being a water sand. This may be because the formation from which it came was dry or nearly so or simply because conditions were right for the quick coloring of the sand by the mud. Whether a given porous formation is a water stratum or not can only be determined by testing. It is only in rare cases that the hydrostatic pressure is sufficient to cause the thinning of the mud in a rotary hole. While drilling in a dry hole with the cable tool, it is known at once how prolific a porous stratum is.

A sand containing no cementing material nor clay very often caves badly in the hole. If this sand settles with such rapidity as to threaten to stick the tools, it is designated quicksand. Such a free sand may, on the other hand, have such properties that it seems to tend to float. It not only caves but fills the hole above its original horizon, sometimes heaving clear to the surface. This sand is called a heaving sand. The presence of gas or a high hydrostatic head often accounts for the heaving of the sand.

An oil sand is a sand containing oil. There is no particular sand which is associated with oil; any porous stratum might contain oil. A porous stratum containing oil is often called a sand although it may actually be a limestone.

A gas sand is any sand containing gas; even a hard limestone is sometimes designated as a gas sand.

Boulders and Gravel.—True boulder formations are rarely encountered in drilling for oil. They are encountered above the Trenton in Ohio and Indiana and occasionally in California. Concretions are often encountered which fall into the hole and follow the bit for some time and are reported as boulders. A green rotary driller will report boulders when he is drilling in sticky gumbo, which causes the bit to jump excessively.

Gravel is also quite rare and as it is a question what is coarse sand and what is gravel, a report of gravel may mean coarse sand. The cable tools will bring up gravel so that it may be recognized. Loose shale or oyster shells may be reported as gravel by the rotary driller.

Shale.—Shale, to many drillers, is only that kind of true shale which appears in the overflow, or bailer, in flakes, that is, laminated shale with well-defined bedding. Other drillers include formations that are sedimentary in character and are consolidated enough to appear in the overflow, or bailer, in pieces as large as a pea or larger. They usually call a shale too hard to scratch with the finger nail rock, particularly in rotary territory. The rotary driller finds it hard to differentiate between hard shale and soft limestone.

Rock Classification Summary

General Class	Rotary-drillers' Term	Use in Rotary System	Cable-drillers' Term	Use in Cable-tool System	Technical Equivalent
Sands.........	Sand	Any uncemented sand.	Sand	Any uncemented sand; also many slightly cemented sands or very porous formations.	Sand
	Water sand	Sands, the samples of which appear clean and bright. Sands tested and found to produce water.	Water sand	Sands producing water.	Sand
	Quicksand	Sands that cave and settle rapidly.	Quicksand.	Sands that cave and settle rapidly.	Sand
	Heaving sand	Sands that cave and are forced up the hole.	Heaving sand	Sands that cave and are forced up the hole.	Sand
	Oil sand	Sands or other porous formations containing oil.	Oil sand	Sand or other porous formation containing oil.	Oil sand
	Gas sand	Sands or other porous formations containing oil.	Gas sand	Sand or other porous formation containing gas.	Gas sand
Gravel, boulders........	Gravel	Any formation having the feel of gravel while drilling.	Gravel	Correctly used.	Gravel
Clay, shale.....	Boulders	Large loose ... of any formation.	Boulders	Correctly used.	Boulders
	Clay	Clay or soft shale; usually not sticky.	Clay	Correctly used.	Clay, or sandy clay
Consolidated formations.	Shale	Soft sticky clay.	Shale	Soft sticky clay.	Clay
		Formations having parallel bedding.	Shale	Consolidated clays.	Shale
	Rock	Any consolidated formation.	Rock	Term not used.	Rock
	Gas rock	Any rock formation containing gas.	Gas rock	Term not used.	Rock
	Chalk rock	Applied to light-colored chalk only.	Chalk rock	Correctly used.	Chalk
	Sand rock and sand- stone	Terms used interchangeably for all cemented sand.	Sandstone	Correctly used.	Sandstone
	Packed sand	Loosely cemented sand.	Packed sand	Correctly used.	Sandstone
	Shell	Thin layer of hard material.	Shell	Thin layer of hard material.	Rock
	Shell rock	Any consolidated formation con- taining fossil shells.	Rock with shells	Formation containing shells.	Rock with shells
	Flint or flinty rock	Any very brittle rock.	Flint or flinty rock	Correctly used.	Flint
	Limestone	lime, also hard shale.	lime	Correctly used.	Limestone
	Lignite	All fossil wood.	Lignite.	Correctly used.	Lignite or fossil wood
	Gypsum	Correctly used when recognized also reported as lime or shale or sticky gumbo.	gym	Correctly used.	Gypsum
Miscellaneous.........	Shells	Fossil shells	Shells	Fossil shells.	Fossil shells

Rock, Gas Rock, Chalk Rock, Sand Rock, Sandstone, Shell, Shell Rock, Flinty Rock, Limestone, Lignite.—When the rotary driller strikes anything hard and does not know what it is, he puts down rock. If this hard substance is a concretion near the surface, it is a rock just the same as the most consolidated formations deeper down. The cable-tool driller has a much better general knowledge and a much better chance to get samples and hunts for some name to apply to the formation.

A gas rock is any rock formation containing gas; the term is applied to both sandstone and limestone.

Chalk rock is usually readily recognized by both rotary and cable-tool drillers. It is usually white or very light in color and quickly changes the rotary mud from its usual dark gray to almost white.

Sand rock, or sandstone, is usually recognized by the rotary driller, except when it is so soft as to be classified as packed sand. The harder formations appear in the overflow in pieces sufficiently large to be readily recognized. The cable-tool driller is able to recognize sandstone and all other hard formations as he finds large fragments in the bailer.

Shell is a very misleading term. If a driller, either rotary or cable-tool, drills from a soft formation into a hard one he gives it what he considers its proper name. If, however, after drilling for a short distance, he goes back into a soft formation again he is liable to put down shell. This shell may be from a few inches to a foot or two in thickness, it means a thin layer or shell of rock.

Shell rock means a rock formation containing fossil shells, unless the driller is very careless or misunderstands the term shell, in which case he may put down shell rock, meaning a thin shell of rock.

Most cable-tool drillers are able to distinguish the characteristic fracture of flint and their report of flint or flinty rock may usually be relied on. Flint is very seldom encountered with the rotary and when reported in a structure in which it is not likely to be found is probably used to designate a very brittle limestone, flinty in character.

The cable-tool driller's report of limestone is usually correct but the rotary driller does not always distinguish between hard shale and limestone.

Lignite is used to designate both the petrified and the bituminous forms of wood found in drilling. Even when the wood has not lost its fibrous character, it is often designated as lignite.

Shells.—In rare instances, solid beds of shells are encountered. They are easily recognized as such with the cable tools but with the rotary they may not be recognized and may be reported as sand or gravel, depending on the feel of the bit while drilling. When mixed with clay or sand, shells usually appear in the log as, "sand with shells" or "clay with shells."

Investigations Concerning Oil-water Emulsion *

By ALEX. W. McCoy, BARTLESVILLE, OKLA., H. R. SHIDEL, EL DORADO, KANS., AND E. A TRAGER, BARTLESVILLE, OKLA.

(Chicago Meeting, September, 1919)

SAMPLING of the fluid from oil wells for percentages of oil, emulsified oil, and water during the last two years has brought out some interesting facts concerning oil-water emulsion. This result led to a laboratory investigation of emulsion, which substantiated the conclusions made from the field observations. The purpose of this paper is to present the information collected, the laboratory experiments, and our interpretation of the same. In order to define emulsified oil exactly, give its synthesis and origin, and to show how and when it is formed in the wells, the work was necessarily divided into two separate lines—laboratory work and field observations. It is hoped that this study may lead to a discussion of such points so that the petroleum engineer, geologist, or technologist may be benefited by its practical bearing on oil-field management. Special credit is due Mr. Everett Carpenter, chief geologist of the Empire Gas & Fuel Co., for his assistance and coöperation in this work.

LABORATORY INVESTIGATIONS ON EMULSIFIED OIL

Laboratory investigations were conducted in an attempt to learn the composition and some of the properties of emulsified oil, or B. S., as it is more commonly called, also to demonstrate, by laboratory methods, how B. S. may be formed under conditions similar to those existing at the time a well is being pumped, and how it may be broken down. Literature bearing on this subject is widely scattered and very limited in scope. Bacon and Hamor define B. S., or bottom settlings, as "earthy matter, inert organic matter, or, in the case of Pennsylvanian petroleum, an emulsion of paraffin wax and water, which accompanies crude oil." In this discussion we will limit the term B. S. to that heavy, dark-brown emulsion, composed of a physical mixture of water, oil, and air with some included inert matter, either organic or inorganic.

Possibly the first step in a description of this product should be a description of its physical properties, but since most operators are quite

* Published through the courtesy of the Empire Gas & Fuel Co   Read before the Tulsa Section, February, 1919.

familiar with emulsified oil, and because the physical discussion will be better understood after one is familiar with the microscopic studies, that side of the investigation will be presented first.

A thin layer of emulsified oil under the microscope appears as a yellowish to brownish green, solid mass of small bubbles, with an occasional larger colorless bubble of water and smaller brownish globules of oil. Fig. 1 shows this relation. All of the large and most of the small, colorless bubbles are composed of water surrounded by an oil film. The dark spots are bubbles of oil. The dark material surrounding and between the bubbles is oil. The few, very bright small bubbles are air.

Careful examination shows that permanently emulsified oil is composed of millions of small bubbles of water that range in diameter from 0.004 to 0.020 mm., the most numerous having a diameter of about 0.016 mm. These bubbles are packed very closely together in a medium of oil, the average distance between them being less than one-half their diameter. Scattered in and among these very small bubbles is a relatively small number of larger bubbles of water, which vary in diameter from 0.034 to 0.070 mm. There is also about one-tenth this number of still larger bubbles of water, which vary in diameter from 0.110 to 0 250 mm. It is about these larger sizes that the very smallest bubbles are concentrated. There are a few bubbles of either water or air, with a diameter of 0.004 mm., scattered among the "groundmass" of small bubbles, which are about 0.016 mm. in diameter; but there are many very small ones located in the oil film that envelops the large water bubbles. This arrangement can be clearly seen by noticing the large bubbles of water in Fig. 1. Nearly all of the small bubbles are filled with water; a few contain air. If the material is heated very slightly, the small bubbles begin a more or less constant motion toward and away from the larger bubbles. The motion is eddying in nature and becomes more rapid as the heat is increased. Occasionally one of the small bubbles drifts away from the current that causes it to move about the larger bubble and moves along the oil passage between the water bubbles of the "groundmass," finally attaching itself to some large bubble.

There are also small globules, or isolated patches of oil, 0.003 to 0.050 mm. in diameter, trapped among the water bubbles. Some of these are perfectly spherical in outline while others have no definite shape. These globules of oil are usually composed of oil free from foreign matter and appear dark reddish brown in color. Some of these may also be seen in Fig. 1.

Air bubbles of any considerable size, that is, over 0.020 mm. in diameter, are rarely found in emulsified oil that has stood for any length of time. This is probably due to the fact that the films surrounding the air, after they reach this size, are easily broken either by mechanical agitation of the mass or by expansion of the air due to heating. The air bubbles are surrounded by a layer or film of oil, a film of water, and a second film of oil.   Fig. 2 shows the oil film on the outside of an air bubble. There appears to be a constant shifting or stretching of these films, which is most easily seen by watching the water film.   The oil films appear to

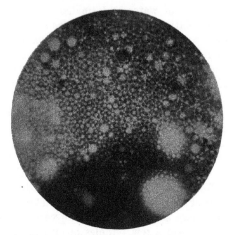

FIG. 1.—PHOTOMICROGRAPH OF TYPICAL B. S.   × 450.

slide about or change their tension, causing streaks to develop in the water film, which appear similar to convection currents or the streams of water that move about on the surface of a soap bubble.   Possibly this

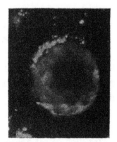

FIG. 2.—OIL FILM AROUND AN AIR   FIG. 3.—WATER FILM BETWEEN THE OIL
BUBBLE.   × 450.   FILMS OF AN AIR BUBBLE.   × 450.

movement is caused by the heating due to the light from the condenser or the microscope, for a large bubble of air seldom lasts over 5 min., in the field of view, without breaking; or it may be due to evaporation of the lighter constituents in the outer oil film, since such bubbles can only be studied by isolating them in a thin layer with the upper part of the bubble exposed to the air. The large bubble shown in Fig. 3 was taken with the water film in focus and will give some idea of the irregularities in this film. The circular shadows visible are from small bubbles on the opposite side of this bubble.

The purity of the oil that fills the spaces between the water bubbles varies widely with different samples. In some cases the oil is practically free from foreign matter, while in others it is very muddy in appearance. The water bubbles, in a sample composed of dirty oil, do not have as definite sizes as they do in samples that are comparatively free from dirt.

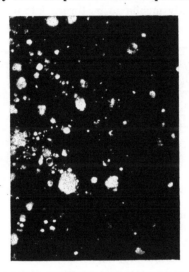

FIG. 4.—B. S. CONTAINING HIGH PERCENTAGE OF FOREIGN MATTER.   × 450.

Fig. 4 shows a sample containing much foreign matter. In general, it was found that the samples that tend to be most permanent are those in which the oil contains a large percentage of suspended matter; and those samples that are easily broken down contain relatively clean oil.

DEFINITION OF ·EMULSIFIED OIL

The appearance of emulsified oil, under the microscope, is so different from that of good crude oil that the two would never be confused by such an examination. Of course, there are gradations from crude oil to emulsified oil. A sample of good oil, under the microscope, appears much the same as it does in a cylinder, with the exception that the foreign matter in suspension is visible and an occasional water bubble will be seen. As the oil approaches emulsification, the water bubbles become closer and closer together, until finally they appear to be touching each other, when examined with a low-power lens. Just when an oil ceases to be crude oil and is to be classified as emulsified depends more on its physical appearance and plastic properties than on characteristics revealed by microscopic examination. It might be said that oil in which the water bubbles are spaced closer together than their diameter should be termed emulsified. This definition would fit most cases, but there would be exceptions because of the importance of suspended matter in forming emulsified oil. An emulsion, containing water bubbles with this spacing, that has a low specific gravity and is free from suspended matter might be sufficiently mobile to be turned into a pipeline run and the entire run treated by a simple heating process. An emulsion having the

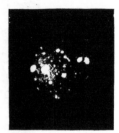

FIG. 5.—B. S. FREE FROM FOREIGN   FIG. 6.—B. S. WITH HIGH PERCENTAGE
MATTER.   OF FOREIGN MATTER.

same spacing of the water bubbles but composed of a heavier oil and containing a considerable amount of suspended matter would be quite viscous and could be treated only by the use of a complex steaming plant. Or, the water bubbles might be twice as far apart and yet the emulsified oil would be more viscous than the first sample, because of the difference in the character of the oil and the amount of foreign matter in suspension. These varying factors make it difficult to establish a dividing line between crude oil and emulsified oil, based on microscopic examination, although usually such an examination will reveal instantly the degree of emulsification, by revealing the spacing of the water bubbles and the amount of suspended matter present.

Figs. 5 to 9 will give some idea of the appearance of different types of emulsified oil under the microscope. Fig. 5 shows a sample of permanent B. S. that is practically free from foreign matter. Fig. 6 shows very heavy and dirty B. S. Fig. 7 shows B. S. that contains very little suspended matter. In Fig. 8, the sample is only partly emulsified; the group of small bubbles in the lower left-hand side marks the place where a large air bubble broke just as the picture was being taken. Fig. 9 shows B. S. that is drying up and shows the coalescing of the water bubbles, which causes the irregular shape of the large bubbles.

The surface tension of an oil-water contact makes it necessary for the water bubbles to be very small in order to have permanent B. S. If the water bubbles are large, say 1 mm. in diameter, the masses of water in two such bubbles have sufficient attraction for each other and

FIG. 7.—DIFFERENT SIZES OF BUBBLES.   × 200 ±.

FIG. 8.—PARTLY EMULSIFIED OIL.   × 450.

sufficient force, in case of an impact, to break the film of oil surrounding them and thus make a larger water bubble. As this process repeats itself and the bubbles increase in size, their weight will cause them to settle through the B. S. until the water collects below the emulsion.   Or an

increase in the temperature may cause sufficient expansion to break
the oil film; if the bubble at this time is in contact with the container
or a water layer below the B. S., the water in the bubble will either
join that below it or will adhere to the container. If it remains attached
to the side of the container, the water from other large bubbles may
be added to it until the large bubble so formed settles to the bottom.

In the case of very small water bubbles, the force of attraction upon
impact is not sufficient for the water to break the oil film, neither will
an increase in the temperature cause sufficient expansion to rupture this
film. It is a simple matter to join two large water bubbles together by
puncturing the oil films surrounding them with a needle; but it is practically impossible to join two small water bubbles, say 0.005 mm. in

FIG. 9.—PHOTOMICROGRAPH OF B. S. WHILE DRYING.   × 450.

diameter, by any amount of patience or skill. Such small bubbles may
be subjected to rather violent impacts and the tendency is to break
into even smaller bubbles, rather than to coalesce.

In a recent article by Harkins, Davies, and Clark,[1] it is stated that
"for the emulsoid particle to be stable, the molecules which make the
transition from the interior of the drop to the dispersion medium, or the
molecules of the 'film' should fit the curvature of the drop. From this
standpoint, the surface tension of very small drops is a function of the
curvature of the surface." Their studies have suggested that small
drops in an emulsion tend to be stable only when the size of the drop is
such that the molecules in the surface film fit the curvature of the sur-

[1] W. D. Harkins, E. C. H. Davies, G. L. Clark: Orientation of Molecules in the Surfaces of Liquids, etc. *Jnl.* Amer. Chem. Soc. (April, 1917) **39**, 541–596.

face. There may be more than one size of bubble in which the number of molecules in the surface fit the curvature of the drop. This tendency for the water bubbles in B. S. to arrange themselves in definite sizes is clearly seen in Figs. 1, 3, 7, and 8.

Physico-chemical Properties of Emulsified Oil

The physico-chemical properties of emulsified oil are quite variable, and each sample is more or less of an individual problem. The color that is most common is a dark, reddish brown, although any color from yellowish or greenish to gray or nearly black may be found. The darker colors generally contain more suspended matter.

Classes of Emulsified Oil

The permanency of emulsified oil may be used as a basis for division into two classes: Temporarily emulsified oil and permanently emulsified oil. The two classes cannot be separated by their appearance. This fact was brought out by two sets of samples sent to the laboratory. When the first set was opened, the glass jars appeared to contain about one-third water and two-thirds crude oil. The sampler's attention was called to this fact, but he insisted that the samples were "the best looking B. S." he had ever seen. A second set came at a later date and about half of these had no microscopic resemblance to B. S. when they reached the laboratory. The oil from these samples was examined under the microscope, and it was found that there were water bubbles present, but they were spaced about ten to twenty times their diameter apart. There were practically no very small water bubbles present, which indicates that this material was not subjected to as violent treatment as is the case with permanent B. S. Investigations in the laboratory demonstrated that oil emulsified by a minimum amount of agitation will mostly settle out in from one to three days. Of course there will be small water bubbles present in the apparently good crude oil remaining, but the percentage will be low.

Permanently Emulsified Oil

Permanently emulsified oil will stand indefinitely and the amount of settling out is negligible. This oil is somewhat more viscous than "fresh" temporarily emulsified oil, and does not contain as many large globules of water, but otherwise the oils appear similar to the unaided eye. The specific gravity of emulsified oil falls within rather narrow limits, 0.95 to 0.995, although the average is about 0.96 (15.8° Bé). The oil that separated from temporarily emulsified oil from Augusta had an average specific gravity of 0.86 (32.8° Bé). The specific gravity of the Augusta crude oil used in the laboratory was 0.849 (34.0° Bé).

Permanently emulsified oil has a very high viscosity. At room temperatures there are all gradations from a thick syrupy consistency to a near-solid. In many samples a hydrometer placed upon the surface will remain there indefinitely. Heat rapidly reduces this viscosity. A sample of the most viscous oil, when heated to 122° F. (50° C.), will readily drop from a glass rod. At 167° F. (75° C.), it has the consistency of a thin syrup, and at or near its boiling point, about 190.4° F. (88° C.), it is almost as mobile as water.

Groups of Permanently Emulsified Oil

In addition to lowering the viscosity, heat also divides permanent B. S. into two groups. In the first group, the B. S. separates into water and oil shortly after it has been heated to the boiling point, the change taking place rather suddenly. Several degrees below the boiling point there is no apparent change in the appearance of the sample, but as soon as this temperature is reached the water settles rapidly.

Samples belonging to the second group may be heated to 221° F. (105° C.) and held at this temperature until all the water and part of the oil have been distilled over, and at no time will there be any signs of separation of the oil and water in the still. However, in some of the samples the water and oil did separate, on standing from 24 to 48 hr., after being heated for 1 hr. at 221° F. To heat above this temperature in an open vessel is impossible under ordinary conditions, for the material begins to froth and boil vigorously between 221° F. and 230° F.

The size and number of the water bubbles in these two groups of B. S. is approximately the same, but just what is the cause of this marked difference in behavior is not fully understood. However, the gravity of the oil from which the B. S. was made, and the per cent. of foreign material present, appear to be the factors that control this behavior.

AMOUNT OF WATER IN B. S.

It is the common belief among practical men in the Mid-Continent field that B. S. may contain from 1 to 99 per cent. water. If an entire pipeline run that contains water is all termed B. S., this statement may be true; but if we limit B. S. to that emulsified product which is commonly recognized to be unfit for the refinery without a preliminary treatment to remove the water, this statement is not true. On the other hand, the per cent. of water in true B. S. is its most constant factor. In all types of B. S., excepting temporary, the water content is very nearly 66 per cent. This fact was determined by distilling a number of samples of B. S. from various sources, including samples that were manufactured in the laboratory. An attempt was then made to synthesize B. S. in the laboratory, hoping to learn what are the controlling factors in its formation, and more about its properties.

A 4-oz. oil-sample bottle completely filled with 70 per cent. water and 30 per cent. oil[2] was rotated about a center, at right angles to its length, at a speed of 900 r.p.m. for about 10 min.; at the end of this time there were no signs of emulsification. The bottle was again rotated, this time for 2½ hr., but no emulsification took place. The bottle was

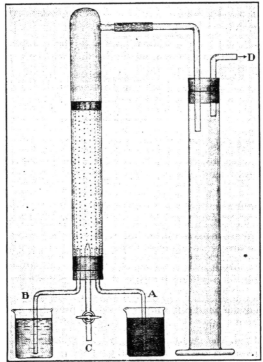

Fig. 10.—Simple Apparatus for Demonstrating Emulsification.

next placed on an automatic shaker and shaken for 3 hr., but no change occurred. Part of the oil and water was then removed and about 10 per cent. of fine sand was placed in the bottle, which was rotated for 3 hr., and then shaken for 3 hr., but negative results were obtained as before.

This experiment was repeated using 70 per cent. water and 30 per cent. oil, but having the bottle only three-fourths full; in just a few minutes after the bottle was placed upon the shaking device emulsification

---

[2] Augusta crude oil was used in all laboratory experiments.

began to take place. In 30 min. the entire contents of the bottle had been converted into permanent B. S. These experiments indicate that air or voids in the fluid, in addition to water and oil, are necessary in the emulsification of oil, the air acting as a catalytic agent.

A simple laboratory apparatus for demonstrating emulsification, under conditions more similar to those existing at the time a well is being pumped, is shown in Fig. 10. One tube A passes into a beaker containing oil, another tube B into a beaker of water, while air could be introduced at will through a third C. The graduated cylinder is used as a reservoir and the "well" is pumped by a vacuum attached to tube D. Water and oil were first drawn through the column of sand; the percentages of each were varied, but under no conditions was more than 1 per cent. of B. S. formed. But when water, oil, and air were drawn through the sand, immediately the percentages of B. S. formed began to rise. The "well" was pumped at different rates of speed and variable amounts of air were introduced; the amount of B. S. formed ranged from 1 to 15 per cent.

If, instead of using fine sand, pebbles, about 0.25 in. (6.35 mm.) in diameter or larger, are used and air under a low pressure is introduced through a large opening, 0.1875 in. (4.7 mm.) diameter or larger, the amount of B. S. formed will be very small. For under these conditions most of the water and oil pass through the gravel without being broken into bubbles small enough to remain permanently emulsified.

Since, to permanently emulsify oil, it is only necessary to break the water present into very small bubbles and then surround these small water bubbles with a film of oil before they have an opportunity to coalesce, B. S. can be made by blowing air bubbles through water and oil in a cylinder. If 67 c.c. of water and 23 c.c. of oil are placed in a 100 c.c. graduated cylinder, and air, under 5 to 8 lb. pressures coming from an opening about 0.5 mm. in diameter, is allowed to bubble through the column of liquid, the entire contents of the cylinder will be converted into permanent B. S. within 3 to 5 min. The amount of B. S. formed, up to a limiting per cent., will depend on the percentages of water and oil present.

Table 1 gives the results of such an experiment, using various combinations of oil and water, and agitating for 5 min. with a current of air under 5 lb. pressure. The first readings were taken after standing for 30 min., the second after standing a week.

The settling out shown in the table is due mostly to large globules of water trapped in the B. S. that later worked their way down through the B. S. to the water level; and to small amounts of oil rising to the top of the B. S. B. S. may also be made by this method if natural gas or steam is substituted for the air.

TABLE 1.—*Percentage of B. S. Formed Upon Five-Minute Agitation, Using Various Combinations of Oil and Water*

Water, Per Cent.	Oil, Per Cent.	B. S. After Standing 30 Minutes, Per Cent.	B. S. After Standing One Week, Per Cent.
90	10	3	4
80	20	32	16
75	25	62	22
70	30	86	66
65	35	82	70
60	40	90	58
50	50	80	60
40	60	66	50
30	70	56	38
20	80	34	24
10	90	18	18
0	100	0	0

Table 2 gives the distillation results from three samples of B. S. No. 1 was taken at a well, No. 2 was received from a steaming plant, and No. 3 is a composite of the samples obtained from the experiment shown in Table 1. In these experiments, the material was heated from 6 to 8 hr. to obtain all the water. The residue is composed of all the oil boiling over ˙105° C., fine particles of shale, sand, and limestone, and a few crystals of salt and pyrite, flakes of mica, etc.

TABLE 2.—*Distillation Results*

Number of Sample	Oil Distilling below 105° C.,* Per Cent.	Water, Per Cent.	Residue and Air, Per Cent.
1	10.5	66.0	24.5
2	10.0	66.5	24.5
3	6.3	66.0	21.0
			7.0 air and loss.

*Distillation runs under 105° C. to determine the per cent. of water in B. S.

FIELD INVESTIGATION ON EMULSIFIED OIL

Oil accumulated at the bottom of a well enters the mechanical parts through the perforated tubing. It is drawn, through the standing valve, into the working barrel, when the ball is raised from its seat, by the vacuum created in the up-stroke. On the down-stroke, the working valve opens and allows the fluid to pass into the tubing. It is then lifted about 3 ft. by each successive up-stroke until it reaches the surface and flows from the well through a lead line to a receiving tank (see Fig. 11).

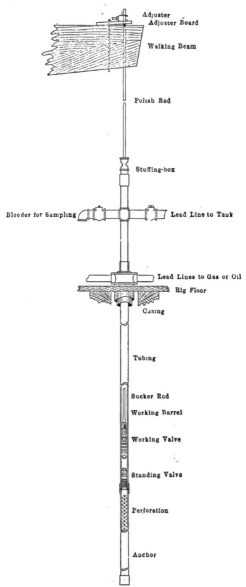

FIG. 11.—DIAGRAM OF WORKING PARTS OF PRODUCING OIL WELL.

When a sample is to be taken, the check valve on the tank lead line is closed so that the fluid coming from the tubing is not affected by the back pressure from the tank.   The valve on the short lead line is opened and the sample is caught in a bucket (see Fig. 12) and allowed to settle for about 30 sec.   The vertical cylinder is then placed in the guides of the bucket, separating a representative column measuring about 90 or 100 c.c. of fluid, which is drawn off through the small valve in the bottom of the bucket.   The sample is then centrifuged, the oil, B. S., and water separating out.   By plotting the results of such samples, taken every 10 min.,

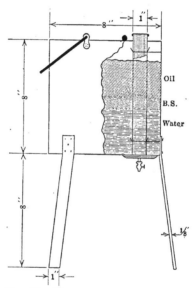

FIG. 12.—DIAGRAM OF SAMPLE BUCKET.

many irregular conditions have been noted.   All producing oil wells do not perform in the same manner; some of the different conditions are shown by the accompanying charts.

Fig. 13 shows the graph of a well producing a high percentage of water; there is no emulsion in the fluid.   The oil and water remains in about the same ratio over a period of 8 hr.   The well was pumping about 200 bbl. per day when this test was made.   Fig. 14 also shows a graph of a well making a large percentage of water; there is, however, a small percentage of emulsion plotted through the entire time of the test.   Fig. 15 represents a well producing a large percentage of water and a comparatively large percentage of B. S.

Fig. 16 shows the graph of a well producing a low percentage of water, a high percentage of oil, and a comparatively high percentage of emulsion.

Fig. 17 illustrates a well that has just pumped-off a head of water, after which the percentage of oil suddenly rose and remained about the same for some time then gradually decreased.   At the time of the increase in the

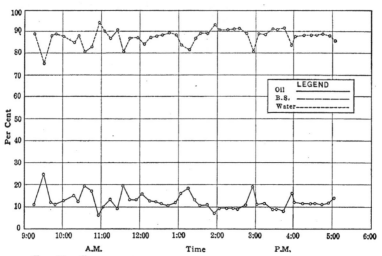

FIG. 13.—GRAPH OF WELL PRODUCING HIGH PERCENTAGE OF WATER.

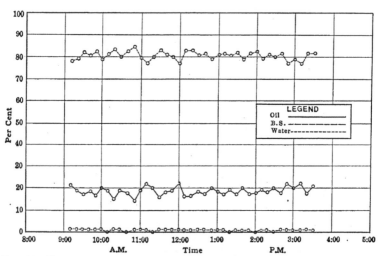

FIG. 14.—GRAPH OF WELL PRODUCING LARGE PERCENTAGE OF WATER AND SMALL PERCENTAGE OF EMULSION.

percentage of oil, the percentage of B. S. rose and continued to increase at about the same rate as the oil dropped later.   The fluid came from the tubing much more slowly as pumping progressed.   Fig. 18 shows a greater increase in the percentage of B. S.   The amount of fluid produced during the last hour was 25 per cent. of the amount pumped during the first hour of the test.

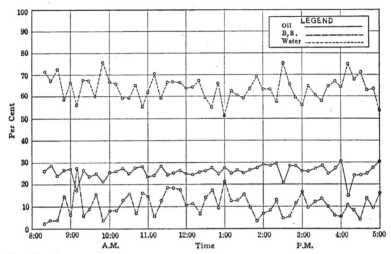

FIG. 15.—GRAPH OF WELL PRODUCING LARGE PERCENTAGE OF WATER AND COMPARATIVELY LARGE PERCENTAGE OF B. S.

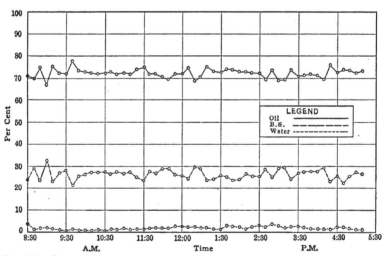

FIG. 16.—GRAPH OF WELL PRODUCING LOW PERCENTAGE OF WATER, HIGH PERCENTAGE OF OIL, AND COMPARATIVELY HIGH PERCENTAGE OF EMULSION.

Fig. 19 represents the behavior of a well pumping all water for several hours; after this was exhausted the percentage of oil increased rapidly. The water pumped during the early hours of the test was the water that separated out of the fluid behind the tubing.   As the head of fluid was reduced, the oil finally came into the tubing.   The percentage of B. S. was practically nil.

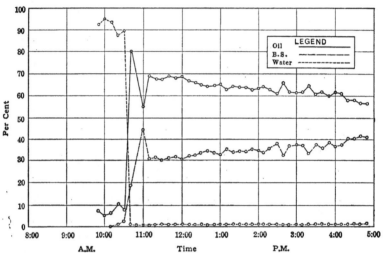

FIG. 17.—GRAPH OF WELL THAT HAS JUST PUMPED OFF HEAD OF WATER.

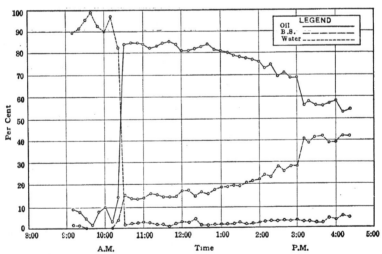

FIG. 18.—SIMILAR TO FIG. 17 BUT SHOWING GREATER INCREASE IN PERCENTAGE OF B. S.

Fig. 20 is the graph of a well that pumped only 6 hr. during the day. The fluid in the tubing was composed entirely of oil and B. S., which did not separate out.   The high percentage of water following was probably the accumulation of water behind the tubing, which passed into the hole during the shutdown. When the water was about exhausted, the percentages of oil and B. S. were about the same as when the well was started.

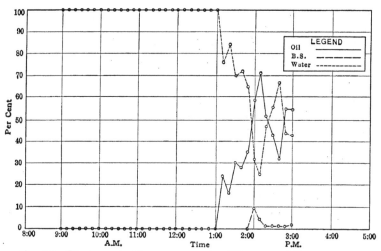

FIG. 19.—BEHAVIOR OF WELL PUMPING ALL WATER FOR SEVERAL HOURS.

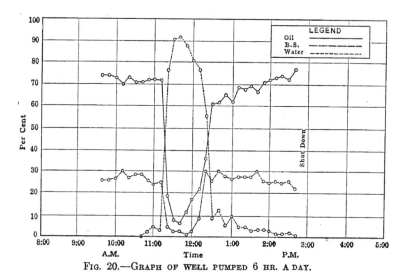

FIG. 20.—GRAPH OF WELL PUMPED 6 HR. A DAY.

A well of this order, when pumping continually, would not show such an erratic condition, as there would be no chance for a large head of water to collect in the hole.

Fig. 21 shows the action of a well producing an extremely high percentage of B. S. with small percentages of oil and water. The well was shut down for 3½ hr. When pumping was started again, it first pro-

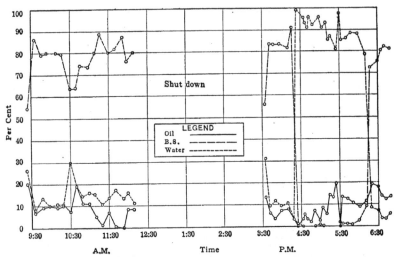

Fig. 21.—Graph of well producing extremely high percentage of B. S. and small percentages of oil and water.

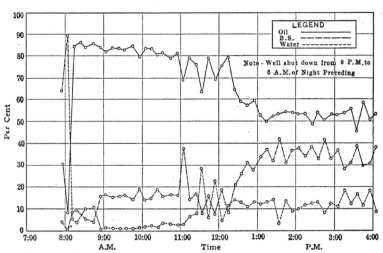

Fig. 22.—Graph of well pumping large amount of fluid at start.

duced the same percentage of B. S. as before. Within 40 min., though, the high percentage of B. S. dropped to nothing with a great increase of water. At this time, the percentage of oil increased some. In 2 hr.

the water was exhausted and the well started to operate with a more regular flow. The influx of water was due to the accumulation of water behind the tubing, during the shutdown.

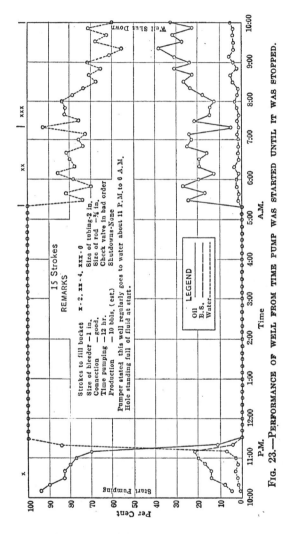

FIG. 23.—PERFORMANCE OF WELL FROM TIME PUMP WAS STARTED UNTIL IT WAS STOPPED.

Fig. 22 shows a well that was pumping a large amount of fluid at the start, the greater percentage of which was oil. As pumping continued, the percentage of oil decreased and that of the B. S. increased. There was a perceptible rise in the percentage of water too. At the end of the test, the well was producing about one-third the amount of fluid that it did at first.

Fig. 23 illustrates the performance of a well from the time the pump was started until it was stopped. At first, the well produced a good percentage of oil, which rapidly decreased to nothing. This percentage represents the oil that had settled in the tubing during the 12-hr. shutdown previous to starting. After this was pumped out, the well produced nothing but water for several hours; oil was then noted. By this time, the water in the tubing, behind the tubing, and some that had probably been backed up in the oil sand had been pumped off. The oil that had separated out to the top of the water outside of the tubing was being pumped. Less fluid was pumped at this time than when only water was being produced. During the next 2 hr., the well was producing at about the same rate. Undoubtedly, some of this fluid was partly accumulated during the shutdown and partly coming into the well during the pumping. After this, a decided drop in the amount of production was noted, with a decrease in percentage of water and an increase in the percentage of oil. The fluid from this time on was probably coming direct from the sand.

Fig. 24 shows the performance of an oil well over a period of 33 hr. During that time several experiments were tried and the results noted. At 9:30 A.M., the pump was started, showing a high percentage of oil, which dropped materially within 10 min.; this was the oil that had settled out during the shutdown. The well produced a large percentage of B. S. for 1½ hr. From 11:10 A.M. to 11:30 A.M., there was a large increase in the percentage of oil followed by a sudden and greater drop than before. This was the oil that had risen to the top of the fluid around the tubing during the period of shutdown. From 12:20 P.M. until 6:20 P.M., the well was pumping about as fast as oil and water were coming from sand; gradually a big increase of B. S. was noted. From 8:30 P.M. until 11:30 P.M., the well was shut down. When it started to pump again, the first test reported a high percentage of oil, which immediately dropped and the well produced fluid in about the same percentages as that before the shutdown until 1:50 A.M. At this time an increase in the percentage of oil was noted, followed immediately by the correspondingly large increase of water.

The fluctuations following are the results of quantities of oil and water getting into the working barrel in separate bodies. From 4 until 8 o'clock, the well was operating with about an average percentage of oil, B. S. and water. The well was shut down from 8:40 A.M. until 9:20 A.M. and from then until 11:20 A.M., it operated about the same as it did from 4 A.M. until 8 A.M. At this time there was a noted increase of water. The fluctuation of oil and water was due to the separation around the tubing during the shutdown. From 12:10 until 5:30, the well operated with a regular, or normal, flow.

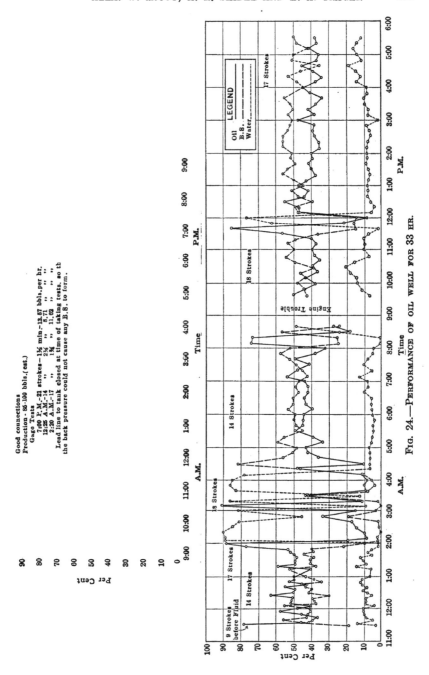

FIG. 24.—PERFORMANCE OF OIL WELL FOR 33 HR.

An interesting feature of the curve is the fact that after the well had been standing idle for a while it produced with a regular and then an irregular flow. The average production of the well was about 40 per cent. oil, 50 per cent. B. S., and 10 per cent. water. This mixture in the tubing did not settle out readily during the shutdown so that when the well started producing again it pumped approximately this ratio of fluid for the first 2 hr., clearing the tubing of the fluid left there during the idle period. While the well was not operating, oil and water filled up behind the tubing from the sand, which was pumped largely as clear oil and water with comparatively little B. S. after the fluid in the tubing was exhausted. When the accumulated head behind the tubing was reduced, the normal production returned. This series of conditions followed each shutdown. The B. S. content is only important when there is no large head of fluid behind the tubing. As shown by the graph, the B. S. content increased materially from 6 until 8 P.M., after the normal flow had gone on for 5 hr. Pumping was no doubt going on at a faster rate than the production from the sand so that gas, air, or voids in the fluid column were admitted and the oil emulsified to a greater amount.

Referring again to Fig. 23, the following computation was made to determine the amount of fluid at different times during the test. During the first hour (10:10 to 11:10 A.M.) of pumping the average of oil content was 83 per cent. The beam was making 15 strokes per minute and it took two strokes to fill a bucket of 7 qt. (6.6 l.) capacity. Consequently, this is equal to 786 gal. (2975 l.) per hour or about 18.7 bbl. per hour. If the oil content is 83 per cent. of this fluid, the amount of oil pumped during this period is 15.52 bbl. For the next 6½ hr., the well produced practically no oil.

From 5:20 to 7:20 P.M., the well was producing about 20 per cent. oil. The b eam wasmaking 15 strokes per minute and it took four strokes to fill a bucket of 7 qt. capacity. This is equal to 393.6 gal. (1490 l.) or 9.37 bbl. of fluid per hour. If the oil content is 20 per cent. of this fluid, the amount of oil pumped during this period is 3.75 barrels.

From 7:20 to 10:00 P.M., the well was producing about 25 per cent. oil. The beam was making 15 strokes per minute and it took six strokes to fill a bucket of 7 qt. capacity. This amount is equal to 262.2 gal. (992 l.) or 6.24 bbl. of fluid per hour. If the oil content is 25 per cent. of this fluid, the amount of oil pumped during this period is 4.16 barrels.

The total amount of oil produced is: $15.52 + 3.75 + 4.16 = 23.43$ bbl. The total amount of B. S. and water produced is: $115.38 + 14.99 + 12.48 = 146.04$ bbl. The total amount of fluid is: $23.43 + 146.04 = 169.47$ bbl. During the 12-hr. shutdown, the oil and water were allowed to accumulate in the well. The tubing remained full of fluid as it was when pumping was started. The fluid entering the well filled up behind the tubing.

The following computation shows the amount of time required for the raising of the fluid at the bottom of the well to the top: Area of 3-in. tubing is 7.06 sq. in. Area of ¾-in. rods is 0.44 sq. in.; difference, 6.62 sq. in.   The number of cubic inches in 1 ft. of 3-in. (76-mm.) tubing is equal to 6.62 by 12 in. or 79.44; then 2.91 ft. of 3-in. tubing contains 1 gal. of fluid. The depth of the well, 2425 ft. (739 m.), divided by 2.91 is equal to the number of gallons of fluid in the tubing, which is 833.3 gal. or 19.84 bbl. Pumping at the rate of 18.71 bbl. per hour, the time that it would require to empty the tubing would be $\frac{19.84}{18.71}$ which is equal to 1.06 hr. or 1 hr. 4 min. It will be noted by the graph that the big influx of water came in 1 hr. 10 min. after the well started to pump.

Referring to Fig. 24, the following computation has been made to show the amount of time required in this well for the raising of the fluid from the bottom to the top.   The depth of the well (2475 ft.) divided by 2.91 is equal to the number of gallons of fluid in the tubing, which is 850 gal., or 20.2 bbl.   Pumping at the rate of 8.71 bbl. per hour, the time required to empty the tubing would be $\frac{20.2}{8.71}$ which is equal to 2.32 hr. or 2 hr. 19 min.   The well was pumped at the above rate beginning at 11:30 P.M. and the time required for the first big fluctuation to occur was 2 hr. 20 min.   These figures give an idea of the time required to pump the fluid from the tubing and the variation of the same in different wells.

The question naturally arises as to whether or not the separation of oil and water while passing through the tubing is sufficient to cause a discrepancy in the ratios of oil and water in each unit volume as it flows from the bleeder, and the ratio of the fluids as they enter the perforations.   In other words, after the well has pumped the full column of fluid in and behind the tubing, are the proportions of oil and water at the bleeder the same as they are in entering from the sand?   The rate of separation of oil globules in a water column depends on the difference in the specific gravity of the two liquids, the temperature of the same, and the size of the globules.   If the full column of fluid is lifted 2500 ft. (762 m.) in 1 or 2 hr., certainly that time is sufficient for considerable separation if the fluid remains quiet.   However, it has been noted by experiment that a slight stirring will prevent any separation of the fluids, and since the rods are constantly flapping through the fluid column, it seems that any tendency to separate while pumping would be greatly if not altogether reduced. Moreover, if the water from each unit volume should be constantly descending to the next unit volume etc. all the way down the column, the bleeder would still receive something of the same ratio, only apparently at later time.

From a number of the curves, it will be noted that after a well has been pumped for several hours, the ratios of oil, B. S., and water tend to remain nearly constant, without large or rapid fluctuations. This may continue for a long time, only after the fluid head behind the tubing has been reduced. For that and the above reasons, we have considered the ratios at the bleeder when the fluid head is once reduced to be the same as the ratios of water and oil entering from the sand and have called this the "normal flow."

Conclusion

Permanent B. S. is an emulsion of very small water bubbles in oil having a diameter generally less than 0.5 mm. The oil may be relatively clean or it may contain variable amounts of suspended matter. There are generally a few air bubbles present.

The behavior of B. S. on heating may be used as an economically important basis for division into two groups. In the first group, the water separates from the oil rapidly with a small amount of heating. In the second group, the water can be removed only by distillation.

To form B. S., it is necessary to have present, in addition to oil and water, either air, a gas, or voids in the continuity of the fluid, i.e., a break in the fluid.

The percentages of oil, B. S., and water vary in the individual wells; each well is a problem in itself.

A small steady amount of B. S. is probably due to bad valves and cups. Percentages of B. S. are increased as the column of fluid around the tubing is exhausted; such a condition allows air to enter the working barrel or a break to occur in the column of fluid. This condition is responsible for large amounts of B. S. The bubbles of the different liquids and gases are made smaller and consequently more stable by the whipping of the rods.

The maximum efficiency of a pumping well, which is producing both water and oil, is obtained when the fluid level is kept above the perforated tubing and below the point where the accumulated head of water would stop the flow of oil into the hole, and when the fluid is pumped at the same rate that it comes from the sand. Such conditions can only be determined by a special test of the individual well.

DISCUSSION

A. W. AMBROSE, Washington, D. C.—Did you make any analysis of the amount of emulsion at the well and after you flowed it through a lead line to the storage tank?

E. A. TRAGER.—B. S. can be formed in passing through a lead line by the friction due to the roughness of the pipe and the irregularities at the joints.

R. W. MOORE.—Did you find the percentage of water to be limited to the percentage of oil in the emulsion which formed?

E. A. TRAGER.—Yes, the percentage is about 67 per cent. water and 23 per cent. oil.

R. W. MOORE.—If you added more water would the emulsion be permanent?

E. A. TRAGER.—Yes, it would be permanent, but the excess water would separate out.

THE CHAIRMAN (C. W. WASHBURNE, New York, N. Y.).—Did you use hot or cold water in these experiments?

E. A. TRAGER.—It makes no difference which is used. We tried to determine whether the composition of B. S. formed in the presence of excess water would differ from that formed in the presence of excess oil. The percentage composition in each case appears to be the same.

F. G. COTTRELL, Washington, D. C.—In electrical demulsification experiments in the West a number of years ago, we found no lower limit to the size of globules in an emulsion that could be dealt with, and I believe this has been borne out in the operation of the commercial plants that grew out of this work and are in operation today. I am therefore surprised at the results that Mr. Trager has secured, and am inclined to think that he may not have applied the treatment in the same way, because it was with those very fine emulsions that we were working in our experiments.

E. A. TRAGER.—The chemical laboratory worked on this same subject and tried using a high voltage current to break down the emulsion, but the results were not commercially practical for it was found necessary to treat fine emulsions several times before they were completely broken down.

F. G. COTTRELL.—Do you know the details of the experiments—the voltages and conditions?

CHAIRMAN WASHBURNE.—I believe you used high voltages in your experiments did you not, and alternating current?

F. G. COTTRELL.—Yes.

CHAIRMAN WASHBURNE.—It seems to me, since there is no doubt that every globule must have its charge of static electricity, the smaller the globule, the easier it would be moved by electrical currents and discharges. The normal static charges will be of like kind and proportional to the surface area of the globules of oil which will vary with the square of the radius, while the volume to be moved will be proportional to the cube of the radius. It is very evident, from the consideration of squares versus cubes, that it must be easier to combine large drops than small ones, because the smaller they are, the easier it is for these little static charges to keep the globules from quite touching each other. These are all technical questions and of value in the manipulation of oil emulsions. In the geological sense, there can be no emulsion. In unlimited time, the globules must come together and coalesce into larger bodies, thereby destroying the emulsion.

E. A. TRAGER.—I had a discussion with Dr. Born (chief chemist) on the subject of electrical treatment of emulsions and the following are the conclusions arrived at: The smaller bubbles, as Mr. Washburne says, move toward the electrode with greater speed, and when two such bubbles collide or when these small bubbles strike the electrode, the tendency is rather to break down into even smaller bubbles than to coalesce. The larger bubbles apparently move more slowly and when two meet they coalesce quite readily.

F. G. COTTRELL.—There are two entirely distinct technical processes which are often confused with one another. One is the electrical precipitation of suspended particles out of a gas with a direct or at least unidirectional current, and the other is the demulsification of liquid mixtures using an alternating current. The fundamental phenomena on which these are based are quite different as they are actually carried out.

In the first case, the suspended particle takes a charge by convection from one electrode, and then is driven over and deposited on the other electrode. In the case of the demulsification of the oil and water mixtures with the alternating current, however, there is no steady migration toward either electrode. The field is continually reversing so the only tendency is for the irregularly distributed globules of water to arrange themselves along the shortest lines between the electrodes. With a very fine emulsion, you may easily observe this through a microscope, the globules forming chains and gradually coalescing along these chains. In all probability, the apparent contact is not directly between the actual oil in the globules but is a contact of a film of impurities projected to the surface of the globules. With perfectly pure paraffin-oil and water, it is very difficult, if not impossible, to make a reasonably permanent emulsion, but by adding a trace of some resin or similar substance to the oil, the emulsion becomes stable at once. In crude oils there are varying amounts of such material. Large drops tend to flow together and break through that film by the force of gravity. As the size of the globules decreases, a limit is reached where that force is no longer sufficient to press the globules together sufficiently to break through these films, but if the globules are polarized by being brought between the electrodes, it may be possible to puncture that film enough to make them coalesce. That is the picture of the process I have formed from watching it under the microscope and from the general action I have seen in the electric treaters. In the case of the demulsifying process, it is not a matter of electricity being actually discharged from one electrode to the globule, but of the water being a better conductor, and of the consequent tendency for water bubbles to arrange themselves in the oil along the shortest lines between the electrodes. This finally brings the globules into contact and causes their coalescence.

R. W. Moore.—Did you use distilled water, and what type of oil?

E. A. Trager.—In these experiments we used ordinary city water which comes from the river and contains considerable inorganic matter. The oil was Augusta crude.

R. W. Moore.—Were any chemical means taken to bring down the emulsion, such as treating the emulsion with salts?

E. A. Trager.—We found nothing that would treat all types of emulsion and do it economically.

Chairman Washburne.—Were any of these experiments repeated with different oils? Sometimes one emulsified oil will act very differently from another, although both come from the same field.

E. A. Trager.—We used oil from eight or ten wells, but all the wells were in Kansas.

Chairman Washburne.—Can you tell us anything about the chemical means of separating emulsions?

E. A. Trager.—I believe a process is now being used in Oklahoma that employs sodium salts and various other compounds (preparations similar to soft soap), but I do not know whether or not they are commercially successful.

R. W. Moore.—Where the oil is emulsified in the water, heat under pressure was worked out very nicely in some of the European products, particularly in lubricating oil. There is a very rapid separation, so, if a man is treating lubricating oil under a heavy pressure, he can throw that in his tanks and get a very rapid separation by purely, we may say, mechanical and not chemical means.

Chairman Washburne.—Is pressure an essential part in that operation?

R. W. Moore.—I do not know. It is claimed that under ordinary conditions of heating they got no separation but with the oil and water emulsion under about 60 lb. pressure they did.

R. E. COLLOM,* Washington, D. C. (written discussion).—The writer disagrees with the definition and use of certain terms in the paper. The second paragraph says: "Laboratory investigations were conducted in an attempt to learn the composition and some of the properties of emulsified oil, or B. S., as it is more commonly called. . . . ·In this discussion we will limit the term B. S. to that heavy, dark-brown emulsion composed of a physical mixture of water, oil, and air with some included inert matter, either organic or inorganic."

The abbreviation B. S., in oil-field practice, is never properly applied to an emulsion. B. S. may contain some emulsion in the form of sludge, which is a mixture of mud—derived from clay or shale—and emulsified fluid. But B. S. means "bottom sediments" or "bottom settlings" and such sediments are entirely different and distinct from oil-water emulsions. Bottom sediments contain certain definite ingredients of oil-well production that have no commercial value. They include sand, mud, sludge, and other semisolid material. Oil-well emulsions, when properly treated by electric dehydrators or other means, give up certain quantities of valuable oil. The term B. S. certainly excludes the greater bulk of emulsions, which are nothing more or less than mechanical mixtures of oil and water. The writer prefers the use of the word "sludge," rather than the abbreviation B. S., for the particular physical mixture—in bottom sediments—containing emulsion.

Emulsified fluids vary in their combined proportions of oil and water. The gravities of the oil undoubtedly control the proportion of oil and water in emulsified mixtures. Light oil will carry less free water in suspension than heavy oil but an emulsion of light oil and water will show a higher water content than one of heavy oil and water. If the Baumé gravity of the pure oil in emulsion is known, a fairly close figure for the percentage of water in the emulsion may be determined in the following manner. Baumé gravities are proportional to volumes. The Baumé gravity of water is 10. The Baumé gravity of each fluid, multiplied by the respective percentage of volume of each fluid and divided by the sum of percentages of volume, or 1, equals the Baumé gravity of the emulsion. That is, where

p = gravity of pure oil;
w = gravity of water;
e = gravity of emulsion;
x = per cent volume of pure oil;
y = per cent volume of water;

$$\frac{px + wy}{x + y} = e \qquad x + y = 1.0 \qquad x = 1 - y$$

$$p - py + wy = e \qquad p - y(p - w) = e$$

e, p, and w are known, solve for y.

---

* Petroleum Technologist, U. S. Bureau of Mines.

Drilling and Production Technique in the Baku Oil Fields

By Arthur Knapp, M. E., Shreveport, La.

(New York Meeting, February, 1920)

No oil territory in the world has been so rich in large producing wells, in a comparatively small area, as the Baku field. Particularly is this true of the Bibi Eibat field, which formerly produced millions of "poods" of "gusher," or as it is called in Russia, "fountain" oil. The Bibi Eibat and Balachany fields have been exhausted of gas and ruined by water, but the Surachany and Benegadi fields are still fountain territories and many outlying districts that have only been prospected produce rich fountains.

The method of controlling fountains, or gushers, is the result of growth, along with the Russian system of drilling, where large diameters and riveted casing have been in vogue. The screw casing is seldom used except to exclude water. Formerly the method of finishing wells and the condition of the casing at the top of the well would not permit the use of gates, manifolds, and connections as is standard elsewhere.

The life of the flowing wells is very short, particularly those 10 in. (25 cm.) or more in diameter, which produce large quantities of sand and often flow for but a few days and are then a complete loss. More than 1,000,000 poods in 24 hr. have been claimed in several instances, but in no case was the flow for more than a few days.

The oil sands of this district are free uncemented sands and vary in thickness from paper thin to a maximum of 10 ft. (3 m.). The sands are interlaid with strata of soft clay. In spite of this, the practice has been to drill into such sands and produce from the open hole without screen or liners. Sometimes the casing is set below the oil sand but in this case holes from $2\frac{1}{2}$ to 3 in. in diameter are drilled opposite the oil sand, which would not have the effect of a screen.

Fig. 1 is an outline of the fountain shield ready for the control of a fountain. It is composed of an inner and an outer covering made from rough boards. The framework is made of rough round poles. When a light flow is expected, only the inner lining is built; and when the fountain comes in unexpectedly, it is often possible to build only the outer cover. The bridge is for the purpose of renewing the blocks as they become worn by the flow. The lower block, as here shown, is made of hardwood and is bolted to the crossbeams with brass bolts. The grain end is toward the flow. The upper block, as shown here, is made of cast steel and is also bolted to the crossbeam with brass bolts.

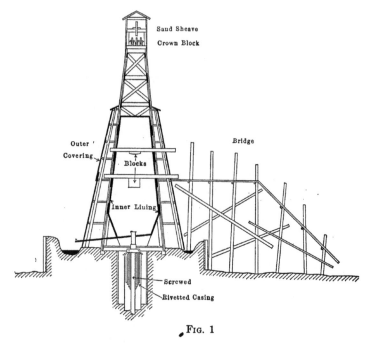

FIG. 1

FIG 2

Both blocks may be wood or steel, depending on the fancy of the engineer in charge.

Valves, tees, and lead lines are sometimes put on the well to prevent its flowing before the shield is complete and to control the well in case of fire. No attempt is made to control the production or to direct the flow when the well is put to producing; the well is always allowed to flow to capacity against the blocks. Usually the valves are cut out

FIG. 3.—INTERIOR OF DERRICK SHOWING END OF WALKING BEAM AND TEMPER-SCREW

and rendered useless soon after the flow commences. The wear of the blocks depends on the flow; sometimes they must be renewed daily. This arrangement of fountain control is not always effective. The top of the derrick, blocks and all, is sometimes lifted entirely off and the well flows wild until it sands up or the flow has weakened sufficiently to allow the blocks to be replaced. The derricks are usually set on embankments from 6 to 10 ft. high, these embankments being reinforced with pilings lined with planks. This bank is necessary to allow the oil to flow away from the well and to give space for the handling of the large amount of sand that some of the wells produce.

Fig. 2 is an exact copy of the record of six wells in the Balachany field. It shows the average condition of producing wells drilled by the Russian method. In general, the formation to 1100 ft. (335 m.) is clay, shale, and sandy shale. Below 1100 ft., there is also soft limestone, soft sandstone, and hard shale. There are no thick layers of hard rock. Sand, entirely uncemented, is seldom found except in thin layers between layers of shale.

FIG 4.—WELL BLOWING OUT THROUGH 26-IN. CASING. MAKING GAS, WATER AND LARGE QUANTITIES OF SAND.

The Russian riveted casing is far from water-tight, which accounts for the large amount of cement put in around the casing. Sometimes it is necessary to fill the casing with clay before the liquid cement will remain behind the casing long enough to set. The cement is put in with pumps, 1 or 1½-in. (2.5 or 3.8-cm.) pipes being inserted between the casings to convey it to its place.

The Russian method of drilling makes use of steel poles to actuate the tools but differs from the Canadian system, which also uses poles of wood or iron, in that a "free-fall" is used above the tools instead of the tools dropping with the rods. This free-fall picks up the tools at the bottom of the stroke and releases them at the top of the stroke, allowing them to fall free to the bottom of the hole. Thus the fall is limited to the length of the stroke of the walking beam and differs from the American cable tool system where the fall, due to the elasticity of the drilling cable, may be several times the stroke of the beam.

The Russian drilling machine is a slow ponderous machine, very heavy and very hard to handle, transport, and install. It cannot be said in reality that this machine drills; it manages to worry a hole in the ground. The results, as shown in the tables, prove this. The large starting diameters are necessary because of the large number of strings

FIG. 5.—TYPICAL VIEW IN BALACHANY.

of casing used and the large oil string necessary for producing by bailing. Also, there are several producing horizons, and wells are drilled large into the first horizon so that later they may be deepened. Casing, both Russian riveted and screwed, is lowered by means of clamps somewhat on the style of the American casing clamp. The Russians have never developed nor learned to use elevators, spiders and slips, casing tongs, or other modern oil-well tools, except when these tools have been brought in with American machinery and used by the American drillers. The casing is always carried with the tools in the Russian system. The bit is seldom advanced more than 20 ft. (6 m.) beyond the shoe. Under-reaming is practised to a large extent.

Tables 1 and 2 are chronological records of two wells shown in Fig. 2. They give, in detail, the time required for various operations under ordinary circumstances. When one string of casing has been carried

TABLE 1.—*Casing Record Shown in Well 5, Fig. 2*

Days	Depth, Feet	Advance, Feet	Average per Day, Feet	Diameter of Casing, Inches	Operation
10	133	133	13	42	Drilling.
2					Lowering 36-in. casing, 64 ft. per day.
6	230	87	15	36	Drilling.
2					Shut down, no boiler water.
2	260	30	15	36	Drilling.
10					Shut down, no boiler water.
5	370	110	22	36	Drilling.
8					Shut down, no boiler water.
10	585	215	21	36	Drilling.
2					Shut down, no boiler water.
6	714	129	21	36	Drilling.
8					Lowering 34-in. casing, 90 ft. per day.
17	1000	286	17	34	Drilling.
11					Lowering 32-in. casing, 90 ft. per day.
19	1250	250	15	32	Drilling.
7					Working casing.
6					Lowering 26-in. casing, 210 ft. per day
80					Cementing.
9					Lowering 20-in. casing, 135 ft. per day.
13	1315	65	5	20	Drilling.
6					Taking out 20-in. riveted casing, 220 ft. per day.
13					Lowering 20 in. screwed pipe, 100 ft. per day
143					Cementing.
11					Lowering 18-in. casing, 120 ft. per day.
2	1336	21	10	18	Drilling.
6					Shut down for repairs.
11	1449	113	10	18	Drilling.
10					Testing casing.
75					Cutting off 18-in., lowering 16-in., and cementing.
27	1687	238	9	16	Drilling.
165					Lowering 10-in. casing and cementing.
30					Waiting for cement to set.
60					Bringing well in.

Days of work, 730; days drilling, 128 or 18.2 per cent.; days lowering casing, 98 or 14.0 per cent.; days idle, 28 or 0.4 per cent. Average advance per day of actual drilling 13.2 ft. Days of work do not include the last 60 days testing for oil and bringing the well in.

Days	Depth, Feet	Advance, Feet	Average per Day Feet	Diameter of Casing Inches	Operation
16	136	136	8	42	Drilling
4					Lowering 36-in. casing, 34 ft. per day.
2	150	14	7	36	Drilling.
1					Testing well for plumbness.
5	246	96	19	36	Drilling.
3					Machinery repairs.
6	373	127	21	36	Drilling.
1					Machinery repairs.
20	735	362	18	36	Drilling.
14					Lowering 34-in. casing, 29 ft. per day.
20	987	252	12	34	Drilling.
6					Waiting for material.
9					Lowering 28-in. casing, 102 ft. per day.
19	1166	179	9	28	Drilling.
23					Cementing.
10					Lowering 26-in. casing, 116 ft. per day.
23	1340	134	6	26	Drilling.
19					Working casing.
8					Lowering 24-in. casing, 166 ft. per day.
6					No steam.
9	1393	53	6	24	Drilling.
14					Working casing, etc.
9					Lowering 22-in. casing, 154 ft. per day.
5	1414	21	4	22	Drilling.
52					Working casing, etc.
10					Lowering 20-in. casing, 154 ft. per day.
27	1484	70	2½	20	Drilling, also working casing.
19	1575	91	5	20	Drilling.
8					Working casing, etc.
12					Lowering 18-in. casing, 146 ft. per day.
13	1676	101	8	18	Drilling.
39					Freeing and repairing casing.
14	1701	25	2	18	Drilling.
17					Freeing and repairing casing.
95					Lowering 16-in. screwed casing, waiting orders.
25	1911	210	8	16	Drilling.
12					Working casing.
27					Lowering 14-in. screwed casing, 71 ft. per day.
4					General repairs.
5	1946	35	7	14	Drilling.
12					Shut down, labor troubles.
14	2044	98	7	14	Drilling.
8 months waiting result of an offset well					
15					Lowering 12-in. screwed casing, 130 ft. per day.
9					Repairs.
26	2184	140	5½	14	Drilling.

Days of work, 737; days drilling, 268 or 36.3 per cent.; days lowering casing, 130 or 17.5 per cent.; days idle, 29 or 0.4 per cent. Average advance per day of actual drilling 8.15 ft. Average amount of casing lowered per day, 140 ft.

as far as possible, a smaller string must be lowered before drilling can be continued. The time required for this is a large item and appears in detail in the column marked Operations. The general character of the Russian method accounts for most of the slow progress, together with poor tools, material, and labor. Wells are usually drilled by contractors who are paid per linear foot on a sliding scale, depending on the depth. They are paid a fixed sum per day while fishing and are not liable for casing lost.

Days of work does not include the 8 months waiting on the result of another well. Working casing includes cementing, testing for water, raising and lowering casing to free it, cutting off, etc. Days drilling includes time of raising and lowering tools and the lowering of casing as the drilling proceeds.

DISCUSSION

I. N. KNAPP, Philadelphia, Pa. (written discussion).—I had considerable correspondence with the author of this paper during the two years he was engaged at Baku, from July, 1914, to August, 1916, and since his return I have had many interesting talks with him on his Russian experience. I think, therefore, that instead of making a strictly technical discussion of the paper, it will be more interesting to include some personal details.

The author was employed to advise on American methods of drilling and operating and particularly to introduce American methods of pumping oil from wells. Fully one-third of the oil then being produced around Baku was used as fuel for bailing the production.

On arriving at his destination he was not allowed by the local Russian management to go upon the properties of the company that had employed him. This gave him an opportunity to study the Russian language under competent teachers; in 6 mo. he was able to do business over the telephone in that language. In talking of this I said "It was extremely fortunate that you were able to learn Russian in such a way that you did not have to differentiate the common 'cuss words' of the oil fields from polite language." To this he replied that Russian is not commonly spoken by the workmen of the Baku oil fields. The principal language used is Tartaric with a mixture of Persian, Armenian, and some Georgian.

After several months of idleness, he was given a practically abandoned well to put to pumping. The Russian manager seemed to expect him to tear down the Russian rig and build one in American fashion, which could easily be made to take a year's time. He chose rather to repair and line up the old Russian rig and machine already at the well and on running the tools found a couple of bailers stuck in the hole. The workmen were surprised to see an American engineer who was not afraid to repair one of their machines and operate it.   The author soon found that some of the workmen were good oil men in their way, so after he had run an impression block, had some fishing tools shaped up, and had cleared the junk out of the hole, they became willing and helpful workers.   He found blacksmiths and machinists in the field that could do surprisingly good work, considering the facilities they had.   The workmen had never used American elevators, modern pipe tongs, monkey or Stillson wrenches, so it was necessary to show them how to use such tools efficiently.   The several large properties of his company had no tools of this kind until the shipment of American goods arrived.

The first well was soon got on the beam and put to pumping.   The author assures me that there is no more difficulty in pumping a properly screened well from the Baku sands than from the Midway, Calif., sands where he has had experience in both drilling for and pumping oil.   Also, he says that many places in Louisiana and in Trinidad present greater difficulties in drilling and pumping than Baku.

A second practically abandoned well was turned over to him to be put to pumping but instead of the good American tubing got for the purpose a lot of junk tubing that would drop apart before 1000 ft. was run in the hole was substituted.   Each break made a fishing job that would last some time.   After awhile the Russian management forgot about trying to pump wells.

The author was given an opportunity to become familiar with the Russian free-fall system of drilling and, for a time, had charge of the operation of a Holland rig, or European water-flush system, and an American rotary.   He understood that it was compulsory for each oil property to be managed by a qualified engineer with Russian diplomas and such managers commonly opposed any innovations on general principles.

In some of the Government-owned pools, operations are restricted to hand-dug wells not to exceed a maximum depth of 420 ft.   The well diggers employed are skilled in their trade, which has been carried on in that region since time immemorial.   They are very loyal to their mates. Men, when digging, are frequently overcome with gas and the workmen are adepts at resuscitation in such cases.   The laws of the country have been made extremely drastic on deaths caused by asphyxiation.   When the workers conclude that a man is really dead from this cause, they pound him on the head with a rock or kick in his ribs so as to claim that he died from an accident and, of course, the authorities decide in such cases that nobody is to blame.

He further said that the rules laid down for operations in the Baku field were perhaps made in good faith but many lacked practicability. For instance, there was a rule that only 60 lb. of steam was to be allowed ordinarily on any boiler in the field.   American rotary rigs are designed for at least 100 lb. steam pressure and were hard to operate at the low pressure on account of the small steam cylinders. Some boilers sent with such rigs were built under the Burmah specification and really had a fair factor of safety at 200 lb. steam pressure. But notwithstanding all this every boiler in the field must have two safety valves, one of which is set at 60 lb. pressure and sealed by a Government engineer. Admittedly the water used is bad but not the slightest regard is paid to the thickness of the sheets or the workmanship of the boiler. Gradually the author was permitted to examine all the geological and drilling records of his company from which the data given in the paper were taken.

There were a few native Russians in the Baku district who had studied in American and English colleges and occasionally one was employed in the oil fields. Partly through the influence of these men and partly because of his ability to speak Russian fluently and to make sketches of American methods of drilling and operating he was invited, during the last few months of his stay at Baku, to attend and take part in the proceedings of the weekly meetings held by the Government engineers in general charge of the Baku oil fields. He says that it is generally recognized that the days of the rich shallow gushers, or fountains, at Baku have passed and less expensive methods of drilling and operating must be adopted, such as is offered by the American method of rotary drilling, cementing in the casing, screening off the sand, and pumping the wells.

So far as I can see, the first great step toward progress would be to do away with the former misdirected Governmental interference of all kinds. Let the investigations be directed by men skilled in the oil business, and not by the impractical scientist whose findings only serve to entrench the administrative and bureaucratic machines in the strangle hold that smothers initiative, progress, and real conservation.

Determination of Pore Space of Oil and Gas Sands*

BY A. F. MELCHER,† M. S., WASHINGTON, D. C.

(Lake Superior Meeting, August, 1920)

THE present paper is a progress report on an investigation of the physical factors of oil and gas and especially of their sands,[1] such as pore space, size of pores or permeability, retentivity, viscosity of the oil, temperature, pressure, thickness and area of the pay sand, water relations, and capillarity. The purpose is to determine as many of these physical factors as possible, and to ascertain the relations existing, directly or indirectly, between these physical factors and the production of oil and gas. As yet only pore space[2] and size of grains of pay sands have been investigated, although an apparatus has been designed to determine the permeability of a sand to oil, water, or gas under definite drops of pressure between the entrance face and exit face of the sample.

Messrs. E. W. Shaw, R. Van A. Mills, D. Dale Condit, G. B. Richardson, G. C. Matson, and C. H. Wegemann collected the samples upon which the physical determinations were made. Acknowledgment is made to my colleagues of the United States Geological Survey for many suggestions and criticisms; to Mr. Mills for the production data given of the oil wells from which some of the samples were collected; to Mr. A. W. McCoy of the Empire Gas and Fuel Co., Bartlesville, Okla., for many ideas.

DETERMINING PORE SPACE OF OIL AND GAS SANDS '

In selecting a method for the determination of pore space, two objects were kept in mind: First, the method must not only be sufficiently

---

* Published by permission of the Director, U. S. Geological Survey.

† Associate Physical Geologist, U. S. Geological Survey.

[1] Sand is used in this paper with the meaning of oil and gas pay sands as they exist in nature, either coherent or incoherent. The sand samples tested in this paper were coherent.

[2] The pore space of a rock can be divided into two kinds, the total pore space and the effective pore space. The total pore space is the total interstitial space and includes not only the communicating pores, but any isolated pores that may exist. The effective pore space, on the other hand, is relative, depending on such factors as the constitution of the liquid, the size of the pores, the material of the rock, temperature, pressure, and other conditions. It is apparent that the total pore space is a maximum limit for the effective pore space. The method described in this paper determines the total pore space.

accurate to be of a truly scientific nature but must also be rapid enough to justify its commercial use. Second, the method, to have as large a range as possible, must permit the determination of pore space of many types of samples of different composition and structure, with great range of size. The method selected is based on the principle that the volume of the fragment of the sand minus the volume of its individual grains equals the volume of the pore space. The volume of the pore space divided by the volume of the fragment gives the per cent. pore space by volume. The volume of a fragment of sand is chosen because it is a constant factor. Density and weight of a fragment of sand are not constant unless all substances are removed from its pore space, but will vary with the quantity and kind of material in the pores of the stone.

To determine the pore space of an oil or gas sand, it is quite necessary to have unbroken fragments or parts of the pay sand as it existed in nature—free from cracks or cleavage planes and its surface free from foreign material. Disintegrated sand is not so valuable in the determination of pore space, as it would be impossible to place the separate grains in their original positions in order that their true interstitial space might be found. It is better to have several samples from each well, beginning at the top of the pay sand, or even at the top of the cap-rock and extending to the bottom of the pay sand, as often the productive sand is within another sand. The core-drill method of obtaining samples is the best. Samples are obtained when the well is shot and from drill cuttings. Sometimes they come up with the bailer and with oil and gas when oil and gas come out of the well under considerable pressure. Samples are often obtained from outcrops and, in some cases, where mine shafts penetrate the pay sand. The fragments are sometimes irregular and quite small, weighing less than 1 gram.

Dipping Samples in Paraffin

Sometimes the texture of the samples is so loose that it is difficult to keep the grains of sand from rubbing off while handling them; other fragments are firmer and more compact. It was because of this looseness of texture and the small size of some of the samples that the method of dipping in paraffin[3] was adopted. After the surface of a sample was

---

[3] Julius Hirschwald ("Die Prüfung der Natürlichen Bausteine auf ihre Wetterbeständigkeit." Berlin, 1908. W. Ernst und Sohn) describes a method of dipping the specimens in paraffin, which he used to determine the specific gravity of building stones. The volumenometer was employed instead of weighing the sample in water to find the specific gravity. He determined the absolute pore space from a comparison of the specific gravity of the powdered stone with that of a greater fragment of the stone. By the method of finding the pore space by a comparison of specific gravities, Hirschwald eliminates the difficulties of saturating the sample with water, but in choosing the specific gravity instead of volume he retains the difficulties of removing foreign material from the pores of his fragment specimen.

thoroughly cleaned of foreign material with an assay brush and loose particles brushed off, it was broken into two parts; one part was used for finding the volume of the fragment and the other was used for finding the volume of the individual grains making up the fragment.

The pieces that were to be used for finding the volume of the fragment were weighed and then dipped into paraffin heated to a temperature a little above its melting point. The layer of paraffin around the sample was then examined for air bubbles and pinholes. If any were found, they were removed by remelting the paraffin at that point with the end of a hot wire.

Fig. 1.—Samples of oil- and gas-bearing sands dipped in paraffin; the scale reads in centimeters.

The fragments are best dipped by holding them with the fingers. First, the half of the sample opposite the fingers is dipped, then the sample is turned around and the other half is dipped. The samples should never remain in the melted paraffin longer than 2 or 3 sec., and very small samples or very porous ones should be immersed for shorter periods. Bubbles should not be permitted to come out of the samples as they usually indicate that the paraffin is beginning to enter the pores. If there is any doubt about the paraffin entering the pores of the sample, the specimen may be broken, after it is weighed, in distilled water and examined with

a hand lens or microscope, depending on the size of the pores. It will
be found that, after a little practice, if the samples are cold, there will
not be much difficulty in dipping them so that the paraffin will not enter
the pores, as the paraffin almost immediately hardens when it comes into
contact with the cold surface of the sand. When the paraffin cools,
the sample with its coating is weighed to determine the weight of the
paraffin.

DETERMINING VOLUME OF FRAGMENT

The sample with the coating of paraffin is suspended in distilled
water by a No. 30 B. & S. gage platinum wire and weighed; a fine wire
is used so that the error due to surface tension will be as small as possible.
The water should have been boiled and its temperature taken to one-
tenth of a degree at the time of the weighing. The sample is then
removed from the water, dried by pressing the surface against bibulous
paper or a smooth towel and weighed in air. This weighing is made to
see whether the sample absorbed any water. If an appreciable quantity
of water is absorbed, a correction can be made to the weight of water
displaced from the difference between the last weighing and the former
weighing of the sample plus the paraffin in air.

From the weight of the water displaced, its temperature, and density, the volume of the sample plus the volume of the paraffin can be
obtained. The tables by P. Chappuis[4] on the change of density with the
temperature of pure water free from air were used. From a previous
determination of the density of paraffin, which in this case is 0.906, and
the weight of the paraffin covering the sample, its volume can be obtained. Subtracting this volume from the total volume of the sample,
plus the volume of the paraffin, gives the volume of the fragment of stone
used.

DETERMINING VOLUME OF INDIVIDUAL GRAINS

The second part of the sample is weighed and crushed in an agate
crucible into its separate particles; or, in the case of a very fine sand,
until it will pass through a 100-mesh sieve. It is again weighed and
thoroughly dried in an electric oven, or better in the Steiger toluene[5]
bath at from 100° to 150° C. for 30 min. to 1 hr.; a lower temperature
is used when there is danger of driving off an appreciable quantity of
combined water. It is then placed in a dessicator to cool. After the
particles have cooled, the sample is weighed and exposed to the air to
take up moisture. After the particles have reached a constant weight,

---

[4] Bureau International des Poids et Mesures, *Travaux et Memoirs* (1907) **13**;
U. S. Bureau of Standards, *Circular* 19, 5th ed., Table 27.
[5] U. S. Geol. Survey *Bull.* 422 (1910) 75–76.

or nearly so, they are again weighed to correct for hydroscopic water. The particles of sand are then transferred to the pycnometer, using glazed paper. The pycnometer plus the sample are weighed to correct for the loss in transfer. The pycnometers used are of the type designed by John Johnston and L. H. Adams,[6] of the Carnegie Institution.

The advantages of this type of pycnometer over others are: (1) There is no appreciable loss by evaporation of the liquid from the pycnometer, the pycnometer can, therefore, be allowed to stand in the balance case

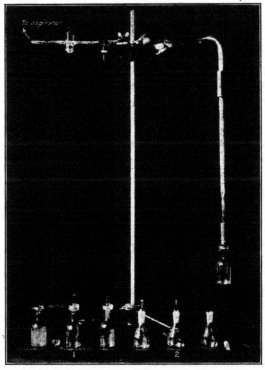

FIG. 2.—APPARATUS FOR REMOVING THE AIR FROM DISINTEGRATED SAND IN PYCNOMETER, AND TWO TYPES OF PYCNOMETERS, THE JOHNSTON & ADAMS PLANE-JOINT PYCNOMETERS, NO. 1, AND COMMON PYCNOMETERS, NO. 2.

until temperature and moisture equilibrium is attained; (2) there is no error due to grease, which is necessary in other pycnometers where the stopper fits into flask; (3) any particle of grit or dirt can be easily wiped from the joint between the stopper and flask. It is about as easy to obtain an accuracy in density of 2 in the fourth decimal place by this pycnometer as it is of 2 in the third decimal place for the ordinary type of pycnometer, where the stopper fits inside the flask. The device of G.

[6]*Jnl.* Amer. Chem. Soc. (1912) **34,** 566.

E. Moore,[7] slightly modified by Day and Allen,[8] was used for the evacua-
tion of the air from the ground particles. Fig. 2 shows this apparatus
with pycnometers of two types—the Johnston & Adams plane joint
pycnometer and the common pycnometer, in which the stopper fits inside
the neck of the flask.

After the pycnometer is nearly filled with boiled distilled water, the
aspirator is removed and the pycnometer is placed in a constant-tem-
perature thermostat regulated to 0.1° C. The filling of the pycnometer
is completed from distilled water taken from another vessel in the thermostat. The pycnometer is then removed from the thermostat and weighed
after its outside surface has been dried with a towel. From a previous
calibration of the pycnometer, which gives the weight of the water necessary to fill the pycnometer, the weight of water that the crushed
sample displaced is found. The volume of the ground particles in the
pycnometer is found from the weight of water displaced and the table
of densities of water at the temperature of the thermostat.

By proportion, the total volume of grains in the fragment dipped in
paraffin is determined. Then the volume of the fragment dipped in
paraffin minus the volume of its grains is equal to the volume of the pore
space. This volume divided by the volume of the fragment gives the
per cent. pore space by volume.

DETERMINING PORE SPACE OF VERY SMALL SAMPLES

In case the sample is too small to break into two parts, the whole
sample can be dipped into paraffin and the paraffin burned off, if the
grains of the sample are of sufficiently pure quartz not to be appreciably
changed in volume or weight by the burning. In many cases the paraffin
can easily be shaved and brushed off with a knife and assay brush and
a new weighing made to determine the loss of weight of particles brushed
off. In case there is oil in the fragment that is crushed, the oil is either
burned out by placing the crushed sample in a platinum crucible or it is
dissolved by a solvent, as petroleum ether or carbon tetrachloride.
It was possible to burn out the oil in nearly all cases as most of the samples
consisted of practically pure quartz.

BASIC PRINCIPLE OF METHOD

This method is based upon the principle that the volume of the fragment minus the volume of its grains equals the volume of the pore space.
Let V = volume of fragment;

V_{tg} = volume of grains of that fragment free from moisture;

V_p = volume of pore space;

$$V_p = V - V_{tg} \tag{1}$$

[7] *Am. Jnl. Sci.* [3] (1872) **3,** 41.

[8] Carnegie Institution of Washington *Pub.* 31; U. S. Geol. Survey *Bull.* 422
(1910) 48–50

The per cent. pore space is

$$P = \frac{V - V_{tg}}{V}\, 100 = 100\left(1 - \frac{V_{tg}}{V}\right) \qquad (2)$$

Now, the problem is to find V and V_{tg} in known quantities, and quantities that can easily be obtained experimentally. In order to do this, the sample is broken into two parts, one to be dipped in paraffin and the other to be used for the determination of the volume of the grains.

Let W and W_1 = weights, respectively, of the two pieces;
$\qquad$ W_p = weight of one of fragments dipped in paraffin;
$W_p - W$ = weight of paraffin.

The volume of the paraffin is

$$V_{p1} = \frac{W_p - W}{0.906},$$

wherein 0.906 = density of paraffin at $20.4°$ C.

Let W_{p1} = weight of fragment dipped in paraffin plus wire carrier in boiled distilled water;
$\qquad$ W_c = weight of wire carrier immersed an equal distance in water as when fragment was attached.

The weight of the water displaced by the fragment plus its coating of paraffin is $W_p - (W_{p1} - W_c)$, and the volume of the fragment is

$$V = \frac{W_p - (W_{p1} - W_c)}{D_t} - V_{p1}$$

where D_t = density of water taken from the density tables at the temperature of the water when the weighing was made. Substituting for V_{p1} its value,

$$V = \frac{W_p - (W_{p1} - W_c)}{D_t} - \frac{W_p - W}{0.906} \qquad (3)$$

If W_g = weight of grains immediately after sample is crushed, $W_1 - W_g$ = weight lost or gained in breaking up sample into its separate grains or so that the material will pass through a 100-mesh sieve in case the sand is very fine. The difference ($W_1 - W_g$) is usually quite small. It can be made a negligible quantity by first using the Ellis crucible[9] for breaking the fragments into coarse particles and then using the agate crucible for the finer grinding. In real fine material, there is an error due to the particles taking up moisture, but in this work the error is inappreciable or the above quantity can be used as a correction.

---

[9] U. S. Geol. Survey *Bull.* 422, 50–51.

Let

W_{g1} = weight of crushed sample, after drying, at 100° to 150° C. in an electric oven for about 1 hr.;

W_k = weight of pycnometer;

W_{k1} = weight of water content of pycnometer at standardized temperature t_1;

W_{k2} = weight of pycnometer with crushed sample filled with water;

D_{t1} = density of water at temperature t_1.

The weight of water displaced by the crushed sample is,

$$W_{k3} = W_{k1} - [W_{k2} - (W_k + W_{g1})]$$

and the volume of the grains is

$$V_g = \frac{W_{k3}}{D_{t1}} = \frac{W_{k1} - [W_{k2} - (W_k + W_{g1})]}{D_{t1}} \qquad (4)$$

Then the total volume of the grains V_{tg} in the fragment that was coated with paraffin is found from the proportion $W_g : W = V_g : V_{tg}$; or, expressed in the form of an equation,

$$V_{tg} = \frac{W V_g}{W_g}.$$

Substituting for V_g its value in equation (4),

$$V_{tg} = \frac{W \{W_{k1} - [W_{k2} - (W_k + W_{g1})]\}}{W_g D_{t1}}$$

Substituting in equation (2) for V and V_{tg} their values,

$$P = 100 \left[1 - \frac{D_t W \{W_{k1} - [W_{k2} - (W_k + W_{g1})]\}}{D_{t1} W_g \left[W_p - (W_{p1} - W_c) - \dfrac{W_p - W}{0.906} D_t \right]} \right]$$

in which W_k, W_{k1}, W_c are experimental constants and D_t and D_{t1} are constants found from the tables on density of water free from air. These leave six quantities, W, W_g, W_{k2}, W_{g1}, W_p, and W_{p1} to be found by weighing. The density of the grains free from moisture, or specific gravity of the grains referred to water at 4° C. as unity is,

$$D = \frac{W_{g1}}{V_g}$$

For very accurate determination of pore space, it is necessary to add a correction to some of the weighings for buoyancy of the air. In this method it is not necessary to dry the fragments. In a number of cases the percentages of different diameters (of grains) were found by sieving.

ADVANTAGES OF METHODS ADOPTED

The method used will determine the pore space of oil- and gas-bearing sands in about one-tenth the time it takes by the water absorption method, and is more accurate. In most cases it would be impossible to determine the pore space of the samples by water absorption with sufficient accuracy even for commercial use, on account of the small size and lack of solidity of most of the available fragments. Pore space determinations can be made of chunk samples that weigh 0.1 oz. with an error of less than ±1 per cent. The pore space of a chunk sample weighing 0.05 oz., the grains of which will pass through a No. 20 mesh sieve (and most grains of oil and gas sands will pass through a mesh of this size) can be determined with sufficient accuracy for commercial use.

One of the chief sources of error in determining the pore space of a small sample by water absorption is that the quantity of water that may be taken from the pores or left on the surface in drying the sample may be a large percentage of the total amount of water absorbed. It is also quite difficult to clean thoroughly oil- and gas-bearing fragments of sands from their oil, water, and gas, as well as to saturate them with water after they have been cleaned. Another difficulty met in loosely connected grains of a sample is that the sample may disintegrate when it is placed in water. These sources of error and difficulties are eliminated by the method that has been described.

POROSITY TESTS ON BUILDING STONES

The following are some determinations of pore space by Julius Hirschwald[10] and show the variations in the value of porosity of different methods of water absorption on the same sample. These porosity tests were made on building stones.

Percentage Proportion of Water Absorption to Total Pore Space*

Number of Sample	By Method of Quick Immersion	By Method of Gradual Immersion	By Method of Gradual Immersion in Vacuum	By Method of Pressure 50–150 Atmospheres
1	53.0	61.3	85.5	100.0
17	45.9	52.2	61.1	100.0
11	71.3	81.2	81.5	100.0
2	60.9	63.0	99.4	100.0
4	72.6	77.2	81.0	100.0
18	53.3	54.6	99.5	100.0
16	47.6	49.7	96.1	100.0

* The total pore space in these seven cases is the same as the pore space determined by the method of applying a pressure of 50–150 atmospheres to the sample under water immediately after the method of gradual immersion in water under a vacuum has been completed.

[10] Julius Hirschwald: Die Prüfung der Natürlichen Bausteine auf ihre Wetterbeständigkeit. Berlin, 1908. W. Ernst und Sohn.

RESULTS OBTAINED FROM PORE-SPACE DETERMINATIONS

Pore-space determinations have been made from 107 chunk samples of oil and gas sands, cap-rocks, and shales collected from Pennsylvania, West Virginia, New York, Ohio, Kentucky, Oklahoma, Texas, Louisiana, Wyoming, and Montana. The distribution of diameter of grains of 36 of these samples have been determined. The pore space with density and distribution of diameters of grains of oil- and gas-bearing sands and associated rocks are given in the accompanying tables. None of the pay sands in which oil was found that had a porosity less than 10.5 per cent. were producing sands. The most probable explanation for this fact is that there are sufficient fine grains, including cementing material, between the larger grains in these samples to reduce the interstitial openings to a size sufficiently close to the subcapillary[11] size so that the oil, on account of the resistance it meets under existing pressure and temperature, will not move rapidly enough to produce in commercial quantities. A pore in an ideal sand, in which the grains are uniform spheres, does not have a constant diameter throughout its length, but varies in diameter and cross-section, passing continuously from a minimum to a maximum cross-section.

Professor Slichter[12] has shown that the flow of water through a sand may be reckoned as passing through an ideal sand, the pores of which are continuous tubes of the minimum size. This reduction of the cross-section of the pore to the minimum for the flow of the oil would make the size of the pore approach much closer to the subcapillary than at first it would appear from the diameter of the grains. On the other hand, the grains of sand can be of such a shape and laid down in such a way that the width, or diameter, of the pores at places are sufficiently close to the subcapillary to interfere materially with production. The production in such a case might not be sufficient for commercial quantities, even when the well is repeatedly shot.

In Ohio, there are four wells of which production or non-production are given. Plots of the sands of these wells are shown in Fig. 3, in which the percentages, by weight, are plotted as ordinates and the diameters of grains are plotted as abscissas. The depths of the sands from which samples 5, 7, and 10 were obtained are about the same, the depth of the sand from which sample 16 was procured is not given, but is probably about the same as the others. Sample 10 has a pore space of 16.9 per cent., the grains of its maximum column are larger in diameter than

---

[11] A subcapillary tube is one in which molecular attraction extends across the tube; the average size of such a tube, as determined experimentally by different physicists, is about 0.00002 mm. in diameter.

[12] C. S. Slichter: Theoretical Investigation of the Motion of Groundwater. U. S. Geol. Survey, 19th *Ann. Rept.* (1899) Pt. 2, 305–323.

the grains of the maximum column of samples 7 and 16, and are about the same diameter as the grains of the maximum column of sample 5. The well from which sample 10 was collected had an initial production of 400 bbl. Sample 5 had a pore space of 13.1 per cent. and the well from which this sample was obtained had an initial production of 80 bbl. Sample 16 had a pore space of 16.8 per cent. and has its maximum column at a much smaller diameter of grain than samples 5 and 10, and the well from which this sample was collected had an initial production of 100 bbl. Sample 7 had a pore space of 4.7 per cent. and its maximum column had a small diameter of grain; the well from which this sample was taken was non-productive.

RELATION OF PORE SPACE TO PRODUCTIVITY OF POOL

Pore space is undoubtedly one of the several factors that control production from an oil or gas pool. Professor Slichter[13] also has shown that if two samples of the same sand are packed, one sample so that its porosity is 26 per cent. and the other sample so that its porosity is 47 per cent., the flow through the latter sample will be more than seven times the flow through the former. If the two samples of the same sand had been packed so that their porosities had been 30 per cent. and 40 per cent., respectively, the flow through the latter sample would have been about 2.6 times the flow through the former. He states that "These facts should make clear the enormous influence of porosity on flow, and the inadequacy of a formula of flow that does not take it into account."

Comparison of the production of one oil or gas pool with another or of one oil well with another, from a comparison of their physical constants and factors, is very similar to the comparison of two unknown quantities, each of which is made up of an equal number of factors. It is at once apparent that the more known factors there are of each, the more nearly can they be compared or estimated. In this same manner can the production of an oil or gas pool, or oil and gas well, be estimated or compared, and the more factors known the closer can the production be estimated or compared with a known production of a pool or well of known physical factors. In three out of four samples where the quantity of combustible matter burned out of the sample of sand amounted to 3 per cent. or less by volume, the well from which the sample was taken produced salt water with the oil; see Table 1.

CONCLUSIONS

A method has been established that will determine the pore space of very small fragments of oil and gas sands and determine the pore space accurately. None of the pay sands in which oil was found, if the pore

---

[13] C. S. Slichter: *Op. cit.*, 323.

TABLE 1.—*Total Pore Space and Per Cent., by Volume, of Combustible Matter Burned Out, with Density and Percentage Distribution of Diameters of Grains*

	Developers Oil & Gas Co., Petrolia, Tex.	(1) Berea Sand From Well No. 1, on M. O. Huth Farm, Near Woodsfield, Ohio. Sample From Productive Oil Well	(4) Keener Sand From Woodsfield, Ohio, Near Huth Farm. Fragment From Bed Yielding Gas, Oil, and Salt Water	(5) "Big Injun" Sand, Well No. 2, A. C. Weber, Lewisville, Ohio. Fragment From Bed Yielding Gas, Oil, and Salt Water. Initial Production 30 Bbl.	(7) Berea Sand From Chaseville Field, Ohio. Fragment From Non-productive Bed	(10) Keener Sand, Well No. 2, J. R. Scott, Jerusalem, Ohio. Fragment From Well of Initial Production of 400 Bbl.	(11) Berea Sand From Well No. 5 on the Henry Herdershot Farm. Specimen From Bed Yielding Both Oil and Gas	(15) Berea Sand From Well No. 4, Taylor Heirs Farm, near Woodsfield, Ohio. Fragment From Bed Producing Oil and Very Little Salt Water	(16) Berea Sand, Well on Shepherd Farm, Armstrong Mills, Ohio. Fragment From Productive Bed. Initial Production 100 Bbl.	Coal River No. 4, Pay Sand, Dawes, W. Va.	Cabin Creek No. 92, Pay Sand, Dawes, W. Va.	Bradford Pay Sand, Minald Run Oil Co., Custer City, Pa.
Pore space, per cent. by volume	18.5	11.2, 10.8 }11	12.7	13.1	4.7	16.9	10.5	14.8	16.8	19.3	18.7	17.8
Total weight, in grams	20.114	53.239, 31.974	40.819	41.120	16.238	41.482	28.591	36.412	60.867	5.030	30.933	26.616
Density of grains free from moisture or specific gravity referred to water at 4° C. as unity	2.76	2.647	2.646	2.654	2.665	2.653	2.649	2.693	2.682	2.672	2.664	2.663
Per cent. by volume of combustible matter burned out of sample*	11.3	3	3	2	0	8	5	3	4	9	6	13.6
Diameters greater than 1.651 mm., per cent.				20.4		1.9						
Diameters from 0.833–1.651 mm., per cent.	0.2						4.6		0.3			0.3
0.417–0.833 mm., per cent.	8.2	48.0	2.4	56.4	36.0	41.1	29.6	0.4	25.4		11.0	19.6
0.295–0.417 mm., per cent.	45.8	36.8	29.4	14.7	38.2	47.2	30.1	56.8	54.3	20.6	39.7	39.6
0.208–0.295 mm., per cent.	29.5	4.9	53.3	2.5	9.6	3.4	10.0	20.0	11.4	17.1	28.4	26.9
0.147–0.208 mm., per cent.	12.5		10.1	1.6		2.7	10.3		2.9	14.9	13.7	
0.104–0.147 mm., per cent.	1.8		1.7	1.1		1.7					3.7	
0.074–0.104 mm., per cent.				0.7		0.5						
Diameters less than 0.074 mm., per cent.	2.0	10.3	3.2	0.9	16.2	0.6	15.5	22.8	5.8	47.4	4.1	13.6
Total, per cent.	100.0	100.1	100.1	100.0	100.0	100.1	100.1	100.0	100.1	100.1	100.1	100.0

* 0.84 was used as the density of the combustible matter burned out, as nearly all of the combustible matter burned out was oil.

DIAMETER OF GRAINS IN MILLIMETERS

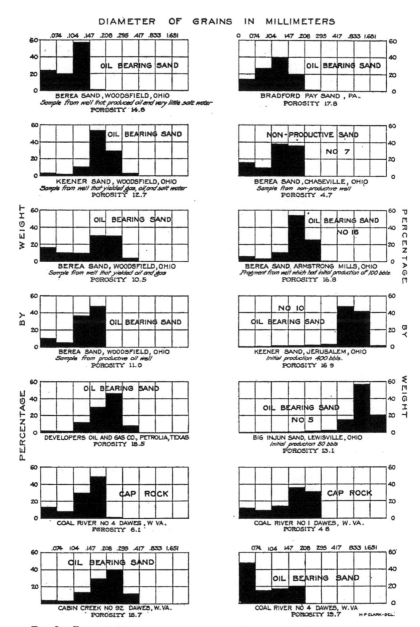

FIG. 3.—PERCENTAGE DISTRIBUTION OF DIAMETER OF GRAINS OF ELEVEN OIL-
BEARING SANDS, ONE NON PRODUCTIVE SAND AND TWO CAP-ROCKS.

space was less than 10.5 per cent., were producing sands. A study of samples 10, 16, 5, and 7 in Fig. 3, in connection with the corresponding pore spaces and diameters of grains in the maximum columns of the different samples suggests the conclusion that production is dependent on both pore space and size of grains, other factors being equal.

Data already obtained indicate that results of physical experiments involving the constants and factors stated in this paper are not only of scientific value but can also be used in connection with the most efficient methods of recovery of oil and gas; namely, in the valuation of oil and gas fields, in the possible application of various methods of oil extraction in fields where the normal flow is not sufficient to justify financially the continuation of the well, in plugging off water, in spacing and rate of pumping of wells, in avoiding interference of wells with one another, in recognizing the nature and texture of oil-bearing beds, which will respond to shooting, and where shooting will be detrimental.

TABLE 2.—*Total Pore Space with Density and Percentage Distribution of Diameters of Grains of Gas-bearing Sands from Mexia-Groesbeck Gas Field, Limestone Co., Tex.**

	Rawls No. 2, Specimen No. 1	Rawls No. 2, Specimen No. 2	Welsh No. 2	Clark No. 1, Specimen No. 1	Clark No. 1, Specimen No. 2	Kendrick No. 1, Specimen No. 1	Kendrick No. 1, Specimen No. 2	Cargile No. 1
Pore space, per cent. by volume.	13.2	10.7	16.6	34.2	37.7	25.7	22.8	34.4
Total weight, in grams.........	0.847	1.557	0.310	1.258	0.843	4.766	0.845	3.274
Density of grains free from moisture, or specific gravity referred to water at 4° C. as unity......................	2.76		2.73	2.70		2.68	-	2.73
Diameters from 0.833–0.417, mm., per cent..............	0.4		2.4	0.3	0.3	3.5		10.8
0.417–0.295 mm., per cent......	1.4		8.1	2.3	2.2	10.4		20.1
0.295–0.208 mm., per cent......	9.2		21.5	5.6	5.4	16.2		17.1
0.208–0.147 mm., per cent......	33.3		29.2	39.0	38.9	26.6		13.6
0.147–0.104 mm., per cent......	40.1		29.2	40.2	42.0	25.3		14.1
0.104–0.074 mm., per cent......	6.2		4.8	5.9	4.7	9.7		8.5
0.074–0.000 mm., per cent......	9.5		4.8	6.6	6.4	8.4		15.8
Total.....................	100.1		100.0	99.9	99.9	100.1		100.0

* These samples were collected by Mr. George C. Matson.

In Table 2, the Clark No. 1 sample contains a small amount of iron. The distribution of diameters of the Rawls, Clark, and Kendrick sands was determined from a mixture of the two samples. The two determinations of Clark No. 1 are duplicate measurements of the same sample. Cargile No. 1 consists largely of very small particles cemented together;

nearly all of this material would pass through the 200-mesh sieve (0.074-mm. diameter of grain) without breaking the particles. Kendrick Nos. 1 and 2 contain a small percentage of cemented material, but the remaining portion of Kendrick Nos. 1 and 2, and all of Rawls, Welsh, and Clark, consist of well-defined individual grains.

TABLE 3.—*Total Pore Space with Density and Percentage Distribution of Diameters of Grains of Gas and Oil-bearing Sands from Developers Oil & Gas Co., Petrolia, Tex.**

	Well No. 5, Oil-bearing Sand	Beatty Well No. 1 Gas Sand	Byerswell Specimen No. 1 Gas Sand	Byerswell Specimen No. 2 Gas Sand
Pore space, per cent. by volume	18.5	21.7	26.6	24.9
Total weight, in grams.........	17.725	8.522	22.798	19.897
Density of grains free from moisture, or specific gravity referred to water at 4° C. as unity	2.65	2.64		2.66
Diameters from 0.833–0.417 mm., per cent...............	0.2	0.01		0.01
0.417–0.295 mm., per cent......	8.2	0.3		0.02
0.295–0.208 mm., per cent......	45.8	23.0		0.1
0.208–0.147 mm., per cent......	29.5	55.6		31.5
0.147–0.104 mm., per cent......	12.5	15.8		50.0
0.104–0.074 mm., per cent......	1.8	2.6		9.6
0.074–0.000 mm., per cent......	2.0	2.7		8.8
Total....................	100.0	100.0		100.0

* These samples were collected by Mr. E. W. Shaw.

TABLE 4.—*Total Pore Space and Density of Grains of Oil Sands and Associated Rocks from Butler and Zelienople Quadrangles, Pennsylvania**

	No. 1	No. 2	No. 3	No. 4	No. 5	No. 6	No. 7	No. 8
Pore space, per cent. by volume.	8.0	8.5	22.2	14.5	10.4	4.5	7.3	5.5
Total weight of sample, in grams	48.281	14.840	5.770	9.331	21.994	9.637	9 811	6.493
Density of grains free from moisture, or specific gravity referred to water at 4° C. as unity.......	2.68	2.65	2.67	2.67	2.66	2.66	2.65	2.66

* These samples were collected by Mr. G. B. Richardson.

In Table 3, the distribution of diameters óf the Byerswell sand was determined from a mixture of the two samples. Both samples of the Developers Oil & Gas Co. Well No. 5, and Beatty Well No. 1, contained a small quantity of magnetite. The Developers Well No. 5 consisted of the largest size grains, a maximum percentage being of a diameter between 0.208 mm. and 0.295 mm. The Beatty sample gave a maximum percentage of grains between 0.147 and 0.208 mm. The Byerswell sand consisted of the smallest grains of the three samples, the maximum percentage being between 0.104 mm. and 0.147 mm. All three samples consisted of well-defined individual grains.

TABLE 5.—*Total Pore Space in Oil and Gas-bearing Sands and Associ-*

	No. 1	No. 2	No. 3	No. 4	No. 5	No. 6
Pore space, per cent. by volume.................	11.2 10.8	7.0	12.7	12.7	13.1	11.3
Total weight, in grams.........................	53.239 31.974	10.733	17.706	40.819	41.120	8.970
Density of grains free from moisture, or specific gravity referred to water at 4° C. as unity......	2.647	2.675	2.659	2.646	2.654	2.651
Diameters greater than 1.651 mm., per cent.....					20.4	0.6
Diameters from 1.651–0.833 mm., per cent.......					56.4	31.8
0.833–0.417 mm., per cent.....................					14.7	45.2
0.417–0.295 mm., per cent.....................				2.4	2.5	10.3
0.295–0.208 mm., per cent.....................		8.3		29.4	1.7	5.0
0.208–0.147 mm., per cent.....................	48.0	50.6	13.1	53.3	1.6	3.6
0.147–0.104 mm., per cent.....................	36.8	26.1	63.3	10.1	1.1	1.7
0.104–0.074 mm., per cent.....................	4.9	3.7	8.6	1.7	0.7	0.4
0.074–0.000 mm., per cent.....................	10.3	11.3	15.0	3.2	0.9	1.2
Total...	100.0	100.0	100.0	100.1	100.0	99.8

Only one sample, No. 4, showed the presence of oil. It is quite certain that none of the other samples belong to the oil-bearing sands.

No. 1, Fourth Sand, Speechley Pool. Concord Township, Butler Co., Butler Quadrangle.

No. 2, Third Sand, Evans City Pool, Forward Township, Butler Co., Zelienople Quadrangle.

No. 3, Snee Sand, Petersville Pool, Conoquenessing Township, Butler Co., Zelienople Quadrangle.

No. 4, "Clover seed," Top of Fourth Sand, Haysville Pool, Fairview Township, Butler Co., Butler Quadrangle.

No. 5, Fourth Sand (hard), Haysville Pool, Fairview Township, Butler Co., Butler Quadrangle.

No. 6, 100-ft. Sand, Evans City Pool, Forward Township, Butler Co., Butler Quadrangle.

No. 7, 100-ft. Sand, Evans City Pool, Forward Township, Butler Co., Butler Quadrangle.

No. 8, Fourth Sand, Haysville Pool, Fairview Township, Butler Co., Butler Quadrangle.

No. 1. Berea sand; depth 2050± ft.; from Well No. 1 on the M. O. Huth farm near Woodsfield, Center Twp., Monroe County, Ohio; fragment from a productive oil well.

No. 2. Berea sand; depth 1807–1829 ft.; from Well No. 2 on the N. H. Burkhead heirs farm near Woodsfield, Center Twp., Monroe County, Ohio; hard densely cemented fragment from a Berea gas well.

No 3. Berea sand; depth 1440± ft. sand sent by Larrick Bros. from Well No. 2, J. W. Steel farm, near Chaseville, Sec. 26, Buffalo Twp., Noble County, Ohio; hard densely cemented fragment from a non-productive well.

No. 4. Keener sand; depth 1539–1570 ft.; near the Huth farm, Woodsfield, Center Twp., Monroe County, Ohio; from a bed that yielded gas, oil, and salt water.

No. 5. "Big Injun" sand; depth 1460–1470± ft.; well No. 2, A. C. Weber, Lewisville, Summit Twp., Monroe County, Ohio; from a bed that yielded gas, oil, and salt water. Initial daily yield of well was 80 bbl. of oil.

No. 6. Keener sand; depth 1220–1259 ft.; from well No. 5 on the G. W. Kysor farm near Coats Station, Center Twp., Monroe County, Ohio; fragment of sandstone from a bed from which oil and salt water were pumped.

ated Rocks, with Diameter and Density of the Component Grains *

No. 7	No. 8	No. 9	No. 10			No. 11	No. 12	No. 13	No. 14	No. 15	No. 16
4.7	9.7	11.0	17.7	14.5	18.4	10.5	11.0	13.0	11.3	14.7	17.1
							12.4			15.0	15.9
											17.4
16.238	14.090	5.765	41.482	19.693	8.446	28.591	5.057	3.819	1.956	36.412	60.867
							4.309			28.900	43.213
											30.979
2.665	2.727	2.705	2.653	2.649	2.647	2.649	2.662	2.731	2.658	2.698	2.682
		0.8	1.9	23.5	0.2						
		0.8	41.1	34.4	2.6			8.5			
		16.6	47.2	30.1	78.9			70.8			
		18.3	3.4	5.1	12.4	4.6		12.9	30.3		0.3
		13.8	2.7	3.6	2.6	29.6		3.2	28.8		25.4
36.0		11.2	1.7	1.7	1.4	30.1	18.7	1.4	20.5	0.4	54.3
38.2	37.6	11.0	1.0	0.9	0.9	10.0	48.6	1.0	10.1	56.8	11.4
9.6	26.0	12.0	0.5	0.4	0.4	10.3	12.6	0.4	4.1	20.0	2.9
16.2	36.4	15.4	0.6	0.3	0.6	15.5	20.1	1.8	6.2	22.8	5.8
100.0	100.0	99.9	100.1	100.0	100.0	100.1	100.0	100.0	100.0	100.0	100.1

* These samples were collected in Ohio by Mr. R. Van A. Mills and Mr. D. Dale Condit.

No. 7. Berea sand; depth 1535± ft.; Chaseville, Seneca Twp., Noble County, Ohio; hard, densely cemented fragment from a non-productive well.

No. 8. Hard, bluish-gray shale overlying the Keener sand; depth 1445 ft.; J. R. Scott farm, Jerusalem, Sunsbury Twp., Monroe County, Ohio.

No. 9. Light bluish-gray shale, overlying the Keener sand; depth 1475± ft.; Hinderlong farm, Miltonsburg, Malaga Twp., Monroe County, Ohio.

No. 10. Keener sand; depth 1451–1469 ft.; Well No. 2; J. R. Scott farm, Jerusalem, Sunsbury Twp., Monroe County, Ohio; loosely cemented sandstone from productive bed, initial daily production of oil from well was approximately 400 bbl

No. 11. Berea sand; depth 1890–1920± ft.; Well No. 5; Henry Herdershot farm, Monroe County, Ohio; fragment from productive bed; well yielded both gas and oil.

No. 12. Berea sand; depth 1500± ft.; McLaughlin Well No. 5; Chaseville, Seneca Twp., Noble County, Ohio; fragment from productive bed; well yielded both oil and gas, with practically no salt water.

No. 13. Gas sand; depth 1350–1355 ft.; George Reem-Schneider farm, Sec. 11, Malaga Twp., Monroe County, Ohio; fragment from a bed that yielded gas.

No. 14. "Big Injun" sand; depth 1460–1500± ft.; Well No. 1, Ben Butts farm, Lewisville, Summit Twp.; Monroe County, Ohio; fragment from a bed that yielded oil and salt water.

No. 15. Berea sand; depth 2140–2160± ft.; Well No. 4, Taylor heirs farm, Woodsfield, Center Twp., Monroe County, Ohio; fragment from productive bed; well yielded oil with very little salt water.

No. 16. Berea sand; Shepherd farm, Armstrong Mills, Belmont County, Ohio; fragment from a productive bed; initial daily production of well was 100 bbl.

TABLE 6.—*Physical Properties of Chattanooga Black Shale from the Irvine Oil Field, Irvine, Ky.* *

	Pore Space, Per Cent. by Volume	Total Weight of Sample, in grams	Density of Grains Free from Moisture, or Specific Gravity Referred to Water at 4° C. as Unity
Trial No. 1..............	7.6	34.646	2.57
Trial No. 2..............	7.4	45.670	2.57

* These determinations of porosity have been previously published. See Eugene W. Shaw: The Irvine Oil Field. U. S. Geol. Survey *Bull.* 661-D, 190. Mr. Shaw collected the samples.

TABLE 7.—*Total Pore Space and Density of Grains of Oil and Gas Sands and Associated Rocks from Wyoming and Montana* *

Name and Location of Sand		Pore Space Per Cent. by Volume	Density of Grains Free From Moisture, or Specific Gravity Referred to Water at 4° C. as Unity
1. Wall Creek sand, Pine Mts., Wyo..........	Trial 1	3.6	2.659
	Trial 2	3.2	
2. Peay sand, Stump triangle station, Big Horn Mts..................................	Trial 1	5.1	2.651
	Trial 2	5.1	
3. Wall Creek (Peay) S. S. Jack Creek, Mont...		5.0	2.710
4. Torchlight S. S. about 2½ mi. east of Greybull, Wyo.............................	Trial 1	28.6	2.634
	Trial 2	30.1	
		28.9	
5. Wall Creek, S. S., Calcareous layer. From east side Powder River House, Wyo., 10 mi. west Salt Creek...		7.6	2.675
6. The main oil sand of Salt Creek field, Wyo..	Trial 1	25.8	2.640
	Trial 2	25.8	
7. Shannon sandstone, ¾ mi. northeast of town of Salt Creek, Wyo.....................	Trial 1	26.9	2.667
	Trial 2	26.6	
8. Wall Creek S. S., 10 mi. west of Salt Creek, Wyo. Lower Ledge (full of cleavage planes)	Trial 1	19.9	2.650
	Trial 2	20.9	
9. Peay S. S. Jack Creek, Mont..............	Trial 1	18.7	2.639
	Trial 2	18.5	
10. Peay S. S. Greybull oil field, Wyo..........	Trial 1	28.6	2.655
	Trial 2	28.6	

* These samples were collected by Mr. Carroll H. Wegemann.

TABLE 8.—*Total Pore Space with Density and Percentage Distribution of Diameters of Grains of Oil- and Gas-bearing Sands and Associated Rocks from Dawes, W. Va.* *

Ohio Cities Gas Co.	No. 1 Coal River No. 1, Cap Rock	No. 2 Coal River No. 4, Cap Rock	No. 3 Coal River, No. 4 Pay Sand				No. 4, Cabin Creek No. 92 Pay Sand		No. 5 Kelley Creek, W. Va. Gray (Wier) Sand. Only Well on this Creek
			Specimen 1	Specimen 2	Specimen 3	Specimen 4	Specimen 1	Specimen 2	
Pore space, per cent. by volume..................	4.8	6.1	21.7	20.1	18.8	16.6	18.0	19.4	13.7
Total weight, in grams.....	12.707	6.730	5.030	3.479	4.172	5.964	30.933	38.285	1.347
Density of grains free from moisture, or specific gravity referred to water at 4° C. as unity..................	2.656	2.636	2.672				2.664		2.663
Diameters from 0.295–0.417 mm., per cent...........							11.0		
0.208–0.295 mm., per cent...	30.6	0.5					39.7		3.2
0.147–0.208 mm., per cent...	34.7	48.7	20.6				28.4		25.3
0.104–0.147 mm., per cent...	13.7	30.3	17.1				13.1		25.8
0.074–0.104 mm., per cent...	8.8	7.7	14.9				3.7		15.1
Diameters less than 0.074 mm., per cent...........	12.1	12.8	47.4				4.1		30.6
Total.................	99.9	100.0	100.0				100.0		100.0

* These samples were collected by Mr. E. W. Shaw.

TABLE 9.—*Total Pore Space and Density of Grains of Sand from Bartlesville, Okla.* *

	(1) Bartlesville Pay Sand, Bartlesville, Okla., Skelton-Moore Well No. 11, Specimen 1	(1) Bartlesville Pay Sand, Bartlesville, Okla., Skelton-Moore Well No. 11, Specimen 2	(2) Outcropping Ledge of Sandstone From Same Section, Specimen 1	(2) Outcropping Ledge of Sandstone From Same Section, Specimen 2
Pore space, per cent. by volume	16.6	16.1	16.4	17.7
Total weight, in grams........	7.258	8.464	9.264	6.079
Density of grains free from moisture, or specific gravity referred to water at 4° C. as unity.....................	2 643		2.672	
Diameters from 0.295–0.417 mm., per cent..............	6.0		5.3	
0.208–0.295 mm., per cent......	12.3		27.9	
0.147–0.208 mm., per cent.....	46.1		39.6	
0 104–0.147 mm., per cent.....	18.3		16.4	
0.074–0.104 mm., per cent......	6.6		4.6	
Diameters less than 0.074 mm., per cent....................	10.0		6.2	
Total....................	99.3		100.0	

* These samples were collected by Mr. G. B. Richardson.

TABLE 10.—*Total Pore Space and Density of Grains of Gas-bearing Sands and Associated Rocks from Shreveport, La.* *

Name and Location of Sand	Pore Space, Per Cent. by Volume	Total Weight of Sample, in Grams	Density of Grains Free From Moisture, or Specific Gravity Referred to Water at 4 C. as Unity
1. Woodbine sand, Butler well.............	17.4	17.512	2.733
2. Tooke and Burke, No. 1..............	29.1	4.068	2.681
3. Pay sand, rare sample, Curtis No. 1.....	24.3	1.942	2.647
4. Greenish shale, just above pay sand of Curtis No. 1........................	22.6	1.628	2.717
5. Reddish shale, just above pay sand of Curtis No. 1........................	20.0	0.870	2.769
6. Flournoy, No. 1......................	37.7	7.300	2.700
7. McCutcheon fee No. 1................	22.2	4.752	2.691
8. Henderson and Hester, two determinations, very fine sand, almost (a) the appearance of shale......... (b)	31.1 32.6	0.808 0.772	 2.725
9. Independent Ice Co., fee No. 1, close to above.............................	36.7	8.705	2.688
10. McCormick fee, Well No. 155..........	25.3	4.876	2.728
11. McCullough fee, No. 1............ (a) (b)	17.8 20.3	37.106 25.582	 3.314
12. Stoer, fee No. 1.....................	14.6	12.003	2.693
13. Vivian field, Conlay No. 5.............	22.5	2.776	2.640
14. May Oil Co. No. 3 on S. W. Gas & Electric Co. No. 2. Two determinations................... (a) (b)	28.5 26.0	2.313 3.897	 2.591
15. Monroe gas pay sand. Smith Nos. 1 and 2.............................	27.2	2.448	2.672
16. Stringfellow fee No. 2................	9.4	3.647	2.705
17. Christian No. 4................ (a) (b)	9.6 8.8	1.160 1.137	 2.662
18. Hodges ward 1, sample of shale, deep gas sand...........................	16.9	6.931	2.752
19. Sample of sandstone containing some shale, same well as No. 18............	19.7	23.348	2.673
20. Sample from same well as No. 18, dark gray sand...........................	23.7	11.812	2.716
21. Sample from same well as No. 18, light reddish gray sand...................	20.4	15.526	2.654

* These samples were collected by E. W. Shaw.

After grinding until they passed through the 100-mesh sieve, the samples were washed in petroleum ether. The sample that gave 7.6 per cent. porosity was washed twelve times; the other was washed eighteen times A washing consisted of covering the sample with petroleum ether, letting it boil for about 15 min., and then decanting off the petroleum ether. Petroleum ether was again poured over the sample and then poured off; the process was then repeated. The final porosity of each sample was found to be 8 per cent.

The separated grains of the shale passed through the 300-mesh sieve. A 3-gm. sample of the shale was passed through a 100-mesh sieve and then boiled for 20 min. in concentrated hydrochloric acid. The porosity determined from the powder thus treated is 8 per cent., the same as was obtained by boiling in petroleum ether   Another 3 gm. sample was passed through the 100-mesh sieve and heated 20 min. in a Bunsen flame; the porosity determined from this final product is 19.6 per cent. The specific gravity of the powder thus treated is 2.59, and a solid cubic foot of it would weigh 161.69 lb. A cubic foot of the shale with the 19.6 per cent. of pore space emptied in the way outlined would weigh 130 lb.

In Table 7, the second determination of sample No. 4 is known to be slightly erroneous. The value, 28.9, is a weighted mean in which the first observation is given a weight equal to four times the second. No oil was found in any of the samples, when they were tested in the flame.

Two tests of each sample have been made. No indications of oil were found by heating specimens 1 and 2 of pay sand No. 1 in a platinum crucible.

After a fragment of the Bradford pay sand, Table 11, weighing 11.817 gm. was heated in a platinum crucible by a Bunsen flame until all organic matter and moisture were expelled it weighed 11.330 gm. The volume of the fragment equaled 4.255 c.c. If 0.80 is the specific gravity of the oil, the volume of the organic matter (mainly oil) is 0.608 c.c. Then the per cent. of the total volume of the fragment burned is 14.3. Taking 0.84 as the density of the oil, the per cent. of the total volume of the fragment burned is 13.6.

TABLE 11.—*Total Pore Space and Density of Grains of Bradford Oil-bearing Sand and Medina Sand* *

	Pore Space, Per Cent. by Volume	Total Weight of Sample, in Grams	Density of Grains Free From Moisture, or Specific Gravity Referred to Water at 4° C. as Unity
Medina sand, Niagara Gorge, Niagara, (a)	7.8	24.363	
N. Y. Two determinations......... (b)	8.0	15.126	2.657
Bradford pay sand, Minaid Run Oil Co., (a)	18.0	19.850	
Custer City, Pa. Two determinations (b)	17.6	26.616	2.663

* These samples were collected by Mr. G. B. Richardson.

DISCUSSION

R. Van A. Mills,* Washington, D. C.—Changes induced in the sands by drilling and operating wells have an important bearing on this paper; the porosities of sands and the sizes of grains and of pores change as the wells produce. Reductions in porosity and sizes of pores are caused by induced cementation, brought about through the infiltration of reactive waters into the wells and through the breaking down of bicarbonates in the oil-field waters incident to the liberation of carbon dioxide when wells are drilled and operated. The resistance to flow increases and the rate of production decreases as the sizes of the pores are reduced. Account must be taken of these facts in order to establish valid relations between the initial rates of production of wells and the porosities of sands that have undergone induced cementation.

The textures and bedding in sandstones are extremely variable and it is doubtful if many of the lumps of sand collected from wells after they are shot are truly representative of the pays. In many sands, the porous, open-textured parts of the pays are so soft and friable as to be disintegrated by shooting, so that most of the remaining lumps represent hard, tight parts of the sands. The collecting of representative samples, together with adequate collateral data, constitutes an important part of the investigation outlined by Mr. Melcher.

W. M. Small, Tulsa, Okla.—Has Mr. Mills any ideas concerning the zone of influence within which this cementation would take place; would it be more pronounced close to the bore hole and how far would it extend into the rock?

R. Van A. Mills.—That is a difficult question to answer at the present stage of the investigation. The greatest deposition of carbonates occurs within or close to the wells, but in some fields there is evidence that the sands become plugged in this way at considerable distances from the wells. Shallow pay sands are frequently calcareous throughout considerable areas, but this may be caused by natural agencies similar to those causing induced cementation. Some new wells in old fields reveal induced cementation by carbonates several hundred feet from the nearest old wells, but how general this may be remains to be determined. Photographs of lumps of sand shot and cleaned from old wells in Butler County, Pennsylvania, were published in Geological Survey Bulletin 693. These photographs show how thoroughly the sands were plugged by carbonates.

In parts of Ohio there is much evidence that declines in production are due largely to induced cementation of the sands. This is indicated not only by the examination of sands from the old wells, but by the high

---

* Petroleum Technologist, U. S. Bureau of Mines.

yields of new wells drilled among the old cemented wells. In one locality where the old wells have declined to average yields of approximately ½ bbl. per day, the initial rates of production of new wells, drilled within 300 ft. of the old wells, run as high as 50 bbl. per day. In many cases, the wells in this field were abandoned, not because the oil was exhausted but because the sands became so cemented that the oil would not pass through. The Bureau of Mines is pursuing field and laboratory experiments upon the removal of carbonates from the pay sands immediately around old oil wells through the use of chemical reagents. These experiments afford considerable promise of successful application.

R. VAN A. MILLS (written discussion*).—The trend of modern petroleum technology is to displace speculation by establishing facts and relationships through which to interpret underground conditions. Conditions in the sands, such as thickness and lenticularity, coarse or fine textures, openly porous or tight sands, initial or induced cementation,[14] initial or depleted rock pressures, the presence or absence of water are mapped as guides to the development and operation of fields. Samples of sand from apparently barren beds penetrated by the drill are examined to determine their possible productivity. Underground conditions that change during the operation of wells, more especially the changes in the textures of pay sands caused by induced cementation, and the movements and rearrangements of the fluids, such as the encroachment of water, are closely observed and recorded on field and office maps. More reliable criteria for the valuation of oil and gas properties are being established; studies of the probable oil and gas content of sands, together with the production records of wells, are being supplemented by studies of the conditions or causes governing the rates of production. In all of this work and in the operation and conservation of individual wells, investigations like those outlined by Mr. Melcher are of primary importance.

The United States Geological Survey Bulletin,[15] from which Mr. Melcher has taken his Table 5, production data, and other field notes, establishes the importance of correlating physical and chemical studies of the reservoir rocks and contained fluids with the production histories of the wells to establish relationships for practical application, but the few production figures published are not adequate for the use Mr. Melcher

---

* Published by permission of the Director, U. S. Bureau of Mines.

[14] The term induced is used to designate the deep-seated effects of man's activities. The cementation of pay sands incident to the drilling and operation of wells is one of the induced effects in oil and gas fields previously described by the writer. See U. S. Geol. Survey *Bull.* 693, 44–55, and 98.

[15] R. V. A. Mills and R. C. Wells: Evaporation and Concentration of Associated with Petroleum and Natural Gas. U. S. Geol. Survey *Bull.* 693

makes of them. They suggest broad relationships that the porosities and sizes of pores bear to the initial rates of production from a few wells, but these relationships might be more definitely established, in the Appalachian fields, by using the large collection of sands and accompanying field notes and production data that the writer and others have contributed to the Geological Survey.

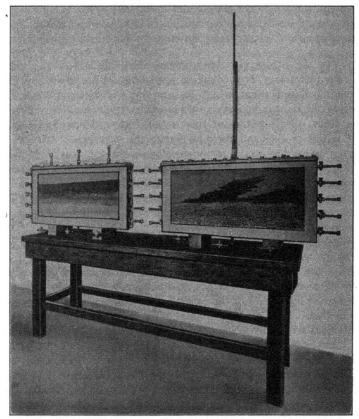

FIG. 4.—APPARATUS FOR STUDYING CAUSES AND EFFECTS OF MIGRATION OF OIL AND WATER THROUGH OIL SAND. (*Plate XXII*, *U. S. Bureau of Mines Bull.* 175.)

In studying subsurface relationships, we are obliged to deal with the summations of effects of many factors whose values are only relative and rarely alike in different localities or at different depths in the same locality. Porosity and size of pores are among these factors. A minimum porosity value, or a minimum size of pores, below which sands are nonproductive in one locality, need not necessarily apply in localities where the bedding of the sands, the modes of occurrence of oil, gas, and water, the viscosities of the oils, the gas pressures, the subsurface temperatures, etc. are different. Consequently, it is imperative that various principles and relationships be studied, in conjunction with porosity tests, through adequate field and laboratory methods for each locality.

The writer supplements field work and porosity tests, such as Mr. Melcher describes, by comparative studies of fluid movements through sands arranged in steel tanks. The tanks are equipped with plate-glass fronts, to facilitate observations and the making of photographic records of experiments. Oils from different fields are used, the number of variables in each experiment is restricted, and the relative values of different factors such as porosity, viscosities of oils, buoyancies of oils and gases in water-saturated sands, expansive forces of compressed gases in sands, capillary forces, and many other factors are definitely established for each set of experimental conditions.

The importance of adequate field methods and notes in the collection of samples of oil- and gas-bearing sands must also be emphasized. It should be understood that a large proportion of the hard, densely cemented fragments shot and cleaned from wells represent so-called shells, breaks, and tight sand, rather than true pay sands. Many of the pay sands are so granular and friable as to render lump samples exceptional, but where lump samples of the "pays" can be obtained, they should be collected before they have been exposed to the weather. It is good practice to collect several lumps together with loose sand from the same well for comparison. Large proportions of the loose sands cleaned from shot cavities in producing wells come from the relatively friable pays. The textures of these loose sands furnish criteria for the identification of lumps from the same parts of the beds. Care should be taken to differentiate between loose sands from shot cavities and drill sludge, which is unsatisfactory because of its pulverized condition.

Owing to the extremely variable nature of beds of sandstone, no part or section of a sandstone may be regarded as truly representative of the bed. But the average result of several porosity tests upon a carefully selected multiple sample from the productive horizon in a well should most nearly represent the porosity of the pay sand at that place. Adequate studies of the variations in texture and porosity of pay sands, and the relationships that these conditions bear to the occurrence and recovery of oil and gas can be made only through intensive field and laboratory work. Samples of the sands should be collected in conjunction with detailed geologic studies of a field or by a resident engineer or geologist during the development and operation of the field. But no matter how the work is done samples of the pay sands, drill logs, and production records from as many wells as possible should be obtained and applied in each field under examination.

To establish the relationships that porosity and size of pores in sands bear to productivity, it is advantageous to collect samples of the non-productive rocks for comparison with pay sands. Pieces of non-productive rocks are occasionally brought to the surface through the shooting of dry holes; also pieces of the cap sands are sometimes ejected in shooting the pays, but for the most part we must depend for these samples on small fragments or chips, of uncertain origin, found in the drill sludge and cavings cleaned out while the wells are being drilled. The use of core drills for sampling oil sands and their associated rocks has long been considered, but the writer believes that a sampling device (working on the same principle as the under reamer) that will break fragments from the walls of a well as it is being drilled should prove advantageous to companies applying physical and chemical studies of oil- and gas-bearing rocks.

The period in the productive history of a well at which a sample of pay sand is collected, together with the water conditions in the well, have much to do with the physical and chemical qualities of the rock and the relationships that these qualtites bear to production. The porosities, sizes of pores, and chemical compositions of water-bearing pay sands frequently undergo marked changes during the operation of wells. The induced cementation of pay sands by carbonates is exceedingly common

Through the coöperation of Mr. C. W. Paine, of Ozark, Mr. George Vandergrift, of Woodsfield, and other operators in that locality, the writer has had the groups of Ohio wells, cited by Mr. Melcher, under surveillance since the summer of 1914. The subsurface geology has been studied in detail[16] and samples of the oils, gases, waters, and reservoir rocks have been collected and examined periodically. Some of these wells have now (April, 1920) ceased to produce because the pay sands are plugged by inorganic deposits from the waters associated with the oil. The porosities and sizes of pores in the pay sands around the wells have been so reduced as to stop production.[17] New wells situated within 300 ft. of the old ones, and drilled after the old wells had been abandoned, yielded oil at initial rates as high as 10 bbl. per day from the same sands; apparently from the same pays, where they had not been plugged.

The causes and effects of induced cementation have already been described.[18] To ignore them in studying the relationships that porosity and sizes of pores bear to the productivity of sands may cause errors. For instance, consider two of the sands represented in Table 1, which together with the other Ohio sands cited by Mr. Melcher, were collected and examined in the preparation of Geological Survey Bulletin 693. The Berea

---

[16] R. Van A. Mills and D. Dale Condit: Unpublished manuscript and maps in the files of the United States Geological Survey.

[17] Figures showing changes in composition and reductions in porosity and sizes of pores will be presented in later papers.

[18] See U. S. Geol. Survey *Bull.* 693, 44–50 and 98.

sand, from Armstrong's Mills, was collected from an oil and gas well that had been producing for ten years. The initial rate of production from the well was 100 bbl. of oil per day, but in 1914, when the sample was collected, the rate of production had declined to about 2 bbl. of oil with a little water. Chemical and petrographic examinations of the sand indicate that it has undergone induced cementation through the deposition of carbonates. Judging from the high proportion of secondary carbonates in the sample the original porosity may have been diminished 7.7 per cent. of the volume of the rock.[19] The sizes of pores and the permeability of the sand have undoubtedly been diminished since the well started to produce. Consequently the porosity and sizes of pores in this sample can bear no valid relation to the initial rate of production of the well. To interpret the loss by ignition of this sample as a loss of combustible matter is erroneous. As shown in the accompanying table, the sample contains 4.23 per cent., by weight, of CO_2 combined with calcium, magnesium, and iron to form carbonates. Part of this CO_2 would probably be lost during ignition of the oil and paraffin contained in the sample. If the hydrocarbons in this sample were removed by ignition prior to the porosity measurements, the value of the porosity measurements themselves were impaired through the breaking down of the carbonate minerals, which constituted an important part of the sample.

TABLE 12.—*Analyses of Sands from Oil Wells**

(*R. C. Wells, Analyst*)

	Keener Sand from Well No. 2, J. R. Scott Farm, Per Cent.	Berea Sand from Well on Shepherd Farm, Per Cent.
SiO_2	93.82	85.90
Fe_2O_3 (all Fe as Fe_2O_3)	1.75	2.82
Al_2O_3	0.73	1.68
CaO	0.19	2.96
MgO	0.25	0.84
P_2O	0.02	0.03
CO_2	Trace	4.23
TiO_3	0.12	0.12
Loss on ignition, less CO_2	2.31	1.35
	99.19	99.93

* U. S. Geol. Survey *Bull.* 693, 17.

[19] The calculation is based on the assumption that the sample, having a total porosity of 16.8 per cent., contained 4.23 per cent. by weight of CO_2 combined with calcium, magnesium, and iron to form carbonates, and also that the sand was free from carbonates when the well was drilled. The examination of a large number of samples of water-bearing pay sands from new wells in new fields in Ohio, Pennsylvania, and West Virginia reveals only traces of carbonates.

The ⎡Keener sand from well No. 2 on the J. R. Scott farm, near Jerusalem, Ohio, though collected after the well had been producing for 11 years, apparently had not undergone induced cementation. The sand contained only a trace of CO_2 and the well has remained the best producer in the field.[20] The injury to the sand that necessitated shooting the well was caused by so-called paraffining of the sand, and the high percentage of combustible matter in the sample was due to the presence of waxy hydrocarbons, which reduced the effective porosity. The examination of this sample, after the hydrocarbons were burned off, furnishes more reliable data for use in establishing the relationships that porosity and sizes of pores bear to initial rates of production.

The relationships that total porosities of coherent sands may bear to the rates of production, as well as to the ultimate productions from such sands, depend largely on relationships between total porosities, effective porosities, and sizes of pores. All of these conditions are related one to the other and all of them influence the retentivity as well as the fluid movements through sands. Cementation, either natural or induced, has played a major role in reducing the total and effective porosities of lithified sands, but it has likewise reduced the sizes of the pores. The most densely cemented, or in other words the least porous, of the lithified sediments generally contain relatively fine pores. In unconsolidated sands, where there has been little cementation, the total porosities, effective porosities, and sizes of pores are not so closely related. The writer's experiments with unconsolidated sands indicate that sizes of pores, and especially the sharp variations in the sizes of pores between different beds or in different parts of the same beds, are factors of primary importance in the movements of oil and gas through water-bearing strata, regardless of total porosities.[21]

The selective or differential permeabilities of sands to waters, oils and gases are also of primary importance in recovery problems. These selective or differential permeabilities depend not only on the porosities and sizes of pores of the sands, and the viscosities, pressures, and temperatures of the fluids, but also on the order and degree with which the sands have become wet or saturated by water or oil. Studies of these relationships, especially the studies of effective porosity and of permeability that Mr. Melcher proposes to make, constitute a new and promising phase of petroleum technology in which real advances can be made only through intensive field work supplemented by systematic and scientific laboratory experimentation.

---

[20] See U. S. Geol. Survey *Bull.* 693, 97.

[21] R. Van A. Mills: Experimental Studies of Subsurface Relationships in Oil and Gas Fields. Manuscript in course of publication.

C. W. WASHBURNE, New York, N. Y. (written discussion).—This paper marks an advance in technical methods. The data indicate that the oil-bearing parts of sands are not more porous than the same sands at their outcrops, a result that does not accord with the prevailing opinion of many geologists, none of whom, however, has made such extensive observations. The figures, though, should be regarded as minima rather than averages, for the observations were made on coherent chunks of oil sand obtained, probably, from the bottom of drill holes without the use of core barrels. Chunks of this kind probably represent the harder, more cemented, and less porous parts of the sand. The greater part of the sand is more friable; it is ground up by the bit and comes out as sand, not as fragments of stone. Chunks obtained from wells, therefore, are not likely to be average samples of the sand.

To get the true average porosity of a sand, cores of the whole sand should be obtained. The cores obtained by the core barrel on the Gulf Coast are too fragile to ship, except in the core barrel itself. The efficient field manager must study these cores before deciding how to case and tube his well. He must therefore break them up immediately at the well. It would give a truer figure of porosity if a method were developed whereby the porosity of these cores could be determined in the field without interfering with the driller's examination.

Water Displacement in Oil and Gas Sands

By Roswell H. Johnson, M. S., Pittsburgh, Pa.

(New York Meeting, February, 1920)

ALL STRATA not yielding oil or gas in commercial quantities or a corresponding amount of water may be called dry in a wide sense. In petroleum geology, however, we may exclude all sands of too low or fine porosity to yield gaseous or fluid contents to the hole drilled in the sand before any original pressure that its contents may be under is disturbed. Most rocks are of this class and they are not reservoirs in our definition; their "dryness" is wholly a matter of course. What are the contents of the pores or what is the exact porosity of such rocks is of almost no concern to us, for economically they are "dry."

What does interest us is the content of a rock having sufficient porosity and the pores of sufficient size to yield oil or gas in commercial quantities, if they were present under original pressure. Dryness of these reservoirs is a matter of supreme practical importance. Three views current as to such dryness seem, to me, to apply in a few cases only. It is the purpose of this paper to give reasons for this position and for believing that, in ordinary sedimentary rocks, there is only rarely a reservoir of competent porosity and undisturbed pressure that is dry in the sense of not yielding water, oil, or gas when first penetrated.

1. Gardner[1] writes of some Kentucky sands, "There has never been present any salt water or other water in the sand." Absence of water cannot demonstrate this position. It is necessary to show that the rocks were not laid down in water, but in air, and that they became so enclosed, while still above the water-table of the ground water, that water has not been able to enter since. Most of these sands, and certainly the productive limestones, were deposited in water; and such sands as have been commercially productive show no reason for believing that the overlying shale or limestone was not laid down progressively from one direction and in water that would have flooded it. No adequate explanation has been offered for this hypothesis, which is so inherently improbable.

2. Reeves[2] urges that "sands originally water filled may have been drained of their water and not filled when later covered." It is difficult

[1] James H. Gardner: Kentucky as an Oil State. *Science*, N. S. (1917) **46**, 279–280.

[2] F. Reeves: Origin of the Natural Brines of Oil Fields. Johns Hopkins Univ. *Circ.*, N. S. (1917) N. 3;

Absence of Water in Certain Sandstones of the Appalachian Field. *Econ. Geol.* (1917) **12**, 354–378.

to see how the presence of the air could prevent the entrance of water where the water overlaps the sand from one side and so has ample opportunity to expel the air.   However, we have an excellent test of whether the sand is dry because air filled, as supposed by Gardner, by merely analyzing this supposed entrapped air.   Instead of the air called for by Reeves' hypothesis we nearly always find methane.   There are very rare occasions where it is mainly nitrogen, probably entrapped air denuded of its oxygen by the oxidizing of materials in contact.   For these occasions, as at Dexter, Reeves' hypothesis is helpful; but its unimportance is measured by the extreme rarity of such cases.

3. Shaw holds that a sand may be adequately porous and hold water and yet not yield it to a drill hole because of lack of expansive force behind it.   In view of the almost universal rule of an increase of pressure with depth in our ordinary sedimentary strata, such as we find in oil fields, such a failure must be excessively rare.

An absence of methane would not be expected in the sedimentary series in which our oil and gas fields are found, because these rocks are so generally charged with some gas, either free or dissolved in oil, in some part of the reservoir.   Even with no methane, we know that propane and butane are soluble in water to an extent of nearly 3 per cent. so that they could give it expansibility for at least a short time.

DISPLACEMENT AND RESULTING MOVEMENT IN OIL AND GAS SANDS

Concluding, then, that the reservoirs now containing oil and gas originally were water filled and that the gas and oil later entered the reservoir, thereby displacing water, it becomes a matter of interest to postulate the resulting movement of the oil, gas, and water, respectively. We may assume that the oil and gas enter on all sides of the reservoir. If at the bottom they would rise to the top, although in all probability generally deflected en route along some bedding plane.   Having reached the roof of the reservoir, since this is ordinarily elongated and pitching, they would move along the inclined plane until they formed an oil and gas accumulation at the upper end.

The matter of especial interest to us is the action as it finds minor dome-like irregularities.   These will necessarily be filled if there is enough oil and gas to fill them.   If more than enough oil and gas reach these local catchments, the oil and gas will resume their movement up dip. However, as this movement continues, the proportion of the gas in these catchments will increase.   Indeed, the oil may nearly all be forced down into the general stream and so move on up to the highest oil and gas mass.   In this motion upward along the crest of the reservoir, the path would not be a broad one.   Any "bulge" in the roof to one side of such a "path" would not be fed with oil and gas, except such as was caught by direct upward movement to it by side paths flowing on the way to the ridge.   If the crest of the reservoir was very flat and broad, we might possibly have a series of braided paths, such as one finds in some rivers of broad bed.   In the top mass of oil and gas to which the paths lead, the percentage of oil to gas should be higher than any bulge below because of the excessive proportion of gas held below. This selective action explains some of the differences in relative percentage of gas and oil in different pools.   Suppose now the reservoir as a whole is arched, each flank is then working as before suggested but the oil-gas mass is held at the crest instead of by the termination of the reservoir.

So far as the upward motion of the oil and gas has been discussed, we have assumed that there are no obstacles to the free motion of any molecule of oil and gas, as directed by gravitation.   However, one serious obstacle, surface tension, leads to the oil or gas rounding off into a bubble, which thereby offers great resistance to motion in sandstone as fine as we generally find it.   A bubble forms in each "chink" between grains, but its oil cannot move until the bubble grows so large as to extend as a bud through one of the larger passageways into the adjoining chink between grains.   Only a continuous invasion can make progress.   It is a mistake to think of a passage of a series of bubbles as such.   The resistance in that case would be so great that gravitation at least would be impotent except with very coarse deposits.

The water must have a motion away from the upper part of the reservoir as the movement of the oil and gas upward along the roof drives the water, in part, back into the shale and, in part, down the reservoir to the lower end.   Again, we must consider the effect of depressions in the floor (whether depositional or deformational) on the water as it recedes to the lower end of the reservoirs.   The water would fill each depression and spill over its oil in the general movement down the reservoir.   It retains a disproportionate share of water after all the oil and water have passed this depression going down dip.   Some of the water may be forced out through the floor of the reservoir, but it would usually leave the water in excess until the gas accumulation was quite large.   Therefore we conclude that these depressions are less favorable points for oil and that most of the oil will accumulate at the lowest part of the reservoir, assuming that the displacement continues that long.   The lowest part of the reservoir being so frequently a matter of lateral variation or "tailing" of the bed, this place is more difficult to locate.   Hence the search for oil in sands without water is more difficult than in those carrying much water.   It is not a case of mere reversal, seeking anticlines in one case and synclines in the other.   Structure is, then, of still less help in the waterless sands than would otherwise be supposed.

DISCUSSION

DAVID WHITE,* Washington, D. C.—This is a most interesting point concerning the genesis and distribution of oil, gas, and water in rocks. According to common acceptance, a dry sand is one from which oil, water or gas will not exude when it is penetrated by the drill or the mine shaft. However, strictly speaking, there is no arenaceous sediment or clastic, not excepting eolian sands, which has not been laid down in water or has not later been submerged beneath and filled with water before any sealing cap-rock has been laid down. All sands have at some time been full of water. _ The expulsion of the water under varying conditions is a topic not yet adequately discussed. It does not seem to have been generally recognized that the essential reason why oil does not flow from the sand when resistance is removed by perforation by the drill or the mine shaft probably lies in the fact that former pressures have been reduced to the point where capillary resistance prevents the outflow into the void. There is one more question: Does the deformation occur while the oil, gas, and water are in process of migration, or do these migrate after the deformation occurs? Deformation takes a long time. The migration also ought to require a long period. Is not the migration in progress when the deformation is developed?

G. H. ASHLEY, Harrisburg, Pa.—Within the past few months there has been, in the McKeesport gas pool in western Pennsylvania, a development that, if it has been properly interpreted, has some bearing on this problem. The principal gas reservoir is the so-called Speechley sand, found at a depth of about 2900 ft. (884 m.). Between 400 and 500 ft. above that is the Elizabeth sand. The first big well contained too much gas to be carried off by the 6-in. main that had been laid, so a valve was placed in the main to allow the escape of gas above a pressure of 430 lb. Mr. Tonkins, of the Peoples Gas Co., suggests that as a result of the back pressure thus generated in that big well the gas from the lower sand entered the upper, or Elizabeth, sand and enriched it, as indicated by the fact that other wells put down to the upper sand have increased their flow and later wells have obtained an enlarged flow from that upper sand. If that is true, it indicates that the Elizabeth sand · was dry, not because nothing would flow out of it or into it, nor because of closeness of grain, for otherwise the sand would not have taken up gas.

SIDNEY PAIGE,† Washington, D. C.—You say the back pressure; could the back pressure have been any greater than the original pressure before the oil was tapped? How would this new movement have occurred? It is not clear to me.

---

* Chief Geologist, U. S. Geol. Survey.
† Geologist, U. S. Geol. Survey.

G. H. ASHLEY.—Before the tapping of the lower sand, there was no connection between the upper and lower sands.

SIDNEY PAIGE.—It came up along the pipe?

G. H. ASHLEY.—It came up along the pipe; there was no tubing or piping between the sands. The 100-ft. sand was the last one that was cut off.

R. H. JOHNSON.—I should say that hydrocarbons are still coming in while deformation is going on. The main reason for that is that the deformation is particularly active in making hydrocarbons, as David White's work has well shown. Most of the hydrocarbons must come into the reservoirs quite a little later than was formerly thought.

May I add a point in connection with this well at McKeesport? At the Elk City gas field, the other prominent gas field we have had recently, the pressure started to decline at a rather rapid rate, but when the pressure reached a certain point, the decline, although we were taking out still more gas, was not so rapid. In explanation, it was said that the well was tubed to a place above the productive sands, so that there was an open hole of several feet. This sand, when first struck, I would suggest therefore was feeding in there just as the Elizabeth was being fed at Mc-Keesport, so the pressure dropped fairly rapidly during this period of underground wastage; but after this sand had been fed to its capacity, apparently the pressure declined more slowly. I suspect that something very similar happened at McKeesport.

If we could have had pressures on that well right along, we could have learned something about the feeding situation. The Elizabeth sand was fed until it would take no more. From then on, of course, it was not as serious a source of underground wastage except as the gas might go through other wells than those of the owner.

These Elizabeth sands are not as large as they really ought to be, considering the magnificent chance of being charged by this gigantic well, which seems to be the result of a lower porosity. The sands yielded a small amount of gas before this feeding process and the amount since is only moderate compared with the great wells; I should say that was because it did not have the capacity to receive much of that gas.

E. W. SHAW,[*] Washington, D. C.—In the Caddo, Elm Grove, and Monroe fields, Louisiana, we have such extensive underground migration of gas that after some of the big wells have been completed but not successfully cased the country all around sizzles. The gas creeps from one sand to another and sometimes blows out the surface as much as $\frac{1}{4}$ mile from the well where it left its natural reservoir.

---

* Geologist, U. S. Geol. Survey.

I do not see the bearing of this on the question of dry sands, concerning which there seems to be a good deal of difference of opinion, for the reason that when the gas rises from a lower sand, where the pressure is high, to a higher sand, where the pressure is low, it is not essential, and it is not to be inferred that the pores in the higher sand are empty or even free from liquid contents. All that is required is that the gas or liquid move off somewhere else or accommodate itself in smaller quarters.

I was much interested in Mr. White's remark that we are all agreed that pores are filled with something. If we can agree on this we have made a real step in advance. The following step to be taken is longer and more difficult, but it is a step that we must take sooner or later. This step is to recognize that most dry sands are myths.

R. H. JOHNSON.—The question of the helium in the Kansas and some of the Texas gases, I think, has a bearing on Reeves' hypothesis of entrapped air. Those gases have more helium in proportion to nitrogen than the air.

In this paper, I have accepted the notion that we might have entrapped air to explain these nitrogen reservoirs. Since writing that, I have become more skeptical. We can easily explain away the lack of oxygen; that can be taken up to make carbonate, but why this super-atmospheric amount of helium? These helium gases may have a deep-seated origin over faults that do not come to the surface. May they not be gases of a cosmic nature—gases that have been extruded from original earth stuff from still greater depths, that have worked along some faults and have not been able to get closer to the surface?

Do not think that that means I am inclining toward any inorganic origin of hydrocarbons, but if we do not accept that hypothesis, we have difficulty in getting that much helium because the air must have been entrapped, and it is utterly unreasonable to suppose there was more helium in the air then than there is now. I dare say that higher up in the air, there is a greater amount of helium, but that will not help us because these gases were laid down close to the earth's surface, and the gravitational contrast was as great then as it is now.

H. W. HIXON, New York, N. Y.—That question of helium in the gases goes back, I believe, to the origin of the hydrocarbons; and while Mr. Johnson evidently does not believe in the inorganic origin of oil and natural gas, I most decidedly do. If you assume that the earth had an origin, it must have been either according to the planetesimal hypothesis or a gaseo-molten condition. Taking the latter view, a planet above its critical temperature is all gaseous. Under that condition, by applying the law of the diffusion of gases, you have each gas occupying the whole space of the body of the planet as if the other gases were not there.

Gravitational compression will produce a condition of density greater than that of the solids at sufficient depth, so that when such a planet cools, the solid material, being lighter than the highly compressed gases, will act just as if it were a solid throughout. You still have, in the body of the planet, some of each of the gases that were present in the original planet when it was all gaseous.

As regards the origin of petroleum and natural gas, there is just the same reason for the hydrocarbons being in that gaseous interior as any of the other gases. That is the reason why, from volcanoes, all the known gases of the atmosphere and others are extruded. So the origin of helium goes back to the original gaseous planet, like the origin of the hydrocarbons. I take that stand, knowing that nearly all petroleum engineers and geologists believe that petroleum and natural gas are of organic origin.

I first became interested in this matter when I heard Mr. Eugene Coste speak on the subject. He did not, however, go back as far as that and simply denies that fossils or organic matter produce oil. I can see how from the application of the law of diffusion of gases to a gaseous planet, where all of these things would come about in that way, the oil and gas would be entirely of inorganic origin. In the question whether the dome is the cause of the accumulation of gas or the gas the cause of the dome, I think you have the cart before the horse. I think the domes are caused by the accumulation of gas, the gas causing the dome or the anticline or both.

DAVID WHITE.—The origin of the helium in such large amounts in the natural gas of parts of Ohio, Kansas, northern Oklahoma, and Texas is a geological problem of great interest and importance that is yet to be solved, and it is greatly to be hoped that the oil- and gas-field geologists will find the key to the situation. There is some circumstantial evidence pointing toward the occurrence of the helium-rich gas of Kansas and Oklahoma over areas of deep-seated faults or disturbance. The same may be true of the north Texas region. But the singular fact that the helium now occurs, in general, in the shallow sands, and is present only in relatively small amounts or not at all in the deep sands in most areas is baffling. Apparently the Ohio area, Hocking and Vinton Counties, in which the helium is found in the Clinton as well as in the Berea, offers no exception. One does not look for badly disturbed rocks in the center of the basin in southeastern Ohio, although the unexpected frequently happens, and it may have happened in this case.

Composition of Petroleum and Its Relation to Industrial Use

By Charles F. Mabery,* S. D., Cleveland, Ohio

(New York Meeting, February, 1920) "

So far as the elementary composition of petroleum is known, it may be briefly stated. Petroleum consists principally of a few series of hydrocarbons, with admixtures of sulfur, nitrogen, and oxygen derivatives in comparatively minute proportions, which may be regarded as impurities to be removed in the preparation of commercial products. But as each series is represented by many homologs, in the aggregate, crude petroleum is an extremely complex mixture of hydrocarbons and their derivatives. In part, these hydrocarbons individually conform in structure to the system of synthetic hydrocarbons whose structure is well defined and represented by the typical series C_nH_{2n+2}, C_nH_{2n}, the series C_nH_{2n-2}, C_nH_{2n-4}, the members of which have not been sufficiently studied to establish their structure, and the series C_nH_{2n-6} composed of the aromatic group, benzene and its homologs. Hydrocarbons of greater density contained in the portions of petroleum that cannot be distilled without decomposition doubtless have less hydrogen than is represented by these formulas. Since to every hydrocarbon there is a definite temperature, even in vacuum, at which its constituent carbon and hydrogen atoms fall apart, and since for the heavier bodies this temperature is not much above 360° C. in vacuum, it is evident that some other method than distillation must be devised for their separation if anything further is to be learned concerning their individual constitution.

CLASSIFICATION OF PETROLEUMS

There is such a wide variation in the composition of petroleum from different fields, it would seem possible to make a classification on this basis were it not that no single variety is entirely free from hydrocarbons contained in others. Such a classification has been suggested of the exceptionally pure Pennsylvania petroleum, the sulfur oil from Trenton limestone and other sources, the California oil with its large amount of nitrogen (quinoline) derivatives, and the Russian oil, composed chiefly of the naphthene hydrocarbons.[1] A commercial distinction is made

---

* Emeritus Professor of Chemistry, Case School of Applied Science.
[1] S. F. Peckham: *Jnl.* Frank. Inst. (1896) **141**, 219; C. Engler: "Das Erdol," **1**, 228. Leipzig, 1913.

between oils with a paraffine base, of which Pennsylvania crude is typical, and oils with an asphaltic base, typical Texas and California crudes, the heavier varieties; but this distinction cannot be sharply drawn since there are oils that contain both constituents.

Both theoretically and commercially, there is a corresponding difference in quality between such light oils as those of the Appalachian fields, some of them composed to the extent of 50 per cent. or more of gasoline and kerosene hydrocarbons, and almost entirely of the hydrocarbons C_nH_{2n+2}, including paraffine, and the Texas Gulf oils which contain no hydrocarbons of this series but are composed of heavy members of the series C_nH_{2n-2}, and C_nH_{2n-4}, besides the still heavier asphaltic bodies. But even here there is a connecting link in the hydrocarbons of the series C_nH_{2n-2} that form the light lubricants of the Pennsylvania oil. Such interrelations have been verified in all American petroleum. From petroleum of many fields containing sulfur derivatives, such as that of the Ohio Trenton limestone, of the Illinois fields and even of Canada with large sulfur content, there are good yields of gasoline, kerosene, and paraffine. The great fields of Oklahoma, Kansas, Wyoming, and the lighter crudes of Texas and Louisiana with a variable composition between the Appalachian and the asphaltic crudes also fall within this category.

Petroleum from oil territory in other parts of the world does not differ materially in composition from that of the American fields. The principal foreign fields are those of Galicia, Russia, Rumania, Japan, and the East Indian Islands. They contain, in variable amounts, paraffine, gasoline, and kerosene hydrocarbons, but not of the same series as those of American gasoline and kerosene. They all contain sulfur and nitrogen derivatives. Rumanian and Japanese oils are both composed to a large extent of the naphthenes, to be more fully described later, as is also Russian oil to the extent of 80 per cent. or more, in which these hydrocarbons were first identified. Large amounts of Russian oil have been sold here for medicinal purposes, but, no doubt, some varieties of American petroleum are fully its equal in this field.

Basic Series of Hydrocarbons

Referring again to the basic series of hydrocarbons alluded to above as constituting the main body of American petroleum, the series C_nH_{2n+2}, commonly known as the methane, or marsh-gas, series for it begins with methane or, marsh gas, CH_4, the principal component of natural gas, is the most comprehensive for it includes the main portions of gasoline, kerosene, and paraffine, and is often alluded to as the paraffine series and its members as paraffine hydrocarbons. The latter increase in unit order, by the increment CH_2, through the more volatile gasolines with boiling points from 30° to 150° C., C_5H_{12} to C_9H_{20}, and next through kerosene, with boiling points from 150° to 325° C., C_9H_{20} to $C_{19}H_{40}$, when they soon begin to solidify as paraffine composed of the crystalline hydrocarbons from $C_{20}H_{42}$ to $C_{35}H_{72}$ and distilling in vacuum as high as 350° C. In practical use, these hydrocarbons include paraffine for candles, kerosene for illumination, and the lower members for motor fuels, and various minor uses, such as cleansers and solvents. They are extremely inert, entirely devoid of lubricating quality, easily decomposed by heat (cracked) into lower members of the same series or into unsaturated hydrocarbons. Such decompositions, which include also other heavier hydrocarbons, are the basis of the numerous cracking processes, in which heavy oils are converted into more volatile forms for use as motor fuels.

ETHYLENE AND NAPHTHENE SERIES OF HYDROCARBONS

Continuing with our scheme of the hydrocarbons, the next paralle series C_nH_{2n}, the ethylene unsaturated series, is present in small amounts in most petroleum. The oil that separates by dilution of acid sludge is composed to a considerable extent of these hydrocarbons, for they are dissolved by the acid in refining the crude distillate. It was formerly thought that these bodies formed a large proportion of American crude oil, but they have since been shown to be another series of the same empiric composition and formula, C_nH_{2n}, but altogether different in properties; they are cyclic, or closed-chain, hydrocarbons with the name naphthene, proposed by Markownikoff, who first discovered them in Russian petroleum. These naphthene hydrocarbons are probably present in all petroleum to a certain extent, in small amounts in the light Appalachian oils and in larger proportions in the heavy sulfur and asphaltic varieties. They form a considerable part of light American gasolines, and Russian burning oil of superior luminosity is composed altogether of these bodies. Like the hydrocarbons of the methane series, they are devoid of lubricating quality, but the lower members form good motor fuels.

HYDROCARBONS HAVING SOME VISCOSITY

The next series of hydrocarbons, of the general formula, C_nH_{2n-2}, is found in all petroleum. Collecting in the fractions above 300° C. and having some viscosity, they form the lubricants in Appalachian petroleum that are prepared for sewing machines, typewriter machines, and for other similar light lubrication. The higher members of this series are also constituents of the heavy motor-car lubricants. Heavy petroleum, in general, is composed to a large extent of these hydrocarbons; but although in such general use, their structure has not yet been ascertained.

HYDROCARBONS POSSESSING HIGH VISCOSITY

Next in order is the series C_nH_{2n-4}, made up of hydrocarbons possessing a high viscosity; $C_{25}H_{46}$ is one of them. These hydrocarbons form the constituents of the best lubricants it is possible to prepare from petroleum. Heavy petroleum with an asphaltic base contains these hydrocarbons in large proportion, and lighter varieties in smaller amounts. With boiling points above 250° C. in vacuum, the decomposition, when distilled with dry heat, is partly prevented by the use of steam in the still or, better, by excluding air and reducing the boiling points by exhaustion when these hydrocarbons may be distilled repeatedly with but slight decomposition. Straight petroleum lubricants are, therefore, made up mainly of a few viscous hydrocarbons of the last two series mentioned, and they are graded by varying the mixtures to provide for the kind of lubrication desired.

AROMATIC HYDROCARBONS

The last series of hydrocarbons in petroleum, concerning which anything is definitely known, is represented by the general formula, C_nH_{2n-6} or the so-called aromatic series, beginning with benzene, C_6H_6. These hydrocarbons are contained in all varieties of petroleum so far as known, but in only minute proportions in light grades, such as those of the Appalachian fields. Some heavier grades, especially those of California, contain large amounts of the aromatic hydrocarbons—benzene, toluene, the xylenes, mesitylene, and naphthalene has been observed. But these bodies are rather a detriment in petroleum to be removed in the processes of refining. They are closely related to the cyclic naphthenes in structure, the latter partaking of the properties of both the methane, or paraffine, series and the aromatic series. For instance, by the addition of hydrogen, benzene unites with six atoms to form hexahydro-benzene C_6H_{12}, and from the latter by proper treatment the six atoms of hydrogen may be removed to form the same benzene. The same relation holds for all the homologs of benzene and their hexahydro derivatives.

OXYGEN COMPOUNDS OF PETROLEUM HYDROCARBONS

Of the oxygen compounds of the petroleum hydrocarbons, phenols are found in some heavy varieties, such as California oil, and the naphthene acids first discovered in Russian oil, which contains them in considerable amounts, are generally to be found. But they have no influence on commercial products for they are removed by proper refining, although it is probable that they have something to do with the formation of emulsions.

Nitrogen Bases

The nitrogen bases, the quinolines, are contained in all petroleum, in some varieties in large proportions; it has been estimated that some California petroleums contain as much as 10 to 20 per cent.; but they also are completely removed in refining. These bases are of especial interest in their bearing on the origin of petroleum as indicating its evolution from organic remains, animal or vegetable.

Sulfur in Petroleum

Sulfur is the most undesirable impurity in petroleum, and it is pretty nearly everywhere present, except in the Appalachian oils and in certain heavy oils from shallow wells. In general, it appears that the proportion of sulfur has considerably diminished as compared with the quantities contained in the earlier development of oil territory. The largest proportion that has come under my observation is 2.75 per cent. in the early Humble crude, about one-third free sulfur in solution, nearly all that the crude oil can hold, and two-thirds combined. Formerly, the free sulfur often crystallized out in the tank cars during transportation; now the amount in this oil is less than 1 per cent. Sulfur was first observed in Canadian oil at Petrolia, which carried 1 per cent., next in Ohio Trenton limestone oil in the late eighties, containing 1 per cent. or less; and more recently in the fields of Illinois, Oklahoma, Louisiana, Kansas, and Wyoming containing variable proportions below 0.5 per cent. In combination with the hydrocarbons, sulfur derivatives are of the form $C_nH_{2n}S$, such as the individual $C_{10}H_{20}S$, unstable when heated in contact with air, but distilled without decomposition in vacuum. Their structure' is uncertain but probably cyclic with sulfur the connection link. Since, as in Texas, where wells are often drilled through beds of sulfur with which the oil has long been in contact, it is not difficult to understand its mechanical solution.

In the ordinary refining of petroleum, sulfur is removed only in part, necessitating the use of special methods for its removal to the extent that it should contain not more than 0.05 per cent. in burning oil to avoid SO_2 in the atmosphere of the compartment, and not in excess of 0.1 per cent. in lubricants, to avoid corrosion of metals. Distillation over copper oxide or metallic iron is the usual method of removal where the amount of sulfur is large. The presence of combined sulfur in petroleum has an especial interest to the geologist, for it is doubtless associated with the primary formation of the heavier varieties. Such large amounts as petroleum contain could not have had an origin in vegetable or animal matter; it must have been the result of secondary changes, in which the oil came in contact with beds of sulfur, the latter having been formed from sulfates in underground sulfate water by reduction of organic matter. When heated with sulfur, the hydrocarbons readily give off hydrogen sulfide and under proper conditions the sulfur combines with the hydrocarbons. These changes no doubt explain in part the principal difference between the light oils of the Appalachian region, which have never been in contact with sulfur, and the heavier varieties of the middle west and south, which have always been associated with sulfate waters. The former are nearly pure mixtures of hydrocarbons, the lighter individuals predominating, and of the most stable series, such as is known to be experimentally formed from the decomposition of vegetable or animal matter, containing only small amounts of sulfur. Derived from vegetable matter in the Appalachian region, far removed from the organic remains of the ancient sea that left the great saline beds of the middle west between the Appalachian and the Rocky Mountains and far away from contact with the sulfur or sulfates of those deposits, this petroleum may be regarded as the typically pure product of vegetable organic decay with exclusion of air.

The conditions were very different in the formation of the heavier varieties of Ohio, Illinois, Oklahoma, and Kansas in the great sea bed of this region. Concurrent with the decay of sea life yielding oil, or subsequently it may be, and with an increase in temperature, came the action of sulfur removing hydrogen, increasing the density of the oil and introducing sulfur in combination. This sharp demarcation between the formation of the Appalachian and middle west petroleum is sufficient to account for these differences in composition and properties. Changes subsequent to the formation of Appalachian oil, of moderate temperature, pressure, possible transference or infiltration through different strata in many periods of decomposition, elevation and folding, all combined to produce an oil unlike in purity that of any other field. The heavier quality of Trenton limestone petroleum, doubtless due in part to its origin from the same source as the lime rock, was increased by the action of sulfur in the formation of the compounds it now contains.

For the original formation of California petroleum the records are plainly written in the great beds of marine shell life, asserted by Doctor Dickenson to be an adequate source of all petroleum in those extensive fields. As in other oil territory of similar origin, most of this petroleum is thick and heavy, lacking altogether the lighter constituents of deposits derived from a vegetable source. It contains much sulfur, indicating that this element had something to do in the formation of its heavy condition, much nitrogen in the form of quinolines, and a large proportion of heavy asphaltic hydrocarbons—the asphalt base. That it contains much organic matter not fully converted into the petroleum hydrocarbons is shown by the maggoty condition of some of the oil pools. On the other hand, there is evidence that this petroleum has been subject to none of the secondary changes that have contributed to the clarifying effects of the eastern deposits—that it has been changed little, if at all, in the location of its origin. A possible contribution to the formation of California petroleum, and it may be to other petroleum, is suggested by what has taken place in the Rancho La Brea asphalt pits, and the interesting collection in the museum at Los Angeles of animal skeletons representing all the extinct mammalian fauna of that region caught in those pits in the glacial epoch that terminated 25,000 years ago after a probable duration of 500,000 years.

HEAVY OHIO OIL

Besides the varieties of petroleum already described, there is another essentially different in its composition and quality, in fact almost a class by itself. As Appalachian petroleum, composed of a pure mixture of hydrocarbons, stands at the end of a series with the lighter individuals predominating, so this oil may be regarded as the other end of the series, also a pure mixture of hydrocarbons but of the least volatile end. It has none of the gasoline nor kerosene constituents, none of the naphthenes, no paraffine nor asphaltic base. This Ohio oil is, doubtless, of more recent origin than most other petroleum, and it has never been in contact with sulfur. It has been found in three localities not far removed. One is a depression on the Mahone River, in quartz sand 150 ft. (45 m.) deep, the oil overlying a pool of brine and closely adjacent to large beds of coal. Just when the commercial development of this oil territory had begun, the entire area was flooded with water by a water company.

A second field of similar character is the ancient Mecca district, one of the first in this country to be operated on a commercial scale on account of its use as a natural lubricant. This oil has also been long known for its medicinal quality. But since it contains neither gasoline nor kerosene hydrocarbons its output has been much restricted. The greatly increased demand for lubricants has again attracted attention to this oil, and it is now being systematically pumped for the manufacture of high-grade lubricants. The wells are shallow, 70 to 100 ft. (21 to 30 m.) deep, and the oil is taken from a surface of brine.

A third field of the same general type is near Middlebranch, Ohio, likewise in a shallow depression of a few hundred acres; the oil is here reached in wells about 700 ft. deep, also above salt brine. For some time this field yielded a large supply of gas, which is still utilized in considerable quantity. Both these oils, like the Mahone, are of more recent origin than those of other fields, and they have undergone no other metamorphosis by the influence of sulfur or changes in location than the apparent evaporation of the volatile end—gasoline and kerosene hydrocarbons. Thus in nature's laboratory through long periods of time this oil, com-

512 COMPOSITION OF PETROLEUM

posed of a few hydrocarbons of maximum viscosity, has been formed and preserved, and now with proper treatment it yields lubricants of the best quality it is possible to prepare from petroleum. Since the crude oil has a viscosity of 3000 sec., a specific gravity of 0.90 at 20° C., and all the hydrocarbons it contains, except 5 per cent. of the lighter end, having marked viscosity, in the treatment of the oil in refining, it is only necessary to select the hydrocarbons for the viscosity desired, without decomposition, and to give the resulting oil a proper finish. The lubricant value of this petroleum is explained by its composition. Containing none of the paraffine hydrocarbons, none of the naphthenes, it is composed chiefly of the two series C_nH_{2n-2} and C_nH_{2n-4}, both of high lubricant quality. As to the composition of the hydrocarbons beyond the range of distillation without decomposition, nothing is known; these residues still retain their viscosity without the ready formation during distillation of asphaltic products common to most heavy petroleum.

For a more complete résumé of the composition, geology, occurrence, genesis, and technology of American petroleum reference is made to a paper by Clifford Richardson,[2] a paper by C. F. Mabery[3] and the most complete work on American petroleum industry that has appeared, by Bacon and Hamor.[4] In 1915, David White,[5] of the U. S. Geological Survey, gave a very complete review of the data from extensive observations and their bearing on the relations in formation of coal and petroleum.

PREPARATION OF COMMERCIAL PRODUCTS FROM PETROLEUM

There has been little fundamental change in the refining of petroleum since the early days of this industry. The first stage in the process is distillation, to separate the cuts, or distillates, that are to be used for gasoline, kerosene, and lubricants. Until recently, these cuts were made by specific gravity of the distillate at the end of the condensers; now pyrometers set into the stills give a fairly good separation by recording temperatures. To avoid the decompositions of outside heat alone, live or superheated steam is now freely used within the still.

There is always a certain amount of decomposition products in the distillates, besides the natural impurities in the crude oil, so the next stage has always been to agitate with concentrated sulfuric acid, which removes these bodies as a heavy acid sludge that is drawn off after standing some time to settle. Another process, which has found limited use, consists

---

[2] *Jnl.* Frank. Inst. (1906) **162,** 57, 81.
[3] Mabery: *Jnl.* Amer. Chem. Soc. (1906) **28,** 415.
[4] R. F. Bacon and W. A. Hamor: "American Petroleum Industry." N. Y., 1916, McGraw-Hill.
[5] *Jnl.* Wash. Acad. Sci. (1915).

in agitating with liquid sulfurous acid, but this process has not been generally adopted.

For the complete removal of the acid sludge and acid compounds in solution, the next operation consists of agitation with caustic soda in sufficient excess to neutralize the acid. Since a very slight excess of the caustic causes an emulsion of the oil, this stage of the treatment demands the best skill and care on the part of the man in charge of the treating house, with the aid of the works chemist. It is not possible to work by definite formula, because the wide difference in the distillates from crudes of different fields requires varying amounts of caustic. The formation of emulsions is, and always has been, the worst trouble with which the refiner must contend, for it means loss of oil, besides the additional labor, and a darkening of the finished oil through the application of heat, which alone will break up an emulsion. In such oil emulsions, minute particles of aqueous alkali are completely enclosed within films of oil and retained almost indefinitely at ordinary temperatures. Washing the emulsion merely increases its volume by the absorption of more water. On the other hand, if caustic is not used in sufficient excess to remove the sludge, there is danger of an objectionable color as well as an acid condition in the finished oil. There is more danger of emulsions in finishing heavy distillates. In the last stage of refining, the dry oil is passed through Fuller's earth to lighten its color.

PRODUCTION AND USE OF GASOLINE

At first, kerosene was the principal product refined from petroleum, with a limited use of gasoline, as a solvent and for cleansing, and of the heavy distillates. Later, with the adaptation of acetylene and the cheapening of electricity for both city and country lighting, the demand for kerosene diminished to such an extent that, just before the war, the refiner informed the seller of gasoline that he must take a certain proportion of kerosene with his gasoline. The use of gasoline had already rapidly increased, on account of the adaptation of the stationary gasoline engine for power; and when the economic efficiency of the gasoline engine was so perfected that it could be used for motive power in the automobile and a popular demand for motor cars was established, the consumption increased to such an extent that the output of crude oil, although very greatly enlarged, could not meet the demand. Then appeared numerous attempts and many patents were obtained for the production of motor oil by cracking the higher hydrocarbons into lighter oils that could be used in motor engines.

Even now it appears that the production cannot keep pace with consumption, and that, as reported, reserve supplies are being drawn upon to maintain a demand that, in large part, serves no economic nor

prevent such friction and that is to avoid contact, but it is possible of control within the limits of economic mechanical operation by the insertion of a third body capable of bearing the moving weight. Such a body is known as a lubricant and the lubricating materials are restricted to solids, the softer metals, graphite and certain other unctuous substances like talc, and some oils and greases. A hard metal bearing on a softer metal may be lubricated to some extent by the softer metal, but the nearest approach to an ideal solid lubricant is pure graphite, which forms a veneer on a hard surface, closing the pores and, by means of its highly unctuous quality, reducing friction to the lowest possible limit. Of oil lubricants, the undecomposed petroleum hydrocarbons with impurities removed possess the best wearing quality. They lubricate until the last molecule is used up. Certain vegetable and animal oils have the requisite viscosity, but they are less stable, gum and corrode by decomposition, and are inferior in durability.

In the preparation of petroleum lubricants, the grade must be selected with reference to the work that it is expected to perform, first in the cuts of the distillation and then in the combination of the hydrocarbons for the quality desired. The principal means of control are specific gravity, viscosity, and the heat quality, as represented by the tests of flash and fire. As factors of safety, the fire tests and especially the flash test, must be closely controlled in oils designed for motor-engine lubrication and made to conform to established safety limits—for water-cooled engines 250° F. (120° C.) and for air-cooled engines 350° F. These temperatures of the cylinders should be exceeded by the flash points of the lubricant oils by at least 50° F. For steam cylinders, lubricants must have a flash of 500° to 650° F.

The actual value of a lubricating oil is based on its viscosity—the peculiar quality of oiliness or greasiness that holds the molecules together with sufficient force to maintain the pressure of the surfaces they hold apart. The viscous quality is wanting in the paraffine hydrocarbons C_nH_{2n+2} and in the naphthenes C_nH_{2n}. It appears in the series C_nH_{2n-2}; and of the distilled lubricants, reaches its highest value in the series C_nH_{2n-4}. In heavy crudes, such as those of Texas and California, the lubricating quality ends with the distillates from the asphaltic residues, and only partly appears in the paraffine residue of the lighter crudes. But in the heavy Mecca oil, all but a few hydrocarbons of the first distillate, not more than 5 per cent., are decidedly viscous, the viscosity increasing rapidly and continuing throughout the entire mass of the oil, such that the residue of vacuum distillation has an extremely high viscosity. Thus, it is possible to prepare from this crude oil a wide range of lubricants; beginning with the light oils needed for sewing machines and typewriter machines, watch and clock oils, through the various grades of motor-oil lubricants, heavy-engine and steam-cylinder oils.

Next to the production of power in a motor car, the most important detail in its operation, and one that is too much neglected, is lubrication. Too often the car owner has not the slightest knowledge as to what sort of lubricant is best adapted to his car; he uses what is given him or what he is advised to use, which is often too low in viscosity. An oil that seems very oily at ordinary temperature may become as thin as water when exposed to the great heat of the cylinders. Excepting perhaps the lightest cars, all others should be run on lubricants with a viscosity of at least 1000 to 1200 sec., Universal viscosimeter, and the oil should not be used until it becomes too thin. There is doubtless greater unnecessary wear in motor cars from lack of lubrication than from any other neglect.

Of the many annoyances in the operation of a motor car, one of the most serious is the necessity for frequent removal of carbon from the cylinders, on account of its interference with the passage of the spark for the explosion and its deadening effect on the resulting power. Deposits of carbon may be formed from the lubricant and from the gasoline. With the use of normal gasoline, a lubricant properly refined and adapted to the size of the engine, to its load, and the conditions of its use, it is safe to say that, with proper manipulation, smoking exhausts should disappear and carbon deposits be reduced to a minimum, or easily removed by such simple expedients as pouring gasoline into the engine while hot.

The common use of too light lubricants in motor cars is poor economy, as well as the use of the same grade in all cars, irrespective of their weight. No doubt the quality of the lubricant as well as the quality of the gasoline has much to do with carbonization. Cracked gasoline, consisting of partly decomposed hydrocarbons, more readily escapes complete combustion, sending forth dense fumes, a sure indication of excessive carbon deposits. Yet with a sufficient excess of air and an adequate temperature for complete combustion in the cylinders, even these hydrocarbons may be completely burned, leaving behind little carbon.

An exaggerated importance is attached to what is termed "free carbon" in motor lubricants, a misleading term, based on the differences in the carbon residue of a loose method of determination, which consists in evaporating the oil until a mixture of free carbon and hydrocarbons remain, the latter not completely expelled by heat in this manner. It is assumed that the results indicate the comparative extent to which different oils form carbon in an engine. But on account of other elements of operation, as well as the fact that carbon deposits consist to a considerable extent of mineral matter, sometimes as much as 70 per cent., the tendency to leave carbon in analytical determinations has little bearing on the comparative value of lubricants with reference to carbon deposition. Some of the best lubricants, as regards carbon deposits in the cylinders, give higher percentage of "free carbon" in the analytical determinations than others that carbonize more freely in the engine.

UTILIZATION OF OTHER PETROLEUM SOURCES

With the present abundant and convenient supply of petroleum, it is not yet necessary to earnestly cast about for other forms of bitumen as sources of commercial products most in demand. But we are assured that the great deposits of rich carboniferous shales in the west only await a serious falling off in petroleum production to provide a practically unlimited output of motor-engine and lubricating oils, not perhaps of the equivalent value of petroleum products but good substitutes.

Closely related to petroleum as to their origin and oils they yield by distillation, are certain other varieties of bitumen found in Colorado and Utah—Gilsonite and Grahamite.[6] These natural asphaltic bitumens still contain a considerable proportion of volatile oils, the lighter portions having evaporated during their slow formation from petroleum leaving these brittle solids resembling coals. But unlike coals containing the inorganic residue of their primary vegetable formation, Gilsonite and Grahamite, owing their genesis through secondary phases to petroleum, are free from all inorganic residues.

As recently as twelve years ago, petroleum that could not produce kerosene, such as that of Texas and California, was of lower commercial value and in demand only for fuel, or for a limited production of lubricants. Very soon the rapid development of the automobile industry, stimulating a demand for gasoline and, consequently, for lubricants, gave greater prominence to the heavy crudes, both as a source of lubricants and of motor fuels by cracking the higher hydrocarbons. But with production pushed to the utmost, the older fields alone could not have prevented a gasoline famine during the war, with serious results. When in 1917 the English Admiralty was confronted by defeat within three months unless it could halt the submarine destruction, the situation was saved by the contributory influence of the great increase in the output of motor fuel from the new fields of Illinois, Oklahoma, Kansas, Wyoming, and Mexico, all but the latter yielding large amounts of normal gasoline. With the material falling off in production of the Appalachian fields, the new territory came in at an opportune moment to meet the demands of the war and the great expansion of the automobile industry.

VALUE OF SCIENTIFIC WORK

Much of the loss in the early days of the petroleum industry due to haphazard prospecting and drilling was later avoided by the scientific investigations of geologists, especially by Hunt, Orton, Winchell, Newberry, and Hoefer. From the prospector's point of view, the most important of all was the theory of Hunt suggesting the storage of oil in an anti-

---

[6] Mabery: *Jnl.* Amer. Chem. Soc. (1917) **39²**, 2025.

clinal and synclinal system, and referring the great rock pressure on oil and gas to an extensive underground hydraulic system. In the later application and expansion of this theory, the storage of oil in all the great fields of the world was found to be in well-defined anticlines. When the Trenton limestone was identified by Orton as the reservoir of the immense deposits of oil in Ohio and Indiana, he established such an extended system of anticlinal storage of gas and oil that the direction of the dip could be traced with sufficient accuracy to direct the prospector in his drilling operations. Furthermore, Orton identified the dolomitic nature of the oil-bearing rock and, by means of outcropping rock formation, indicated the underlying strata that could be relied on as sources of gas and oil storage.

What geological science has done in placing the exploitation of petroleum territory on a practical and successful economic basis has its partial counterpart in the results of chemical research. What is known concerning the chemistry of petroleum is the result of independent investigations carried on in the limited facilities of the chemical laboratory; altogether unlike the opportunities of the geologist who always had the advantage of unrestricted observations at fundamental sources. If, in the early development of the petroleum industry, there had been established a properly organized refinery with adequate funds and an adequate working force to ascertain the complete composition of crude oil from the producing fields then known, and to take up thorough investigation of newly discovered territory as it came into commercial prominence, the gain to the present and future industry would have been beyond calculation. Even now, before oil territory is too far exhausted and abandoned, such an organization should place on record a great accumulation of facts as to the constituents of petroleum concerning which little or nothing is known, and which should incidentally be of service to the petroleum industry.

DISCUSSION

SAMUEL P. SADTLER, Philadelphia, Pa. (written discussion).—I have read, with great interest, this discussion on the individual series of hydrocarbons that are found to be represented in natural petroleums. One of the subjects of very great practical interest is, what hydrocarbons possess special viscosity. Doctor Mabery very properly calls attention to the class of hydrocarbons that seem to be characteristic of the lubricants prepared from Appalachian petroleum. These, he states, are higher members of the series C_nH_{2n-2}. He does not particularize as to which kind of hydrocarbons of this general formula he means. It is obvious that they are not the hydrocarbons of the acetylene series, but of what are termed unsaturated cyclic hydrocarbons, also possessing this formula.

He calls attention to the hydrocarbons of the series C_nH_{2n-4} as possessing high viscosity. Here again, it is proper to understand that reference is made to unsaturated cyclic hydrocarbons, as distinguished from aliphatic hydrocarbons, and he practically awards the whole value for lubricating power in prepared oils to hydrocarbons belonging to these series.

He specifically states also, on page 507, of the naphthenes, first recognized in Russian petroleum, and now known to be present in most American petroleums, that, "like the hydrocarbons of the methane series, they are devoid of lubricating qualities." This rather positive statement of Doctor Mabery's is probably not a matter that is, as yet, universally agreed upon. The studies that have been made upon this subject in recent years, largely by the aid of the "formolite" reaction, do not as yet give conclusive evidence on this subject.

Engler, in his work on the chemistry of petroleum, says (1, 387): "The controversial question as to which group of hydrocarbons are the chief bearers of the viscosity, which has recently been especially studied on the one hand by Nastjukoff, as well as Herr, who take the view that the viscosity belongs to the unsaturated oils, and on the other hand by Charitschkoff, who attributes it also to the saturated naphthenes, is not yet definitely decided. That the unsaturated cyclic hydrocarbons of high molecular weight are also highly viscous is settled beyond doubt, as is conceded also by Charitschkoff. It is, however, not yet shown that the high molecular saturated cyclic hydrocarbons are not also very viscous."

On page 558, Engler quotes a series of relatively recent results by Marcusson, which were obtained by the study of both American and Russian oils. These results seemed to conform with the view of Charitschkoff, that the nonformolite-forming constituents (other than the unsaturated cyclic hydrocarbons) are, not only in relative quantity but in their viscous quality, the chief representative elements in the lubricating power, and these include the paraffins, the naphthenes, the poly-naphthenes, and the olefines. Engler discusses these results of Marcusson at some length, and calls attention to the particular care with which the formolite reaction must be carried out to insure accurate results, and intimates that many of the discrepancies in the results of previous experimenters may have been due to the overlooking of necessary precautions in the carrying out of this reaction, and apparently expresses himself as satisfied with these latest results of Marcusson, based on the use of formolite reaction.

To sum up, therefore, I would merely say that it is desirable to consider this question as not yet so definitely settled as seemed to be expressed by the statements of Doctor Mabery. These studies of the lubricating oils and the relation of their composition to viscosity are, of course, of the greatest interest and importance, but we must not draw sharp deductions based largely on a reaction that may be carried out in such a way as to give varying results.

Doctor Mabery gives a very interesting account of the special class of Ohio oils in which this viscosity is particularly developed.

He gives an outline of the general methods of preparing commercial products from petroleum, covering the general methods of refining, the production of gasoline in increasing amount by cracking processes, although he does not refer particularly to the cracking methods used (in which there is considerable variation) and takes up particularly the matter of lubricants and their uses. Some notice might have been given of the very large use of clay filtration, which is an important part, particularly in the preparation of high-grade lubricants. Many of the special grades of lubricants with an exceptional low cold test and consequent availability for lubrication, under conditions of low temperature, are prepared exclusively by these methods of filtration with fuller's earth or special grades of clay.

B. F. TILLSON,* Franklin, N. J.—The subject of lubrication received considerable attention at the last annual meetings of the American Society of Mechanical Engineers and of the Society of Automotive Engineers, and interest in it seems to be spreading so that there is a tendency towards coördination of research along the lines of what should be the properties of a lubricant and a bearing in order properly to utilize a lubricant. The opinion seems to be quite general that oil grooves in bearings are, in the main, detrimental and, as far as possible, should be removed; if used, the edges of the groove should be tapered so as not to form a sharp surface that will wipe away the oil from the moving shaft or body that rests on the bearing.

But I wonder whether there is not the same agreement that, in general, high viscosities are not at all necessary properties of an excellent lubricant, that the lower the viscosity of the lubricant it is possible to use, the less is the frictional loss; that the internal friction of the molecules moving in oils of high viscosities is a considerable power loss?

In general, the rule seems to be that the viscosity lessens as the temperature increases; but some instances seem to indicate a reverse condition. I have heard that waxes or paraffins with a low melting point that have been separated and left in a refrigerator for some time become liquid at lower temperatures; that some soaps, if left on a window sill in cold weather, change from solid to liquid form. Do not such examples indicate that research concerning the colloidal conditions of the elements that form the oils we are using is greatly needed?

Further research may show that if they do not crack readily or deposit their carbons, oils of much lower viscosities than the present practice indicates may be used in both automotive and general mechanical engineering design.

---

* Min. Engr., New Jersey Zinc Co.

CHARLES F. MABERY (author's reply to discussion).—In reply to Mr. Tillson's question, as to the relation of viscosity to the quality of lubricants, and the influence of temperature, I think that oils with the lowest viscosity to meet the frictional conditions should be selected. The influence of temperature on lubricants does not receive the attention it should. The great falling off in viscosity by even slight raise in temperature is a direct measure of the loss in power of the lubricant to keep the bearing surfaces apart.

In reply to Doctor Sadtler's question, the lubricant hydrocarbons represented by the formula $C_n H_{2n-2}$, are not unsaturated in the same manner as the acetylenes or ethylenes, and in only one or two instances has the structure been made out. But evidently a double bonded structure is necessary to account for the smaller number of hydrogen atoms. Concerning the series $C_n H_{2n-4}$, these hydrocarbons are the least volatile portions of petroleum that can be distilled without decomposition.

Doctor Sadtler alludes to my ommission of the use of fuller's earth in refining. For some time I have been connected with the preparation of low-test and high-viscosity lubricants, and have had abundant opportunities to become familiar with the usefulness of clay filtration in finishing these products.

Carbon Ratios of Coals in West Virginia Oil Fields

BY DAVID B. REGER,* MORGANTOWN, W. VA.

(New York Meeting, February, 1921)

THE value of carbon ratios in determining the boundaries of possible oil deposits appears to have passed the hypothetical stage. The theory that the ratio of fixed carbon in pure coals is an invariable index of incipient metamorphism in both surface and underground rocks and that it may be applied in defining the limits of petroleum, advanced by David White,[1] has been received with keen interest by many petroleum geologists. Detailed isocarb maps have been prepared of the Pennsylvanian area of North Texas and Eastern Oklahoma by M. L. Fuller.[2] A similar map of the coal-bearing area of West Virginia is given here. .

An isocarb[3] is a line showing an equal fixed-carbon percentage, pure coal basis; the term has been proposed by David White to supersede a less expressive nomenclature.

The term carbon ratio is applied to the percentage of carbon in pure coal after water and ash have been eliminated. As a comparatively small number of analyses have been made on this basis, it is usually necessary to compute the ratio by dividing the fixed carbon of the proximate analysis by the sum of the fixed carbon and volatile matter of the same analysis.

Many thousands of proximate analyses have been made by the West Virginia Geological Survey, covering nearly every county in which coal is found. Numerous others have been made by the U. S. Geological Survey and the U. S. Bureau of Mines, but as uniformity of results is best secured by adhering to one set of analyses, the tests of the West Virginia Survey have been exclusively used.

---

* Assistant Geologist, West Virginia Geological Survey.

[1] Some Relations in Origin between Coal and Petroleum. *Wash. Acad. Sci.* (March 19, 1915) **5**, 189–212; Late Theories Regarding the Origin of Oil. *Bull.* Geol. Soc. Am. (Sept. 30, 1917) **28**, 727–734.

[2] Relation of Oil to Carbon Ratios of Pennsylvanian Coals in North Texas. *Econ. Geol.* (November, 1919) **14**, 536–542; Carbon Ratios in Carboniferous Coals of Oklahoma, and Their Relation to Petroleum. *Econ. Geol.* (April–May, 1920) **15**, 225–235.

[3] The term isocarb has been suggested by David White as more accurate than the word isovol originally used by him. The term isovolve, used by Fuller, appears to be a corruption of isovol.

In preparing the isocarb map, it has been necessary to use analyses ranging from the Dunkard (Permo-Carboniferous) coals down to, and including, the Pocahontas Group of the Pottsville (Pennsylvanian), as, with certain exceptions, there is a progressive rise of strata from the Appalachian geosyncline southeastward to the Alleghany Mountains, where the coals disappear above the summits. The Dunkard coals, being much different in character from those of the Pennsylvanian, have been used in only one county (Tyler). With certain minor exceptions, in each county analyses from the oldest coal seams available have been employed, in order to secure the nearest possible approach to under-

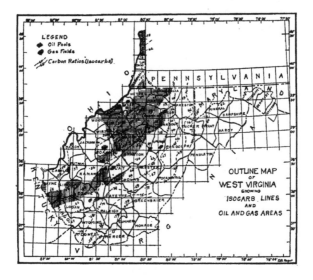

ground conditions. In carrying out this rule two or more seams have been used for different portions of several counties where the pitch of the measures is large.

The table shows in detail the various coals used in preparing the m ap few analyses of coals above the Pittsburgh have been employed.

ISOCARB MAP

The map shows the main productive oil and gas fields of the state, together with isocarb lines for the coal-bearing area, as plotted from the preceding data. The dots with accompanying figures show the approximate localities represented by each average determination. The average carbon ratio falls below 55 in parts of Roane, Calhoun, Gilmer, Doddridge, Lewis, and Harrison counties; this area lies just southeast of the Appalachian geosyncline, the axis of which extends roughly from the southwest corner of Pennsylvania to the Kentucky state line in the northern half of Wayne County.

It is apparent from the map that the main oil pools lie within the limit of isocarb 60, the most notable exceptions being the Cabin creek pool of Kanawha and Boone Counties, the southwestern limit of which is not yet fully defined, and certain pools in eastern Kanawha and Clay Counties. Oil also occurs in quantity in Brooke County, of the northern panhandle, where the carbon ratio is 62. Recent we'ls are reported from Mingo County near the point where isocarb 65 crosses the Kentucky line. Gas occurs in several localities on the high side of isocarb 65 in Raleigh, Fayette and Nicholas Counties; it has also been reported in an uncompleted well in Northern Wyoming, where the average carbon ratio is 69.

PROBABLE LIMIT OF OIL AND GAS

It would seem, from the record of numerous wells drilled on the high side of isocarb 60, that dry holes or gas will be the main result of tests in such territory (certain exceptions have been noted above) and that wells drilled on the high side of isocarb 65 can hope for only occasional occurrences of gas.

It would seem, from the map, that new production in the western portion of the state is most likely in Wayne, Cabell, Putnam, Kanawha, Mason, Jackson, Roane, Wirt, Wood, Marshall and Ohio Counties. In some of these, however, the strata have not been sufficiently disturbed to afford gravitational segregation of oil, gas, and water, and in various regions of these counties where favorable synclines occur the sands are known to be saturated with water, contrary to the general rule throughout the state. In spite of these two unfavorable features, the writer believes that several new pools will be developed in some of these counties.

In the central belt, certain undeveloped portions of Nicholas, Braxton, Lewis, Upshur, Barbour, and Marion Counties lie on the low side of isocarb 60, so that production may reasonably be expected from some of these. As the structure of this region is largely monoclinal, the main hope of oil will depend on terraces. Inasmuch as there is a rapid southeastward expansion of the Pottsville, Mauch Chunk, and Greenbrier Series structure maps based on surface strata do not fully reveal such terraces and their location cannot be made until sufficient holes have been drilled to afford the necessary subsurface data.

COMPARISON WITH OTHER STATES

The researches of M. L. Fuller show that in both Oklahoma and northern Texas the main producing fields lie on the low side of isocarb 55,

Table of Coal Analyses and Carbon Ratios

County	Locality	Coal Seam and Group	Number of Analyses	Volatile Matter, Average, Per Cent.	Fixed Carbon, Average, Per Cent.	F. C. / (F. C. + V. M.) Nearest Per Cent.
Hancock.........		L. Kittanning (Ca)	3	37.91	53.00	58
Brooke..........		Pittsburgh (Cm)	15	34.20	56.08	62
Ohio............		Pittsburgh (Cm)	6	35.92	55.17	60
Marshall........		Pittsburgh (Cm)	7	38.40	50.97	57
Wetzel.........		Uniontown (Cm)	1	37.07	53.07	59
Tyler...........		Washington (Cd)	3	34.27	47.92	58
Pleasants........						
Wood...........						
Jackson..........						
Mason...........		Pittsburgh (Cm)	5	38.79	49.33	56
Cabell..........		Little Pittsburgh (Ccm)	1	39.82	46.83	54
Wayne..........		No. 2 Gas (Ck)	1	40.02	55.42	58
Lincoln.........		No. 2 Gas (Ck)	1	39.29	55.81	58
Putnam.........		Pittsburgh (Cm)	6	38.53	51.40	57
Roane...........		Brush Creek (Ccm)	1	37.20	44.12	54
Wirt............		Bakerstown (Ccm)	1	33.43	42.75	56
Ritchie.........		Pittsburgh (Cm)	1	37.45	52.50	57
Doddridge.......		Uniontown (Cm)	3	39.36	45.29	53
Monongalia......	East	L. Kittanning (Ca)	4	30.47	57.53	65
	West	Pittsburgh (Cm)	8	36.47	54.86	60
Marion..........	East	L. Kittanning (Ca)	2	33.90	49.83	59
	West	Pittsburgh (Cm)	43	36.62	55.73	60
Harrison........	East	Harlem (Ccm)	5	35.81	49.70	58
	West	Pittsburgh (Cm)	56	38.24	53.53	58
Lewis...........	North	Pittsburgh (Cm)	2	41.92	51.08	55
	South	L. Kittanning (Ca)	8	34.52	53.41	60
Gilmer..........		Pittsburgh (Cm)	21	41.84	49.49	54
Calhoun.........		Pittsburgh (Cm)	2	34.99	56.95	61
Clay...........		Coalburg (Ck)	6	34.05	56.10	62
Kanawha........	Northwest	Pittsburgh (Cm)	14	40.06	51.48	56
	Southeast	Eagle (Ck)	7	31.22	63.09	66
Boone...........	Northwest	No. 5 Block (Ca)	13	37.31	55.73	59
	Southeast	Eagle (Ck)	6	34.98	57.55	61
Logan...........		Campbell Creek (Ck)	9	34.92	59.27	62
Mingo...........		Eagle (Ck)	5	30.45	63.06	67
McDowell........	West	Eagle (Ck)	3	31.29	62.44	66
	East	No. 3 Pocohantas (Cp)	14	17.90	76.13	80
Wyoming........	West	Gilbert (Ck)	4	27.84	63.61	69
	East	Sewell (Cnr)	5	23.31	72.76	75
Raleigh.........	West	Eagle (Ck)	23	29.29	65.02	70
	East	Fire Creek (Cnr)	5	20.56	77.37	79
Fayette.........	West	Eagle (Ck)	16	28.25	66.61	70
	East	Fire Creek (Cnr)	18	19.29	75.42	79
Nicholas........	North	L. Kittanning (Ca)	17	35.33	53.97	60
	Central	Eagle (Ck)	24	33.43	59.74	64
	South	Sewell (Cnr)	29	29.10	65.73	69
Braxton.........	North	Pittsburgh (Cm)	7	39.68	51.78	56
	South	L. Kittanning (Ca)	3	37.02	56.02	60
Upshur..........	North	Redstone (Cm)	12	38.36	54.35	58
	South	L. Kittanning (Ca)	5	35.36	51.78	59
Barbour.........	Northwest	Pittsburgh (Cm)	10	36.90	55.63	60
	Southeast	L. Kittanning (Ca)	18	30.77	56.85	64
Taylor..........	West	Pittsburgh (Cm)	10	37.01	55.19	59
	East	L. Kittanning (Ca)	5	29.90	58.96	66
Preston.........	West	U. Freeport (Ca)	43	27.25	62.52	69
	East	L. Kittanning (Ca)	12	29.49	59.42	66
Tucker..........	North	U. Freeport (Ca)	6	22.05	70.20	76
	South	Sewell (Cnr)	1	24.81	68.18	73
Randolph........	Northwest	M. & L. Kittanning (Ca)	8	32 14	55.63	63
	Southeast	Sewell (Cnr)	15	30.35	61.80	67
	Central	Sewell (Cnr)	9	26.59	60.84	69
Webster.........	North	L. Kittanning (Ca)	7	32.47	54.08	62
	Central	Eagle (Ck)	6	35.45	54.04	60
	South	Sewell (Cnr)	17	30.07	62.08	67
Greenbrier.......	West	Sewell (Cnr)	1	26.96	66.77	71
Summers.........						
Mercer..........		No. 3 Pocahontas (Cp)	17	19.85	66.48	77
Pocahontas.......	West	Gilbert (Ck)	1	29.73	56.21	65
Grant...........		U. Freeport (Ca)	1	18.46	68.64	78
Mineral.........		U. Freeport (Ca)	4	15.50	73.46	82

NOTE.—Group abbreviations are as follows: Cd-Dunkard (Permo-Carboniferous); Cm, Monongahela; Ccm, Conemaugh; Ca, Allegheny; Ck, Kanawha; Cnr, New River; and Cp, Pocahontas (last six Pennsylvanian).

there being important exceptions in the former and smaller deviations in the latter state. In neither state are oil pools of importance noted above isocarb 60. In West Virginia, however, many of the large pools lie above isocarb 55, and some large pools occur above isocarb 60, while gas extends to still higher limits. The occurrence of oil at high carbon levels in the latter state is reflected by its quality, since it is generally of higher Baumé gravity than those of the other states, and commands a higher market price, indicating that the process of natural distillation is more nearly complete.

DISCUSSION

JOSEPH T. SINGEWALD, JR.,* Baltimore, Md.—The utility of geology in locating and developing oil fields is as much negative as it is positive; it is just as important and valuable to eliminate large areas in which you cannot expect to find oil or gas, as it is to select certain smaller and restricted areas in which you may expect to find it. The processes are really the same. If the isocarbs are a criterion for eliminating large areas as not being likely fields of oil and gas, they are of immense practical value. We cannot always apply this criterion as we do not always have coal seams in an oil and gas region. Although first announced in 1915, for the Appalachian field, this criterion has since been tested only in Oklahoma and northern Texas. The Pennsylvania State Survey is now making an isocarb map of that state. West Virginia offers the best opportunity for testing it, in that we have many more analyses of its coals than those of other states. The map of Mr. Reger shows a distinct line running northeast and southwest, southeast of which there are no oil or gas fields. In Maryland we now have an opportunity to test this method. The western boundary of the state lies to the east of the line on Mr. Reger's map. They are drilling along the Potomac River and are about to commence drilling in Garrett County, the extreme western part of the state. Stratigraphically, one might consider conditions as favorable there as in West Virginia and, at first sight, structurally they might be considered more favorable, in that the structure is more pronounced. The folding, thinking of it only in terms of the angles of dip, would appear no more intense than in some of the western states. So, if the degree of metamorphism is not taken into consideration, we might be inclined to recommend drilling in Garrett County. But if the degree of metamorphism is tested by the isocarbs we ought to say not to drill. As yet, however, the theory has been tested in so few places that we cannot say "Do not drill," with absolute confidence.

So much for the grosser application of the isocarbs in ruling out certain areas as not being possible localities for the occurrence of oil and

---

* Associate Professor, Economic Geology, Johns Hopkins University.

gas. It seems to me that the theory might also be applied to great advantage in more detail. Take the case of West Virginia, where we have a great many analyses in individual counties. If we draw in greater detail on a larger scale, the curves of isocarbs and compare the irregularities in those curves with the physical and chemical character of the oil, the theory may give us valuable conclusions as to the geologic history of the oil itself.

We know that a certain oil has certain physical and chemical properties, but we do not know to what extent that oil has its own peculiar properties on account of the original composition of the material from which it has been derived, and to what extent it has those porperties on account of the geological history through which it has gone. The details of the isocarbs would appear to have within them the possibility of throwing light on that subject. It is a line of investigation worthy of more attention.

General Notes on the Production, Marine Transportation and Taxation of Mexican Petroleums

BY VALENTIN R. GARFIAS,* NEW YORK, N. Y.

(New York Meeting, February, 1921)

PRODUCTION AND MARINE TRANSPORTATION

ALTHOUGH the work on which this paper is based was carried on by the writer as Special Commissioner of the Petroleum Department of the Mexican Government, the opinions expressed are only his personal views for which the Mexican Government can in no way be held responsible. Notwithstanding that the appointment covers all phases of the so-called petroleum question, as the one question of immediate importance was that relating to taxation, it was decided to devote all the time to this phase of the work.

The present report has been divided into two closely related parts, the first dealing with the production and marine transportation of Mexican petroleum, and the second, with the Mexican taxation on petroleum and its products.

The writer wishes to acknowledge the loyal coöperation rendered by Mr. M. C. Ehlen, who had charge of the statistical work, and by Miss S. Stern and Messrs. E. P. Heiles and J. E. Morrissey.

The aim throughout the work has been to pave the way for a thorough understanding between the Mexican Government and the oil operators and to present a true and clear statement of facts relating to the questions at issue.

MEXICAN OIL FIELDS DEVELOPMENT

The salient features in the development of the Mexican oil fields may be summarized as follows:

1901.   First commercially productive well in Ebano field.

1902.   Tehuantepec field came in; Diaz Government granted concession to Pearson & Son on practically all Federal lands along the Gulf of Mexico.

1907.   First producing well in the Furbero field.

1908.   Dos Bocas well caught fire, advertising tremendous productivity of Southern oil fields of Mexico.

---

* Manager of Foreign Oil Department, Henry L. Doherty & Co.

1909. Tanhuijo field, discovery well.

1910. Discovery of oil in Juan Casiano, Potrero, Panuco, and Topila, placed Mexico as a leader in oil production.

1911. First shipment of Mexican oil made to United States, on May 25.

1913. Alamo field of Penn-Mex Co. discovered.

1917. New Constitution of Mexico, establishing national ownership of subsoil rights was enacted, May 1.

· 1917. Cerro Azul field discovered.

1918. Tepetate and Naranjos fields came in.

1919. Casiano, Tepetate, and Potrero fields went to water.

1920. Chinampa field showed water in August.

1920. Zacamixtle field came in on Oct. 8.

1921. Naranjos field showed water, Feb.

Table 2 shows the holding companies and subsidiaries at present operating in the fields and marketing Mexican oils.

TABLE 1.—*Relative Daily Production per Well in Mexican and American Oil Fields*

Field and State	Producing Wells	Drilling Wells	Well Locations	Abandoned Wells	Production Daily	Production Well Daily
Mid-continent.						
Oklahoma.................	47,574	1,691	594		302,567	6.40
North and Central Texas....	9,693	2,814	1,159		192,533	19.90
North Louisiana............	2,694	488	333		103,867	38.50
Kansas....................	13,708	368	113	In 1913 = 7163 wells.	117,166	8.50
Gulf Coast.................	3,252	462	184	In 1914 = 5607 wells.	68,067	20.90
Illinois.	15,692	49	4	In 1915 = 6029 wells.	31,033	1.90
Lima–Indiana.				In 1916 = 6017 wells		
Lima.....................	13,975	66	10	In 1917 = 6542 wells	4,467	0.32
Indiana...................	512	55	7		2,666	5.20
Kentucky–Tennessee..........	8,796	932	13		25,633	2.90
Appalachian.						
Pennsylvania and New York.	71,101	246	115		24,467	0.34
West Virginia..............	17,302	252	147		23,067	1.30
South, East and Central Ohio	14,178	116	75		15,433	1.10
Rocky Mountain..............	868	381	181		54,767	63.00
California....................	9,357	422	66		273,000	29.50
Total....................	228,702	8,342	3,001		1,240,633	5.4
Mexico.....................	(a) 61	26	12		237,884	3900.00
	(b) 80			567	296,305	3710.00
	(b) 249				296,305	1190.00

1. The figures relating to the American fields are those for the month of June, 1920, published by the U. S. Geological Survey.

2. The wells abandoned in the American fields are listed by years, as given by the U. S. Geological Survey. The figures (567) for wells abandoned in the Mexican fields include the total abandonments to date.

3. The well statistics for the Mexican fields (a) refer to conditions on June, 1919, as given by *Boletín del Petróleo:* b) refer to conditions during December, 1919, as given in *Oil and Gas Journal.* The number of actually producing wells (80) during December, 1919 is estimated.

TABLE 2.—*Foreign Companies, Producing and Marketing, at Present Operating in Mexico*

Atlantic, Gulf and West Indies S. S. Co.
  Cia Petrolera de Tepetate.
  Agwi Pipeline Co.
  Agwi Refining Co.
  Agwi Terminal Co.

Cities Service Co.
  Cia de Gas y Combustible Imperio, S. A.
  Cia Terminal Imperio, S. A.
  Cia Emmex de Petroleo y Gas, S. A.
  Empire Transportation and Oil Corpn.
    Gúlf Coast Corpn.
    Lagunita Oil Co.
    National Petroleum Corpn.
    Southern Fuel & Refining Co.
    Tampascas Oil Co.
    Cia Mexicana de Oleoductos Imperio, S. A

Compania Terminal Union.
  Hispano Mejicana Oil Co.
  Hispano Cubana Oil Co.

East Coast Oil Co., S. A.
  Southern Pacific Railroad.

General Petroleum Company of California.
  Continental Mexican Oil Co.

Gulf Oil Corpn.
  Mexican Gulf Oil Corpn.

Island Oil and Transport Corpn.
  Antillian Corpn.
  Capuchinas Oil Co.
  Colombia Petroleum Syndicate, Ltd.
  Cia Metropolitana de Oleoductos.
  Cia Mexicana de Petroleo La Libertad, S. A.
  Cia Petrolera Nayarit, S. A.
  Esfuerzo Tampiqueno, S. A.
  Island Refining Corpn.
  Metropolitan Petroleum Corpn.

Interocean Oil Co.
  U. S. Asphalt Refining Co.
  Mexican Crude Oil & Asphalt Product Co.

Mexican Petroleum Co., Ltd.
  Pan-American Petroleum & Transport Corpn.
  Caloric Co.
  Huasteca Petroleum Co.
  Cia Naviera Transportadora de Petroleo, S. A.
  Tamiahua Petroleum Co.
  Tuxpam Petroleum Co.
  Chiconcillo Petroleum Co.
  Doheny & Bridge.

National Oil Co.
  National Shipbuilding Co.
  Gomales Oil Co.
  Cia Exploradora del Petroleo, S. A.

New England Oil Corpn.
  Cochrane, Harper and Co.
  Canada Mexico Oil Co.
  France and Canada Oil Transport Co.
  New England Exploration Co.
  New England Oil Refining Co.
  (See also Magnolia Petroleum Co.)

Pierce Oil Corpn.
  Cia Mexicana de Combustibles, S. A.

Anglo-Dutch Interests.
  La Corona Petroleum Co.
    Chijoles Oil, Ltd.
    Cia Mexicana de Petroleo La Corona.
    Tampico Panuco Petroleum Co.
  Mexican Eagle Oil Co., (Aguila).
    Eagle Oil Transport Co., Ltd.
    Oilfields of Mexico, S. A.
  Scottish American Oil and Transport Corpn.
    Southern Oil & Transport Corpn.
    Fuel Oil Distribution Corpn.
    Tampico Navigation Co.
    Tampico Shipbuilding Corpn.
    Tal Vez Oil Co.
    Scottish Mexican Oil Co.

Sinclair Consolidated Oil Corpn.
  Sinclair Gulf Corpn.
  Sinclair Mexican Petroleum Co.
  Freeport and Tampico Fuel Oil Corpn.
  Freeport and Mexican Fuel Oil Corpn.
  Freeport and Tampico Fuel Oil Transp. Corpn.
  Mexican Seaboard Oil Co.
  International Petroleum Co.

Standard Oil Interests.
  Atlantic Lobos Oil Co.
    Port Lobos Petroleum Corpn.
    Cortes Oil Corpn.
    Atlantic Petroleum Producing & Refining Co. of Mexico, S. A.
    Atlantic Oil Co.
    Panuco Boston Oil Co.
    Producers Terminal Corpn.
  Magnolia Petroleum Co.
    Azteca Petroleum Co.
    Cia Inversiones de Aztlan, S. A.
    New England Fuel Oil Corpn.
  Compania Transcontinental de Petroleo, S. A.
    Panuco Excelsior Oil Co.
    Vera Cruz Mexican Oil Co.
    Penn-Mex Fuel Co.

Texas Company of Mexico.
  Panuco Transportation Company of Mexico.
  Tex-Mex Oil Co.

Tidewater Petroleum Co.
  Tide-Mex Oil Co.

Union Oil Company of California.
  Otontepec Petroleum Co.
  California Investment Co.

TABLE 3.—*Comparison of Mexican and American Standards for Measuring Petroleum and Its Products*

Volume Relations

1 U. S. barrel	=	42.00 U. S. gal.	1 U. S. gallon	=	231.00 cu. in.
	=	5.6145 cu. ft.		=	0.13368 cu. ft.
	=	158.985 li.		=	3.7853 li.
	=	0.158985 cu. m.		=	0.0037853 cu. m.
1 cubic foot	=	7.4807 U. S. gal.	1 cubic meter	=	1000.00 li.
	=	0.17811 U. S. bbl.		=	35.31445 cu. ft.
	=	28.317 li.		=	264.1775 U. S. gal.
				=	6.2899 U. S. bbl.

$$1 \text{ liter} = 0.03531 \text{ cu. ft.}$$
$$= 0.26418 \text{ U. S. gal.}$$
$$= 0.0062899 \text{ U. S. bbl.}$$

The American standard weight of water is taken with water at 60° F.; the Mexican standard, with water at 4° C. (39.2° F.).

Volume Weight Relations of Water at 60° F.

8.32823 lb. per U. S. gal.

349.78566 lb. per U. S. bbl.

Metric ton (2204.622 lb.) per 6.3028 U. S. bbl.

Long ton (2240 lb.) per 6.4039 U. S. bbl.

The specific gravities of oils are expressed in the Mexican and American standards as follows:

Mexican Standard

$$\text{Specific gravity} = \frac{\text{Weight of volume of oil at 20° C.}}{\text{Weight of same volume of water at 4° C.}}$$

American Standard

$$\text{Specific gravity} = \frac{\text{Weight of volume of oil at 60° F.}}{\text{Weight of same volume of water at 60° F.}}$$

Density is the weight per unit volume.

The formulas for converting Baumé degrees into specific gravity at 60°/60° F. and vice versa, are as follows:

$$\text{Baumé degrees} = \frac{140}{\text{Sp. Gr.} \times 60°/60° \text{ F.}} - 130$$

$$\text{Specific gravity} = \frac{140}{130 + \text{Baumé-} 60°/60° \text{ F.}}$$

To convert specific gravity of oil from the American to the Mexican standard multiply the specific gravity as shown on the American hydrometer by 0.001061 and subtract the result from the American hydrometer reading; vice versa, to correct from Mexican into American reading, multiply the specific gravity shown by the Mexican hydrometer by 0.001062, and add the result to the Mexican hydrometer reading.

The coefficient of expansion per degree Fahrenheit for oils varies with the specific gravity (60°/60° F.) as follows:

0.67 specific gravity	0.000728	0.87 specific gravity	0.000419
0.72 specific gravity	0.000627	0.91 specific gravity	0.000392
0.77 specific gravity	0.000540	0.96 specific gravity	0.000368
0.82 specific gravity	0.000470	1.00 specific gravity	0.000356

To ascertain the number of U. S. barrels of oil at 60° F. in 1 metric ton of 2204.622 lb., divide the specific gravity at 60° F. into 6.3028; for barrels per long ton, divide specific gravity at 60° F. into 6.4039.

TABLE 4.—*Relation Between World's and United States' Petroleum Production and Mexican Production and Exports in Barrels of 42 U. S. Gallons*

Year	World's Production‡	United States Production		Mexican Production		Mexican Exports		
		Total	Per Cent. of World	Total†	Per cent of World	Total†	To U. S. A.‡	Per Cent. to U. S. A.
1901	167,434,434	69,389,194	41.4	10,345	0.006			
1902	182,006,076	88,766,916	38.7	40,200	0.02			
1903	194,879,669	100,461,337	51.3	75,375	0.04			
1904	218,204,391	117,080,960	53.6	125,625	0.06			
1905	215,292,167	134,717,580	62.6	251,250	0.12			
1906	213,415,360	126,493,936	59.2	502,500	0.23			
1907	264,245,419	166,095,335	62.8	1,005,000	0.38			
1908	285,552,746	178,527,355	62.5	3,932,900	1.38			
1909	298,616,405	183,170,874	61.4	2,713,500	0.91			
1910	327,937,629	209,557,248	63.9	3,634,080	1.11			
1911	344,174,355	220,449,391	64.1	12,552,798	3.65	893,709		
1912	352,446,598	222,935,044	65.0	16,558,215	4.70	7,627,795	7,383,229	96.9
1913	383,547,399	248,446,230	64.8	25,696,291	6.70	20,915,928	17,809,058	85.2
1914	403,745,342	265,762,535	65.8	26,235,403	6.50	22,880,530	16,245,975	71.0
1915	427,740,129	281,104,104	65.8	32,910,508	7.70	24,279,375	17,478,472	72.0
1916	461,493,226	300,767,158	65.0	40,545,712	8.79	26,746,432	20,125,657	75.2
1917	506,702,902	335,315,601	66.2	55,292,770	10.91	42,545,843	29,933,516	70.4
1918	514,538,716	355,927,716	69.2	63,828,000	12.40	51,768,010	40,819,870*	78.9
1919	557,500,000°	377,719,000	67.8	92,402,055*	16.56	77,703,289*	57,618,589*	74.1
1920	660,000,000	443,402,000	67.0	185,000,000	28.0	147,204,000	112,374,000	76.0

* *Oil and Gas Journal.* ‡ U. S. Geological Survey.
† *Boletin del Petroleo* (Mexico City). ° Estimated.

TABLE 5.—*Summary of Mexican Oil Production, Exports, Stocks, and Domestic and Field Consumption During 1917 to 1920 in Barrels of 42 U. S. Gallons*

	1917	1918	1919	1920
Exports....................	42,545,843	51,780,479	75,812,760	147,204,000*
Bunker fuel...............	12,746,927	2,336,768	1,890,529 ⎫	
Domestic consumption........		9,218,491	14,098,766 ⎬	33,000,000
Loss in refining............		492,588	600,000 ⎭	
Total net production......	55,292,770	63,828,326	92,402,055	180,204,000*
Field consumption and losses		4,781,509	5,000,000	
Oil in steel storage..........	10,000,000	14,526,559	12,528,518	13,000,000

NOTE.—1917–1918 figures taken from *Boletin del Petroleo.* 1919 figures taken from the *Oil and Gas Journal.* Field consumption for 1918 estimated by Mexican Petroleum Department. 1919 figures also estimated.

* Estimated.

TABLE 6.—*Mexican Oil Exports from Tampico, Tuxpan, and Port Lobos to Destinations in Barrels of 42 U. S. Gallons*

	1917		1918		1919		1920	
	Exports	Per Cent.	Exports	Per Cent.	Exports	Per Cent.	Exports	Per Cent.
From Tampico........	32,537,821	70.6	37,176,008	65.5	44,092,135	55.0	89,909,213	58.7
Tuxpan.............	13,516,337	29.4	15,849,998	27.9	17,166,714	21.0	18,786,904	12.2
Port Lobos.........			3,739,390	6.6	20,070,993	24.0	44,579,183	29.1
	46,054,158	100.0	56,765,396	100.0	81,329,842	100.0	153,275,300	100.0
Destination:								
United States........	35,386,242	76.7	40,819,870	71.9	57,618,589	70.8	112,373,795	73.3
Canada.............	630,579	1.38	681,177	1.2	2,558,496	3.1	2,010,958	1.3
South America......	4,801,536	10.60	5,557,827	9.7	6,642,985	8.2	13,087,007	8.5
Central America.....	535,367	1.16	377,394	0.6	553,692	0.7	593,252	0.4
West Indies.........			885,483	1.6	1,964,282	2.4	5,754,903	3.8
Great Britain........	679,157	1.48	2,600,806	4.6	3,054,357	3.8	5,493,533	3.7
Netherlands.........					239,985	0.3	1,101,448	0.7
France.............					277,591	0.3	579,496	0.4
Portugal............					110,509	0.1	107,341	0.1
Mediterranean Ports.	49,836	0.11			66,767	0.1	159,130	0.1
Gibraltar...........							144,624	0.1
Malta..............					95,794	0.1	41,952	0.0
Tunis..............							58,433	0.0
Egypt..............					116,716	0.2	463,179	0.3
Algiers.............							60,538	0.0
Italy..............							101,729	0.1
Suez...............					72,504	0.1	132,630	0.1
Mexican coastwise ..:	3,971,441	8.60	4,689,774	8.3	5,996,982	7.4	6,070,439	3.9
Bunker fuel			1,153,065	2.1	1,960,593	2.4	4,895,265	3.2
Local deliveries							45,648	0.0
	46,054,158	100.00	56,765,396	100.0	81,329,842	100.0	153,275,300	100.0

TABLE 7.—*Mexican Oil Exports to United States Harbors from January 1917. In Barrels of 42 U. S. Gallons*

	1917		1918		1919		1920	
	Exports	Per Cent.	Exports	Per Cent.	Exports	Per Cent.	Exports	Per Cent.
Texas Ports..........	9,023,492	25.5	9,878,409	24.2	15,787,493	27.4	30,941,986	27.5
New Orleans.........	7,254,180	20.5	8,082,334	19.8	8,181,840	14.2	14,824,281	13.1
Florida Ports.........	3,680,169	10.4	2,571,652	6.3	1,959,032	3.4	6,087,332	5.4
New York............	13,552,930	38.3	17,144,345	42.0	26,965,499	46.8	48,626,183	43.3
New England Ports....	1,875,471	5.3	3,143,130	7.7	3,975,683	6.9	11,098,771	9.9
California............					749,042	1.3	795,242	0.8
Total to United States	35,386,242	100.0	40,819,870	100.0	57,618,589	100.0	112,373,795	100.0

TABLE 8.—*Mexican Oil Exports by Companies from January, 1917. In Barrels of 42 U. S. Gallons*

Company	1917	1918	1919	1920
Standard Oil Group				
Standard Oil Co. of New York ⎫				
Standard Oil Co. of New Jersey ⎬	5,035,774	7,645,671	6,970,927	21,502,886
Transcontinental Petroleum Co. ⎭				
Penn-Mex Fuel Co....................	3,451,226	7,007,833	8,495,047	3,176,963
Cortez Oil Corpn		1,935,360	9,096,435	7,960,959
Panuco-Boston Oil Co.................			48,777	385,996
Magnolia Petroleum Co. (New England)..				
Mexican Petroleum Co.				
Huasteca Petroleum Co................	12,236,388	11,708,109	12,651,974	29,280,421
Royal Dutch Shell and British Interests				
Mexican Eagle Oil Co. (El Aguila)	8,567,299	8,583,258	12,570,492	17,266,692
La Corona Petroleum Corpn.............			524,626	2,895,587
Tal Vez Oil Co.......................	24,368	98,896	398,889	504,993
The Texas Co.........................	1,955,146	1,256,128	6,814,084	12,355,082
Sinclair Consolidated Oil Co.				
Freeport & Mexican Fuel Oil Corpn......	3,626,917	3,939,756	4,753,862	8,300,045
Gulf Oil Corpn.				
Mexican Gulf Oil Co...................	1,121,236	1,734,191	4,574,520	10,573,622
East Coast Oil Co.....................	3,390,939	3,398,459	4,639,513	5,542,820
Island Oil and Transport Corpn...........			6,212,915	12,410,323
Pierce Oil Corpn.......................	636,469	1,253,133	977,730	2,312,039
Union Oil Company of California..........	1,622,131	2,002,453		68,811
Cities Service Co.				
National Petroleum Corpn..............	258,894	543,791	489,159	792,050
New England Fuel Co...................		166,567	218,244	1,126,967
Cochrane and Harper....................			335,571	1,187,915
Inter-ocean Oil Co.....................	619,056	492,511	635,296	438,754
Compania Terminal Union...............			293,719	400,094
National Oil Co.......................				1,602,134
Atlantic Gulf Oil Co.				
Compania Refinadora del Agwi..........				6,403,967
Total.............................	42,545,843	51,766,116	80,701,780*	146,489,120

* Includes 2,998,491 bbl. Mexican coastwise shipments that were consumed domestically, making net exports 77,703,289 bbl.

TABLE 9.—*Tank Steamers in Operation and Under Construction. by Companies Exporting Mexican Oils*

Company	In Operation			Under Construction		
	Number of Tankers	Total Dead-weight Tons	Maximum Tonnage per Tanker	Number of Tankers	Total Dead-weight Tons	Maximum tonnage per Tanker
Atlantic Gulf Oil Co................				14	160,400	12,000
Eagle Oil and Transport............:..	16	221,000	18,100	7	126,000	20,000
East Coast Oil Co..................	4	25,000	8,000			
Gulf Refining Co....................	15	106,777	12,777	4	30,270	10,300
National Petroleum Corpn..........	1	5,100				
Pan American Petroleum & Transp...	18	145,325	12,350	11	150,000	12,000
Pierce Oil Corpn*..................	2	10,000	5,000			
Shell Transport Co.†...............		263,000				
Sinclair Consolidated Oil Corpn.......	11	50,884	7,500	12	116,600	10,500
Standard Oil Group						
Standard Oil Co. of N. Y.†.........	19	109,039	12,650	15	159,960	12,650
Standard Oil Co. of N. J.†.........	45	449,166	15,000	17	225,000	20,000
The Texas Co......................	15	100,000	9,500		80,000	10,000
Union Oil Company of California†....	12	82,875	11,000	8	89,000	12,000

* It is acknowledged that the information on this table is incomplete.
† Only a small number of these tankers are in the Mexican trade.

PRODUCTION AND EXPORTS

Mexican and American Oil-measuring Standards

The standards for measuring oil in Mexico are based on the metric system and so the weight of water, which is the basis for comparison with oil, is taken with water at a temperature of 4° C. and the specific gravity of oil is based on oil at 20° C.; the American standards are taken on the basis of the relative densities of oil and water at 60° F. (17° C.). It is therefore evident that oil having a specific gravity of, say, 0.982 under the Mexican standards is not an oil of 0.982 specific gravity under the American standards. In order to establish the relation between both systems, Table 3 has been compiled; this gives the specific gravities and Baumé degrees in the American standard and the corresponding specific gravities in the Mexican standard, and also the volume-weight relation showing the number of barrels per metric ton.

The Mexican Government levies the tax on the weight of the oil, pesos-per-metric-ton basis, while the American operator sells the product by volume on the dollars-per-barrel basis.

Relation between World, United States, and Mexican Production

A clear idea of the Mexican oil production and exports may be obtained from Table 4 and Fig. 1, which show that for the first seven years, Mexico's yearly production did not reach 1 per cent. of the world's total; that the increase in production was gradual until the light-oil fields were discovered, in 1910, from which date there has been a rapid increase, until the Mexican production aggregates about 28 per cent. of the world's total. The greatest portion of the Mexican production is exported, the exports began in 1911 and reached an important amount in 1913. About three-fourths of the exports go to the United States, in 1920 this amounted to over 112,000,000 bbl. During 1920, the United States and Mexico produced on an aggregate close to 95 per cent. of the total world's output.

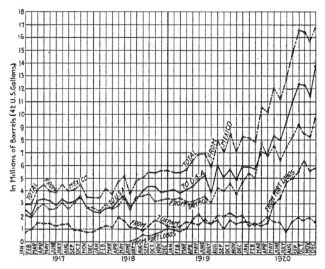

FIG. 1.—MEXICAN OIL EXPORTS, BY PORTS FROM WHICH EXPORTED, AND TOTAL TO THE UNITED STATES.

Summary of Mexican Production

Table 5 gives an analysis of Mexican production, exports, bunker fuel, stocks, and domestic and field consumption during the last four years. It shows that production has increased three and one-half times during this period, domestic consumption has likewise increased, while the volume of oil stocks or in storage has remained practically constant. The figures relating to domestic consumption, losses, and storage shown in the table are admittedly only approximately correct.

The following table gives the storage capacity in the Mexican fields during 1919 and in August, 1920:

STORAGE CAPACITY, IN U. S. BARRELS

	1919	AUGUST, 1920
Steel tanks.........................	26,355,000	31,455,000
Concrete tanks......................	275,000	275,000
Earthen reservoirs	22,005,000	22,060,000
Concrete reservoirs..................	865,000	860,000
Total.........................	49,500,000	54,650,000

The steel storage facilities of four of the large companies is as follows:

	BARRELS
Mexican Eagle	7,120,000
Mexican Petroleum	6,500,000
La Corona	2,500,000
Transcontinental	2,200,000

Exports by Destination

Table 6 and Fig. 2 show that by far the greatest bulk of the oil exported from Mexico goes to the United States, South American harbors

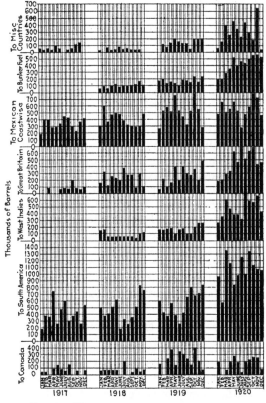

Fig. 2.—MEXICAN OIL EXPORTS BY DESTINATIONS.

coming next; the balance of the exports, a comparatively small sum, go to widely scattered ports in the West Indies, Great Britain, the Mediterranean, and elsewhere.

The amount of oil used as bunker fuel is increasing at a rapid rate, for the first six months of 1920 being about equal to the total for the preceding year. The Mexican coastwise movements include oil shipped from Tampico, Tuxpam, and Port Lobos to Mexican harbors and to the Aguila company's refinery at Minatitlan. The harbor of Tampico still retains the leadership in oil exports with 59 per cent., Port Lobos comes second with 29 per cent., and Tuxpam third with 12 per cent.

The oil shipped to Puerto Mexico is refined at Minatitlan and thence marketed in Mexico or foreign countries by the Mexican Eagle Oil Co. A small amount of oil produced in the Tehuantepec region is likewise refined at Minatitlan, it being difficult to differentiate between this production and the crude from Tuxpam or Tampico, included under "Coastwise shipments."

The yearly over-all exports from Puerto Mexico have been approximately as follows:

	BARRELS		BARRELS
1913	1,003,000	1917	1,401,000
1914	1,846,000	1918	1,010,000
1915	1,933,000	1919	1,882,000
1916	1,536,000	1920	2,300,000 (estimated)

Table 7 shows that since January, 1917, most of the oil exports had gone to New York, Baltimore, Philadelphia, and neighboring ports; over one-half of the Mexican oil exported to the United States going to ports on the Atlantic seaboard. The exports to Texas ports have aggregated about one-fourth of the total exports to the United States, this figure being kept more or less constant since 1917, while the exports to New Orleans have gradually decreased from 20 to about 13 per cent. The exports to Florida ports have likewise decreased from 10 to 5 per cent. and those to New England ports have had a correspondingly gradual increase. The exports to California harbors aggregate a fractional percentage of the total and consist for the most part of about 5 to 15 shiploads during 1919 and the first half of 1920.

Exports by Companies

Table 8 shows that the Standard Oil group easily lead at the present time, the Mexican Petroleum being second; the exports by the Anglo-Dutch interests are third, but they have approximately only one-half the exports of the first group of companies. A number of independent companies exported from 8,300,000 to 68,000 bbl. each during the year 1920.

It is interesting to note that only about sixteen companies, or rather interests, are at present exporting Mexican oil, the transporting and marketing of Mexican oil being thus narrowed down to the well-established oil interests.

Exports by Grades of Oils

It is difficult to obtain figures of exports of Mexican oils by grades; those on which Fig. 3 is based are more or less approximate. However, this chart shows that the exports of light crude have increased tremendously during the last eight months, and that there has been but little increase in the exports of heavy crude, crude gasoline, and fuel oil. This would seem to indicate that refinery facilities have not been increased during that time, the increase in exports being primarily due to the larger volume of the light crude now produced in the Chinampa-Naranjos-Alazan pool.

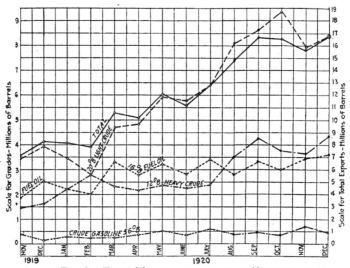

FIG. 3.—TOTAL MEXICAN EXPORTS SINCE NOVEMBER, 1919.

COST OF TRANSPORTING MEXICAN OIL IN TANK STEAMERS

Tables 6 and 7 show that since January, 1917, from 70 to 76 per cent. of the total Mexican exports have gone to the United States; in fact, were Mexican coastwise shipping, bunker fuel, and local deliveries excluded, the net percentage shipped to the United States harbors would be well over 75 per cent.

Table 7 shows that over half of the exports to the United States go to Atlantic seaboard harbors, New York harbor and vicinity leading with 43.3 per cent. Approximately one-fourth goes to Texas ports, 13.1

per cent. to New Orleans, the remaining 15 per cent., or so, is distributed between the Florida and New England ports; the amount being shipped to California is almost negligible. It is therefore evident that in studying the cost of transporting Mexican oils in tank steamers, it is necessary to analyze conditions governing the transportation between the Mexican harbors and Texas ports, New Orleans, Florida ports, New York and New England, and primarily between Mexico and these last two mentioned.

Net Carrying Capacity of Tank Steamers

The net carrying capacity of tank steamers plying between Mexican and American ports has been compiled in Figs. 4 and 5, which show that the larger tankers are being used between Mexico and New York and New England harbors, the smallest being used for the short runs to

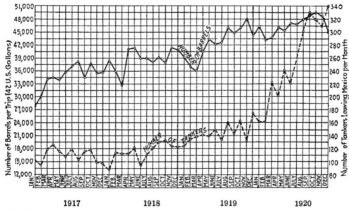

FIG. 4.—TRANSPORTATION OF MEXICAN OILS IN TANK STEAMERS; NUMBER OF TANKERS PER MONTH AND AVERAGE NUMBER OF BARRELS PER TRIP TO ALL PORTS.

Texas and Florida ports. Somewhat larger tankers are used from Tampico to New Orleans.

In a general way, it may be stated that the average tanker plying between Mexican harbors and New England or New York is a 10,000-deadweight-ton tanker or larger, able to carry 60,000 bbl. and more per trip; that the average tank steamers plying to New Orleans have a dead-weight tonnage of about 8000 tons with a carrying capacity of about 45,000 bbl.; the smaller tankers of 3000 to 5000 tons and oil barges make the run between Tampico and Florida and Texas ports.

Fig. 4 gives the average number of barrels transported per tank-steamer-trip and indicates that this has increased from about 28,000 in January, 1917, to 49,000 in October, 1920, showing that larger units are being constantly put into service. This figure also shows that the carrying capacity decreases during the winter months.

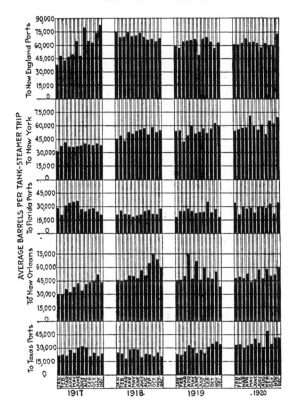

FIG. 5.—AVERAGE NET CARRYING CAPACITY PER TANK STEAMER TRIP FROM MEXICAN
TO AMERICAN PORTS.

Distance from Tampico to American Ports and Time Required for Round Trip

The distance from Tampico to American and other ports and the average number of days required to make a round trip by a tanker, with an average speed of 10 mi. per hr., allowances being made for days lost in repairs, dry-docking, etc. are as follows:

	DISTANCE, MILES	TIME, DAYS		DISTANCE, MILES	TIME, DAYS
Antofagasta, Chile	3,668	38	Callao, Peru	2,874	32
Baltimore, Md	1,951	24	Canal Zone	1,485	20
Bayonne, N. J.	2,030	25	Freeport, Tex	474	12
Beaumont, Tex	475	12	Fall River, Mass	2,131	26
Boston, Mass	2,276	27	Galveston, Tex	473	12
Buenos Aires, Argentine	5,518	54	Houston, Tex	473	12

	DISTANCE, MILES	TIME, DAYS		DISTANCE, MILES	TIME, DAYS
Jacksonville, Fla........	1,361	19	Philadelphia, Pa........	2,000	25
Key West, Fla..........	907	16	Pt. Arthur, Tex.........	473	12
Kingston, Jamaica.......	1,252	19	Portland, Me...........	2,275	27
Liverpool, England......	4,905	49	Providence, R. I........	2,131	26
London, England........	5,201	51	Rio de Janeiro..........	5,417	53
Marcus Hook, Pa........	2,000	25	St. Thomas, W. I.......	1,905	23
Maurer, N. J...........:....	2,025	25	San Francisco, Calif.....	4,150	42
Miami, Fla.............	1,048	16	Savannah, Ga..........	1,439	20
Mobile, Ala............	721	15	Southampton, England..	5,013	50
Montreal, Canada.......	3,301	37	Sparrows' Point.........	1,950	24
New Orleans, La........	721	15	Tampa, Fla.............	921	16
New York..............	2,030	25	Texas City, Tex.........	475	12
Norfolk, Va	1,829	23	Valpariso, Chile.........	4,144	42
Pensacola, Fla..........	759	15	Warner's, N. J..........	2,025	25

The shortest trip, to Texas ports, requires an average of twelve days, while fifteen days are allowed tankers making the New Orleans route. The round trip to New York harbor and vicinity requires twenty-five days; for the New England ports one or two days more are needed.

Cost of Tank Steamers

In pre-war days, the price of a 10,000-ton tanker averaged close to $70 per ton; some of the larger oil companies purchased these steamers for less. During the war, the price reached $200 per ton and higher; but some months ago, a downward tendency began and it is possible to contract for tankers of 10,000 tons and over for between $140 and $150 per ton.

Although the transportation costs are based, in this report, on steamers rated at 10,000 d.w. tons, the tendency is to increase the capacity of the boats to 15,000 and 20,000 tons, as shown on Table 9. The Eagle Oil and Transport, and the Standard Oil of New Jersey have under construction several 20,000-ton boats.

As a general rule, a boat built for a tanker should carry, in barrels, on an average six times its deadweight tonnage; for example, a 10,000-ton tanker should average at least 60,000 bbl. of oil per trip. Naturally, the exact figures depend, among other factors, on the weight of the oil, design of tanker, amount of space needed for tanker's fuel (more bunker space will be needed on longer trips) season of the year, draft of boat as compared to the depth of water in the loading and unloading harbors, etc. Five-thousand-ton tankers will carry somewhat over 30,000 bbl.; 7000-ton tankers about 45,000 bbl.; 10,000 d.w. ton tankers from 60,000 to 65,000 bbl.; 15,000-ton tankers about 95,000 bbl.; and 20,000-ton tankers about 120,000 barrels.

As a general rule, it has been found more advantageous to equip tankers with steam engines using fuel oil for steam generation; boats equipped with Diesel type engines have not given as reliable service as the steamers.

Cost of Transportation

Taking as a unit a 10,000 d.w. ton tanker, able to carry 60,000 bbl. of oil and upward per trip, and costing $200 per deadweight ton, which is higher than the present average cost, and assuming other equally conservative figures, the cost of transporting oil for round trips taking from 12 to 30 days is shown in Fig. 6.

Thus the cost per barrel for transporting oil in a 10,000 d.w. ton tanker, from Tampico to Texas ports, 12 days round trip will be 42.5c.; to New Orleans, 15 days, 53c.; to Florida ports, 16 days, 57c.; and to New York, 25 days, 88 cents.

If the tanker only cost $100 per ton, the correction factors in the lower left-hand corner of Fig. 6 should be used; thus, the New Orleans trip will cost 39.64c. per bbl. and not 53 cents.

In corroboration of the figures given by the chart for the cost of transporting oil, the following is quoted from the *Journal of Commerce* of July 29, 1920: "In closing the contracts, the United States Shipping Board has agreed to charter sufficient tank ships for its transportation at the Government rate of $6.50 per deadweight ton per month."

Allowing an average of 6 bbl. per deadweight and assuming that the oil is to be transported from Tampico to New Orleans in a 10,000-ton tanker costing $200 per deadweight, making two round trips per month, the Shipping Board charter rate would give a transportation cost of 54.2c. per bbl. against the 53c. per bbl. obtained from Fig. 6.

It is the opinion of competent authorities that for round trips taking from 12 to 30 days, the various charges for the shorter and longer trips will about balance each other, leaving a fairly uniform ratio for cost of transporting oil per barrel for the long and short trip.

Conclusions

1. The United States and Mexico will produce on an aggregate, in 1920, close to 90 per cent. of the total world's output of petroleum.

2. The Mexican production, in 1920, will aggregate over 25 per cent. of the world's total.

3. About 75 per cent. of the Mexican exports estimated at 108,000,-000 bbl. in 1920 go to the United States, and this represents about 25 per cent. of the United States production.

4. Of the oil exported to the United States, about 52 per cent. is shipped to New York and North Atlantic ports; 27.5 per cent. to Texas ports; the remaining 18.5 per cent. to New Orleans and Florida ports.

5. Although there has been a gradual increase in the exports of fuel

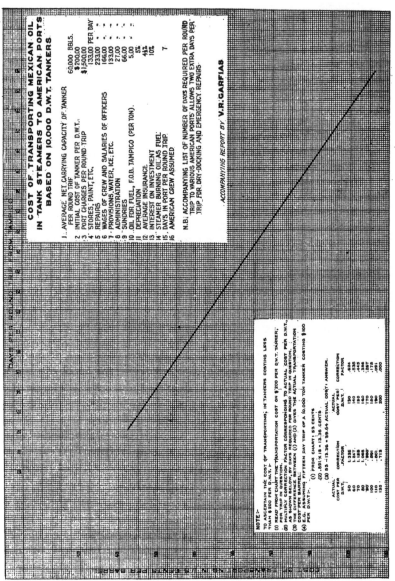

FIG. 6.

oil, heavy crude, and crude gasoline, the great increase is due almost entirely to the greater volumes of light crude exported.

6. The average load per tank steamer trip from Mexican harbors has increased from 28,000 to 48,000 bbl. from January, 1917, to August, 1920. The average load to North Atlantic and New York harbors is close to 60,000 bbl. (10,000-ton tankers); about 45,000 bbl. to New Orleans (8000-ton tankers); 30,000 bbl. to Texas, and Florida ports (5000-ton tankers). Tank steamers under construction, 12,000, 15,000, and 20,000 tons, when placed in operation, should increase the average net carrying capacity per tanker trip.

7. The transportation costs given in Fig. 6, based on 10,000 d.w. ton tankers, are very conservative; in fact, appreciably lower figures are fully justified at present. The figures given apply to boats owned by the operating company, not to chartered boats.

8. No reliable figures of Mexican oil in storage are available.

MEXICAN TAXATION ON PETROLEUM AND ITS PRODUCTS

Although the following analysis of Mexican taxation on petroleum was made by the writer when acting as Special Commissioner of the Petroleum Department of the Mexican Government, the conclusions drawn represent his own views, for which the Mexican Government can in no way be held responsible.

The Mexican Government at present levies on petroleum and its products not utilized in the Republic, the so-called Export Stamp Tax, which is based on certain percentages, varying with the grades of oil, of the prices of the exported commodity. These prices may be determined: (1) as those prevailing within Mexico; (2) prices in New York, or other American harbors, less marine transportation costs; (3) the prices, anywhere in the United States, of similar petroleums as regards physical properties.

The amount of the present taxes that depend on oil prices, which in turn can be interpreted in three ways, has been, and is the source of misunderstandings between some of the operators and the government whenever the government and the companies' manner of evaluating the oils disagree. The aim of the writer is to pave the way for the removal of these causes for controversies and to suggest changes that will make for a more definite and clear basis for taxation.

Taxes Prior to May, 1917

With the exception of the usual stamp tax on documents, and such other minor contributions, the oil companies operating in Mexico did not pay taxes to the government on about 24,800,000 bbl. produced prior to

1912. In fact, companies like the Aguila, Huasteca, and the Standard Oil of New Jersey were exempt from the usual import taxes on machinery, etc. On July 1, 1912, during the administration of President Madero, an export tax of 20 centavos per metric ton, approximately 1.54 U. S. cents per barrel of oil exported, was charged: this tax, which was applied irrespective of the quality of the oil, was in force until November, 1913, when it was increased to 75 centavos per metric ton, about 5.77 U. S cents per barrel, during the Huerta administration. This tax, like the preceding one, was applied to oil exported, irrespective of its quality, and was reduced on May 1, 1914, to 60 centavos per metric ton, approximately 4.62 U. S. cents per barrel, during the Carranza administration and was in force until May 1, 1917, when the present tax, based on a certain per cent. of the value of the oils was established. The exports from 1912 to April, 1917, inclusive, aggregating about 111,700,000 bbl., were taxed, therefore, approximately $4,355,000, or about 3.9 U. S. cents per barrel.

TAXES FROM MAY, 1917 TO DATE

It should be understood that the tax, called in this paper the export stamp tax, applies exclusively to petroleum and its products and is independent of any other former or subsequent tax that applies to petroleum as well as to other exports. Under this heading may be included the paper redemption tax (infalsificable), bar dues for oil shipped from Tampico, etc. It is clear, therefore, that when one speaks of Mexican taxation on oil, one should differentiate between the export stamp tax, to which the decree of Apr. 13, 1917, applies and which is based on a certain percentage of the value of the oils, and any other taxes not inherent to the petroleum industry. The additional taxes that are not properly oil taxes are small, compared with the export stamp tax.

THE EXPORT STAMP TAX

The decree of Apr. 13, 1917, on which this tax is based, reads in part as follows:

. . . that it being of very diversified quality, the petroleum produced in the Republic, and for the same reason of different commercial value, the tax should have as a basis the value of each product, in order that it be reasonable and equitable; that a considerable quantity of this liquid is not utilized because the necessary precautions are not taken in the exploration work and its daily handling, this circumstance occasioning frequent losses, not only to the interested companies, but also to the Government, on account of the taxes that it fails to collect.

In view of the foregoing, I (the President) have enacted the following decree:

Article 1.—All crude petroleum of national production, its derivatives and the gas from the wells, from the moment that it flows from the ground or leaves the storage deposits, are subject to a special stamp tax under the following terms:

(a) Crude and fuel oil........................ 10 per cent. of assigned value.
    Refined gasoline........................... 3 per cent. of assigned value.
    Crude gasoline............................ 6 per cent. of assigned value.
    Refined kerosene.......................... 3 per cent. of assigned value.
    Lubricating oils........................... ¼c. per liter.
    Asphalt................................... $1.50 per ton.
    Gas...................................... 5 per cent. ad valorem.

(For up-to-date rates and changes see Table 1.)

(b) The crude petroleum and its derivatives, when wasted in any quantity, whether for lack of care or not complying with the legal regulations, will pay a tax double the one corresponding to similar products.

The products derived from the natural gas of the wells, when it is wasted from the same reasons, will pay 10 per cent. of its commercial value.

Article 2.—(Exempts from tax, oil consumed in Mexico.)

Article 3.—(Defines "crude oil," "refined oils," etc.)

Article 4.—In order to be able to establish the tax, which, in accordance with fraction (a) of Article 1, corresponds to each one of the products derived from petroleum, the Secretary of Hacienda will fix every two months the prices of said articles at the shipping ports, taking the average of the values reached in the previous month. The manifestations or bills that the companies present regarding sales of the same articles, in the interior in Mexico, will serve as a base for making the estimate referred to.

In case that no operations of sales take place in the interior, the average value which these products had in New York the previous month, or in the harbors of the United States, will be taken, deducting the value of transportation of said products, from the Mexican to the foreign harbors. If there are no available data to make the previous calculations, an equal price will be assigned to that which similar articles have, in regard to physical properties, in the United States, fixing on this price the respective tax.

The Grades of Mexican Oils Exported

Although it is difficult to obtain accurate information regarding the various grades of oils exported from Mexico, enough data has been obtained to show that, at present and for some time past, the bulk of the exports can be divided into four classes:

    Light or southern crude................ 0.9333 sp. gr. (20° Bé.)
    Heavy or Panuco crude................ 0.9859 sp. gr. (12° Bé.)
    Fuel oil............................. 0.9589 sp. gr. (16° Bé.)
    Crude kerosene or tops................ 0.7527 sp. gr. (56° Bé.)

The light crude is exported in large quantities and is also partly refined in topping plants, which produce a low-flash fuel oil and light tops, or crude gasoline. The heavy crude is not refined, being used exclusively for fuel purposes in power plants; its low flash point and high viscosity prevent its extensive use for marine purposes.

Mexican Taxes on Various Grades of Oils

The export stamp taxes from May, 1917, to January, 1921, for the four main grades of oils, have been computed and listed in Tables 10 and

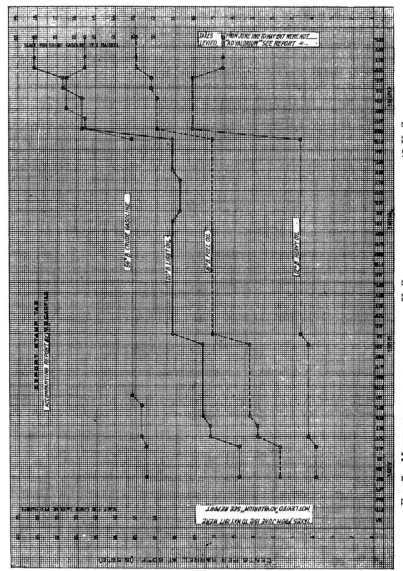

FIG. 7.—MEXICAN EXPORT STAMP TAX IN U. S. CENTS PER BARREL OF 42 U. S. GALLONS.

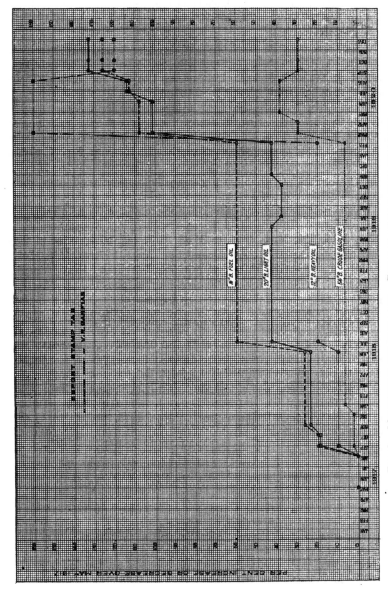

FIG. 8.—MEXICAN EXPORT STAMP TAX.   PER CENT. INCREASE OR DECREASE.

13, on the basis of U. S. cents per barrel. These tables show that the lowest tax, 3.90c. per bbl., was levied on heavy crude in 1917, the highest, 72.7c. per bbl., or about 1.73c. per gal., being levied on crude gasoline for several months in 1920.

In a general way the taxes on light crude have gradually increased from 7.8c., in 1917, to 18.2c., in September, 1920; fuel oil tax has increased from 5.7 to 13c.; heavy crude from about 4 to 10c. and back to 8.6c.; and crude gasoline, from 1.2 to 1.7c. a gallon.

The monthly fluctuation in taxes for the four grades of oils are graphically shown in Fig. 7, which illustrates the comparatively few changes that have occurred from May, 1917, to March, 1920. The percentage of increase in the taxes on the various grades is shown in Fig. 8, which illustrates the abnormal increase of the tax on heavy crude, on March, 1920, as well as the lack of uniformity in the fluctuations of all taxes from March, 1920, to date.

The average export stamp taxes, in U. S. cents per barrel, paid from 1912 to 1917 were: July, 1912, to November, 1913, $1\frac{1}{2}$; November, 1913, to May, 1914, $5\frac{1}{2}$; May, 1914, to May, 1917, $4\frac{1}{2}$; irrespective of the quality of the oils, while the average from May, 1917, to December, 1920, for the four main grades of products has been: Heavy crude, 5; light crude, 11; fuel oil, 9; crude gasoline, 56c. per bbl., or $1\frac{3}{8}$c. per gallon.

Table 11 shows the total taxes on petroleum and its products, which include, besides the export stamp tax, others not exclusively applicable to the oil industry. This table shows the infalsificable, or paper redemption tax (one paper peso be paid for each metal peso paid in taxes) figured on the uniform ratio of 10 to 1 for the relative values of the paper and metal peso, which ratio undoubtedly gives larger figures than have been actually paid. Bar dues have been calculated on the assumption that all the oil exported paid these bar dues, while, as a matter of fact, these dues are applicable only to the oil shipped from the harbor of Tampico.

The total taxes herein listed represent practically all the returns the Mexican Government obtains from the oil industry, inasmuch as no income nor excess profit or similar taxes are in operation in Mexico.

THE VALUE OF MEXICAN OIL

The statement has been often made that the export stamp tax, which by law should represent 10 per cent. of the price of the oil, actually amounts to 40 per cent., but the absurdity of such statements can be realized by analyzing the average taxes from 1917 to date. On the assumption that these taxes represented 40 per cent. of the prices of the oils, we would have:

	40 Per Cent. Tax	Price in Mexico U. S. Cents
Heavy crude	5c.	12½ per bbl.
Light crude	11c.	27½ per bbl.
Fuel oil	9c.	22½ per bbl.
Crude gasoline	1⅛c. per gal.	3⅓ per gal.

The computed prices in this case fall far below the market price, as any one familiar with conditions can certify. This comparison illustrates, further, the difficulties encountered in ascertaining whether the tax in question is, or is not, the exact percentage marked by law of a price that, according to the law, can be computed in three ways, none of which is clearly enough defined to eliminate possibilities of misunderstandings.

It has been advanced by representatives of some companies that the only proper basis for arriving at the true value of Mexican oils, say at Tampico, will be found in the selling contracts made between companies, or between an oil company and the U. S. Shipping Board, which stipulate the price at the Mexican harbor. In support of this contention, contracts are exhibited showing the prices of Mexican oils varying within wide limits, but as a rule well below what might be considered a fair market value. It is further stated, by the supporters of this method of appraising Mexican oils, that account should be taken of long term contracts which net the companies relatively low figures at the present time.

On the other hand, it should be evident to any one familiar with intercompanies' oil contracts, that they do not offer the best means of ascertaining the fair market price, as shown by the following examples: Producing company A sells its oil at Tampico to transportation company B at a price that will net little or no profit to the producing company. Company B, in turn, sells the oil in the United States to marketing company C for a price that will allow the transportation company to operate its boats at a fair profit, leaving to the marketing company the big margin of profits in disposing of the products to consumers in the United States. It certainly would be unfair to claim, in this case, that the inter-company contract price between A and B should be taken as the fair price of the oil at Tampico.

A second case will give another view of this same question. Producing company A, when in great need of financial assistance, was forced to sell its production to outside company B on a long-term contract at a price that is now considerably lower than the market price; it is decidedly unfair to claim that the contract price in question represents the actual market conditions for the duration of the contract.

There is also the case of a long-term contract made under profitable terms in years past, but with poor judgment as to future prices of Mexican oils. It would be unfair to claim that the prices stipulated in these contracts always represent actual market conditions when the net result is only to shift profits from one company to another.

· The U. S. Shipping Board has made contracts, principally for oils that the Shipping Board could not use without refining, and the price of the oil transported was one of the many clauses in the contracts. These contracts often included certain trading agreements for fuel oils that could be utilized as bunker fuel, the price of, say, the light crude contracted, refining of the crude, preferential rights for additional transportation facilities, etc. Here again, the contract price might well not reveal the actual market price of the crude.

Were the letter and not the spirit of the 1917 law followed a tax on gasoline of about 2½ U. S. cents per gallon would be justified, in place of the present tax of 1.7 U. S. cents per gallon, if the current Tampico price of gasoline were taken into account.

Summarizing the foregoing, it may be safely concluded that as long as the export stamp tax is based on the prices of oils in Mexico, as defined in the decree of April, 1917, the result will be endless controversies between the Mexican Government and the operating companies.

RELATION BETWEEN PRICES OF AMERICAN AND MEXICAN OILS

It should be clearly understood that by the following analysis the writer does not intend to establish, for instance, a direct ratio between the prices or values of Mid-Continent crude at the well and those of Mexican petroleum, nor that the composition of Mexican light crude corresponds to that of Gulf Coast, Mid-Continent, or Californian crudes, nor that the price of Mexican oil be established by comparison with the fluctuations in prices of one or all of the American oils mentioned. The endeavor is: (1) To analyze the fluctuations in price of the bulk of American oils, viz: Mid-Continent, Californian and Gulf Coast, which aggregate about 85 per cent. of the total production of the United States; (2) to establish the history of market fluctuations of these oils and such other closely related products as bituminous coal so that the average would represent a fairly stable picture of over-all market fluctuations, independent of the control of any one interest (official or otherwise) and solely related to the laws of supply and demand; (3), once this bench mark is established, to ascertain the relation between the Mexican ad valorem taxes levied from beginning to date and these average prices, not with the view of deciding what the price of Mexican oils has been, but in an effort to ascertain what relation has existed between Mexican taxes and such independent standard on which future taxation could be based thus eliminating past controversies between the Mexican Government and the operating companies. The writer wishes, therefore, to emphasize at this time that what results are given are not offered as the solution of the

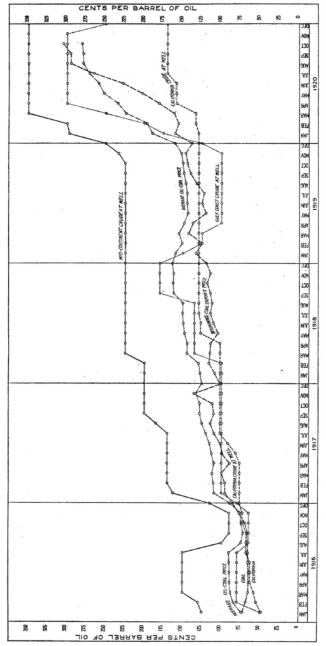

FIG. 9.—COMPARISON OF VARIOUS CRUDE OILS AND BITUMINOUS COAL PRICES IN UNITED STATES.

question as to what really has been or is the value of Mexican oils in American harbors, but are only presented as offering a new and impartial basis for taxing the Mexican oils exported.

As the law of April, 1917, provided that the value of oils in the United States may be taken into account after proper allowances are made for the cost of transporting the oil from the Mexican to the American harbors, realizing the many difficulties encountered in reaching satisfactory results by using the prices in Mexico as per companies' contracts, etc., the writer compiled detailed information on the fluctuation of oil values in the United States from 1917 to date; first, to ascertain whether there has been over-all market conditions uniformly affecting the value of American oils in the western, southern and central fields, and, second, to ascertain what relation, if any, exists between the values of the American and Mexican products.

The average oil prices listed related to the production of the Mid-Continent, California and Gulf Coast fields and therefore represent over-all market conditions. Fig. 9 and Table 12 show that there exists an over-all market condition uniformly regulating fluctuations in prices of fuels. As about 50 per cent. of the Mexican exports are delivered to the Atlantic seaboard, it was thought advisable to include in the analysis the export price of bituminous coal, with which Mexican oil comes directly, or indirectly, in competition, the ratio of 1 ton of coal to 3½ bbl. of oil, which is the generally accepted equivalent, being decided upon, and the export price of coal being converted to the barrel-of-oil basis. The fluctuations of prices of bituminous coal, as shown in Fig. 9, are more uniform than the oil prices, and more closely follow the average market conditions.

The Mexican exports to the United States, which are 75 per cent. of the total exports, equal about 25 per cent. of the American oil production, so when it is sold on the Atlantic or Gulf Coast seaboards, it has to compete with the United States petroleums; therefore, the price of the Mexican oil is controlled by that of the home product. Fig. 9 shows that from the end of 1919 to date, there has been a sharp increase in the prices of fuels, both liquid and solid, throughout the United States, and it is inconceivable that the prices of Mexican oils, the bulk of which is marketed in the United States, did not follow these over-all market fluctuations of values.

RELATIVE COST OF OPERATING IN MEXICAN AND AMERICAN
OIL FIELDS

The claim is often made by some operators that the Mexican oil taxes should be reduced because of the high cost of development compared with this cost in the United States. But it has been proved that the over-all costs are lower in the Mexican than in the American fields, for the average depth of wells in the Mexican fields is less than 2500 ft., which is no greater than that in most American fields, and while the cost of drilling is somewhat greater, it is certainly not much in excess of drilling wells of the same depth in American fields where conditions are similar.

In some American fields, the production cost, made up mostly of the cost of bringing the oil from the underground reservoirs to the surface, is about 40c. per bbl., and as the life of the well decreases, the production cost materially increases. On the other hand, the production cost in Mexican fields is exceedingly low, because practically all the wells are gushers that flow marketable oil, necessitating no dehydration. The cost of pipe lines in the Mexican fields is not materially greater than in some United States fields; the Mexican pipe lines, as a rule, are a good deal shorter than the average lines from the American oil fields to sea-board.

But the main reason for the lower operating cost in the Mexican fields can be found in Table 1. In order to produce, in round figures, 100,000,000 bbl. per year, it is necessary in California to pump about 9400 wells, while in Mexico 250 wells produce a greater amount by natural flow. In fact, the number of wells actually producing in Mexico is much nearer 100 than 250. The California well averages about $29\frac{1}{2}$ bbl. per day, while the productivity of the Mexican wells ranges (according to whether we class as producers every well capable of producing or only those actually producing) between 1190 bbl. and 3000 bbl. per day. California conditions are well above the average in the United States, as the 228,000 wells producing in the country only average about $5\frac{1}{2}$ bbl. per well per day, as compared with $29\frac{1}{2}$ bbl. for the California wells. Besides these producing wells, in the United States and Mexico, many dry holes have been drilled; about 6000 wells have been abandoned each year from 1913 to 1917 inclusive in the United States while the total number of wells abandoned in Mexico to date is less than 600.

These statistics prove the low cost, everything considered, of operating in the Mexican oil fields; a closer analysis discloses the fact that in no other oil field have such economical conditions for operations prevailed as in the Mexican fields to date.

RELATION BETWEEN AVERAGE OIL-COAL PRICE AND MEXICAN TAXES

The average prices of American petroleums and bituminous coal, which represents market conditions of these commodities in the United States, are given in Tables 12 and 13 and are graphically shown in Figs. 7 and 10. These records show a gradual increase in the average price,

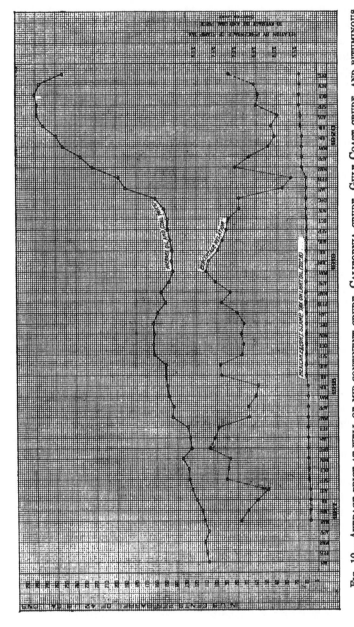

FIG. 10.—AVERAGE PRICE AT WELL OF MID-CONTINENT CRUDE, CALIFORNIA CRUDE, GULF COAST CRUDE, AND BITUMINOUS COAL EXPORT PRICE (CALCULATED PER BARREL OF OIL) COMPARED WITH MEXICAN EXPORT STAMP TAX ON LIGHT OIL (20° BÉ.) PER BARREL OF 42 U. S. GALLONS.

from $1.10 per bbl. in January, 1917, to $1.60 per bbl. in December, 1919, followed by an increase during 1920, from $1.60 to $2.77. The average price increased, therefore, 45.5 per cent. during the years of 1917 to 1919 inclusive, and 73 per cent. in the first eleven months of 1920; the over-all fluctuations from January, 1917, to November, 1920, represent close to 15.2 per cent.

Table 13 shows the per cent. relations, by months, between the average oil-coal price and the export stamp tax on the four grades of oils exported; the figures indicating that the average tax on light crude corresponds approximately to 7 per cent. of the type price (see Fig. 10), the percentage during August, 1920, of 5.98 being about the lowest recorded; the tax on fuel oil represents on an average, 5.4 per cent. of the oil-coal price, the tax for August, 1920 being in proportion the lowest so far levied.

This analysis, based on facts, clearly shows that the Mexican export stamp taxes, with the possible exception of that on heavy crude, are lower in relation to the average market conditions, at the present time, than when initiated in May, 1917.

TAX CONTROVERSIES BETWEEN MEXICAN GOVERNMENT AND OIL COMPANIES

Although a number of important foreign companies have always worked in harmony with the Mexican Government, as was stated to the writer by their representatives in the course of this investigation, other companies have questioned any increase in taxation with the resulting controversies between these companies and the government whenever such changes occurred.

It is undoubtedly true that the principle on which the present Mexican tax operates has not been successful, nor has it met in its application with the full approval of most of the operating companies. This is due not so much to the amount of taxes actually paid as to the inability of the operators to foretell when or what increases will take place, thus preventing sellers and purchasers from taking proper care of these changes at the time of fixing contract prices that extend for considerable time. This has raised difficulties between buyer and seller, the former in some cases being willing only to agree to pay the prevailing tax when the contract is made, thus leaving the seller unable to collect any additional amount in case the tax is increased before the expiration of the contract.

It appears, therefore, that although the operators are not justified in asking for a reduction of the present tax, in order to safeguard the interests of bona fide marketers the Mexican tax should be revamped to conform with the usual business transactions between buyer and seller.

TABLE 10.—*Mexican Government Oil Valuations as Basis for Taxation from May, 1917*

Base of Taxes.—For oils having a density greater than 0.91 sp. gr., decrease valuation 0.20 pesos for each increase in sp. gr. of 0.01 up to 0.97, above which point no credit can be taken.

For each increase of 0.01 in sp. gr. below 0.91, increase valuation 0.40 pesos.

"x" = Pesos per metric ton.
"o" = Pesos per liter.

Year	Month	Grade	Mexican Standard at 68°/39.2° F. Sp. Gr. 20°/4° C. Sp. Gr.	U.S. Standard at 60°/60° F.—15.56/15.56° C. Sp. Gr.	U.S. Standard Degrees, Baumé	Mexican Valuation of Grades in Pesos	Mexican Stamp Tax in Per Cent. on Valuation	Mexican Stamp Tax Per Metric Ton in Pesos	In U.S. Cents per Bbl. of 42 U.S. Gal.
1917	May-June	Fuel oil	0.91	0.911	23.68	8.50	10	0.85 (x)	
		Gas oil	0.91			8.50	10	0.85 (x)	
		Crude	0.97	0.971	14.18	5.00	10	0.50 (x)	
		Crude	0.91	0.911	23.68	11.00	3	1.10 (x)	
		Refined gasoline					6	0.005 (o)	
		Crude gasoline				0.166	3	0.010 (o)	
		Refined kerosene					6	0.0025 (o)	
		Crude kerosene				0.0840		0.0050 (o)	
		Lubricants						0.0025 (o)	
		Asphalt						1.50 (x)	
		Gas					5 ad val.		
		Crude	0.9323	0.9333	20	10.554	10	1.0554	7.8135
		Crude	0.9849	0.9859	12	5.00	10	0.50	3.9105
		Fuel oil	0.9579	0.9589	16	7.542	6	0.7542	5.7370
		Crude gasoline	0.7519	0.7527	56	0.166		13.246	79.0935
	July-Aug.	Fuel oil	0.91	0.911		8.50	10	0.85	
		Gas oil	0.91	0.911		8.50	10	0.85	
		Crude	0.91	0.971		11.00	10	1.10	
		Crude	0.97			5.00	10	0.50	
		Gasoline, bulk				0.11	6		
		Gasoline, cans				0.29	3	0.0066	
		Kerosene, bulk				0.12	6		
		Kerosene, bbl.				0.0554	6		
		Kerosene, cans				0.675	3	0.00332	
		Crude	0.9323	0.9333	20	10.554	10	1.0554	7.8135
		Crude	0.9849	0.9859	12	5.00	10	0.50	3.9105
		Fuel oil	0.9579	0.9589	16	7.542	6	0.7542	5.7370
		Crude gasoline	0.7519	0.7527	56	0.11		8.7777	52.4125

Year	Period	Product	Density	Bbl/ton	°Bé				
1917	Sept.-Oct.	Fuel oil	0.91			10.00	10	1.00	
		Crude	0.91			13.00	10	1.30	
		Crude	0.97			10.00	10	0.55	
		Gas oil	0.91			10.00		1.00	
		Refined gasoline				0.12	3	0.00675	
		Crude gasoline				0.11¼	6	0.00180	
		Kerosene, crude or refined				0.03			
		Crude	0.9323	0.9333	20	12.554	10	1.2554	9.2945
		Crude	0.9849	0.9859	12	5.50	10	0.55	4.3015
		Fuel oil	0.9579	0.9589	16	9.042		0.9042	6.8780
		Crude gasoline	0.7519	0.7527	56	0.11¼	6	8.9772	53.6040
	Nov.-Dec.	Fuel oil	0.91			10.50	10	1.05	
		Crude	0.91			13.50	10	1.35	
		Crude	0.97			5.50	10	0.55	
		Gas oil	0.91			10.50		1.05	
		Refined gasoline				0.12	3	0.00675	
		Crude gasoline				0.11¼	6	0.00180	
		Kerosene, crude or refined				0.03			
		Crude	0.9323	0.9333	20	13.054	10	1.3054	9.6650
		Crude	0.9849	0.9859	12	5.50	10	0.55	4.3015
		Fuel oil	0.9579	0.9589	16	9.542		0.9542	7.2585
		Crude gasoline	0.7519	0.7527	56	0.11¼	6	8.9772	53.6040
1918	Jan.-Feb.	Fuel oil	0.91			10.50	10	1.05	
		Crude	0.91			13.50	10	1.35	
		Crude	0.97			5.50	10	0.55	
		Gas oil	0.91			10.50		1.05	
		Refined gasoline				0.12¼	3	0.00705	
		Crude gasoline				0.11¼	6	0.0018	
		Kerosene, crude or refined				0.03			
		Crude	0.9323	0.9333	20	13.054	10	1.3054	9.6650
		Crude	0.9849	0.9859	12	5.50	10	0.55	4.3015
		Fuel oil	0.9579	0.9589	16	9.542		0.9542	7.2585
		Crude gasoline	0.7519	0.7527	56	0.11¼	6	9.3762	55.9865
	Mar.-June	Fuel oil	0.91			10.50	10	1.05	
		Crude	0.91			13.50	10	1.35	
		Crude	0.97			5.50	10	0.55	
		Gas oil	0.91			10.50		1.05	
		Refined gasoline				0.1250	3	0.00705	
		Crude gasoline				0.1175	6	0.0018	
		Kerosene, crude or refined				0.03			
		Crude	0.9323	0.9333	20	13.054	10	1.3054	9.6650
		Crude	0.9849	0.9859	12	5.50	10	0.55	4.3015
		Fuel oil	0.9579	0.9589	16	9.542		0.9542	7.2585
		Crude gasoline	0.7519	0.7527	56	0.1175	6	9.3762	55.9865

Barrels per metric ton
12° Bé.= 6.3929

TABLE 10.—(Continued)

Base of Taxes.—For oils having a density greater than 0.91 sp. gr., decrease valuation 0.20 pesos for each increase in sp. gr. of 0.01 up to 0.97, above which point no credit can be taken.
For each increase of 0.01 in sp. gr. below 0.91, increase valuation 0.40 pesos.
"z" = Pesos per metric ton.
"o" = Pesos per liter.

Year	Month	Grade	Mexican Standard at 08°/39.2° F. sp. gr. 20°/4° C. sp. gr.	U.S. Standard at 60°/60° F.—15.56/15.56° C. Sp. Gr.	U.S. Standard Degrees, Baumé	Mexican Valuation of Grades in Pesos	Mexican Stamp Tax in Per Cent. on Valuation	Mexican Stamp Tax Per Metric Ton in Pesos	Mexican Stamp Tax In U.S. Cents per Bbl. of 42 U.S. Gal.
1918	July–Dec.	Fuel oil	0.91			13.00	10		
		Crude	0.91			15.00	10		
		Crude	0.97			6.00	10		
		Gas oil	0.91			13.00	10		
		Refined gasoline and refined kerosene							
		Crude gasoline				0.1250	3	0.00705	
		Kerosene, crude or refined				0.1175	6	0.00240	
						0.04			
		Crude	0.9323	0.9333	20	15.054	10	1.5064	11.1455
		Crude	0.9849	0.9850	12	6.00	10	0.60	4.6925
		Fuel oil	0.9570	0.9589	16	12.042	10	1.2042	9.1605
		Crude gasoline	0.7519	0.7527	56	0.1175	6	9.8762	55.9865
1919	Jan.–Feb.	Fuel oil	0.91			13.00	10		
		Crude	0.91			15.00	10		
		Crude	0.97			6.00	10		
		Gas oil	0.91			13.00	10		
		Refined gasoline and refined kerosene							
		Crude gasoline				0.1250	3	0.00705	
		Kerosene, crude or refined				0.1175	6	0.00240	
						0.04			
		Crude	0.9323	0.9333	20	15.054	10	1.5064	11.1455
		Crude	0.9849	0.9850	12	6.00	10	0.60	4.6925
		Fuel oil	0.9570	0.9589	16	12.042	10	1.2042	9.1605
		Crude gasoline	0.7519	0.7527	56	0.1175	6	9.8762	55.9865

1910								
Mar.–Apr.	Fuel oil		0.91			13.00		
	Crude		0.91			15.50		
	Crude		0.97			6.00		
	Gas oil		0.91			18.00		
	Refined gasoline					0.1250	100	
	Crude gasoline					0.06	100	
	Kerosene, crude or refined			0.9333 0.9859 0.9579 0.7527			6	0.00705 0.00360
	Crude		0.9333		20	15.054	100	1.5054	11.1455
	Crude		0.9849		12	6.00	100	6.00	4.6925
	Fuel oil		0.9579		16	12.042	8	1.2042	9.1605
	Crude gasoline		0.7519		55	0.1175	6	9.3762	55.9865
May–June	Fuel oil		0.91			18.00	100		
	Crude		0.91			15.50	100		
	Crude		0.97			6.00	100		
	Gas oil		0.91			18.00	8		
	Refined gasoline					0.1250	6		
	Crude gasoline					0.1175		0.00705	
	Kerosene, crude or refined			0.9333 0.9859 0.9589 0.7527		0.0060		0.0089	
	Crude		0.9333		20	15.054	100	1.5054	11.1455
	Crude		0.9849		12	6.00	100	6.00	4.6925
	Fuel oil		0.9579		16	12.042	8	1.2042	9.1605
	Crude gasoline		0.7519		55	0.1175	6	9.3762	55.9865
July–Aug.	Fuel oil		0.91			18.00	100		
	Crude		0.91			15.00	100		
	Crude		0.97			6.00	100		
	Gas oil		0.91			18.00	8		
	Refined gasoline					0.1250	6		
	Crude gasoline					0.0650		0.00705	
	Kerosene, crude or refined			0.9333 0.9859 0.9589 0.7527				0.0089	
	Crude		0.9333		20	14.554	100	1.4554	10.7755
	Crude		0.9849		12	6.00	100	0.00	4.6925
	Fuel oil		0.9579		16	12.042	8	1.2042	9.1605
	Crude gasoline		0.7519		55	0.1175	6	9.3762	55.9865
Sept.–Oct.	Fuel oil		0.91			18.00	100		
	Crude		0.91			15.00	100		
	Crude		0.97			6.00	3		
	Gas oil		0.91			18.00	8		
	Refined gasoline					0.1175	6		
	Crude gasoline					0.0705		0.00705	
	Kerosene, crude or refined			0.9333 0.9859 0.9589 0.7527				0.00450	
	Crude		0.9333		20	14.554	100	1.4554	10.7755
	Crude		0.9849		12	6.00	100	0.50	4.6925
	Fuel oil		0.9579		16	12.042	8	1.2042	9.1605
	Crude gasoline		0.7519		55	0.1175	6	9.3762	55.9865

Barrels per metric ton
12° Bé.– 6.3929
16° Bé.– 6.5729
20° Bé.– 6.7333
55° Bé.– 8.3736cr
1329.96 Liters

TABLE 10.—(Continued)

Base of Taxes.—For oils having a density greater than 0.91 sp. gr. decrease valuation 0.20 pesos for each increase in sp. gr. of 0.01 up to 0.97, above which point no credit can be taken.

For each increase of 0.01 in sp. gr. below 0.91, increase valuation 0.40 pesos.

"x" = Pesos per metric ton.
"o" = Pesos per liter.

Year	Month	Grade	Mexican Standard at 68°/39.2° F. sp. gr. 20°/4° C. sp. gr.	U.S. Standard at 60°/60° F.—15.56/15.56° C. Sp. Gr.	U.S. Standard Degrees, Baumé	Mexican Valuation of Grades in Pesos	Mexican Stamp Tax in Per Cent. on Valuation	Mexican Stamp Tax Per Metric Ton in Pesos	In U.S. Cents Per Bbl. of 42 U.S. Gal.
1919	Nov.–Dec.	Fuel oil	0.91			13.00	10		
		Crude	0.91			15.50	10		
1920	Jan.–Feb.	Crude	0.97			6.00	10		
		Gas oil	0.91			13.00	10		
		Refined gasoline				0.1250	3		
		Crude gasoline				0.1175	6		
		Kerosene, crude or refined				0.0750	6		
		Crude	0.9323	0.9333	20	15.054	10	1.5054	11.1455
		Crude	0.9859	0.9859	12	6.00	10	0.60	4.6925
		Fuel oil	0.9579	0.9589	16	12.042	10	1.2042	9.1605
		Crude gasoline	0.7519	0.7527	56	0.11¾	6	9.3762	55.9865
	Mar.–Apr.	Fuel oil	0.91			16.62	10		
		Fuel oil, heavier than	0.97			13.00	10		
		Crude	0.91			21.67	10		
		Crude	0.965			13.00	10		
		Gas oil	0.91			16.62	10		
		Refined gasoline				0.1500	3		
		Crude gasoline				0.1425	6		
		Kerosene, crude or refined				0.0950	6		
		Crude	0.9323	0.9333	20	21.224	10	2.1224	15.7135
		Crude	0.9859	0.9859	12	13.00	10	1.300	10.1675
		Fuel oil	0.9579	0.9589	16	15.662	10	1.566	11.9125
		Crude gasoline	0.7519	0.7527	56	0.1425	6	11.3712	67.8990

1920	Product								
May–June	Fuel oil.	0.91			16.62	10		
	Fuel oil, heavier than	0.97			13.00	10		
	Crude.	0.91			21.67	10		
	Crude, heavier than	0.965			13.00	10		
	Gas oil.	0.91			16.62			
	Refined gasoline				0.1600(o)	3		
	Crude gasoline				0.1525(o)	6		
	Kerosene, crude or refined..				0.0950(o)			
	Crude.	0.9323	0.9333	20	21.224	10	2.1224	15.7135
	Crude.	0.9849	0.9859	12	13.00	10	1.30	10.1675
	Fuel oil.	0.9579	0.9589	16	15.662	6	1.566	11.9125
	Crude gasoline.	0.7519	0.7527	56	0.1525		12.1691	72.6635
July–Aug.	Fuel oil.	0.91			17.00	10		
	Fuel oil, heavier than	0.97			13.00	10		
	Crude.	0.91			23.00	10		
	Crude, heavier than	0.965			13.00	10		
	Gas oil.	0.91			25.00			
	Refined gasoline				0.1600(o)	3		
	Crude gasoline				0.1525(o)	6		
	Kerosene, crude or refined..				0.0950(o)			
	Crude.	0.9323	0.9333	20	22.554	10	2.2554	16.6985
	Crude.	0.9849	0.9859	12	13.000	10	1.3000	10.1675
	Fuel oil.	0.9579	0.9589	16	16.042	6	1.604	12.201
	Crude gasoline.	0.7519	0.7527	56	0.1525		12.1691	72.6635
Sept.–Oct.	Fuel oil.	0.91			18.00	10		
	Fuel oil, heavier than	0.97			11.00	10		
	Crude.	0.91			25.00	10		
	Crude, heavier than	0.965			11.00	10		
	Gas oil.	0.91			35.00			
	Refined gasoline				0.1500(o)	3		
	Crude gasoline				0.1425(o)	6		
	Kerosene, crude or refined..				0.0800(o)			
Nov.–Dec.	Crude.	0.9323	0.9333	20	24.554	10	2.4554	18.179
	Crude.	0.9849	0.9859	12	11.000	10	1.1000	8.603
	Fuel oil.	0.9579	0.9589	16	17.042	6	1.704	12.9623
	Crude gasoline.	0.7519	0.7527	56	0.1425		11.3712	67.899

Barrels per metric ton
12° Bé.— 6.3929
16° Bé.— 6.5729
20° Bé.— 6.7533
56° Bé.— 8.3736 or
1329.96 liters

CONCLUSIONS AND RECOMMENDATIONS

The foregoing discussion clearly indicates: (1) That the rate of Mexican taxation on oil exported and, in fact, the aggregate of all Mexican taxes affecting the oil industry, far from being burdensome in nature as some operators contend, have been and are, if anything, reasonably low. (2) Mexican oils exported prior to May, 1917, were taxed only about 4 U. S. cents per barrel, and in several cases the exporting companies were and are exempt from paying the customary import duties on machinery and other supplies thus further benefiting from their Mexican operations. (3) That the export stamp taxes levied under the decree of April, 1917, are in proportion, lower in August, 1920, than in April, 1917, when the law was put into effect. (4) That the basing of the tax on the price of Mexican oils in Mexico, as the decree provides, has given rise to endless arguments and dissensions. (5) That the tax as applied creates difficulties between the seller and purchaser of Mexican oils, which can and should be eliminated. (6) That it would be advantageous to apply Mexican standards for measuring oils, within the metric system, on the volume basis, rather than on the weight basis, inasmuch as all Mexican oil is sold by volume.

Keeping clearly in mind the rights of the Mexican Government as well as the just claims of the operators, and realizing that many of the difficulties can be overcome by the establishment of some stable "benchmark" directly related to market conditions, on which to base the value of Mexican oils and therefore the taxes on their products, the writer offers the following recommendations:

1. That the Mexican tax on each grade of exported oils be based on percentages of the average American oil-coal price, as defined in this report.

2. That these percentage relations between the tax on any one grade of oil and the oil-coal price should remain practically constant unless new conditions should develop to make a change imperative.

3. That, if possible, monthly variations of the average oil-coal price be taken into account.

4. That the tax be applied on the volume (cubic meter) rather than on the weight (metric ton) of the oil exported (the average oil-coal price in dollars per barrel converted to pesos per cubic meter is shown on Table 12).

5. That the law of April, 1917, be abrogated and a new law enacted covering, in a general way, the main points herein advanced.

TABLE 11.—*Total Mexican Export Taxes in U. S. Cents Per Barrel of 42 U. S. Gal.*

Year and Month	Export Stamp Tax	Infalsificable = 10 Per Cent. of Export Stamp Tax	Bar Dues = 10 Centavos per Metric Ton	Grand Total
Light Crude 20° Bé.				
May, 1917	7.813	0.781	0.740	9.335
September, 1917	9.294	0.929	0.740	10.964
November, 1917	9.665	0.966	0.740	11.371
July, 1918	11.145	1.114	0.740	13.000
July, 1919	10.775	1.077	0.740	12.593
November, 1919	11.145	1.114	0.740	13.000
March, 1920	15.713	1.571	0.740	18.025
July, 1920	16.698	1.669	0.740	19.108
September, 1920	18.179	1.817	0.740	20.737
December, 1920	18.179	1.817	0.740	20.737
Fuel Oil 16° Bé.				
May, 1917	5.737	0.573	0.760	7.071
September, 1917	6.878	0.687	0.760	8.326
November, 1917	7.258	0.725	0.760	8.745
July, 1918	9.160	0.916	0.760	10.837
March, 1920	11.913	1.191	0.760	13.864
July, 1920	12.201	1.220	0.760	14.181
September, 1920	12.962	1.296	0.760	15.018
December, 1920	12.962	1.296	0.760	15.018
Heavy Crude 12° Bé.				
May, 1917	3.910	0.391	0.782	5.083
September, 1917	4.301	0.430	0.782	5.513
July, 1918	4.692	0.469	0.782	5.943
March, 1920	10.167	1.016	0.782	11.966
September, 1920	8.603	0.860	0.782	10.245
December, 1920	8.603	0.860	0.782	10.245
Crude Gasoline 56° Bé.				
May, 1917	52.412	5.241	0.597	58.250
September, 1917	53.604	5.360	0.597	59.561
January, 1918	55.986	5.598	0.597	62.182
March, 1920	67.899	6.789	0.597	75.286
May, 1920	72.663	7.266	0.597	80.526
September, 1920	67.899	6.789	0.597	75.286
December, 1920	67.899	6.789	0.597	75.286

TABLE 12.—*Relation Between American Fuel Prices and Mexican Export Stamp Taxes on Petroleum and its Products*

NOTE. One Cubic Meter = 6.2899 bbl.   One U. S. dollar = Two Mexican pesos

Month and Year	Dollars per Barrel					Pesos per Cubic Meter				
	Mid-Conti-nent Crude	Calif-ornia Crude	Bitumin-ous Coal Export Price (3.5 bbl. per ton)	Gulf Coast Crude	Average Price	Type Price	Present Stamp Tax			
							20° Bé. 0.9323	12° Bé. 0.9849	56° Bé. 0.7519	16° Bé. 0.9579
1917										
May................	1.70	0.82	0.991	1.00	1.128	14.190	0.983	.492	.593	.722
June................	1.70	0.92	1.011	1.00	1.158	14.567	0.983	.492	.593	.722
July................	1.70	1.02	1.091	1.00	1.203	15.133	0.983	.492	.593	.722
August.............	1.85	1.02	1.140	1.00	1.253	15.762	0.983	0.492	.593	0.722
September...........	2.00	1.02	1.100	1.00	1.280	16.102	1.169	.541	.743	.865
October............	2.00	1.02	1.120	1.00	1.285	16.165	1.169	.541	.743	.865
November..........	2.00	1.02	1.360	1.00	1.345	16.920	1.216	.541	.743	.913
December...........	2.00	1.02	1.017	1.00	1.259	15.838	1.216	.541	8.743	.913
1918										
January............	2.00	1.02	1.086	1.00	1.277	16.064	1.216	0.541	7.043	0.913
February...........	2.00	1.02	1.165	1.00	1.296	16.303	1.216	0.541	7.043	0.913
March..............	2.25	1.02	1.147	1.35	1.442	18.140	1.216	0.541	7.043	0.913
April...............	2.25	1.02	1.136	1.35	1.439	18.102	1.216	0.541	7.043	0.913
May................	2.25	1.27	1.046	1.35	1.479	18.606	1.216	0.541	7.043	0.913
June................	2.25	1.27	1.105	1.35	1.494	18.794	1.216	0.541	7.043	0.913
July................	2.25	1.29	1.142	1.35	1.508	18.970	1.402	0.590	7.043	1.152
August.............	2.25	1.29	1.122	1.35	1.503	18.907	1.402	0.590	7.043	1.152
September..........	2.25	1.29	1.148	1.80	1.622	20.404	1.402	0.590	7.043	1.152
October............	2.25	1.29	1.185	1.80	1.631	20.518	1.402	0.590	7.043	1.152
November..........	2.25	1.29	1.142	1.80	1.621	20.392	1.402	0.590	7.043	1.152
December...........	2.25	1.29	1.194	1.80	1.634	20.555	1.402	0.590	7.043	1.152
1919										
January............	2.25	1.29	1.336	1.50	1.594	20.052	1.402	0.590	7.043	1.152
February...........	2.25	1.29	1.250	1.25	1.510	18.995	1.402	0.590	7.043	1.152
March..............	2.25	1.29	1.428	1.25	1.555	19.562	1.402	0.590	7.043	1.152
April...............	2.25	1.29	1.371	1.00	1.478	18.593	1.402	0.590	7.043	1.152
May................	2.25	1.29	1.200	1.00	1.435	18.052	1.402	0.590	7.043	1.152
June................	2.25	1.29	1.250	1.00	1.448	18.216	1.402	0.590	7.043	1.152
July................	2.25	1.29	1.228	1.00	1.442	18.140	1.355	0.590	7.043	1.152
August.............	2.25	1.29	1.320	1.00	1.465	18.429	1.355	0.590	7.043	1.152
September..........	2.25	1.29	1.400	1.00	1.485	18.681	1.355	0.590	7.043	1.152
October............	2.25	1.29	1.438	1.00	1.495	18.807	1.355	0.590	7.043	1.152
November..........	2.33	1.29	1.465	1.00	1.521	19.134	1.402	0.590	7.043	1.152
December...........	2.50	1.29	1.380	1.25	1.605	20.191	1.402	0.590	7.043	1.152
1920										
January............	2.97	1.29	1.600	1.75	1.903	23.939	1.402	0.590	7.043	1.152
February...........	3.00	1.33	1.560	2.00	1.973	24.820	1.402	0.590	7.043	1.152
March..............	3.50	1.33	1.615	2.50	2.236	28.128	1.977	1.279	8.542	1.499
April...............	3.50	1.58	1.815	3.00	2.349	29.550	1.977	1.279	8.542	1.499
May................	3.50	1.58	2.028	3.00	2.527	31.789	1.977	1.279	9.141	1.499
June................	3.50	1.58	2.280	3.00	2.590	32.582	1.977	1.279	9.141	1.499
July................	3.50	1.70	2.660	3.00	2.715	34.154	2.101	1.279	9.141	1.535
August.............	3.50	1.70	2.957	3.00	2.789	35.085	2.101	1.279	9.141	1.535
September..........	3.50	1.70	2.957	3.00	2.789	35.085	2.287	1.082	8.542	1.631
October............	3.50	1.70	3.050	3.00	2.812	35.375	2.287	1.082	8.542	1.631
November..........	3.50	1.70	2.870	3.00	2.77	34.846	2.287	1.082	8.542	1.631
December...........	3.50	1.70	2.415	2.50	2.529	31.814	2.287	1.082	8.542	1.631

TABLE 13.—*Mexican Export Stamp Tax Showing Percentage Relation to Average Oil and Coal Price from May, 1917 (in U. S. cents per Barrel)*

Year and Month	Average Oil and Coal Price	Light Crude 20° Bé. (0.9323)		Fuel Oil 16° Bé (0.9579)		Heavy Crude 12° Bé. (0.9849)		Crude Gasoline 56° Bé. (0.7519)	
		Tax	Per Cent.	Tax	Per Cent.	Tax	Per Cent.	Tax	Per Cent.
1917									
May	$1.128	$0.078	6.92	$0.057	5.08	$0.039	3.46	$0.524	46.45
June	1.158	0.078	6.73	0.057	4.95	0.039	3.38	0.524	45.26
July	1.203	0.078	6.50	0.057	4.77	0.039	3.25	0.524	43.57
August	1.253	0.078	6.24	0.057	4.58	0.039	3.12	0.524	41.82
September	1.280	0.093	7.26	0.069	5.37	0.043	3.36	0.536	41.88
October	1.285	0.093	7.23	0.069	5.35	0.043	3.35	0.536	41.71
November	1.345	0.097	7.19	0.073	5.40	0.043	3.20	0.536	39.85
December	1.259	0.097	7.68	0.073	5.77	0.043	3.42	0.536	42.58
1918									
January	1.277	0.097	7.56	0.073	5.68	0.043	3.36	0.560	43.84
February	1.296	0.097	7.46	0.073	5.60	0.043	3.32	0.560	42.53
March	1.442	0.097	6.70	0.073	5.03	0.043	2.97	0.560	38.82
April	1.439	0.097	6.72	0.073	5.04	0.043	2.99	0.560	38.91
May	1.479	0.097	6.53	0.073	6.53	0.043	2.91	0.560	37.85
June	1.494	0.097	6.47	0.073	4.86	0.043	2.88	0.560	37.47
July	1.508	0.112	7.39	0.092	6.07	0.047	3.11	0.550	37.13
August	1.503	0.112	7.42	0.092	6.09	0.047	3.12	0.560	37.25
September	1.622	0.112	6.87	0.092	5.64	0.047	2.89	0.560	34.52
October	1.631	0.112	6.83	0.092	5.62	0.047	2.88	0.560	34.32
November	1.621	0.112	6.88	0.092	5.65	0.047	2.88	0.560	34.54
December	1.634	0.112	6.82	0.092	5.61	0.047	2.87	0.560	34.26
1919									
January	1.594	0.112	6.99	0.092	5.75	0.047	2.94	0.560	35.12
February	1.510	0.112	7.38	0.092	6.07	0.047	3.11	0.560	37.08
March	1.555	0.112	7.17	0.092	5.89	0.047	3.02	0.560	36.00
April	1.478	0.112	7.54	0.092	6.20	0.047	3.18	0.560	37.87
May	1.435	0.112	7.77	0.092	6.39	0.047	3.27	0.560	39.01
June	1.448	0.112	7.70	0.092	6.33	0.047	3.24	0.560	38.66
July	1.442	0.108	7.47	0.092	6.35	0.047	3.25	0.560	38.76
August	1.465	0.108	7.36	0.092	6.25	0.047	3.20	0.560	38.22
September	1.485	0.108	7.26	0.092	6.17	0.047	3.16	0.560	37.70
October	1.495	0.108	7.21	0.092	6.13	0.047	3.14	0.560	37.44
November	1.521	0.112	6.94	0.092	6.02	0.047	3.09	0.560	36.80
December	1.605	0.112	6.94	0.092	5.70	0.047	2.92	0.560	34.88
1920									
January	1.903	0.112	5.86	0.092	4.81	0.047	2.47	0.560	29.42
February	1.973	0.112	5.65	0.092	4.64	0.047	2.38	0.560	28.36
March	2.236	0.157	7.03	0.119	5.33	0.102	4.55	0.679	30.36
April	2.349	0.157	6.69	0.119	5.07	0.102	4.33	0.679	28.90
May	2.527	0.157	6.22	0.119	4.72	0.102	4.02	0.727	28.74
June	2.590	0.157	6.06	0.119	4.60	0.102	3.92	0.727	28.06
July	2.715	0.167	6.15	0.122	4.50	0.102	3.74	0.727	26.76
August	2.790	0.167	5.98	0.122	4.38	0.102	3.64	0.727	26.04
September	2.790	0.182	6.53	0.130	4.66	0.086	3.08	0.679	24.35
October	2.812	0.182	6.47	0.130	4.63	0.086	3.06	0.679	24.30
November	2.77	0.182	6.57	0.130	4.70	0.86	3.10	0.679	24.50
December	2.529	0.182	7.2	0.130	5.14	0.86	3.40	0.679	26.85

Efficiency in Use of Oil as Fuel

By W. N. Best, D. Sc., New York, N. Y.

(St. Louis Meeting, September, 1920)

This paper is not intended as a scientific discussion of the combustion of oil but is written from the standpoint of an operator who has the experience and qualifications necessary to guide others in producing the most economical results in the use of liquid fuels. Oil, in this paper, usually means petroleum or its products but incidental reference is made to other liquid and gaseous fuels, so that the term may be considered as referring to all liquid and gaseous hydrocarbons in comparison with solid fuels, as coal and wood. However, only a few of the principal factors in the use of oil as a fuel can be given.

The present, and prospective, high price of coal is causing users of fuel to renew inquiry as to the merits of other forms of fuel for industrial purposes. Crude oil (petroleum) is proving to be one of the world's most valuable mineral resources. The recent discovery that oil underlies a considerable area of the United States, Mexico, and other parts of the world to a greater extent than was formerly believed and the large production of some of the wells in these areas shows the probable quantity of fuel oil that may now be available. Through the energy of Lord Cowdrey, who was one of the pioneers of the oil industry in Mexico, oil has been discovered in England; some prominent geologists believe that it may be found in quantity in Great Britain.

For years, oil has been known to be of great value in the manufacture of metals. It has proved incomparable in forge shops, steel foundries, heat-treating furnaces, and wherever accuracy of temperatures is essential, or where a maximum output is desired as well as quality of metal. In some types of equipment, the output produced with oil as fuel is double that obtained with coal and at a reduction of 50 per cent. in the cost of the fuel. For example, in drop-forging plants, the metal is always waiting for the man when oil is used as fuel, whereas with, coal, the man must wait for the metal to become sufficiently heated.

It has only been since January, 1919, that the oil supply could be relied on for boiler service, owing to the war conditions and the inability to get oil tankers for the delivery of the oil from Mexico to Atlantic ports; but now a constant supply is assured, and many manufacturers are installing it in their power plants. The cost varies with the size of the plant. In New England and along the Atlantic coast, where the boiler horsepower is large, this fuel is very attractive, for one man can fire and water-tend twelve 300-hp. boilers. It raises the general condition of the man firing the boilers, because the burning of oil is an art and necessitates brain rather than brawn. This fuel responds immediately to the will of the operator in meeting peak or fluctuating loads. The fire room is clean and sanitary, dust from coal and ashes being eliminated. There is practically no loss in fuel, as only a small part of the oil in the storage tank is heated, and that just enough for it to be pumped readily from the storage tank to the supply tank. The handling of the fuel is inexpensive; and it is speedily delivered from the oil tank or tanker. There are, however, certain fundamental principles that must always be observed in making crude-oil installations.

TEMPERATURE OF FUEL

The temperature of the fuel and the method of supply are especially vital points. Oil below 20° Bé should be heated to just below its vaporizing point; steam should always be used for this purpose, as it gives a very accurate temperature; the supply is usually obtained from the exhaust of the pump. Numerous efforts have been made to heat the oil, while passing through the pipes, by electric currents and by heat from coke, gas, and oil fires; these methods have always proved inferior to steam.

Thermometers should always be used, for the manufacturer who heats his fuel accurately and uniformly every day is the one who obtains the greatest efficiency from the fuel burned.

SUPPLY LINES

Supply lines should be so laid as to insure the constant circulation of fuel through all the oil-supply pipes from the pump to the burners. A pressure relief valve should always be placed at the farther end of the burner installation, and the overflow pipe should always return the unused fuel to the supply tank. This is imperative especially when using heavy oils, as they must be heated to reduce their viscosity. Many people have put in a large oil main and run laterals from the main to the boilers or furnaces. Then when a boiler must be washed out or a furnace is shut down for repairs, the oil solidifies in the oil pipe or, if it does not solidify, the residuum from the oil collects in these pipes, causing annoyance and unnecessary trouble. The locating of the oil-storage tank and the laying out of the pipe lines are engineering feats, just as much as the equipment of the boilers or furnaces.

The oil pumps should be brass lined. Two should always be provided, one being held in reserve for use in cases of any emergency. Air chambers and pressure gages should be used; the former to reduce the pulsations caused by the displacement of the oil by the piston and the latter to record the oil pressure maintained upon the oil-supply line. The spring of the pressure-relief valve should be very sensitive, in order that it may release quickly without causing a variation of more than ¼ lb. (0.14 kg.) pressure on the oil supply to the burners. Oil meters should be used whenever possible; the foreman of a boiler plant or furnace department provided with these instruments is encouraged to see that the strictest economy in fuel is maintained.

Types of Burners and Their Use

Numerous oil burners are on the market but the three types most common are: The external atomizing type, which is largely used in loco-motive and stationary boilers and in large furnaces; the internal atomizing type, which is chiefly used on small furnaces; and the mechanical type of burner used on ocean-going vessels, which forces the oil at high pressure through a small aperture, thus making a funnel-shaped flame. This type of burner is used on ocean vessels because no steam is required for atomizing, consequently there is no loss of water. This saving in water, however, is accompanied by loss in fuel, for more oil is required to replace a ton of coal while using a mechanical burner than with the exter-nal atomizing burner, because a mechanical burner cannot atomize the fuel. For example, 180 gal. (681 l.) of oil is the equivalent of a long ton of coal (calorific value, 14,000 B.t.u. per lb.) when using a mechanical burner; while with the use of an atomizing burner only 147 gal. of oil is the equivalent.

When purchasing atomizing burners, several points should always be considered.

1. The burner must not carbonize. A burner that carbonizes should be scrapped at once, as it is not dependable, is wasteful of oil, and requires a great deal of care and attention. Such a burner reduces the burning of oil from a science to a continuous hazard and care.

2. The oil and steam orifices should be independent of each other so that excessive oil pressure is not required and so that no cutting effect is produced when burning oil containing residuum or sand.

3. The burner should be so constructed and filed that it will pro-duce a flame of sufficient length and width to fill the combustion chamber of the furnace or firebox of boiler; in fact, just as perfectly as a drawer fits into its opening in a desk.

4. The oil orifice should be large enough to permit free exit of heavy oils and tars therefrom, and the atomizer opening should be as small as possible in order to reduce to a minimum the amount of steam or com-pressed air used for the atomization of the fuel.

For boiler equipments, steam is preferable as an atomizing agent if 20 lb. (9 kg.) pressure or more is carried upon the boiler; but for smaller pressures air should be used. For furnace equipments, air is preferable to steam for atomizing purposes as it reduces to a minimum the amount of moisture in the furnace.

Today boiler settings are demanded that give ample room for combustion. Boilers for 300 degrees overload are being set with a distance of 14 ft. (4.3 m.) from the coal stokers to the elements of the boiler. When burning oil, the larger the combustion chamber (up to a certain limit), the greater is the efficiency obtained from the fuel and the higher is the boiler horsepower rating obtained. Recording CO_2 instruments should be used in order to gage the air supply accurately and prevent loss of fuel through excessive air supply. For furnace equipments, pyrometers are essential.

Mexican oil is high in sulfur, often containing as much as 3.8 per cent. It is therefore necessary that a combustion chamber be used on furnaces so that the atomized oil may be consumed before it reaches the furnace proper. In boilers there is no difficulty because of sulfur, no matter of what material the stock is made. The question is often asked, "Do steel stacks deteriorate from the use of oil containing as high a percentage of sulfur as Mexican oil?" There will be no deterioration unless the stack temperature reaches 850° F. In ordinary boiler practice, there is, therefore, no likelihood of any detrimental effect because the stack temperatures do not reach so high a degree. Many people condemn the use of this fuel in furnaces because their furnaces do not have combustion chambers to consume the sulfur; when this is consumed in the furnace, there is a detrimental effect upon the metal and the odor in the shop causes the men to complain. In open-hearth furnace work, it has been found good practice to use the lighter oils until the charge is brought down and is covered with slag, after which the Mexican oil can be used with no detrimental effect upon the metal.

Cost of Operating with Oil and Coal

Many engineers and manufacturers take the calorific value of the oil and the calorific value of the coal as bases from which to estimate the difference in cost of operating with these two fuels. This should not be done as the figures thereby obtained are incorrect.

In flue-welding furnaces, 58 gal. of oil is the equivalent of a long ton of coal (2240 lb.) due to the fact that in welding with coal, for safe-ending the flue, it is necessary to coke the fire; this not only means a loss of time but also a loss of the volatile hydrogen and hydrocarbon gases, much of the calorific value of the fuel. These gases are utilized in boiler practice; here the economy effected depends largely on the size of the plant, for one man can fire and water-tend a battery of twelve oil-fired boilers almost as easily as he can care for one boiler. With proper equipment, the tonnage of a locomotive is increased 15 per cent. when changed from coal to oil.

The equivalent of one long ton of coal, in the average locomotive service, is 180 gal. oil; in the average stationary boiler practice, 147 gal.; in forging furnaces, 80 gal; in heat-treating furnaces, with low temperatures, 80 gal.; and in heat-treating furnaces with high temperatures and annealing furnaces, 63 gal. In working these figures, it must be noted that, in each instance quoted, the oil has a calorific value of 19,000 B.t.u. per lb. and weighs 7½ lb. per gal. while the coal averages 14,200 B.t.u. per lb. and weighs 2240 lb. per ton.

3¼ bbl. oil (42 gal. per bbl.) is the equivalent of 5000 lb. hickory or 4550 lb. white oak.

6 gal. oil equals 1000 cu. ft. of natural gas of calorific value of 1000 B.t.u. per cu. ft.

3½ gal. oil equals 1000 cu. ft. of commercial or water gas of calorific value of 620 B.t.u. per cu. ft.

2¼ gal. oil equals 1000 cu. ft. byproduct coke-oven gas at 440 B.t.u. per cu. ft.

0.42 gal. oil equals 1000 cu. ft. blast-furnace gas at 90 B.t.u. per cu. ft.

Steel works are now utilizing their blast-furnace gases, which are of low calorific value, being on an average but 90 B.t.u. per cu. ft. For this reason, it is customary, when these gases are used in boilers, large furnaces, etc., to use an auxiliary fuel in combination therewith. This auxiliary fuel is usually coal tar (the byproduct of coke ovens); this makes a fine combination. Usually 10 gal. of coal tar are made from every ton of coal coked in byproduct coke ovens; this tar has a calorific value of 162,000 B.t.u. per gal. When this coal tar is not available, crude oil is used.

Efforts have been made in West Virginia lately to retain within its border all the natural gas produced in that state. If those fostering this movement succeed, within a period of two years there will be scarcely any natural gas used in the states of Indiana, Ohio, and Pennsylvania. The small quantity of natural gas produced in these three states will be used for domestic or household purposes, rather than in furnaces, etc. Oil, therefore, is the fuel that will be used as it is particularly adapted for furnaces in which natural gas was originally used.

DISCUSSION

S. O. ANDROS,[*] Chicago, Ill.—The most important thing in the burning of fuel oil is the design of the furnace. Almost any of the good burners on the market will be efficient, if the furnace is properly designed.

---

[*] Editor, *Oil News.*

Without proper furnace design, it is impossible to get efficient operation of the burner. The burner itself is not so much of an item in the domestic field as the assembling of the system for domestic use.

RALPH R. MATTHEWS,* Wood River, Ill.—In 1911, when connected with the Bureau of Mines, I inspected various installations of fuel oil burners in Seattle, Portland, and San Francisco and found that every engineer had his pet type. The results seemed to show that as long as the oil is atomized properly, the manner in which it is atomized is not of great importance, but that the furnace must be properly designed. When the furnace design is not proper, there is overheating and probably excessive stack temperature due to burning a larger quantity of fuel oil than should be necessary. Conservation of fuel oil is thus closely linked with furnace design.

HENRY P. MUELLER,† St. Louis, Mo.—I am connected with a bras foundry that makes 15 tons of metal daily, burning 1000 gal. of oil. In the last ten years we have tried practically every burner on the market, but the most satisfactory was one we designed. The oil is discharged as a spray under 2½-lb. pressure; the air comes out of the 1½-in. opening and breaks up the oil at the end of the burner, throwing it 18 in. before it enters the furnace.

Some of our burners are running under compressed air, which in some cases is more economical than steam. The oil is run into the tanks under air pressure and is left in circulation; as a result, it is unnecessary to clean a burner or pipe.

The furnace is of the revolving type. The flame does not come into contact with the metal; instead it heats the top of the furnace, then as that is revolved under the charge, the metals are melted with a loss of only 1½ per cent.

The cost of melting metal on a normal market is 14 cents; today we are paying 10 cents per gallon for oil in carload lots, therefore the cost of melting is practically doubled.

JOHN L. HENNING, Lake Charles, La.—In burning oil we have not had much difficulty in getting proper atomization. The furnace design is the most important thing. We used a common burner and tried to get a happy medium between good combustion and long lived furnace. We burned Mexican oil straight without any trouble with that type burner. When the price of kerosene was low, we burned it in the same burners, simply letting it run out, and got as good combustion as with crude oil under the best conditions.

---

* Chief Chemist, Roxana Petroleum Corpn.
† President, Mueller Brass Foundry.

ARTHUR KNAPP, Shreveport, La.—In burning oil in small quantities it is necessary to remember that oil must be brought into a condition to ignite; that is, each particle must be vaporized before it will burn. When burning a small quantity, as trying to fire a 10-hp. boiler or smaller, it is necessary to have a large surface that will radiate sufficient heat to vaporize the oil and ignite the vapor. In large furnaces, large radiating surfaces above or to the side of the burner furnish the required heat.

W. N. BEST (author's reply to discussion).—While it is important to have the furnace properly designed, if the oil burner will not function with the furnace design, there will be inefficient combustion; vice versa, if the furnace is improperly designed, the most efficient and most modern type of burner will be a failure.

Mr. Mathews is perfectly correct in his premises of a proper design of furnace, but it is just as essential to have a burner that will not carbonize; that will atomize any gravity of liquid fuel; that does not require excessive oil pressure. It is very important, in burning heavy oil or tars to use low oil pressure. Oil or tar should never be burned under an oil or tar pressure exceeding 12 lb., using an atomizing burner.

In the melting of brass it is absolutely essential to have a properly designed furnace; to have a combustion chamber of adequate proportions to insure the consumption of the atomized fuel and the reduction of it to heat before it reaches the furnace proper; to have a burner that will make a flame to fit the combustion chamber as perfectly as a drawer fits an opening in a desk. Without the combustion chamber the furnace will not function properly, owing to the fact that there will be an excessive amount of unmixed air entering the furnace, which will result in an excessive loss of metal.

INDEX

[NOTE.—In this Index the names of authors of papers are printed in small capitals, and the titles of papers in italics.]

Accumulation, oil, salt domes, Gulf coastal plain, **319.**

Alabama: coal, carbon ratios. 141, 146.

    map, geological, **141.**

    oil horizons, 143.

    oil possibilities, 140.

    stratigraphy, **141.**

Alagôas, Brazil, oil-shale, 71.

ALBERTSON, M.: *Isostatic Adjustments on a Minor Scale, in their Relation to Oil Domes,* 418.

ALVEY, GLENN H. and FOSTER, ALDEN W.: *Barrel-day Values,* 412.

AMBROSE, A. W.: *Analysis of Oil-field Water Problems,* 245.

    Discussions: *on Investigations Concerning Oil-water Emulsion,* 454

        on Value of American Oil-shales, 235.

Amortization: definition, 374.

    oil property investment, 350.

Analysis: coal, 525.

    oil and gas sands, 495.

    water in oil wells, 253.

Analysis of Oil-field Water Problems (AMBROSE). 245; *Discussion:* (CONKLING), 265, 267; (DEGOLYER), 265, 266; (REILLEY), 266; (MILLS), 266, 267.

ANDROS, S. O.: *Discussion on Efficiency in Use of Oil as Fuel,* 572.

Anglo-Persian Oil Co., 9.

Appalachian oil fields, geology, 151.

Application of Law of Equal Expectations to Oil Production in California (BEAL and NOLAN), 335.

Application of Taxation Regulations to Oil and Gas Properties (COX), 374; *Discussion:* (ARNOLD), 393.

Appraisal, oil property, method, 356.

Appraisal of Oil Properties (OLIVER), 353; *Discussion:* (BEAL), 361; (JOHNSON), 363.

Argentina, oil, see *Oil, Argentina.*

ARNOLD, RALPH: *Discussions: on Application of Taxation Regulations to Oil and Gas Properties,* 393.

    on Oil-shales and Petroleum Prospects in Brazil, 76.

    on Petroleum Industry of Trinidad, 67, 68.

    on Variation in Decline Curves of Various Oil Pools, 373.

ASHLEY, G. H.: *Discussions: on A Résumé of Pennsylvania-New York Oil Field,* **154.**

    on Water Displacement in Oil and Gas Sands, 501, 502.

Asphalt, related hydrocarbons, 217.

Asphaltenes and asphaltites, definitions, 217.

575

Pore space: oil and gas sands: determination: methods, 469.
        paraffin method, 470.
        pycnometers, 473.
        results, 478, 480.
        small samples, 474.
        volume of individual grains, 472.
        volume of sample by weighing, 472.
        water absorption method, 477.
    nature, 469.
    relation to productivity, 479.
  water absorption, 477.
Porosity, oil and gas sands, see *Pore space, oil and gas sands.*
PRATT, W. E.: *Discussions: on Industrial Representation in the Standard Oil Co. (N. J.)*, 240.
  on Nature of Coal, 223.
  on Oil-field Brines, 289.
  on Petroleum in the Philippines, 54.
Price, future, oil, 347.
Prices, oil and coal, comparison, 553.
Proceedings, St. Louis meeting, 1920, v.
Production curves, oil wells, 335, 338, 340, 341, 365.
Provinces, petroliferous, 199, 205, 215.
Pycnometer, Johnston and Adams, 473.

Red Beds, gypsum origin, 274.
Refining, oil, method, 512.
REGER, DAVID B.: *Carbon Ratios of Coals in West Virginia Oil Fields*, 522.
  Discussion on Genetic Problems Affecting Search for New Oil Regions, 196, 197
Regulations, oil and gas property taxation, 374.
Representation, employees', Standard Oil Co., 237.
REQUA, M. L.: *Discussions: on A Foreign Oil Supply for the United States*, 93.
  on Variation in Decline Curves of Various Oil Pools, 370.
Résumé of Pennsylvania-New York Oil Field (JOHNSON and HUNTLEY,) 151; *Discussion:* (ASHLEY), 154.
Rise and Decline in Production of Petroleum in Ohio and Indiana (BOWNOCKER), 108; *Discussion:* (PANYITY), 119.
Rock Classification from the Oil-driller's Standpoint (KNAPP), 424.
Rogers' hypothesis, origin of salt domes, 297.
Russia: geology of oil fields, 21.
  oil, production, 33, 37.
  oil fields, see *Oil fields, Russia.*

SADTLER, SAMUEL P.: *Discussion on Composition of Petroleum and its Relation to Industrial Use*, 518.
ST. CLAIR, STUART: *Irvine Oil District, Kentucky*, 165.
  Discussion on Oil Fields of Kentucky and Tennessee, 137.
St. Louis meeting, 1920, proceedings, vii.
Salt cores: limestone caps, 274.
  origin, 270, 285.
  origin of gypsum, 272, 285.
Salt deposits, bedded, Gulf coastal plain, 298.
Salt domes: European, 303.
  Gulf coastal plain: analogies with European domes, 303.
    oil accumulation, 319.